Mathematical Studies

for the IB DIPLOMA

Mathematical Studies

for the IB DIPLOMA

Ric Pimentel
Terry Wall

Acknowledgements
p.1 *l* © Inmagine/Alamy, *c* © Feng Yu/Fotolia.com, *r* Ingram Publishing; **p.16** *t* TopFoto, *b* © Bettmann/CORBIS; **p.17** NASA; **p.18** The Granger Collection/TopFoto; **p.27** © The London Art Archive/Alamy; **p.60** *t* www.purestockX.com, *b* © The London Art Archive/Alamy; **p.61** © INTERFOTO/Alamy; **p.62** *l* The Granger Collection/TopFoto, *r* The Granger Collection/TopFoto; **p.74** © The Gallery Collection/Corbis; **p.87** The Granger Collection/TopFoto; **p.106** © Pamela Chandler/TopFoto; **p.107** *t* © Rita Jayaraman/iStockphoto.com, *r* GREGORY SAMS/SCIENCE PHOTO LIBRARY, *b* © The Print Collector/Alamy; **p.108** The Granger Collection/TopFoto; **p.109** *l* Wellcome Library, London, *r* via Wikimedia; **p.154** *t* Imagestate Media, *b* Photodisc/Getty Images; **p.155** *l* The Granger Collection/TopFoto, *c* ©Print Collector/HIP/TopFoto, *r* The Granger Collection/TopFoto; **p.180** © Pictorial Press Ltd/Alamy; **p.196** AAA Collection Ltd; **p.230** © Luis Carlos Torres/iStockphoto.com; **p.231** *t* © spe/Fotolia.com, *b* Imagestate Media; **p.298** *t* © Jonathan Barton.iStockphoto.com *b* Official White House Photo by Pete Souza; **p.299** © Classic Image/Alamy; **p.301** *t* PROFESSOR PETER GODDARD/SCIENCE PHOTO LIBRARY, *b* © Mary Evans Picture Library/Alamy; **p.328** Photodisc/Getty Images; **p.329** Photo by Eric Erbe, digital colorization by Christopher Pooley/ARS/USDA; **p.330** TopFoto; **p.331** Photodisc/Getty Images; **p.344** *t* Photodisc/Getty Images, *b* Library of Congress; **p.345** Alastair Grant/AP Photo/Press Association Images.
l = left, *r* = right, *t* = top, *b* = bottom

Orders: please contact Bookpoint Ltd, 130 Milton Park, Abingdon, Oxon OX14 4SB. Telephone: (44) 01235 827720. Fax: (44) 01235 400454. Lines are open 9.00–5.00, Monday to Saturday, with a 24–hour message answering service. Visit our website at www.hoddereducation.co.uk

First published in 2010 by
Hodder Education, an Hachette UK Company,
338 Euston Road
London NW1 3BH

Impression number 5 4 3 2 1
Year 2014 2013 2012 2011 2010

Cover photo © Deco/Alamy and © CONCEPTUA: Michel Andrault–Nicolas Ayoub
Typeset in 10/12pt Goudy by Pantek Arts Ltd, Maidstone, Kent
Printed in Italy

A catalogue record for this title is available from the British Library

ISBN 978 0 340 98544 1

Contents

Answers to the Revision Exercises can be found on the CD-ROM

Introduction

The IB Mathematical Studies student

We are aware that students working from this book will have begun full-time education in the twenty-first century. IB students come from many cultures and have many different first languages. Sometimes, cultural and linguistic differences can be an obstacle to understanding. However, mathematics is largely free from cultural bias. Indeed, mathematics is considered by many to be a universal language. Even a Japanese algebra text book with Japanese characters will include recognisable equations using x and y. We are also very aware of the major and often fundamental contributions made to mathematics by non-western cultures, especially those of India, China and Arabia.

The syllabus

This textbook covers fully the IB Mathematical Studies syllabus. The topics covered are in the same order as they appear in the syllabus. However, the IB Diploma Mathematical Studies syllabus is not designed as a teaching syllabus and the order in which the syllabus is presented is not necessarily the order in which it should be taught.

The graphic display calculator

The syllabus places great emphasis on the use of the graphic display calculator (GDC), so many teachers will wish to start with Topic 1. This gives a general overview of the use of the graphic display calculator, which will assist students who may be unfamiliar with its use. Throughout the book we have built upon this foundation work and provided very clear and concise illustrations of exactly how such a calculator is used (not merely showing how a graph might look.) We also refer to other computer software (Autograph and GeoGebra) where appropriate.

Each topic also has explanations, examples, exercises and, at the end of the topic, a number of Student assessments to reinforce learning and to point out areas of weakness.

Presumed knowledge

We are aware of the difficulties presented to teachers by students who are taking this course and come from a variety of backgrounds and with widely differing levels of previous mathematical knowledge. The syllabus refers to presumed knowledge. As revision, in each area of study – number, algebra, geometry, trigonometry and statistics – we have included Presumed knowledge assessments which can be used to identify areas of weakness. These areas can then be studied in more detail with reference to our IGCSE text book.

Multi-cultural and historical references

As mentioned earlier, many cultures have made great contributions to our present understanding of mathematics. Arabic, Indian, Greek and Chinese scholars have learned from each other and provided a basis for the work of later mathematicians. We have tried in our introduction to each topic to make reference to the history of mathematics to give a context to the work which follows.

The people who extended the boundaries of mathematical knowledge are many. Where possible we have referred to the major contributors by name, often with a photograph or other image.

Discussion and theory of knowledge

This should form a vital part of the course. The textbook therefore has a section at the end of each topic for 'Discussion points, project ideas and theory of knowledge'. Students should not underestimate the importance of this facet of the course which is why we have given it a dedicated section at the end of each topic; all too often this information is given in boxes in the main text which, in our experience, are read briefly but often not followed up in depth.

CD-ROM

The material on the accompanying CD-ROM is indicated by a coloured CD icon in the text. There are:

powerpoints,

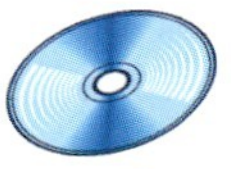

spreadsheets,

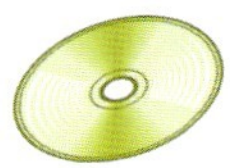

GeoGebra files,

Personal Tutor presentations, step-by-step audio-visual explanations of the harder concepts.

We hope that providing a free copy in every book will enable all students to access this invaluable material. A copy of the GeoGebra installer is also available on the disc.

Internal assessment

A project is chosen by the student and is assessed by their teacher before being externally moderated using assessment criteria that relate to objectives for Group 5 Mathematics and Computer Science. Teachers need to explain these criteria clearly to students.

Project ideas

These can be found at the end of the topics. They are only ideas and are not intended to be project titles. Where it is suggested that students extend their mathematical knowledge as part of the project it is important to discuss this thoroughly with their teacher, both before starting and as the project progresses.

Choice of project

Many teachers feel that too many students choose statistics projects which are too limited in the scope of the mathematics used. Students can look outside the syllabus; it may be advantageous to look at areas of mathematics such as symmetry, topology, optimization, matrices, advanced probability, calculus, linear programming and even mathematics as it applies to art, music and architecture (but be sure that maths is the emphasis). Teachers are allowed to give help and guidance, so students should discuss their ideas before starting their project to be sure that they are feasible.

A textbook is used primarily as preparation for an examination. We have written a book which follows the syllabus and will provide excellent preparation. However, we have also tried to make the book more interesting than that limited aim. We have written a twenty-first century book for students who will probably be alive in the twenty-second century.

Revision exercises

At the end of the book is a revision section, with exercises covering the whole course. It is expected that students will also have access to previous examination papers.

Ric Pimentel and Terry Wall

The authors are both experienced classroom teachers with experience of education in Europe, Turkey, the Far East and the USA. They have also both been teacher trainers and have run courses for English teachers and those from many other countries.

Presumed knowledge assessments

These assessments are intended to identify those concepts from your previous course which may need to be revised. Your teacher will instruct you how these are to be completed.

You may take all the Presumed knowledge assessments 1–8 before you start the course, or it may be better to do assessments 1–4 before Topic 2, assessments 5–7 before Topic 4 and assessment 8 before Topic 5. Your teacher will then, if necessary, refer you to an IGCSE or other textbook.

Student assessment 1: Ordering

1 Copy each of the following statements, and insert one of the symbols $=$, $>$, $<$ into the space to make a true statement.

a $4 \times 2 \ldots 2^3$

b $6^2 \ldots 2^6$

c 850 ml ... 0.5 litres

d number of days in May ... 30 days

2 Illustrate the information in each of the following statements on a number line.

a The temperature during the day reached a maximum of 35 °C.

b There were between 20 and 25 students in a class.

c The world record for the 100 metre sprint is under 10 seconds.

d Doubling a number and subtracting 4 gives an answer greater than 16.

3 Write the information on the following number lines as inequalities.

a

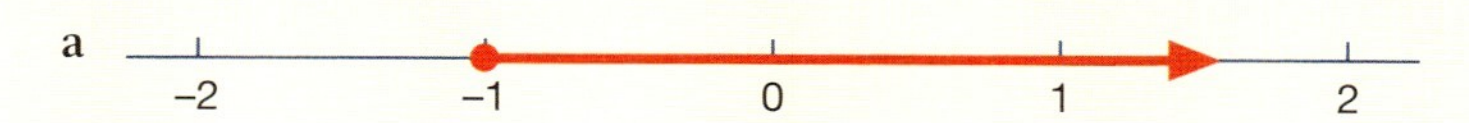

b

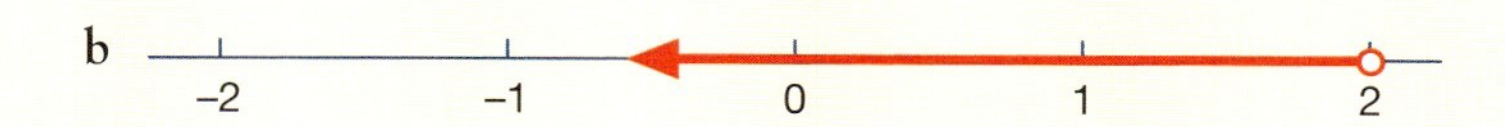

c

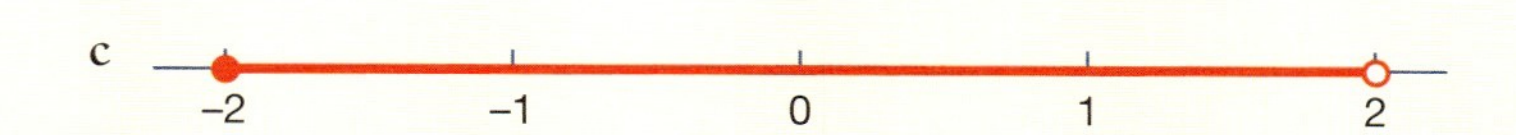

d

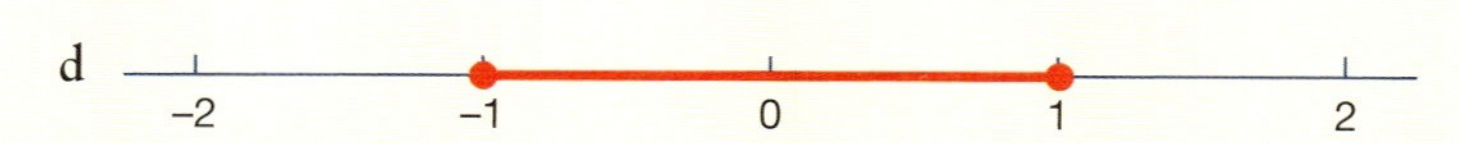

4 Illustrate each of the following inequalities on a number line.

a $x \geq 3$

b $x < 4$

c $0 < x < 4$

d $-3 \leq x < 1$

5 Write the following fractions in order of magnitude, starting with the smallest.

$\frac{4}{7} \quad \frac{3}{14} \quad \frac{9}{10} \quad \frac{1}{2} \quad \frac{2}{5}$

Student assessment 2: The four rules of number

1 Evaluate these.

a $5 + 8 \times 3 - 6$ b $15 + 45 \div 3 - 12$

2 The sum of two numbers is 21 and their product is 90. What are the numbers?

3 How many seconds are there in $2\frac{1}{2}$ hours?

4 Work out 851×27.

5 Work out $6843 \div 19$ giving your answer to one decimal place.

6 Copy these equivalent fractions and fill in the blanks.

$$\frac{8}{18} = \frac{\ }{9} = \frac{16}{\ } = \frac{56}{\ } = \frac{\ }{90}$$

7 Evaluate the following.

a $3\frac{3}{4} - 1\frac{11}{16}$ b $1\frac{4}{5} \div \frac{8}{15}$

8 Change the following fractions to decimals.

a $\frac{2}{5}$ b $1\frac{3}{4}$ c $\frac{9}{11}$ d $1\frac{2}{3}$

9 Change the following decimals to fractions. Give each fraction in its simplest form.

a 4.2 b 0.06 c 1.85 d 2.005

Student assessment 3: Ratio and proportion

1 A piece of wood is cut in the ratio 3 : 7.
   a What fraction of the whole is the longer piece?
   b If the wood is 1.5 m long, how long is the shorter piece?

2 A recipe for two people requires $\frac{1}{4}$kg of rice to 150 g of meat.
   a How much meat would be needed for five people?
   b How much rice would there be in 1 kg of the final dish?

3 The scale of a map is 1 : 10 000.
   a Two rivers are 4.5 cm apart on the map. How far apart are they in real life? Give your answer in metres.
   b Two towns are 8 km apart in real life. How far apart are they on the map? Give your answer in centimetres.

4 a A model train is a $\frac{1}{25}$ scale model. Express this as a ratio.
   b If the length of the model engine is 7 cm, what is the true length of the engine?

5 Divide 3 tonnes in the ratio 2 : 5 : 13.

6 The ratio of the angles of a quadrilateral is 2 : 3 : 3 : 4. Calculate the size of each of the angles.

7 The ratio of the interior angles of a pentagon is 2 : 3 : 4 : 4 : 5. Calculate the size of the largest angle.

8 A large swimming pool takes 36 hours to fill using three identical pumps.
 a How long would it take to fill using eight identical pumps?
 b If the pool needs to be filled in nine hours, how many pumps will be needed?

9 The first triangle is an enlargement of the second. Calculate the size of the missing sides and angles.

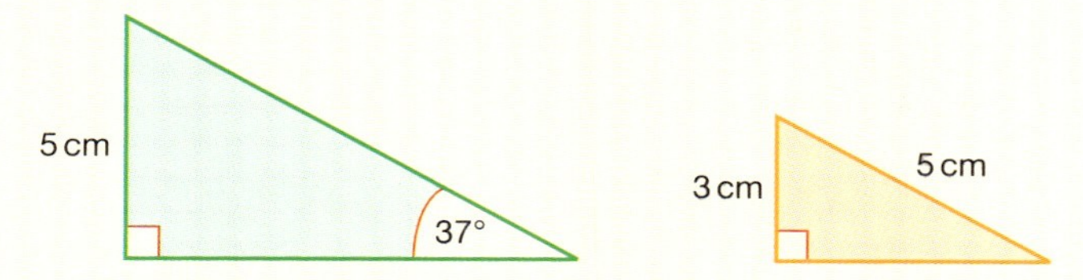

10 A tap issuing water at a rate of 1.2 litres per minute fills a container in 4 minutes.
 a How long would it take to fill the same container if the rate was decreased to 1 litre per minute? Give your answer in minutes and seconds.
 b If the container is to be filled in 3 minutes, calculate the rate at which the water should flow.

Student assessment 4: Percentages

1 Copy the table below and fill in the missing values.

Fraction	Decimal	Percentage
	0.25	
$\frac{3}{5}$		
		$62\frac{1}{2}\%$
$2\frac{1}{4}$		

2 Find 30% of 2500 metres.

3 In a sale a shop reduces its prices by 12.5%. What is the sale price of a desk previously costing 600€?

4 In the last six years the value of a house has increased by 35%. If it cost £72 000 six years ago, what is its value now?

5 Express the first quantity as a percentage of the second.
 a 35 minutes, 2 hours
 b 650 g, 3 kg
 c 5 m, 4 m
 d 15 seconds, 3 minutes
 e 600 kg, 3 tonnes
 f 35 cl, 3.5 l

6 Shares in a company are bought for $600. After a year the same shares are sold for $550. Calculate the percentage depreciation.

7 In a sale, the price of a jacket originally costing 17 000 Japanese yen (¥) is reduced by ¥4000. Any item not sold by the last day of the sale is reduced by a further 50%. If the jacket is sold on the last day of the sale, calculate:

a the price it is finally sold for

b the overall percentage reduction in price.

8 Calculate the original price of each of the following.

Selling price	Profit
$224	12%
$62.50	150%
$660.24	26%
$38.50	285%

9 Calculate the original price of each of the following.

Selling price	Loss
$392.70	15%
$2480	38%
$3937.50	12.5%
$4675	15%

10 In an examination Sarah obtained 87.5% by gaining 105 marks. How many marks did she lose?

11 At the end of a year, a factory has produced 38 500 television sets. If this represents a 10% increase in productivity on last year, calculate the number of sets that were made last year.

12 A computer manufacturer is expected to have produced 24 000 units by the end of this year. If this represents a 4% decrease on last year's output, calculate the number of units produced last year.

13 A farmer increased his yield by 5% each year over the last five years. If he produced 600 tonnes this year, calculate to the nearest tonne his yield five years ago.

Student assessment 5: Algebraic manipulation

1 Expand the following and simplify where possible.

a $3(2x - 3y + 5z)$

b $4p(2m - 7)$

c $-4m(2mn - n^2)$

d $4p^2(5pq - 2q^2 - 2p)$

e $4x - 2(3x + 1)$

f $4x(3x - 2) + 2(5x^2 - 3x)$

g $\frac{1}{5}(15x - 10) - \frac{1}{3}(9x - 12)$

h $\frac{x}{2}(4x - 6) + \frac{x}{4}(2x + 8)$

2 Factorize each of the following.

a $16p - 8q$

b $p^2 - 6pq$

c $5p^2q - 10pq^2$

d $9pq - 6p^2q + 12q^2p$

3 If $a = 4$, $b = 3$ and $c = -2$, evaluate the following.

a $3a - 2b + 3c$

b $5a - 3b^2$

c $a^2 + b^2 + c^2$

d $(a + b)(a - b)$

e $a^2 - b^2$

f $b^3 - c^3$

4 Rearrange the following formulae to make the **bold** letter the subject.

a $p = 4m + \mathbf{n}$
b $4x - 3\mathbf{y} = 5z$
c $2x = \dfrac{3\mathbf{y}}{5p}$
d $m(x + \mathbf{y}) = 3w$
e $\dfrac{pq}{4\mathbf{r}} = \dfrac{mn}{t}$
f $\dfrac{p + \mathbf{q}}{r} = m - n$

5 Factorise each of the following fully.

a $pq - 3rq + pr - 3r^2$
b $1 - t^4$
c $875^2 - 125^2$
d $7.5^2 - 2.5^2$

6 Expand the following and simplify where possible.

a $(x - 4)(x + 2)$
b $(x - 8)^2$
c $(x + y)^2$
d $(x - 11)(x + 11)$
e $(3x - 2)(2x - 3)$
f $(5 - 3x)^2$

7 Factorize each of the following.

a $x^2 - 4x - 77$
b $x^2 - 6x + 9$
c $x^2 - 144$
d $3x^2 + 3x - 18$
e $2x^2 + 5x - 12$
f $4x^2 - 20x + 25$

8 Make the letter in **bold** the subject of the formula.

a $m\mathbf{f}^2 = p$
b $m = 5\mathbf{t}^2$
c $A = \pi r\sqrt{\mathbf{p} + q}$
d $\dfrac{1}{\mathbf{x}} + \dfrac{1}{y} = \dfrac{1}{t}$

9 Simplify the following algebraic fractions.

a $\dfrac{x^7}{x^3}$
b $\dfrac{mn}{p} \times \dfrac{pq}{m}$
c $\dfrac{(y^3)^3}{(y^2)^3}$
d $\dfrac{28pq^2}{7pq^3}$

Student assessment 6: Equations and inequalities

For questions 1–4, solve the equations.

1 a $x + 7 = 16$
b $2x - 9 = 13$
c $8 - 4x = 24$
d $5 - 3x = -13$

2 a $7 - m = 4 + m$
b $5m - 3 = 3m + 11$
c $6m - 1 = 9m - 13$
d $18 - 3p = 6 + p$

3 a $\dfrac{x}{-5} = 2$
b $4 = \frac{1}{3}x$
c $\dfrac{x + 2}{3} = 4$
d $\dfrac{2x - 5}{7} = \dfrac{5}{2}$

4 a $\frac{2}{3}(x - 4) = 8$
b $4(x - 3) = 7(x + 2)$
c $4 = \frac{2}{7}(3x + 8)$
d $\frac{3}{4}(x - 1) = \frac{5}{8}(2x - 4)$

5 Solve the following simultaneous equations.

a $2x + 3y = 16$
$2x - 3y = 4$

b $4x + 2y = 22$
$-2x + 2y = 2$

c $x + y = 9$
$2x + 4y = 26$

d $2x - 3y = 7$
$-3x + 4y = -11$

6 The angles of a triangle are x, $2x$ and $(x + 40)$ degrees.

a Construct an equation in terms of x.
b Solve the equation.
c Calculate the size of each of the three angles.

7 Seven is added to three times a number. The result is then doubled. If the answer is 68, calculate the number.

8 A decagon has six equal exterior angles. Each of the remaining four is fifteen degrees larger than these six angles. Construct an equation and then solve it to find the sizes of the angles.

9 A rectangle is x cm long. The length is 3 cm more than the width. The perimeter of the rectangle is 54 cm.

a Draw a diagram to illustrate the above information.
b Construct an equation in terms of x.
c Solve the equation and hence calculate the length and width of the rectangle.

10 At the end of a football season, the leading goal scorer in a league has scored eight more goals than the second leading goal scorer. The second has scored fifteen more than the third. The total number of goals scored by all three players is 134.

a Write an expression for each of the three scores.
b Form an equation and then solve it to find the number of goals scored by each player.

Student assessment 7: Indices

1 Simplify the following by using indices.

a $2 \times 2 \times 2 \times 5 \times 5$
b $2 \times 2 \times 3 \times 3 \times 3 \times 3 \times 3$

2 Write the following out in full.

a 4^3
b 6^4

3 Work out the value of the following without using a calculator.

a $2^3 \times 10^2$
b $1^4 \times 3^3$

4 Find the value of x in each of the following.

a $2^{(x-2)} = 32$
b $\frac{1}{4^x} = 16$
c $5^{(-x+2)} = 125$
d $8^{-x} = \frac{1}{2}$

5 Using indices, simplify the following.

a $3 \times 2 \times 2 \times 3 \times 27$
b $2 \times 2 \times 4 \times 4 \times 4 \times 2 \times 32$

6 Write the following out in full.

a 6^5
b 2^{-5}

Student assessment 8: Geometry of plane shapes

1 Calculate the circumference and area of each of the following circles. Give your answers to one decimal place.

a

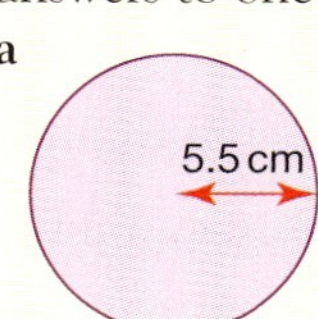

b

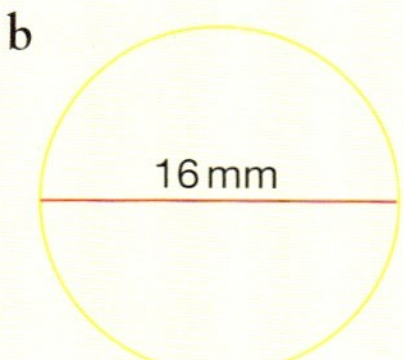

2 A semi-circular shape is cut out of the side of a rectangle as shown. Calculate the shaded area to one decimal place.

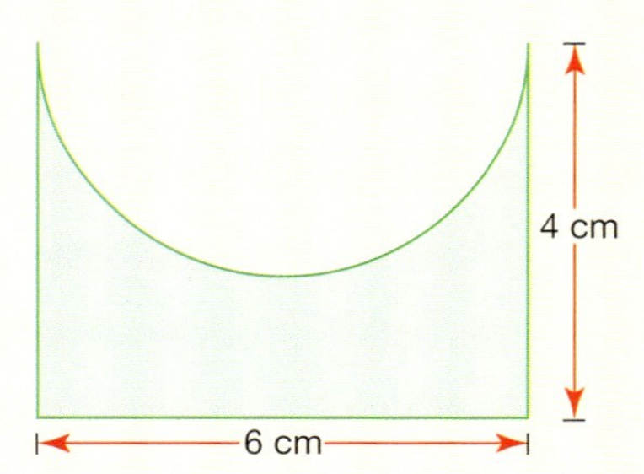

3 For the diagram below, calculate the area of:

a the semi-circle

b the trapezium

c the whole shape.

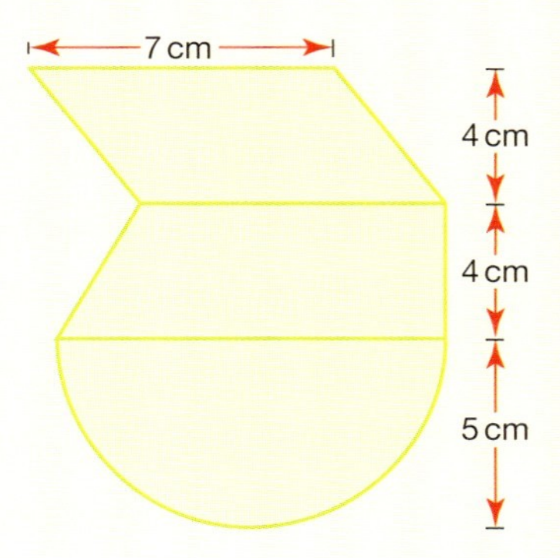

4 Calculate the circumference and area of each of these circles. Give your answers to one decimal place.

a

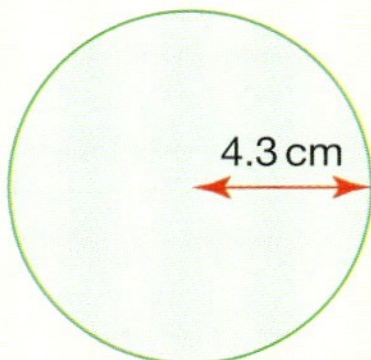

b

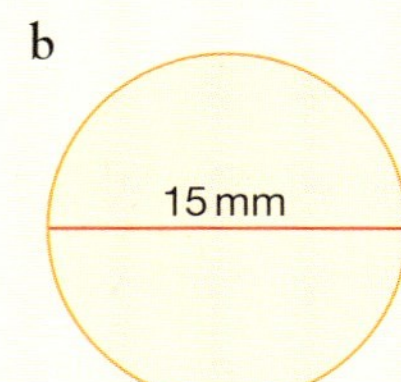

5 A rectangle of length 32 cm and width 20 cm has a semi-circle cut out of two of its sides as shown. Calculate the shaded area to one decimal place.

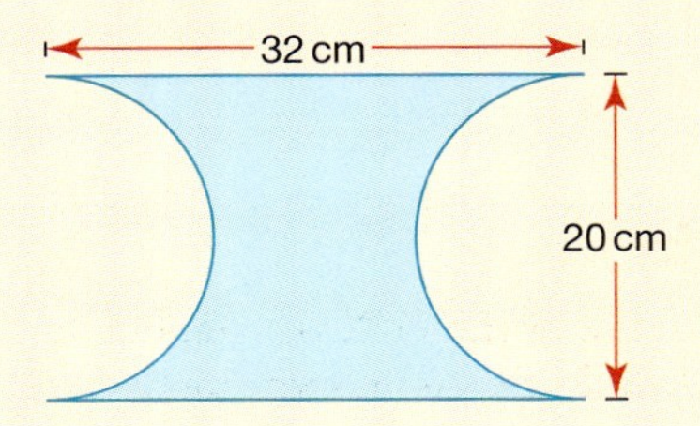

6 Calculate the area of:

a the semi-circle

b the parallelogram

c the whole shape.

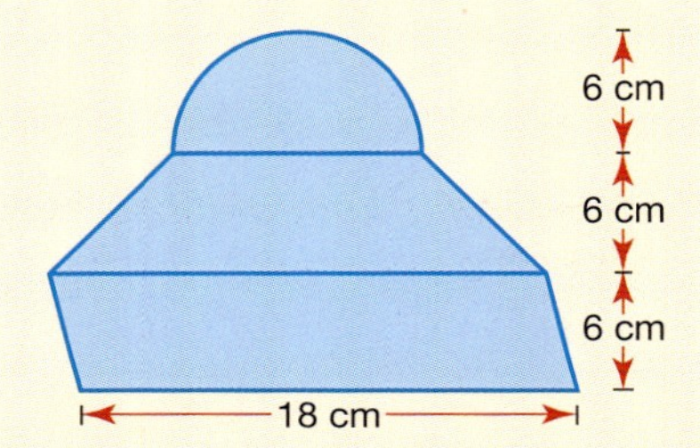

Topic 1

Introduction to the graphic display calculator

Syllabus content

1.1 Arithmetic calculations, use of the GDC to graph a variety of functions.
Appropriate choice of 'window'; use of 'zoom' and 'trace' (or equivalent) to locate points to a given accuracy.
Explanation of commonly used buttons.
Entering data in lists.

Introduction

People have always used devices to help them calculate. Today, basic calculators, scientific calculators and graphic display calculators are all available; they have a long history.

an early abacus

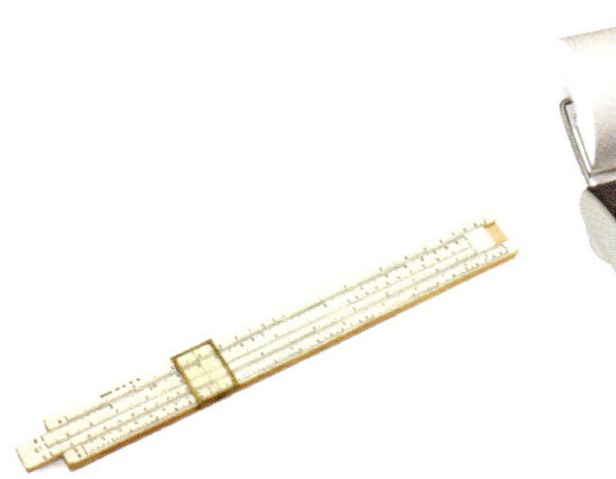

an early slide rule

an early calculator

1.1 Using a graphic display calculator

Graphic display calculators (GDCs) are a powerful tool used for the study of mathematics in the modern classroom. However, as with all tools, their effectiveness is only apparent when used properly. This section will look at some of the key features of the GDC, so that you start to understand some of its potential. The two models used are the Casio *fx*-9860G and the Texas TI-84 Plus. Many GDCs have similar capabilities to the ones shown. However, if your calculator is different, it is important that you take the time to familiarize yourself with it.

Here is the home screen (menu/applications) for each calculator.

Casio	Texas
The modes are selected by using the arrows key and then pressing EXE, or by typing the number/letter that appears in the bottom right-hand corner of each square representing a mode. Brief descriptions of the seven most relevant modes are given below.	The main features are accessed by pressing the appropriate key. Some are explained below.
1 RUN.MAT is used for arithmetic calculations.	 LIST STAT is used for statistical calculations and for drawing graphs of the data entered.
2 STAT is used for statistical calculations and for drawing graphs.	 TEST A MATH is used to access numerical operations.
3 S.SHT is a spreadsheet and can be used for calculations and graphs.	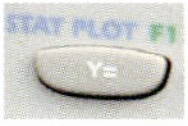 STAT PLOT F1 Y= is used for entering the equations of graphs.
4 GRAPH is used for entering the equations of graphs and plotting them.	 TABLE F5 GRAPH is used for graphing functions.
5 DYNA is a dynamic graph mode that enables a family of curves to be graphed simultaneously.	
6 TABLE is used to generate a table of results from an equation.	
7 EQUA is used to solve different types of equation.	

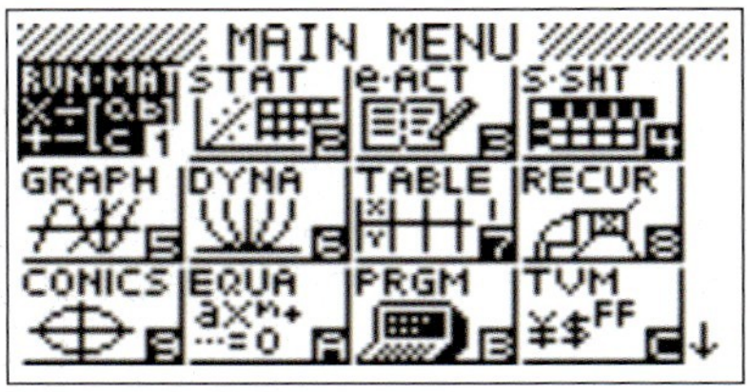

Casio

Texas

Basic calculations

The aim of the following exercise is to familiarize you with some of the buttons on your calculator dealing with basic mathematical operations. It is assumed that you will already be familiar with the mathematical content.

Exercise 1.1.1

Using your GDC, evaluate the following.

Casio	Texas
x^2 (√ r)	x^2 (√ I)

1 a $\sqrt{625}$

b $\sqrt{324}$

c $2\sqrt{8} \times 5\sqrt{2}$

Casio	Texas	
^ ($^x\sqrt{}$ θ)	MATH (TEST A)	5 (L5 U)

2 a $\sqrt[3]{1728}$

b $\sqrt[4]{1296}$

c $\sqrt[5]{3125}$

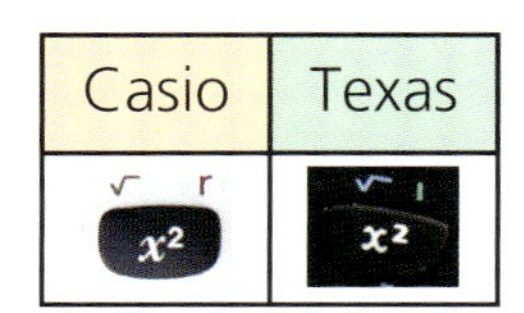

Casio	Texas
x^2 (√ r)	x^2 (√ I)

3 a 13^2

b $8^2 \div 4^2$

c $\sqrt{5^2 + 12^2}$

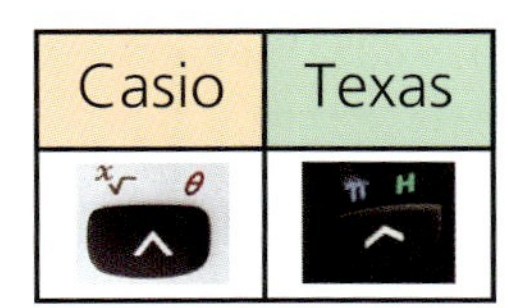

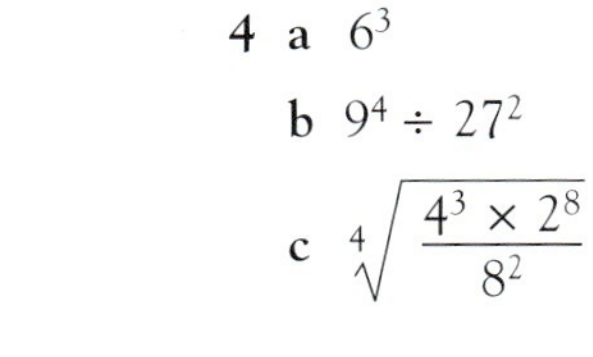

4 a 6^3

b $9^4 \div 27^2$

c $\sqrt[4]{\dfrac{4^3 \times 2^8}{8^2}}$

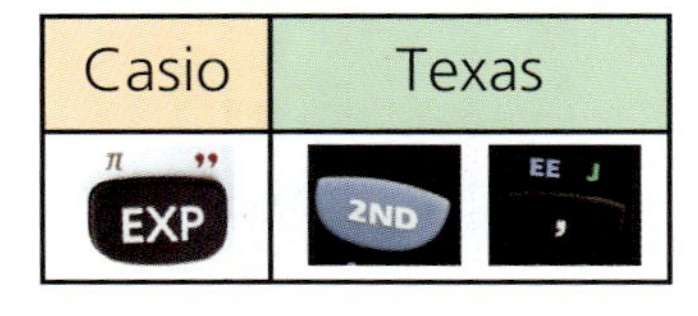

5 a $(2.3 \times 10^3) + (12.1 \times 10^2)$

b $(4.03 \times 10^3) + (15.6 \times 10^4) - (1.05 \times 10^4)$

c $\dfrac{13.95 \times 10^6}{15.5 \times 10^3} - (9 \times 10^2)$

GDCs also have a large number of memory channels. Use these to store answers which are needed for subsequent calculations. This will minimise rounding errors.

Casio	Texas
[→] [ALPHA] followed by a letter of the alphabet	[STO▸] [ALPHA] followed by a letter of the alphabet

Exercise 1.1.2

1 In the following expressions, $a = 5$, $b = 4$ and $c = 2$.
Enter each of these values in memory channels A, B and C of your GDC respectively, and evaluate the following.

a $a + b + c$

b $a - (b + c)$

c $(a + b)^2 - c$

d $\dfrac{2(b + c)^3}{(a - c)}$

e $\dfrac{4\sqrt{a^2 - b^2}}{c}$

f $\dfrac{(ac)^2 + ba^2}{a + b + c}$

2 Circles A, B, C and D have radii 10 cm, 6 cm, 4 cm and 1 cm respectively.

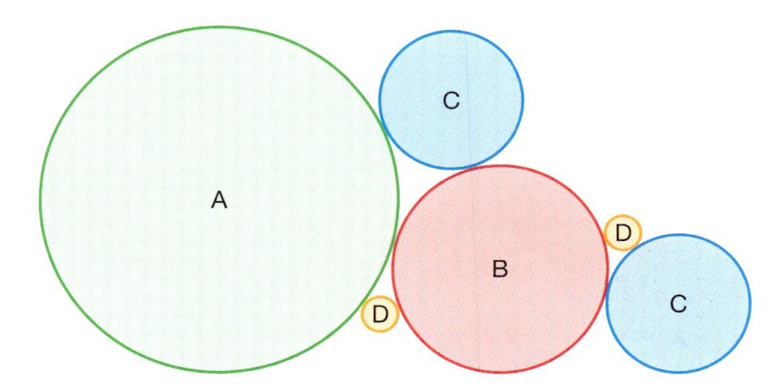

a Calculate the area of circle A and store your answer in memory channel A.
b Calculate the area of circle B and store your answer in memory channel B.
c Calculate the area of each of the circles C and D, storing the answers in memory channels C and D respectively.
d Using your calculator evaluate A + B + 2C + 2D.
e What does the answer to part **d** represent?

3 A child's shape-sorting toy is shown in the diagram. The top consists of a rectangular piece of wood of dimension 30 cm × 12 cm. Four shapes W, X, Y and Z are cut out of it.

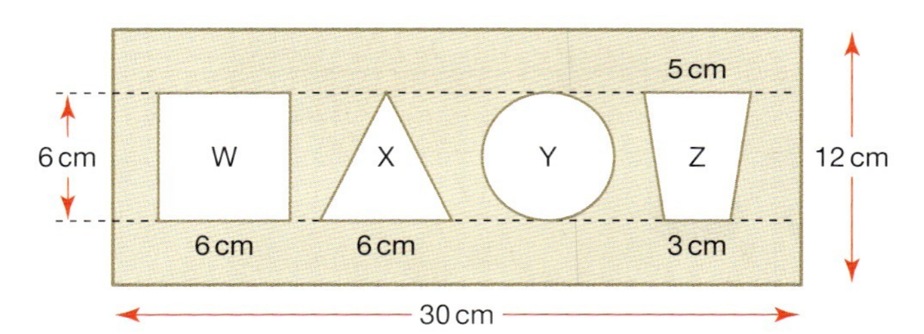

a Calculate the area of the triangle X. Store the answer in your calculator's memory.
b Calculate the area of the trapezium Z. Store the answer in your calculator's memory.
c Calculate the total area of the shapes W, X, Y and Z.
d Calculate the area of the rectangular piece of wood left once the shapes have been cut out.

4 Three balls just fit inside a cylindrical tube as shown. The radius (r) of each ball is 5 cm.

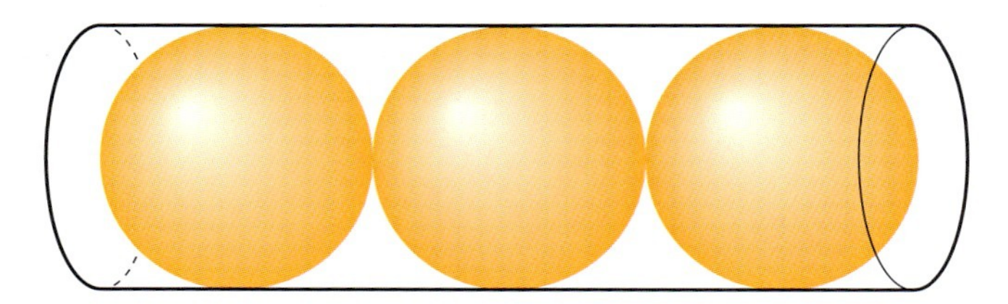

a Using the formula for the volume of a sphere, $V = \frac{4}{3}\pi r^3$, calculate the volume of one of the balls. Store the answer in the memory of your calculator.

b Calculate the volume of the cylinder.

c Calculate the volume of the cylinder *not* occupied by the three balls.

Plotting graphs

One of a GDC's principal features is to plot graphs of functions. This helps to visualize what the function looks like and, later on, it will help solve a number of different types of problem. This section aims to show how to graph a variety of different functions. For example, to plot a graph of the function $y = 2x + 3$, use the following functions on your calculator.

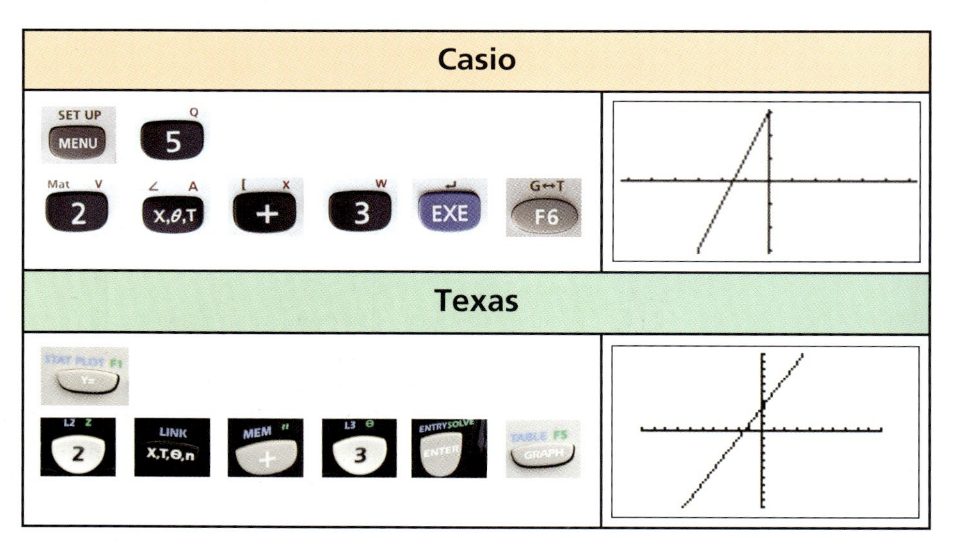

Occasionally, it may be necessary to change the scale on the axes to change how much of the graph, or what part of the graph, can be seen. This can be done in several ways, two of which are described here.

- By using the zoom facility

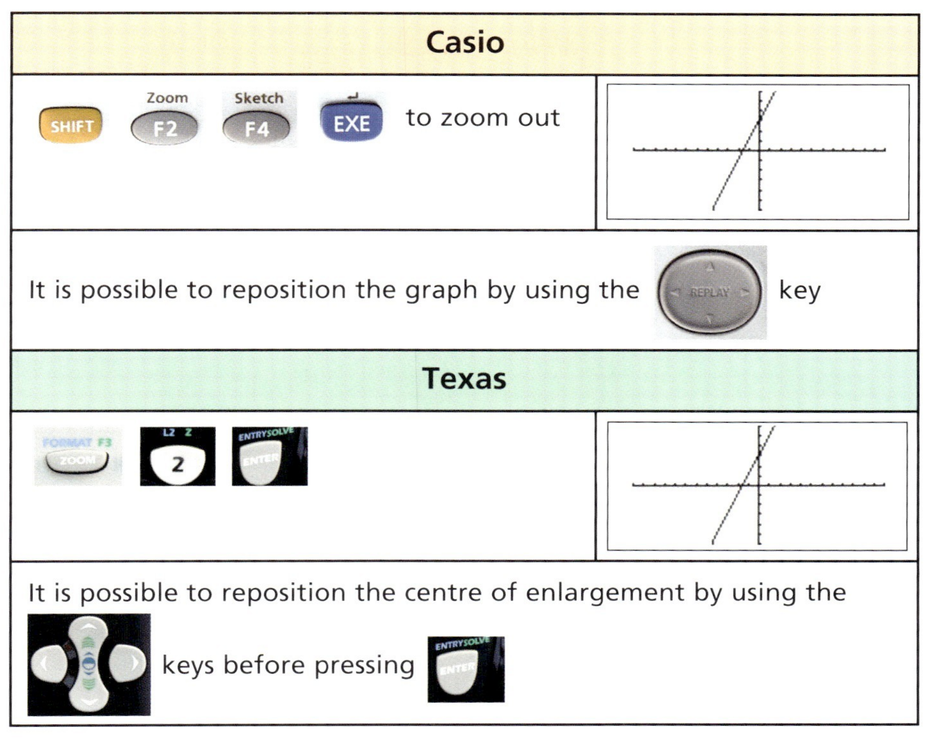

- By changing the scale manually

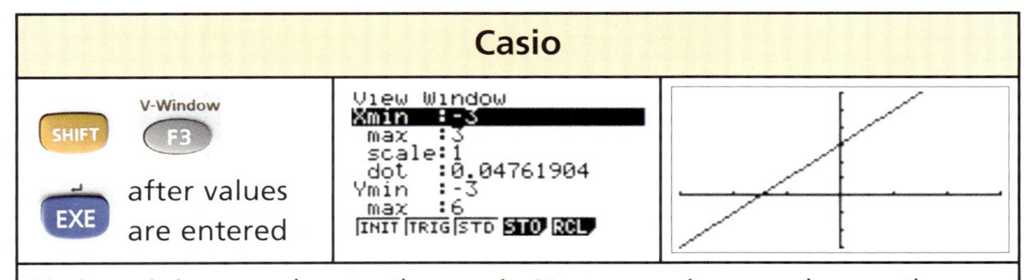

Xmin: minimum value on the *x*-axis, Xmax: maximum value on the *x*-axis, Xscale: spacing of the *x*-axis increments, Xdot: value that relates to one *x*-axis dot (this is set automatically).

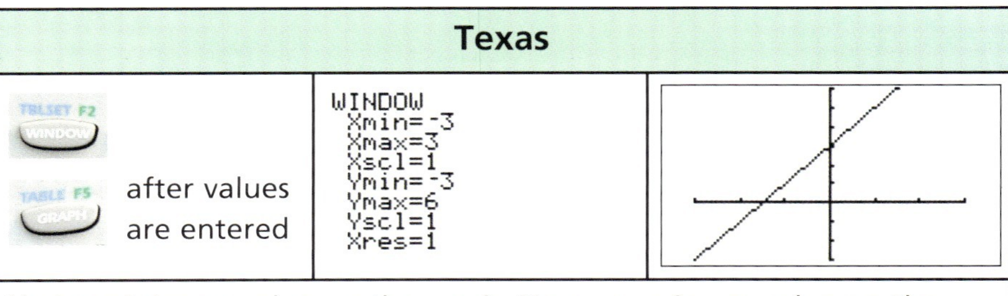

Xmin: minimum value on the *x*-axis, Xmax: maximum value on the *x*-axis, Xscl: spacing of the *x*-axis increments.

Exercise 1.1.3

In the following questions, the axes have been set to their default settings, i.e. Xmin = −10, Xmax = 10, Xscale = 1, Ymin = −10, Ymax = 10, Yscale = 1.

i) 

ii) 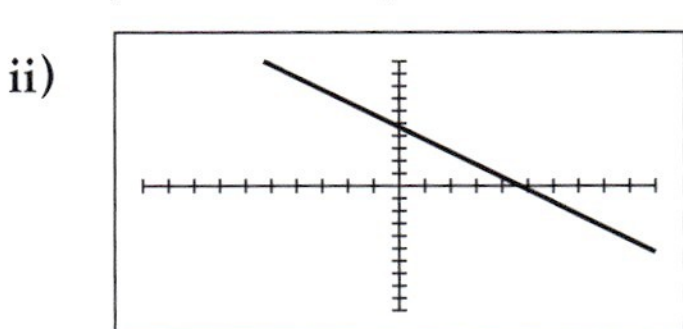

iii) 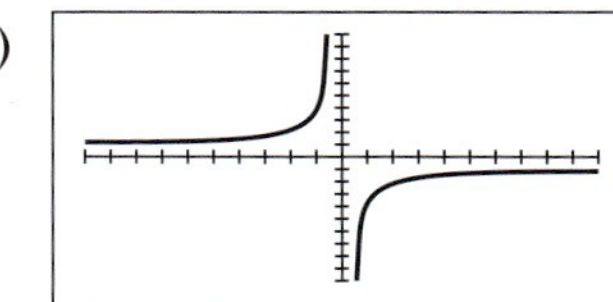

iv) 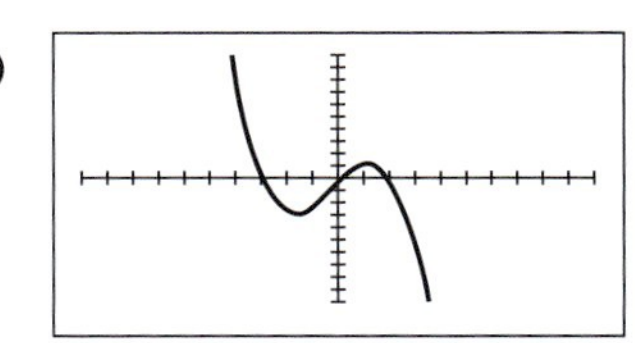

v) 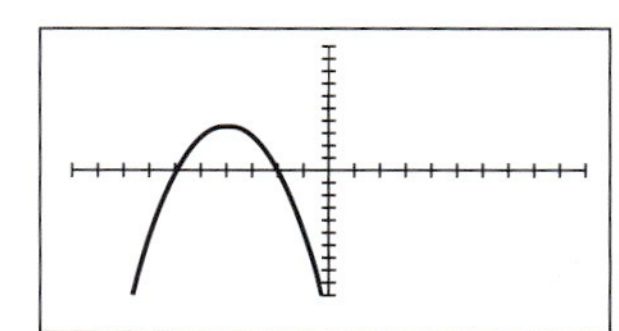

vi) 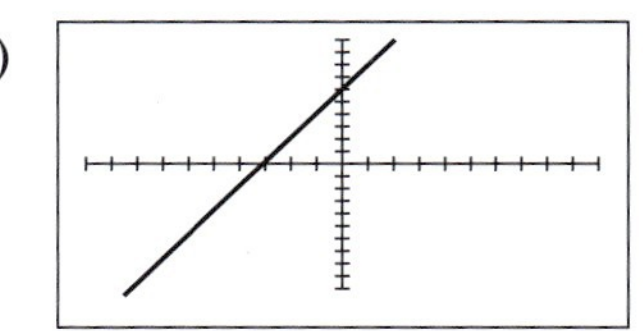

vii) 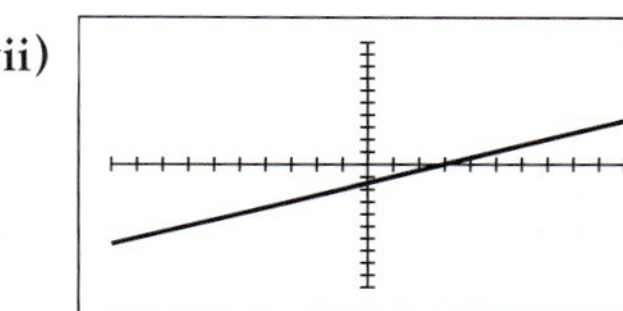

viii) 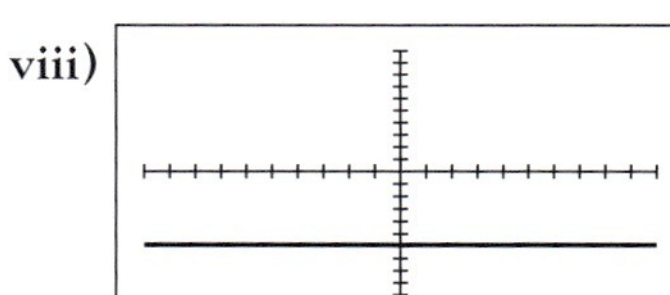

ix) 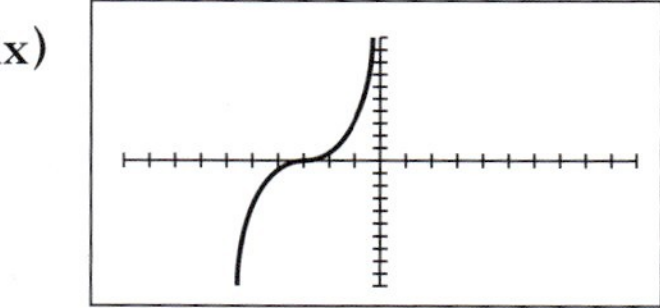

x) 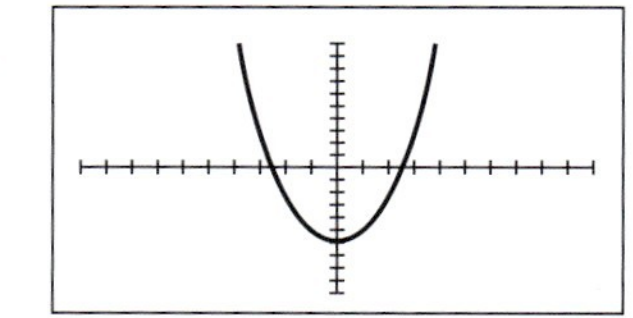

1 By using your GDC to graph the following functions, match each of them to the correct graph.

a $y = 2x + 6$

b $y = \frac{1}{2}x - 2$

c $y = -x + 5$

d $y = -\dfrac{5}{x}$

e $y = x^2 - 6$

f $y = (x - 4)^2$

g $y = -(x + 4)^2 + 4$

h $y = \frac{1}{2}(x + 3)^3$

i $y = -\frac{1}{3}x^3 + 2x - 1$

j $y = -6$

2 In each of the following, a function and its graph are given. Using your GDC, enter a function that produces a reflection of the original function in the x-axis.

a $y = x + 5$

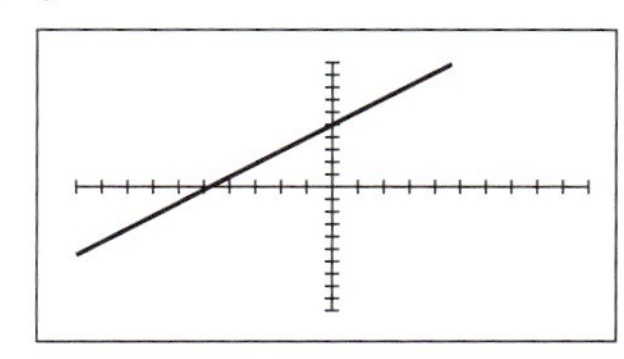

b $y = -2x + 4$

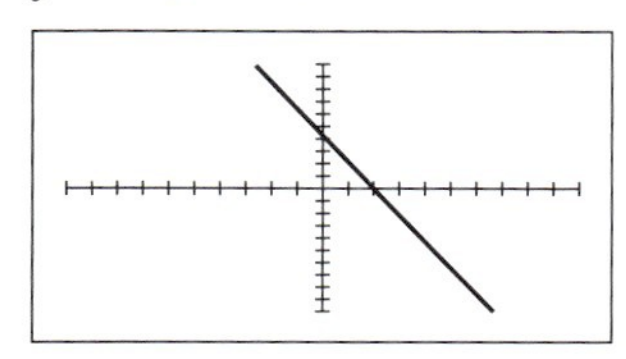

c $y = (x + 5)^2$

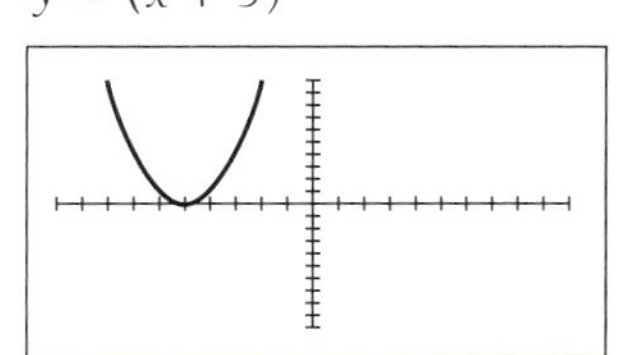

d $y = (x - 5)^2 + 3$

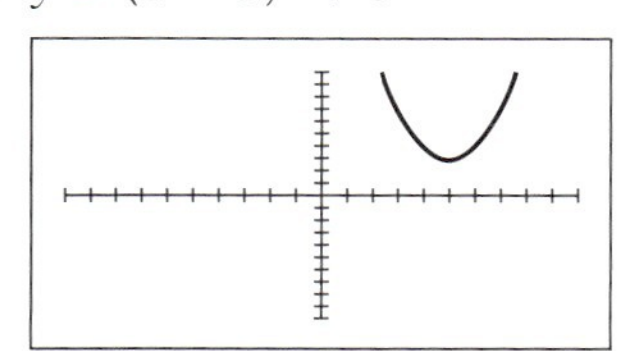

3 Using your GDC, enter a function that produces a reflection in the y-axis of each of the original functions in question 2.

4 By entering appropriate functions into your calculator:

i) make your GDC screen look like the ones shown

ii) write down the functions you used.

a

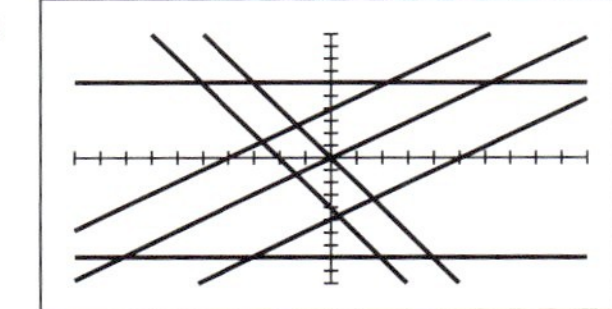

b

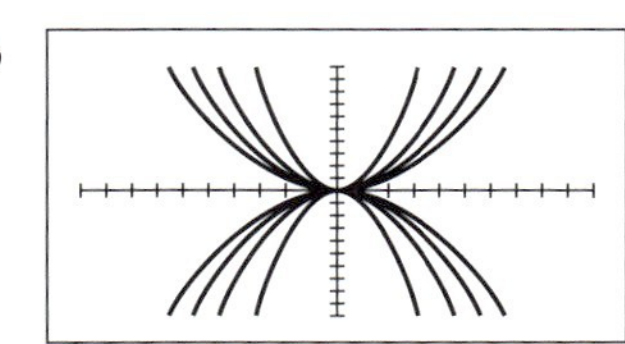

c

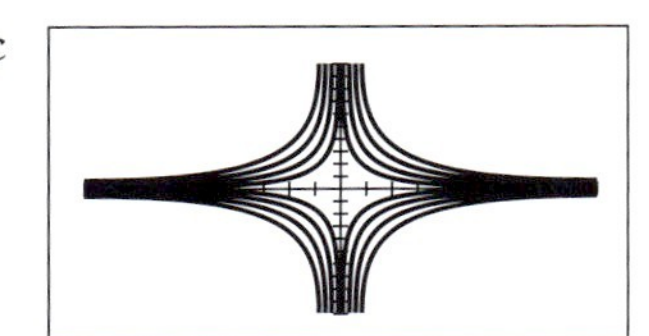

d

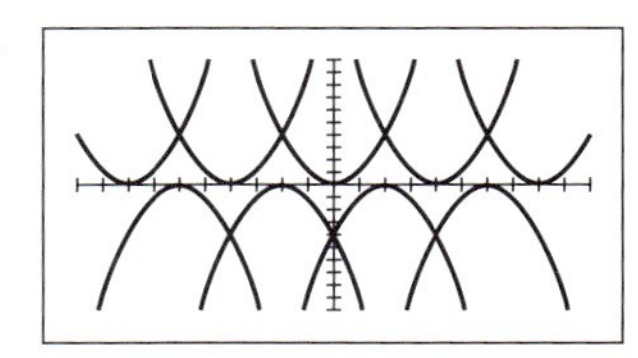

Intersections

When graphing a function it is often necessary to find where it intersects one or both of the axes. If more than one function is graphed simultaneously, it may also be necessary to find where the graphs intersect each other. GDCs have a 'Trace' facility which gives an approximate coordinate of a cursor on the screen. More accurate methods are available and will be introduced in the topics as appropriate.

Worked example

Find where the graph of $y = \frac{1}{5}(x + 3)^3 + 2$ intersects both the x- and y-axes.

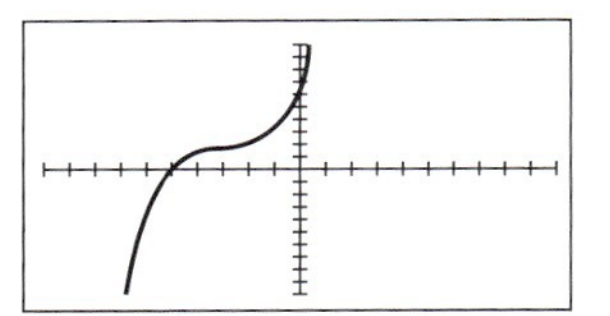

The graph shows that $y = \frac{1}{5}(x + 3)^3 + 2$ intersects each of the axes once.

To find the approximate coordinates of the points of intersection:

Casio

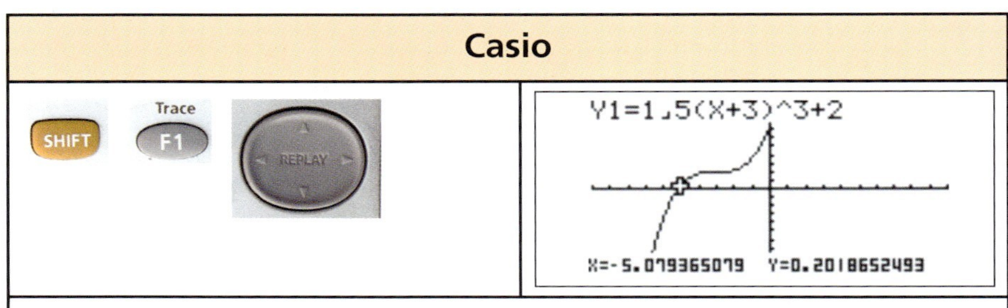

Because of the pixels on the screen, the 'Trace' facility will usually only give an approximate value. The *y*-coordinate of the point of intersection with the *x*-axis will always be zero. However, the calculator's closest result is result is $y = 0.202$. The *x*-value of –5.079 will also therefore be only an approximation. By moving the cursor to the point of intersection with the *x*-axis, values of $x = 0$ and $y = 7.4$ are obtained.

Texas

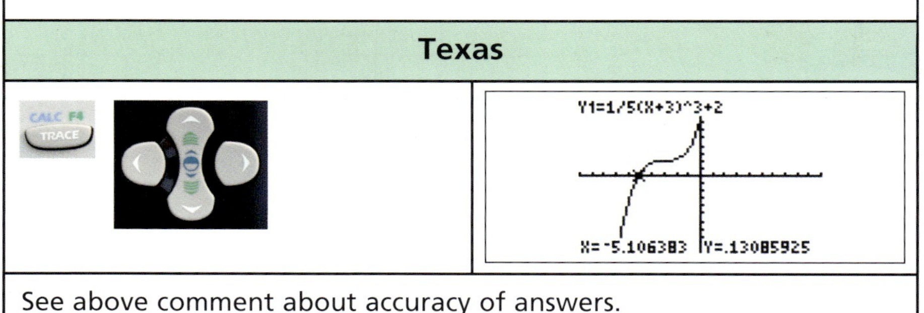

See above comment about accuracy of answers.

Exercise 1.1.4

1 Find an approximate solution to where the following graphs intersect both the *x*- and *y*-axes using your GDC.

a $y = x^2 - 3$

b $y = (x + 3)^2 + 2$

c $y = \frac{1}{2}x^3 - 2x^2 + x + 1$

d $y = \dfrac{-5}{x + 2} + 6$

2 Find the coordinates of the point(s) of intersection of each of the following pairs of equations using your GDC.

a $y = x + 3$ and $y = -2x - 2$

b $y = -x + 1$ and $y = \frac{1}{2}(x^2 - 3)$

c $y = -x^2 + 1$ and $y = \frac{1}{2}(x^2 - 3)$

d $y = -\frac{1}{4}x^3 + 2x^2 - 3$ and $y = \frac{1}{2}x^2 - 2$

Tables of results

A function such as $y = \frac{3}{x} + 2$ implies that there is a relationship between y and x.

To plot the graph manually, the coordinates of several points on the graph need to be calculated and then plotted. GDCs have the facility to produce a table of values giving the coordinates of some of the points on the line.

Worked example

For the function $y = \frac{3}{x} + 2$, complete the following table of values using the table facility of your GDC.

x	–3	–2	–1	0	1	2	3
y							

Casio

Enter function $y = \frac{3}{x} + 2$

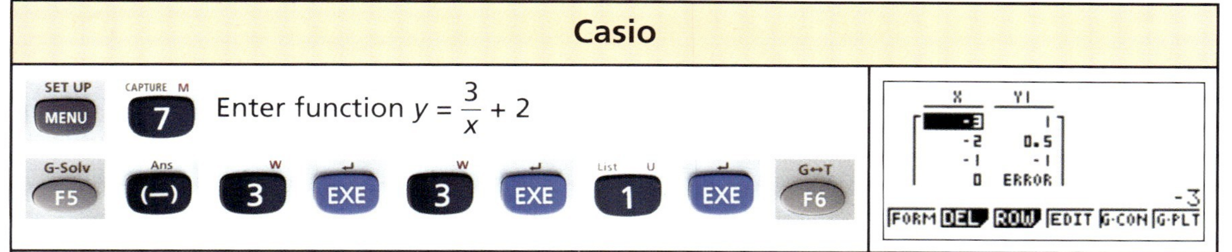

The screen shows that the x-values range from –3 to 3 in increments of 1.

Once the table is displayed, the remaining results can be viewed by using

Texas

Enter function $y = \frac{3}{x} + 2$

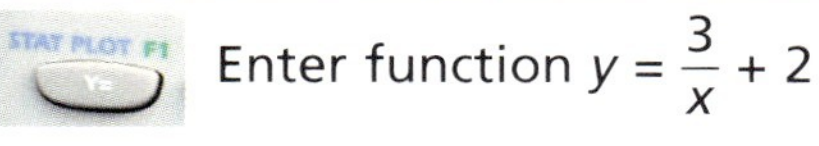

The screen shows that the x-values start at –3 and increase in increments of 1.

Once the table is displayed, further results can be viewed by using

Exercise 1.1.5

1 Copy and complete the tables of values for the following functions using the table facility of your GDC.

a $y = x^2 + x - 4$

x	−3	−2	−1	0	1	2	3
y							

b $y = x^3 + x^2 - 10$

x	−3	−2	−1	0	1	2	3
y							

c $y = \frac{4}{x}$

x	0	0.5	1	1.5	2	2.5	3
y							

d $y = \sqrt{(x+1)}$

x	−1	−0.5	0	0.5	1	1.5	2	2.5	3
y									

2 A car accelerates from rest. Its speed, $y\,\text{m}\,\text{s}^{-1}$, x seconds after starting, is given by the equation $y = 1.8x$.

a Using the table facility of your GDC, calculate the speed of the car every 2 seconds for the first 20 seconds.

b How fast was the car travelling after 10 seconds?

3 A ball is thrown vertically upwards. Its height y metres, x seconds after launch, is given by the equation $y = 15x - 5x^2$.

a Using the table facility of your GDC, calculate the height of the ball each $\frac{1}{2}$ second during the first 4 seconds.

b What is the greatest height reached by the ball?

c How many seconds after its launch did the ball reach its highest point?

d After how many seconds did the ball hit the ground?

e In the context of this question, why can the values for $x = 3.5$ and $x = 4$ be ignored?

Lists

Data is often collected and then analyzed so that observations and conclusions can be made. GDCs have the facility for storing data as lists. Once stored as a list, many different types of calculations can be carried out. This section will explain how to enter data as a list and then how to carry out some simple calculations.

Worked example

An athlete records her time (seconds) in ten races for running 100 m. These are shown below.

12.4 12.7 12.6 12.9 12.4 12.3 12.7 12.4 12.5 13.1

Calculate the mean, median and mode for this set of data using the list facility of your GDC.

Casio

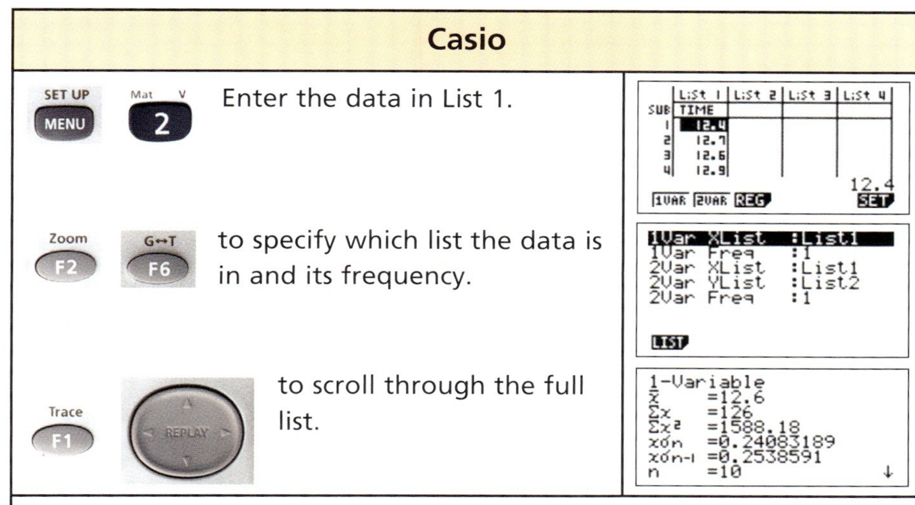

The screen displays various statistical measures.
$\bar{x}$ is the mean, n is the number of data items, Med is the median, Mod is the modal value, Mod: F is the frequency of the modal values.

Texas

Enter the data in List 1.

```
L1      L2      L3     1
12.4    ------  ------
12.7
12.6
12.9
12.4
12.3
12.7
L1(1)=12.4
```

to apply calculations to the data in List 1.

```
1-Var Stats L1
```

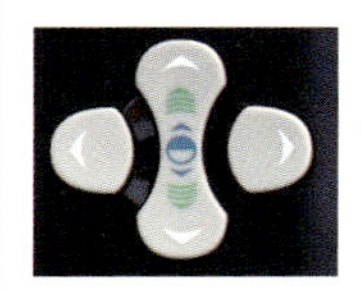

to scroll through the full list.

```
1-Var Stats
 x̄=12.6
 Σx=126
 Σx²=1588.18
 Sx=.2538591035
 σx=.2408318916
↓n=10
```

The screen displays various statistical measures.
$\bar{x}$ is the mean, n is the number of data items, Med is the median. The T1-84 does not display the modal value.

If a lot of data is collected, it is often presented in a frequency table.

Worked example

The numbers of pupils in 30 maths classes are shown in the frequency table.

Number of pupils	Frequency
27	4
28	6
29	9
30	7
31	3
32	1

Calculate the mean, median and mode for this set of data using the list facility of your GDC.

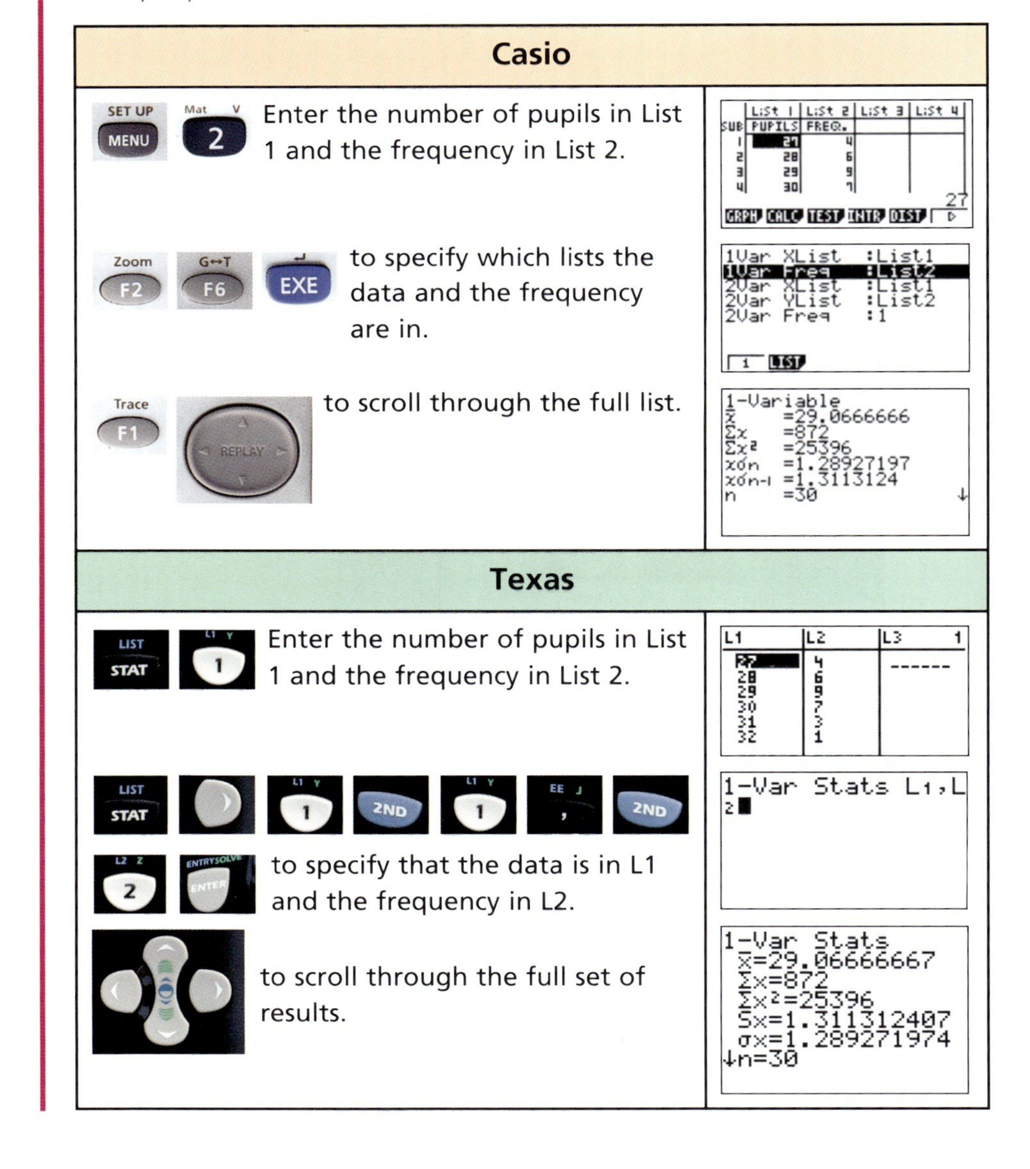

Exercise 1.1.6

1 Find the mean, the median and, if possible, the mode of these sets of numbers using the list facility of your GDC.

a 3, 6, 10, 8, 9, 10, 12, 4, 6, 10, 9, 4

b 12.5, 13.6, 12.2, 14.4, 17.1, 14.8, 20.9, 12.2

2 During a board game, a player makes a note of the numbers he rolls on the dice. These are shown in the frequency table below.

Number on dice	1	2	3	4	5	6
Frequency	3	8	5	2	5	7

Find the mean, the median and, if possible, the modal dice roll using the list facility of your GDC.

3 A class of 30 pupils sat two maths tests. Their scores out of 10 are recorded in the frequency tables below.

Test A										
Score	1	2	3	4	5	6	7	8	9	10
Frequency	3	2	4	3	1	8	3	1	3	2

Test B										
Score	1	2	3	4	5	6	7	8	9	10
Frequency	4	1	0	0	0	24	0	0	0	1

a Find the mean, the median and, if possible, the mode for each test using the list facility of your GDC.

b Comment on any similarities or differences in your answers to part **a**.

c Which test did the class find easiest? Give reasons for your answer.

Discussion points, project ideas and theory of knowledge

1 The use of mathematical tables has declined since the development of scientific calculators. Find examples of both maritime (naval) and other mathematical books of tables and discuss their use.

2 You may be able to get access to an abacus and slide rule. How were they used?

3 Use of devices such as a slide rule and books of calculation tables took time to learn. At what point did using a slide rule justify the time taken to learn its use? Speculate on which professionals might have used such a tool.

4 Investigate what an 'Enigma Machine' was. Where, when and for what purpose was it designed?

5 What are punch cards and how were they used? Investigate simple early computer languages (like basic). This could be a starting point for a project.

6 'Modern graphic display calculators are more powerful than 20-year-old computers.' Discuss this statement.

7 Investigate and learn how to use logarithm and other mathematical tables.

8 'The problem with calculators is that nobody can estimate any more.' Discuss this statement. Design an experiment to test the ability of your classmates to estimate quickly and accurately. This could form the basis of a project.

9 What is meant by 'The law of diminishing returns'? Use your calculator to illustrate it.

10 Explain the difference between 'knowing how' and 'knowing that' with reference to the use of a calculator.

11 'Truth, belief and knowledge are interconnected.' Discuss.

Topic 2 Number and algebra

Syllabus content

2.1 The sets of natural numbers, $\mathbb{N}$; integers, $\mathbb{Z}$; rational number, $\mathbb{Q}$; and real numbers, $\mathbb{R}$.

2.2 Approximation: decimal places; significant figures. Percentage errors. Estimation.

2.3 Expressing numbers in the form $a \times 10^k$ where $1 \leq a < 10$ and $k \in \mathbb{Z}$. Operations with numbers expressed in the form $a \times 10^k$ where $1 \leq a < 10$ and $k \in \mathbb{Z}$.

2.4 SI (*Système International*) and other basic units of measurement: for example, gram (g), metre (m), second (s), litre (l), metre per second (m s^{-1}), Celsius and Fahrenheit scales.

2.5 Arithmetic sequence and series, and their applications. Use of the formulae for the nth term and the sum of n terms.

2.6 Geometric sequences and series, and their applications. Use of the formulae for the nth term and the sum of n terms.

2.7 Solutions of pairs of linear equations in two variables by use of a GDC. Solutions of quadratic equations: by factorising; by use of a GDC.

Introduction

The first written records showing the origin and development of the use of money were found in the city of Eridu in Mesopotamia (modern Iraq). The records were on tablets like the one shown below.

The Sumerians, as the people of this region were known, used a system of recording value, known as 'Cuneiform', five thousand years ago. This writing is now believed to be simply accounts of grain surpluses. This may sound insignificant now, but the change from a hunter–gatherer society to a farming-based society led directly to the kind of sophisticated way of life we have today.

Our present number system has a long history, originating from the Indian Brahmi numerals around 300 BCE.

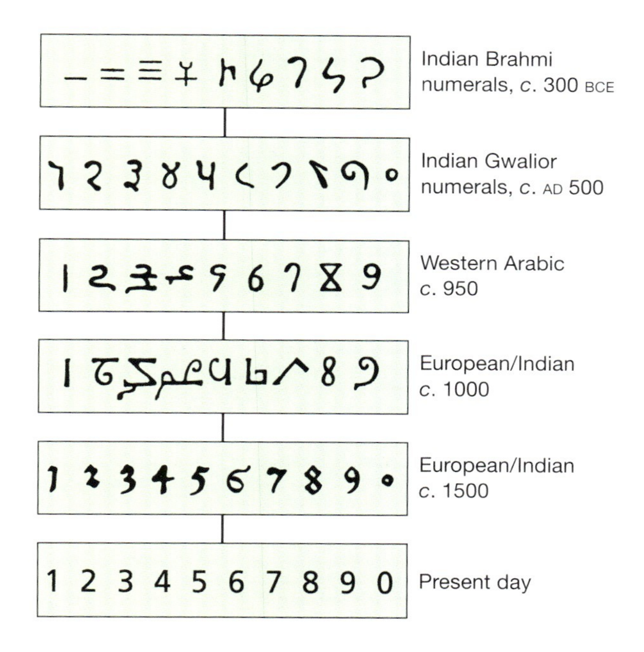

Abu Ja'far Muhammad Ibn Musa Al-Khwarizmi is called 'the father of algebra'. He was born in Baghdad in 790 AD and wrote the book *Hisab al-jabr w'al-muqabala* in 830 AD from which the word algebra (*al-jabr*) is taken.

2.1 Sets of numbers

Natural numbers

A child learns to count: 'one, two, three, four, ...'. These are sometimes called the counting numbers or whole numbers.

The child will say 'I am three', or 'I live at number 73'.

If we include the number 0, then we have the set of numbers called the **natural numbers**. The set of natural numbers $\mathbb{N} = \{0, 1, 2, 3, 4, \ldots\}$.

Integers

On a cold day, the temperature may be 4 °C at 10p.m. If the temperature drops by a further 6 °C, then the temperature is 'below zero'; it is −2 °C.

If you are overdrawn at the bank by £200, this could be shown as −£200.

The set of **integers** $\mathbb{Z} = \{\ldots -3, -2, -1, 0, 1, 2, 3, \ldots\}$.

$\mathbb{Z}$ is therefore an extension of $\mathbb{N}$. Every natural number is an integer.

Rational numbers

A child may say 'I am three'; she may also say 'I am three and a half', or even 'three and a quarter'. $3\frac{1}{2}$ and $3\frac{1}{4}$ are **rational numbers**. All rational numbers can be written as a fraction whose denominator is not zero. All terminating and recurring decimals are rational numbers as they can be written as fractions too, e.g.

$0.2 = \frac{1}{5}$ $\quad 0.3 = \frac{3}{10}$ $\quad 7 = \frac{7}{1}$ $\quad 1.53 = \frac{153}{100}$ $\quad 0.\dot{2} = \frac{2}{9}$

The set of rational numbers $\mathbb{Q}$ is an extension of the set of integers.

Real numbers

Numbers which cannot be expressed as a fraction are not rational numbers; they are **irrational numbers**.

For example, using Pythagoras' rule in the triangle shown below, the length of the hypotenuse AC is found as $\sqrt{2}$:

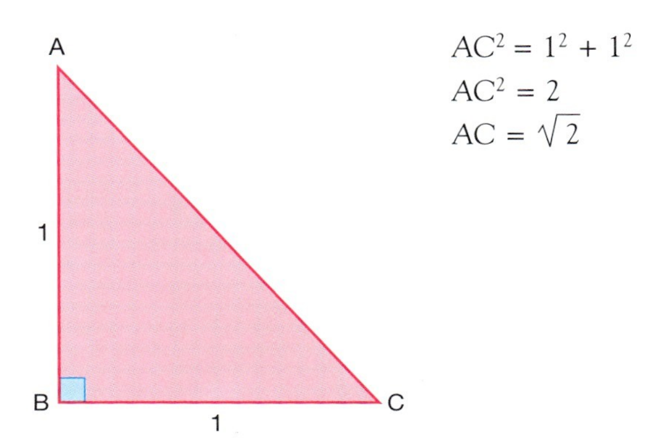

$AC^2 = 1^2 + 1^2$
$AC^2 = 2$
$AC = \sqrt{2}$

$\sqrt{2} = 1.414\,213\,56\ldots$. The digits in this number do not recur or repeat. This is a property of all irrational numbers. Another example of an irrational number which you will come across is π (pi) $= 3.141\,592\,654\ldots$. The set of rational and irrational numbers together form the set of **real numbers** $\mathbb{R}$. There are also numbers, called imaginary numbers, which are not real, but all the numbers that you will come across in this textbook are real numbers.

Exercise 2.1.1

1 State to which of the sets $\mathbb{N}$, $\mathbb{Z}$, $\mathbb{Q}$ and $\mathbb{R}$ these numbers belong.

a 3 b -5 c $\sqrt{3}$ d $11.\dot{3}$

In questions 2–6, state, giving reasons, whether each number is rational or irrational.

2 a 1.3 b $0.\dot{6}$ c $\sqrt{3}$

3 a $-2\frac{2}{5}$ b $\sqrt{25}$ c $\sqrt[3]{8}$

4 a $\sqrt{7}$ b 0.625 c $0.\dot{1}$

5 a $\sqrt{2} \times \sqrt{3}$ b $\sqrt{2} + \sqrt{3}$ c $(\sqrt{2} \times \sqrt{3})^2$

6 a $\dfrac{\sqrt{8}}{\sqrt{2}}$ b $\dfrac{2\sqrt{5}}{\sqrt{20}}$ c $4 + (\sqrt{9} - 4)$

In questions 7–10, state, giving reasons, whether the quantity required is rational or irrational.

7

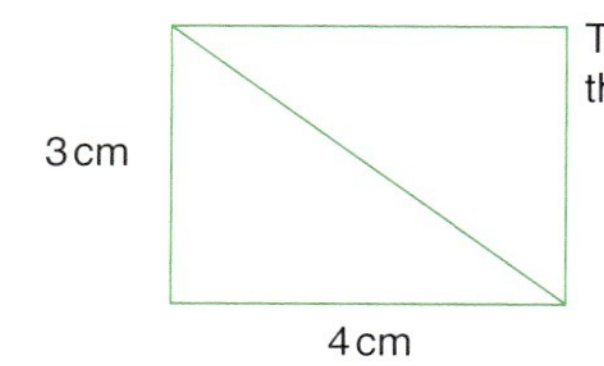

The length of the diagonal

8

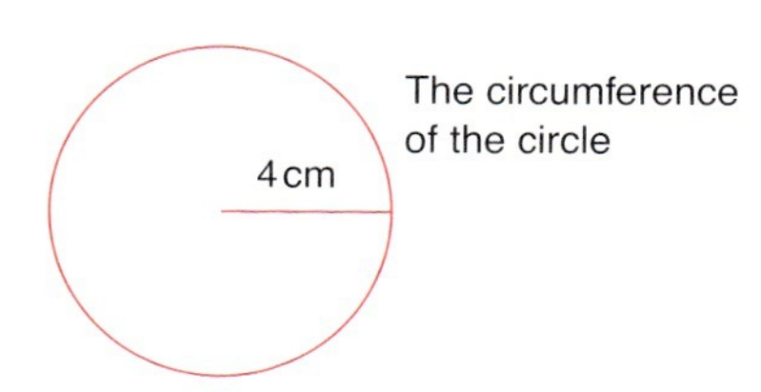

The circumference of the circle

9

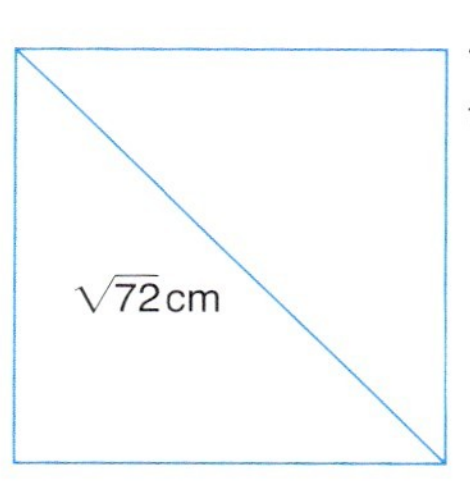

The side length of the square

10

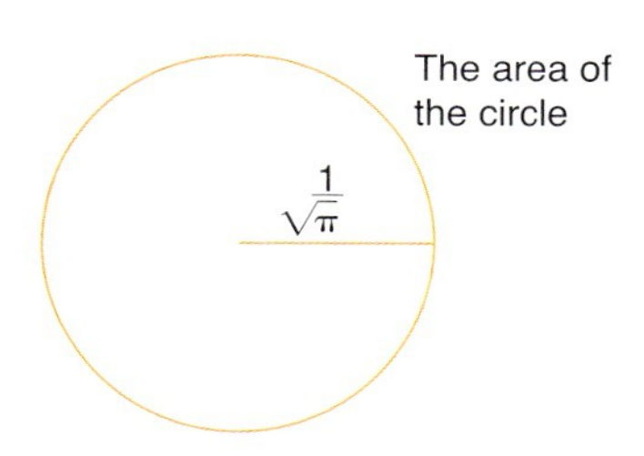

The area of the circle

2.2 Approximation

In many instances exact numbers are not necessary or even desirable. In those circumstances, approximations are given. The approximations can take several forms. The common types of approximation are explained below.

Rounding

If 28 617 people attend a gymnastics competition, this number can be reported to various levels of accuracy.

To the nearest 10 000 the number would be rounded up to 30 000.
To the nearest 1000 the number would be rounded up to 29 000.
To the nearest 100 the number would be rounded down to 28 600.

In this type of situation it is unlikely that the exact number would be reported.

Exercise 2.2.1

1 Round the following numbers to the nearest 1000.

a 68 786   b 74 245   c 89 000
d 4020   e 99 500   f 999 999

2 Round the following numbers to the nearest 100.

a 78 540	b 6858	c 14 099
d 8084	e 950	f 2984

3 Round the following numbers to the nearest 10.

a 485	b 692	c 8847
d 83	e 4	f 997

Decimal places

A number can also be approximated to a given number of decimal places (d.p.). This refers to the number of digits written after a decimal point.

Worked examples

1 Write 7.864 to one decimal place.

The answer needs to be written with one digit after the decimal point. However, to do this, the second digit after the decimal point needs to be considered. If it is 5 or more then the first digit is rounded up. In this case it is 6, so the 8 is rounded up to 9, i.e.

7.864 is written as 7.9 to 1 d.p.

2 Write 5.574 to two decimal places.

The answer here is to be given with two digits after the decimal point. In this case the third digit after the decimal point needs to be considered. As the third digit after the decimal point is less than 5, the second digit is not rounded up, i.e.

5.574 is written as 5.57 to 2 d.p.

Exercise 2.2.2

1 Give the following to one decimal place.

a 5.58	b 0.73	c 11.86
d 157.39	e 4.04	f 15.045
g 2.95	h 0.98	i 12.049

2 Give the following to two decimal places.

a 6.473	b 9.587	c 16.476
d 0.088	e 0.014	f 9.3048
g 99.996	h 0.0048	i 3.0037

Significant figures

Numbers can also be approximated to a given number of significant figures (s.f.). In the number 43.25, the 4 is the most significant figure as it has a value of 40. In contrast, the 5 is the least significant as it has a value of only 5 hundredths. If you are not told otherwise, you are expected to round any answers that are not exact to three significant figures.

Worked examples

1 Write 43.25 to three significant figures.

Only the three most significant figures are written, but the fourth figure needs to be considered to see whether the third figure is to be rounded up or not. Since the fourth figure is 5, the third figure is rounded up, i.e.

43.25 is written as 43.3 to three significant figures.

2 Write 0.0043 to one significant figure.

In this example only two figures have any significance, the 4 and the 3. The 4 is the most significant and therefore is the only one of the two to be written in the answer, i.e.

0.0043 is written as 0.004 to one significant figure.

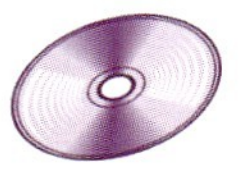

Exercise 2.2.3

1 Write the following to the number of significant figures written in brackets.

a 48 599 (1 s.f.) b 48 599 (3 s.f.) c 6841 (1 s.f.)
d 7538 (2 s.f.) e 483.7 (1 s.f.) f 2.5728 (3 s.f.)
g 990 (1 s.f.) h 2045 (2 s.f.) i 14.952 (3 s.f.)

2 Write the following to the number of significant figures written in brackets.

a 0.085 62 (1 s.f.) b 0.5932 (1 s.f.) c 0.942 (2 s.f.)
d 0.954 (1 s.f.) e 0.954 (2 s.f.) f 0.003 05 (1 s.f.)
g 0.003 05 (2 s.f.) h 0.009 73 (2 s.f.) i 0.009 73 (1 s.f.)

3 Calculate the following, giving your answer to three significant figures.

a 23.456×17.89 b 0.4×12.62 c 18×9.24
d $76.24 \div 3.2$ e 7.6^2 f 16.42^3
g $\frac{2.3 \times 3.37}{4}$ h $\frac{8.31}{2.02}$ i $9.2 \div 4^2$

Estimating answers to calculations

Even though many calculations can be done quickly and effectively on a calculator, an estimate for an answer is often a useful check. This is done by rounding each of the numbers in a way that makes the calculation relatively straightforward.

Worked examples

1 Estimate the answer to 57×246.

Here are two possibilities:
i) $60 \times 200 = 12\,000$
ii) $50 \times 250 = 12\,500$.

2 Estimate the answer to $6386 \div 27$.

$6000 \div 30 = 200$.

Exercise 2.2.4

1 Without using a calculator, estimate the answers to the following.

a 62×19 b 270×12 c 55×60
d 4950×28 e 0.8×0.95 f 0.184×475

2 Without using a calculator, estimate the answers to the following.

a $3946 \div 18$ b $8287 \div 42$ c $906 \div 27$
d $5520 \div 13$ e $48 \div 0.12$ f $610 \div 0.22$

3 Without using a calculator, estimate the answers to the following.

a $78.45 + 51.02$ b $168.3 - 87.09$ c $2.93 \div 3.14$
d $84.2 \div 19.5$ e $\frac{4.3 \times 752}{15.6}$ f $\frac{(9.8)^3}{(2.2)^2}$

4 Using estimation, identify which of the following are definitely incorrect. Explain your reasoning clearly.

a $95 \times 212 = 20\,140$ b $44 \times 17 = 748$
c $689 \times 413 = 28\,457$ d $142\,656 \div 8 = 17\,832$
e $77.9 \times 22.6 = 2512.54$ f $\frac{8.4 \times 46}{0.2} = 19\,366$

5 Estimate the shaded areas of the following shapes. Do *not* work out an exact answer.

a

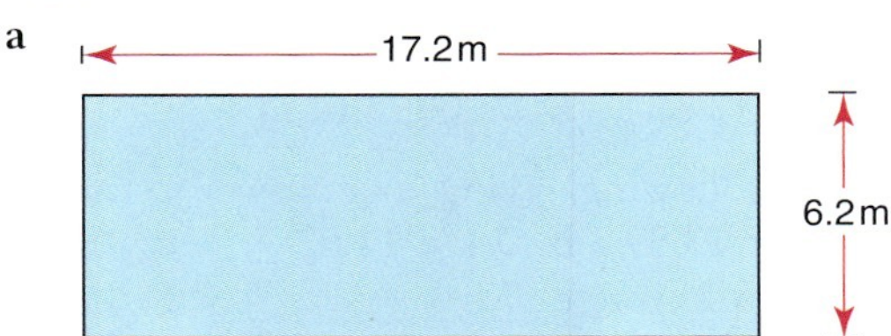

b

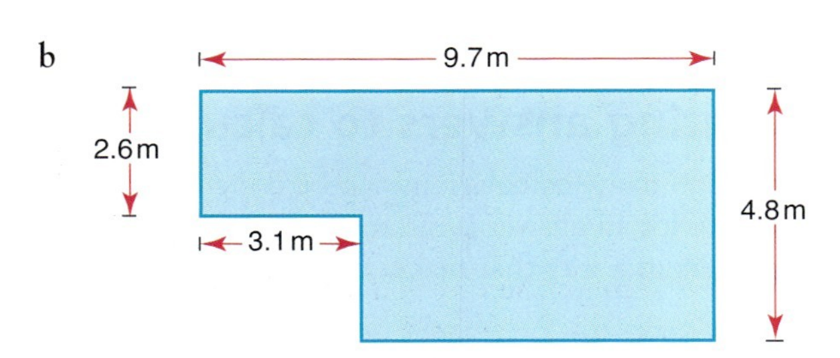

c

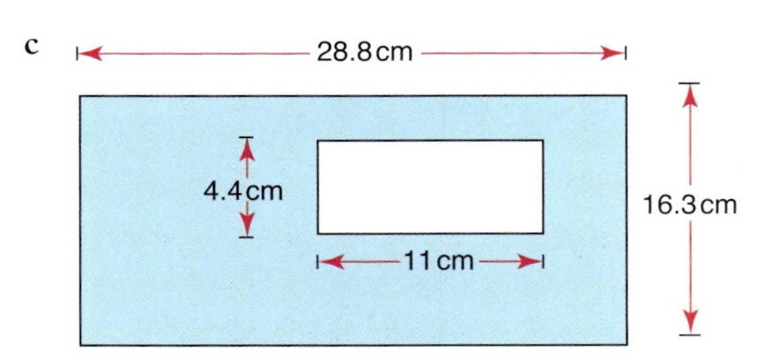

6 Estimate the volume of each of the solids below. Do *not* work out an exact answer.

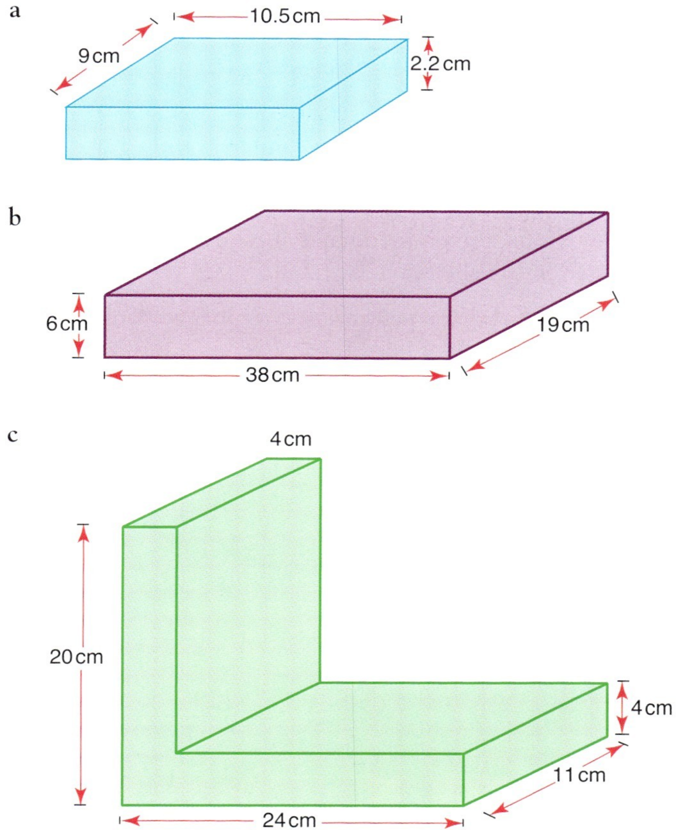

Percentage error

Two golfers are trying to hit a ball to a flag in a hole. The first golfer is 100 m from the hole and lands the ball 4 m short. The second golfer is 20 m from the hole and lands the ball 4 m past it. They have both made a 4 m error.

However, perhaps the first golfer would be happier with his result if the percentage error was calculated.

The percentage error is calculated as follows:

$$\text{Percentage error} = \frac{\text{absolute error}}{\text{total amount}} \times 100$$

Percentage error for golfer 1: $\frac{4}{100} = 4\%$

Percentage error for golfer 2: $\frac{4}{20} = \frac{20}{100} = 20\%$

The first golfer's result now looks much better.

When you are dealing with approximate and exact values you can use this formula.

$$\text{Percentage error} = \frac{\text{absolute error}}{\text{total amount}} \times 100$$

or $$\text{Percentage error, } v = \left|\frac{v_E - v_A}{v_E}\right| \times 100$$

where v_E is the exact value and v_A is the approximate value.

Worked example

Two children estimated their heights. The first child estimated her height at 168 cm, when in fact she was 160 cm tall. The second child estimated her height at 112 cm when in fact she was 120 cm tall.

By calculating the percentage error for each child, calculate which child was better at estimating their height.

$$\text{First child's percentage error} = \left|\frac{v_E - v_A}{v_E}\right| \times 100$$

$$= \left|\frac{160 - 168}{160}\right| \times 100$$

$$= \frac{8}{160} \times 100 = 5\%$$

$$\text{Second child's percentage error} = \left|\frac{v_E - v_A}{v_E}\right| \times 100$$

$$= \left|\frac{120 - 112}{120}\right| \times 100$$

$$= \frac{8}{120} \times 100 = 6\tfrac{2}{3}\%$$

Therefore the first child was better at estimating her height.

Exercise 2.2.5

1 Round the following numbers to two significant figures and calculate the percentage error in doing so.
  a 984 b 2450 c 504

2 Two golfers hit drives. Both estimate that they have hit the ball 250 m. If the first drive is 240 m and the second is 258 m, which player had the smaller percentage error?

3 A plane is flying at 9500 m. If the percentage error is ±2.5%, calculate:
  a the maximum possible height that it is flying
  b the minimum possible height that it is flying.

4 a On the motorway my speedometer reads 120 km h^{-1}, but it has an error of +1.5%. What is my actual speed?
  b At a higher speed, the car shows a speedometer reading of 180 km h^{-1}. What is its percentage error if my true speed is 175 km h^{-1}?

2.3 Standard form

Galileo Galilei (1564–1642)

Galileo was an Italian astronomer and physicist. He was the first person accredited with having used a telescope to study the stars. In 1610 Galileo and a German astronomer, Marius, independently discovered Jupiter's four largest moons: Io, Europa, Ganymede and Callisto. At that time it was believed that the Sun revolved around the Earth. Galileo was one of the few people who believed that the Earth revolved around the Sun. As a result of this, the Church declared that he was a heretic and imprisoned him. It took the Church a further 350 years to officially accept that he was correct; he was pardoned only in 1992.

Facts about Jupiter:

It has a mass of 1 900 000 000 000 000 000 000 000 000 kg
It has a diameter of 142 800 000 m
It has a mean distance from the Sun of 778 000 000 km

If numbers are written in the normal way, as here, they become increasingly difficult to read and laborious to write the larger they become. We can write very large numbers or very small numbers in the form $a \times 10^k$, where a lies in the range $1 \le a < 10$ and the index is a positive or negative integer, which can be expressed as $k \in \mathbb{Z}$. This is known as writing a number in **standard form** or **scientific notation**.

A positive index

$100 = 1 \times 10^2$
$1000 = 1 \times 10^3$
$10\,000 = 1 \times 10^4$
$3000 = 3 \times 10^3$

The number 3100 can be written in many different ways, for example:

3.1×10^3 $\quad 31 \times 10^2$ $\quad 0.31 \times 10^4$ etc.

However, only 3.1×10^3 is written in the form $a \times 10^k$, where $1 \le a < 10$ and $k \in \mathbb{Z}$.

Worked examples

1 Write 72 000 in the form $a \times 10^k$, where $1 \le a < 10$ and $k \in \mathbb{Z}$.

7.2×10^4

2 Of the numbers below, ring those which are written in the form $a \times 10^k$, where $1 \le a < 10$ and $k \in \mathbb{Z}$.

(4.2×10^3) $\quad 0.35 \times 10^2$ $\quad 18 \times 10^5$ $\quad$ (6×10^3) $\quad 0.01 \times 10^1$

3 Multiply the following and write the answer in the form $a \times 10^k$, where $1 \le a < 10$ and $k \in \mathbb{Z}$.

600×4000

600×4000
$= 2\,400\,000$
$= 2.4 \times 10^6$

4 Multiply the following and write the answer in the form $a \times 10^k$, where $1 \leq a < 10$ and $k \in \mathbb{Z}$.

$(2.4 \times 10^4) \times (5 \times 10^7)$

$(2.4 \times 10^4) \times (5 \times 10^7)$
$= (2.4 \times 5) \times 10^{4+7}$
$= 12 \times 10^{11}$
$= 1.2 \times 10^{12}$

5 Divide the following and write the answer in the form $a \times 10^k$, where $1 \leq a < 10$ and $k \in \mathbb{Z}$.

$(6.4 \times 10^7) \div (1.6 \times 10^3)$

$(6.4 \times 10^7) \div (1.6 \times 10^3)$
$= (6.4 \div 1.6) \times 10^{7-3}$
$= 4 \times 10^4$

6 Add the following and write the answer in the form $a \times 10^k$, where $1 \leq a < 10$ and $k \in \mathbb{Z}$.

$(3.8 \times 10^6) + (8.7 \times 10^4)$

Change the indices to the same value, giving the sum:
$(380 \times 10^4) + (8.7 \times 10^4)$
$= 388.7 \times 10^4$
$= 3.887 \times 10^6$

7 Subtract the following and write the answer in the form $a \times 10^k$, where $1 \leq a < 10$ and $k \in \mathbb{Z}$.

$(6.5 \times 10^7) - (9.2 \times 10^5)$

Change the indices to the same value, giving:
$(650 \times 10^5) - (9.2 \times 10^5)$
$= 640.8 \times 10^5$
$= 6.408 \times 10^7$

GDCs have a button which allows you to enter a number in the form $a \times 10^k$, where $1 \leq a < 10$ and $k \in \mathbb{Z}$ and they will also give answers in this form if the answer is very large.

For example, enter the number 8×10^4 into the calculator:

Casio	Texas
CLIP N 8, π ” EXP, CATALOG P 4	8, 2ND, EE J ,, L4 T 4
Note: A number such as 1 000 000 000 000 000 would appear on the screen as 1E + 15. You should write this as 1×10^{15}, not as your calculator displays it.	Note: A number such as 1 000 000 000 000 000 would appear on the screen as 1 E 15. You should write this as 1×10^{15}, not as your calculator displays it.

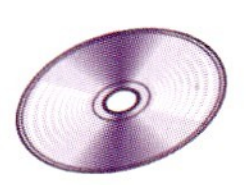

Exercise 2.3.1

1 Which of the following are not in the form $a \times 10^k$, where $1 \le a < 10$ and $k \in \mathbb{Z}$?
  a 6.2×10^5 b 7.834×10^{16} c 8.0×10^5
  d 0.46×10^7 e 82.3×10^6 f 6.75×10^1

2 Write the following numbers in the form $a \times 10^k$, where $1 \le a < 10$ and $k \in \mathbb{Z}$.
  a 600 000 b 48 000 000 c 784 000 000 000
  d 534 000 e 7 million f 8.5 million

3 Write the following in the form $a \times 10^k$, where $1 \le a < 10$ and $k \in \mathbb{Z}$.
  a 68×10^5 b 720×10^6 c 8×10^5
  d 0.75×10^8 e 0.4×10^{10} f 50×10^6

4 Multiply the following and write your answers in the form $a \times 10^k$, where $1 \le a < 10$ and $k \in \mathbb{Z}$.
  a 200×3000 b 6000×4000
  c 7 million $\times$ 20 d 500 $\times$ 6 million
  e 3 million $\times$ 4 million f 4500×4000

5 Light from the Sun takes approximately 8 minutes to reach Earth. If light travels at a speed of $3 \times 10^8\,\text{m s}^{-1}$, calculate to three significant figures (s.f.) the distance from the Sun to the Earth.

6 Find the value of the following and write your answers in the form $a \times 10^k$, where $1 \le a < 10$ and $k \in \mathbb{Z}$.
  a $(4.4 \times 10^3) \times (2 \times 10^5)$ b $(6.8 \times 10^7) \times (3 \times 10^3)$
  c $(4 \times 10^5) \times (8.3 \times 10^5)$ d $(5 \times 10^9) \times (8.4 \times 10^{12})$
  e $(8.5 \times 10^6) \times (6 \times 10^{15})$ f $(5.0 \times 10^{12})^2$

7 Find the value of the following and write your answers in the form $a \times 10^k$, where $1 \le a < 10$ and $k \in \mathbb{Z}$.
  a $(3.8 \times 10^8) \div (1.9 \times 10^6)$ b $(6.75 \times 10^9) \div (2.25 \times 10^4)$
  c $(9.6 \times 10^{11}) \div (2.4 \times 10^5)$ d $\dfrac{1.8 \times 10^{12}}{9.0 \times 10^7}$
  e $\dfrac{2.3 \times 10^{11}}{9.2 \times 10^4}$ f $\dfrac{2.4 \times 10^8}{6.0 \times 10^3}$

8 Find the value of the following and write your answers in the form $a \times 10^k$, where $1 \le a < 10$ and $k \in \mathbb{Z}$.
  a $(3.8 \times 10^5) + (4.6 \times 10^4)$ b $(7.9 \times 10^9) + (5.8 \times 10^8)$
  c $(6.3 \times 10^7) + (8.8 \times 10^5)$ d $(3.15 \times 10^9) + (7.0 \times 10^6)$
  e $(5.3 \times 10^8) - (8.0 \times 10^7)$ f $(6.5 \times 10^7) - (4.9 \times 10^6)$
  g $(8.93 \times 10^{10}) - (7.8 \times 10^9)$ h $(4.07 \times 10^7) - (5.1 \times 10^6)$

9 The following list shows the distance of the planets of the Solar System from the Sun.

Jupiter	778 million kilometres
Mercury	58 million kilometres
Mars	228 million kilometres
Uranus	2870 million kilometres
Venus	108 million kilometres
Neptune	4500 million kilometres
Earth	150 million kilometres
Saturn	1430 million kilometres

Write each of the distances in the form $a \times 10^k$, where $1 \le a < 10$ and $k \in \mathbb{Z}$ and then arrange them in order of magnitude, starting with the distance of the planet closest to the Sun.

A negative index

A negative index is used when writing a number between 0 and 1 in the form $a \times 10^k$, where $1 \le a < 10$ and $k \in \mathbb{Z}$.

e.g.
$100 = 1 \times 10^2$
$10 = 1 \times 10^1$
$1 = 1 \times 10^0$
$0.1 = 1 \times 10^{-1}$
$0.01 = 1 \times 10^{-2}$
$0.001 = 1 \times 10^{-3}$
$0.0001 = 1 \times 10^{-4}$

Note that a must still lie within the range $1 \le a < 10$.

Worked examples

1 Write 0.0032 in the form $a \times 10^k$, where $1 \le a < 10$ and $k \in \mathbb{Z}$.

3.2×10^{-3}

2 Write these numbers in order of magnitude, starting with the largest.

3.6×10^{-3} 5.2×10^{-5} 1×10^{-2} 8.35×10^{-2} 6.08×10^{-8}

8.35×10^{-2} 1×10^{-2} 3.6×10^{-3} 5.2×10^{-5} 6.08×10^{-8}

Exercise 2.3.2

1 Write the following numbers in the form $a \times 10^k$, where $1 \le a < 10$ and $k \in \mathbb{Z}$.
a 0.0006 b 0.000 053 c 0.000 864
d 0.000 000 088 e 0.000 0007 f 0.000 4145

2 Write the following numbers in the form $a \times 10^k$, where $1 \le a < 10$ and $k \in \mathbb{Z}$.
a 68×10^{-5} b 750×10^{-9} c 42×10^{-11}
d 0.08×10^{-7} e 0.057×10^{-9} f 0.4×10^{-10}

3 Deduce the value of k in each of the following cases.

a $0.000\,25 = 2.5 \times 10^k$ **b** $0.00357 = 3.57 \times 10^k$

c $0.000\,000\,06 = 6 \times 10^k$ **d** $0.004^2 = 1.6 \times 10^k$

e $0.000\,65^2 = 4.225 \times 10^k$

4 Write these numbers in order of magnitude, starting with the largest.

3.2×10^{-4} 6.8×10^5 5.57×10^{-9} 6.2×10^3

5.8×10^{-7} 6.741×10^{-4} 8.414×10^2

2.4 SI units of measurement

A soldier in Julius Caesar's army could comfortably march 20 miles in one day, wearing full kit, and then help to build a defensive blockade.

The mile was a unit of length based on 1000 strides of a Roman legionary. The measurement was sufficiently accurate for its purpose but only an approximate distance.

Most measures started as rough estimates. The yard (3 feet or 36 inches) was said to be the distance from the king's nose (reputed to be Henry I of England) to the tip of his extended finger. As it became necessary to have standardization in measurement, the measures themselves became more exact.

In 1791 during the French Revolution, a new unit of measurement, the metre, was defined in France. Originally it was defined as 'one ten-millionth of the length of the quadrant of the Earth's meridian through Paris'. This use of this unit of measurement became law in France in 1795.

However, this measurement was not considered sufficiently accurate and further definitions were required.

In 1927 a metre was defined as the distance between two marks on a given platinum–iridium bar. This bar is kept in Paris.

In 1960 the definition was based on the emission of a krypton-86 lamp.

At the 1983 General Conference on Weights and Measures, the metre was redefined as the length of the path travelled by light in a vacuum in $\frac{1}{299\,792\,488}$ second. This definition, although not very neat, can be considered one of the few 'accurate' measures. Most measures are only to a degree of accuracy.

SI is an abbreviation of *Système International d'Unités*. Its seven base units are listed below.

Quantity	Unit	Symbol
Distance	metre	m
Mass	kilogram	kg
Time	second	s
Electrical current	ampere	A
Temperature	kelvin	K
Substance	mole	mol
Intensity of light	candela	cd

The SI has other derived units. The following questions highlight some of the more common derived units and their relationship to the base units.

Exercise 2.4.1

1 Copy and complete the sentences below.
  - **a** There are ______ centimetres in one metre.
  - **b** A centimetre is ______ part of a metre.
  - **c** There are ______ metres in one kilometre.
  - **d** A metre is ______ part of a kilometre.
  - **e** There are ______ grams in one kilogram.
  - **f** A gram is ______ part of a kilogram.
  - **g** A kilogram is ______ part of a tonne.
  - **h** There are ______ millilitres in one litre.
  - **i** One thousandth of a litre is ______ .
  - **j** There are ______ grams in one tonne.

2 Which of the units below would be used to measure each of the following?

millimetre	centimetre	metre	kilometre
milligram	gram	kilogram	tonne
millilitre	litre		

  - **a** Your mass (weight)
  - **b** The length of your foot
  - **c** Your height
  - **d** The amount of water in a glass
  - **e** The mass of a ship
  - **f** The height of a bus
  - **g** The capacity of a swimming pool
  - **h** The length of a road
  - **i** The capacity of the fuel tank of a truck
  - **j** The size of your waist

3 Draw five lines of different lengths.
  - **a** Estimate the length of each line in millimetres.
  - **b** Measure the length of each line to the nearest millimetre.

4 Write an estimate for each of the following using the correct unit.
  - **a** Your height
  - **b** Your weight (mass)
  - **c** The capacity of a cup
  - **d** The distance to the nearest town
  - **e** The mass of an orange
  - **f** The quantity of blood in the human body
  - **g** The depth of the Pacific Ocean
  - **h** The distance to the moon
  - **i** The mass of a car
  - **j** The capacity of a swimming pool

Converting from one unit to another

Length

1 km is 1000 m, so:

to change from kilometres to metres, multiply by 1000
to change from metres to kilometres, divide by 1000.

Worked examples

1 Change 5.84 km to metres.

1 km = 1000 m so multiply by 1000
$5.84 \times 1000 = 5840$ m

2 Change 3640 mm to metres.

1 m = 1000 mm so divide by 1000
$3640 \div 1000 = 3.64$ m

Exercise 2.4.2

1 Convert these to millimetres.

a 4 cm	b 6.2 cm	c 28 cm	d 1.2 m
e 0.88 m	f 3.65 m	g 0.008 m	h 0.23 cm

2 Convert these to metres.

a 260 cm	b 8900 cm	c 2.3 km	d 0.75 km
e 250 cm	f 0.4 km	g 3.8 km	h 25 km

3 Convert these to kilometres.

a 2000 m	b 26 500 m	c 200 m	d 750 m
e 100 m	f 5000 m	g 15 000 m	h 75 600 m

Mass

1 tonne is 1000 kg, so:

to change from tonnes to kilograms multiply by 1000
to change from kilograms to tonnes divide by 1000.

Worked examples

1 Change 0.872 tonne to kilograms.

1 tonne is 1000 kg so multiply by 1000.
$0.872 \times 1000 = 872$ kg

2 Change 4200 kg to tonnes.

1 tonne = 1000 kg so divide by 1000.
$4200 \div 1000 = 4.2$ tonnes

Capacity

1 litre is 1000 ml, so:

to change from litres to millilitres multiply by 1000
to change from millilitres to litres divide by 1000.

Worked examples

1 Change 2.4 ℓ to millilitres.

1 ℓ is 1000ml so multiply by 1000.
$2.4 \times 1000 = 2400$ ml

2 Change 4500 ml to litres.

1 ℓ = 1000 ml so divide by 1000
$4500 \div 1000 = 4.5$ ℓ

Exercise 2.4.3

1 Convert these to kilograms.
a 2 tonne b 7.2 tonne c 2800 g d 750 g
e 0.45 tonne f 0.003 tonne g 6500 g h 7 000 000 g

2 Convert these to millilitres.
a 2.6 ℓ b 0.7 ℓ c 0.04 ℓ d 0.008 ℓ

3 Convert these to litres.
a 1500 ml b 5280 ml c 750 ml d 25 ml

4 The masses of four containers loaded on a ship are 28 tonnes, 45 tonnes, 16.8 tonnes and 48 500 kg.
a What is the total mass in tonnes?
b What is the total mass in kilograms? Write your answer in the form $a \times 10^k$, where $1 \leq a < 10$ and $k \in \mathbb{Z}$.

5 Three test tubes contain 0.08 ℓ, 0.42 ℓ and 220 ml.
a What is the total in millilitres?
b How many litres of water need to be added to make the solution up to 1.25 ℓ?

Temperature scales

You will have come across three temperature scales in science lessons.

Fahrenheit °F Celsius °C kelvin K

The kelvin is the official SI unit of temperature. It is identical to the Celsius scale (in that a 1 °C change is equivalent to a 1 K change, except that it starts at 0 K, which is equivalent to −273 °C. This temperature is known as absolute zero.

Absolute zero = 0 K = −273 °C

The table shows the conversion between Fahrenheit, Celsius and kelvin scales of temperature.

Scale	Freezing point of water	Boiling point of water
Fahrenheit	32	212
Celsius	0	100
Kelvin	273	373

The number of graduations (scale points of 1 degree) between the freezing and boiling points of water in the Celsius and Fahrenheit scales is 100 and 180 respectively. Therefore:

a change of 100 °C represents a change of 180 °F

a change of 1 °C represents a change of $\frac{180}{100} = \frac{9}{5}$ °F.

Because we know that the freezing point of water is 0 °C and 32 °F, it is possible to derive a formula for converting between Celsius and Fahrenheit.

$T_f = \frac{9}{5}T_c + 32$ which rearranged gives $T_c = \frac{5}{9}(T_f - 32)$

where T_f represents the temperature in degrees Fahrenheit and T_c represents the temperature in degrees Celsius.

An approximate conversion from degrees Celsius to degrees Fahrenheit can be obtained from the formula:

$T_f = 2T_c + 30$

Similarly, because we know that a change on the Celsius scale represents the same change on the kelvin scale and that the freezing point of water is 0 °C and 273 K, it is possible to derive a formula to convert between degrees Celsius and kelvin.

$T_k = T_c + 273$, which rearranged gives $T_c = T_k - 273$

where T_k represents the temperature in kelvin.

Worked example

Convert 40 °C into:

a degrees Fahrenheit
b kelvin.

a $T_f = \frac{9}{5}T_c + 32$

$T_f = \frac{9}{5} \times 40 + 32$

$T_f = 104$

40 °C is 104 °F.

b $T_k = T_c + 273$

$T_k = 40 + 273$

$T_k = 313$

40 °C is 313 K.

Exercise 2.4.4

1 Convert the following to degrees Fahrenheit.

a 20 °C **b** 80 °C **c** 200 °C

2 Convert the following to degrees Celsius.

a 50 °F **b** 275 °F **c** 887 °F

3 Convert the following to kelvin.

a 30 °C **b** −120 °C **c** 95 °F

4 Using the approximate conversion formula, convert the following to degrees Fahrenheit.
a 20 °C b 80 °C c 200 °C

5 Calculate the percentage error in your answers to question 4 above compared with your answers to question 1.

6 At what temperature in degrees Celsius and degrees Fahrenheit does the approximate formula give the same answer as the accurate formula?

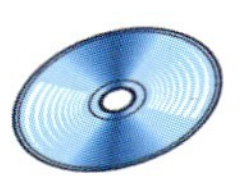

2.5 Arithmetic sequences and series

A **sequence** is a collection of terms arranged in a specific order, where each term is obtained according to a rule. Examples of some simple sequences are given below.

2, 4, 6, 8, 10   1, 4, 9, 16, 25   1, 2, 4, 8, 16

1, 1, 2, 3, 5, 8   1, 8, 27, 64, 125   10, 5, $\frac{5}{2}$, $\frac{5}{4}$, $\frac{5}{8}$

Discuss with another the student the rules involved in producing the sequences above.

The terms of a sequence can be expressed as $u_1, u_2, u_3, \ldots, u_n$ where

u_1 is the first term
u_2 is the second term

u_n is the nth term

Therefore in the sequence 2, 4, 6, 8, 10, $u_1 = 2$, $u_2 = 4$, etc.

Arithmetic sequences

In an **arithmetic sequence** there is a common difference (d) between successive terms, e.g.

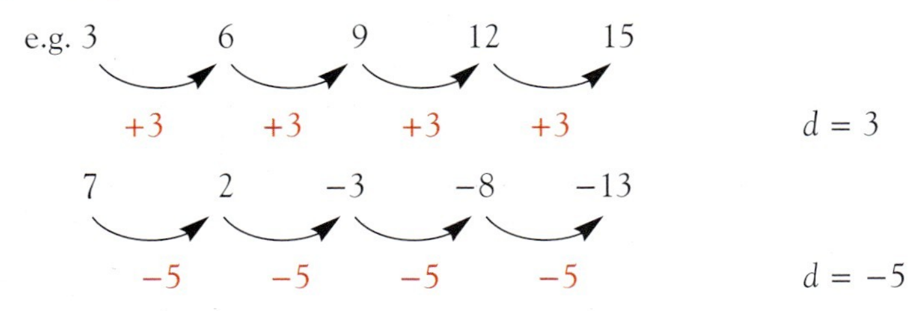

Formulae for the terms of an arithmetic sequence
There are two main ways to describe a sequence.

1 A term-to-term rule, known as a **recurrence relation**.
In the following sequence,

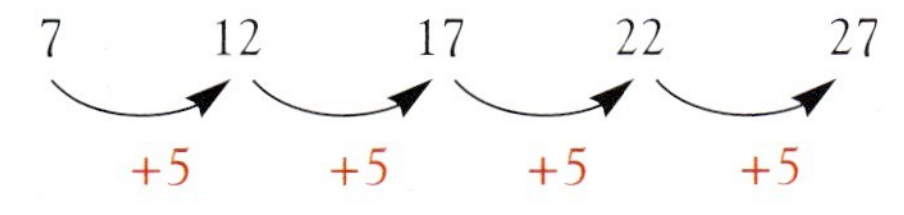

the recurrence relation is +5, i.e. $u_2 = u_1 + 5$, $u_3 = u_2 + 5$, etc.

The general form is therefore written as $u_{n+1} = u_n + 5$, $u_1 = 7$, where u_n is the nth term and u_{n+1} is the term after the nth term.

Note: It is important to give the value of one of the terms, e.g. u_1, so that the exact sequence can be generated.

2 A formula for the nth term of a sequence.
This type of rule links each term to its position in the sequence, e.g.

Position	1	2	3	4	5	n
Term	7	12	17	22	27	

We can deduce from the figures above that each term can be calculated by multiplying its position number by 5 and adding 2. Algebraically this can be written as the formula for the nth term:

$$u_n = 5n + 2$$

A GDC can be used to generate the terms of a sequence from the recurrence relation. For example, to calculate u_2, u_3 and u_4 for the sequence $u_{n+1} = 2u_n - 5$, $u_1 = 1$, the following can be done:

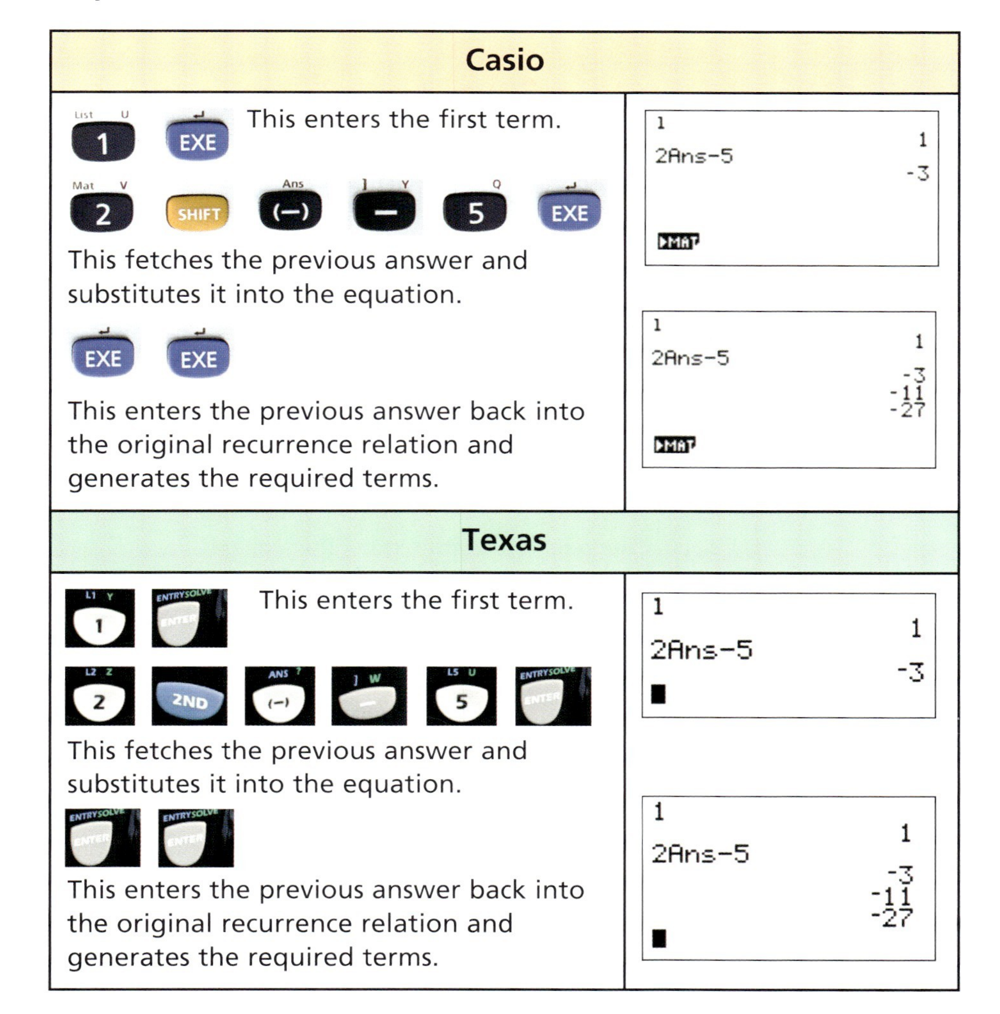

Therefore $u_2 = -3$, $u_3 = -11$ and $u_4 = -27$.

Note: This sequence is not arithmetic as the difference between successive terms is not constant.

With an arithmetic sequence, the rule for the nth term can be easily deduced by looking at the common difference, e.g.

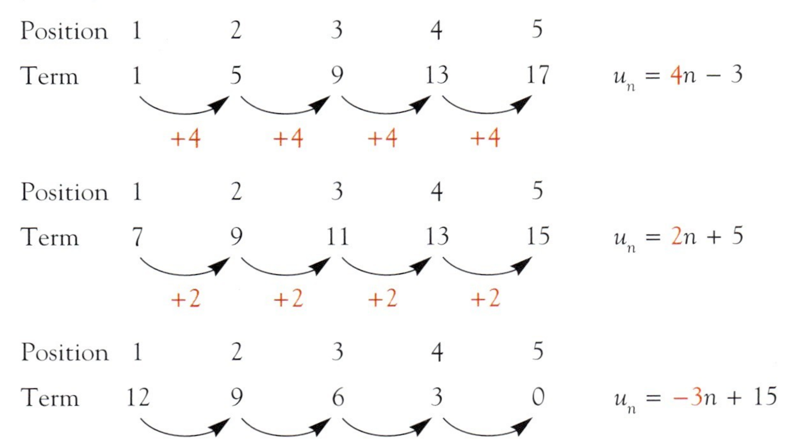

The common difference is the coefficient of n (i.e. the number by which n is multiplied). The constant is then worked out by calculating the number needed to make the term.

Worked examples

1 A sequence is described by the following recurrence relation:

$u_{n+1} = 4u_n - 3$, $u_1 = 2$

Calculate u_2, u_3 and u_4 and state whether the sequence is arithmetic or not.

$u_2 = 4u_1 - 3$ therefore $u_2 = 4 \times 2 - 3 = 5$
$u_3 = 4u_2 - 3$ therefore $u_3 = 4 \times 5 - 3 = 17$
$u_4 = 4u_3 - 3$ therefore $u_4 = 4 \times 17 - 3 = 65$

The sequence is not arithmetic as the difference between terms is not constant.

2 Find the rule for the nth term of the sequence 12, 7, 2, −3, −8....

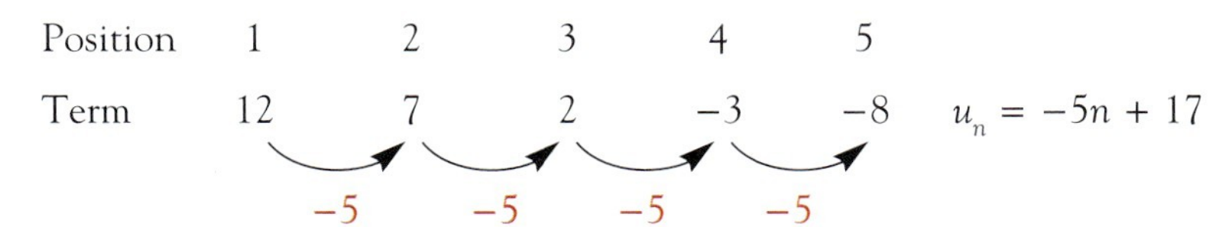

Exercise 2.5.1

1 For each of the following sequences, the recurrence relation and u_1 are given. Calculate u_2, u_3 and u_4 and state whether or not the sequence is arithmetic.

a $u_{n+1} = u_n + 5$, $u_1 = 3$
b $u_{n+1} = 2u_n - 4$, $u_1 = 1$
c $u_{n+1} = -4u_n + 1$, $u_1 = 0$
d $u_{n+1} = 3 - u_n$, $u_1 = 5$
e $u_{n+1} = -4 + u_n$, $u_1 = 8$
f $u_{n+1} = 6 - \frac{1}{3}u_n$, $u_1 = -9$

2 For each of the following sequences:
i) deduce the formula for the nth term
ii) calculate the 10th term.

a 5, 8, 11, 14, 17
b 0, 4, 8, 12, 16
c $\frac{1}{2}, 1\frac{1}{2}, 2\frac{1}{2}, 3\frac{1}{2}, 4\frac{1}{2}$
d 6, 3, 0, −3, −6
e −7, −4, −1, 2, 5
f −9, −13, −17, −21, −25

3 Copy and complete each of the following tables showing arithmetic sequences.

a

Position	1	2	5		50	n
Term				45		$4n - 3$

b

Position	1	2	5			n
Term				59	449	$6n - 1$

c

Position	1				100	n
Term		0	−5	−47		$-n + 3$

d

Position	1	2	3			n
Term	3	0	−3	−24	−294	

e

Position		5	7			n
Term	1	10	16	25	145	

f

Position	1	2	5		50	n
Term	−5.5	−7		−34		

4 For each of the following arithmetic sequences:
i) deduce the common difference d
ii) give the formula for the nth term
iii) calculate the 50th term.

a 5, 9, 13, 17, 21
b 0, __ , 2 , __ , 4
c −10 , __ , __ , __ , 2
d $u_1 = 6, u_9 = 10$
e $u_3 = -50, u_{20} = 18$
f $u_5 = 60, u_{12} = 39$

5 Judi invests £200 in a bond which pays simple interest of 12.5% per year. She receives a cheque each year for the interest. How many years is it before the bond pays for itself?

Series

When the terms of a sequence are added together, the sum is called a **series**. There is specific notation associated with series, with which you will need to be familiar.

Consider the arithmetic sequence 3, 6, 9, 12, 15, …
S_n is used to describe the sum of the first n terms of a sequence.
S_4 therefore denotes the sum of the first four terms of the sequence, i.e.

$S_4 = 3 + 6 + 9 + 12 = 30$

The Greek letter Σ is also used to indicate that the terms of a sequence are to be added. In the sequence above, the formula for the nth term is $u_n = 3n$.

$\sum_1^4 3n$ denotes that the first to the fourth terms of the sequence $u_n = 3n$ are to be added:

$\sum_1^4 3n = 3 + 6 + 9 + 12 = 30$

Similarly $\sum_6^8 3n$ denotes that the sixth to the eight terms of the sequence $u_n = 3n$ are to be added:

$\sum_6^8 3n = 18 + 21 + 24 = 63$

Worked examples

1 Write in full the terms of the series $\sum_1^5 2n - 6$.

$\sum_1^5 2n - 6 = -4 + -2 + 0 + 2 + 4$

2 Evaluate $\sum_3^7 (-n + 4)$.

$\sum_3^7 (-n + 4) = 1 + 0 + -1 + -2 + -3 = -5$

3 Write the series $0 + \frac{1}{2} + 1 + 1\frac{1}{2} + 2 + 2\frac{1}{2}$ using the Σ notation.

The formula for the nth term is $u_n = \frac{1}{2}n - \frac{1}{2}$

Therefore the series can be written as $\sum_1^6 \left(\frac{1}{2}n - \frac{1}{2}\right)$

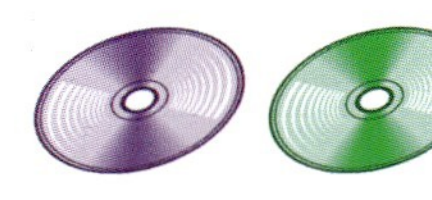

The sum of an arithmetic series

Consider the sum of the first 100 positive integers, i.e. $1 + 2 + 3 + \ldots + 98 + 99 + 100$. There is an efficient way of calculating the sum of the numbers without having to add them all one by one.

$1 + 2 + 3 + 4 + 5 + 6 + \ldots + 95 + 96 + 97 + 98 + 99 + 100$

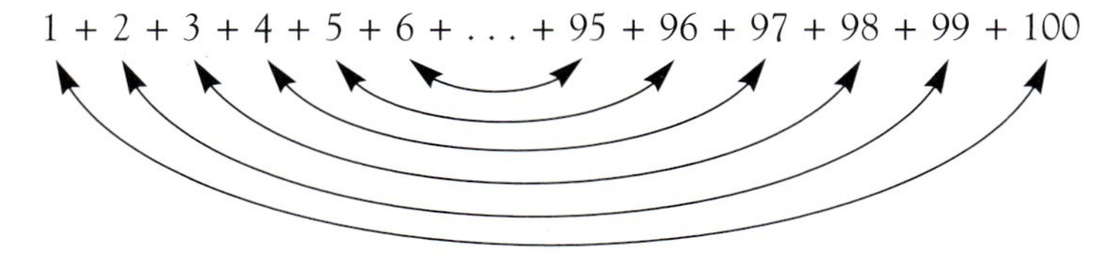

The terms of the series can be paired up as shown above. The sum of each pair is 101 and there are 50 pairs. Therefore the sum of the series is $50 \times 101 = 5050$. This can be generalized for any arithmetic series as

$$S_n = \frac{n}{2}(u_1 + u_n)$$

where u_1 is the first term, u_n the last term in this case and n is the number of terms.

The nth term u_n can also be written in terms of u_1, n and d (the common difference).

Consider the sequence 2, 5, 8, 11, ... To calculate the 20th term, we can use the formula for the nth term, $u_n = 3n - 1$, or we can consider how many times the common difference d needs to be added to the first term in order to reach the 20th term:

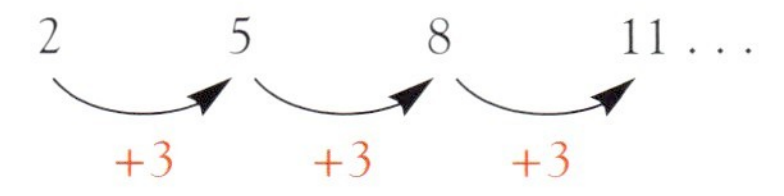

To get to the 2nd term, +3 is added once to the 1st term (2). To get to the 3rd term, +3 is added twice to the 1st term. Therefore to get to the 20th term, +3 will be added 19 times.

$u_{20} = 2 + (19 \times 3) = 59$

In general terms $u_n = u_1 + (n - 1)d$

If this is substituted for u_n in the formula $S_n = \frac{n}{2}(u_1 + u_n)$, then another formula for S_n is derived.

$$S_n = \frac{n}{2}(u_1 + u_1 + (n - 1)d)$$

$$S_n = \frac{n}{2}(2u_1 + (n - 1)d)$$

For the sequence 2, 5, 8, 11, ..., the value of $S_{20} = \frac{20}{2}(2 \times 2 + (20 - 1)3) = 610$.

You can also use your GDC to calculate the sum of an arithmetic sequence. Although you will be expected to show your working in an exam, the GDC will allow you to quickly check your answer.

For example, to evaluate $\sum_{1}^{20}(3n - 1)$:

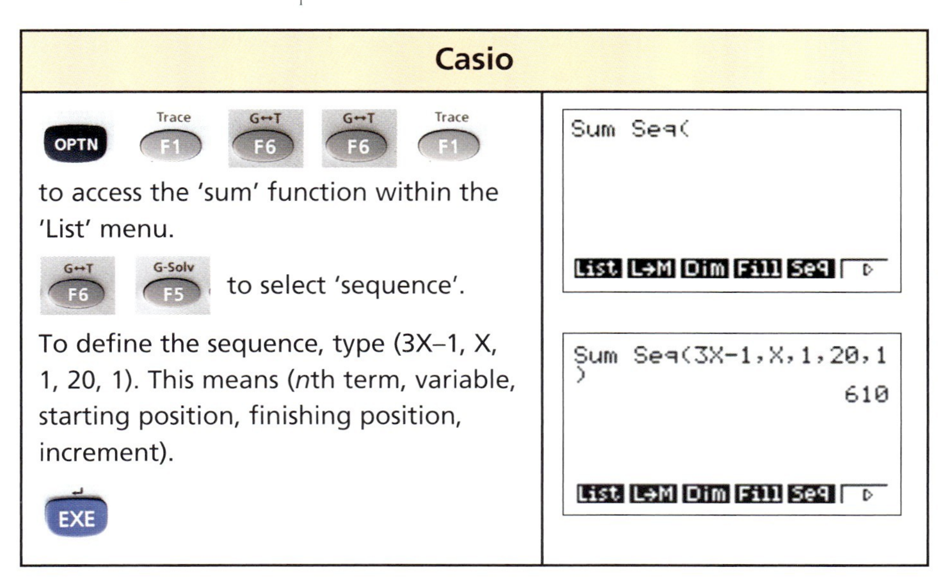

Texas	
  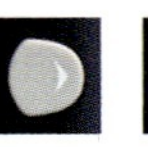  to access the 'sum' function within the 'MATH' menu.	sum(seq(
    to select 'sequence' from the OPS menu. To define the sequence, type (3X – 1, X, 1, 20, 1). This means (*n*th term, variable, starting position, finishing position, increment). 	sum(seq(3X-1,X,1,20,1) 610

Worked examples

1 Evaluate $\sum_{1}^{100} 2n$:

This is the sum of the first 100 even numbers.

$$S_{100} = \frac{100}{2}(2 \times 2 + (100 - 1)2) = 10100$$

2 Evaluate $\sum_{50}^{100}(2n - 8)$.

This is the sum of the 50th to 100th terms of the sequence $u_n = 2n - 8$.
For this sequence the common difference $d = 2$.
The sum can be calculated in two ways.

i) $$\sum_{50}^{100}(2n - 8) = \sum_{1}^{100}(2n - 8) - \sum_{1}^{49}(2n - 8)$$
$$= 9300 - 2058$$
$$= 7242$$

ii) $$\sum_{50}^{100}(2n - 8) = u_{50} + u_{51} + \ldots + u_{100}$$
$$= 92 + 94 + \ldots + 192$$

i.e. $u_1 = 92, n = 51, d = 2$

Therefore $= \sum_{50}^{100}(2n - 8) = \frac{51}{2}(2 \times 92 + (51 - 1)2) = 7242$

3 Four consecutive terms of an arithmetic sequence are $x + 8$, $2x + 6$, $4x - 8$ and $3x + 14$.

a Calculate the sum of the four terms in terms of x.

b Calculate the sum of the four terms.

a $x + 8 + 2x + 6 + 4x - 8 + 3x + 14 = 10x + 20$

b The common difference $d = 2x + 6 - (x + 8) = x - 2$
The common difference d is also $= 4x - 8 - (2x + 6) = 2x - 14$

Therefore $x - 2 = 2x - 14$
$x = 12$

The sum of the four terms $= 10x + 20$
$= 10 \times 12 + 20$
$= 140$

Exercise 2.5.2

1 Find the sum of each of the following arithmetic series.

a $3 + 8 + 13 + \ldots + 53$

b $\frac{1}{2} + 4\frac{1}{2} + 8\frac{1}{2} + \ldots + 60\frac{1}{2}$

c $21 + 17 + 13 + \ldots + -43$

d $100 + 95 + \ldots + -100$

2 Evaluate each of the following.

a $\sum_{1}^{10}(4 - n)$

b $\sum_{1}^{20}\left(\frac{n}{2} - 10\right)$

c $\sum_{10}^{20}(3n - 50)$

d $\sum_{1}^{n}\left(\frac{n}{2} + 4\right)$

3 The second and sixth terms of an arithmetic series are -2 and 10 respectively. Calculate:

a the common difference
b the 1st term
c the 20th term
d S_{20}.

4 The tenth term of an arithmetic series is 18.5. If $S_{10} = 95$, calculate:

a the first term
b the common difference
c S_{50}.

5 The fifth, sixth and seventh terms of an arithmetic series are $3 - 3m$, $m - 9$, $9 - m$ respectively. Calculate:

a the common difference
b the first term
c the sum of the first ten terms.

6 The fourth term of an arithmetic series is twice the first term x.
If the tenth term is 24, calculate:

a the common difference in terms of x
b the first term
c the sum of the first ten terms.

7 The first term of an arithmetic series is 19 and the last term is -51. If the sum of the series is -176, calculate the number of terms in the series.

8 A child builds a triangular structure out of bricks. The top three rows are shown.

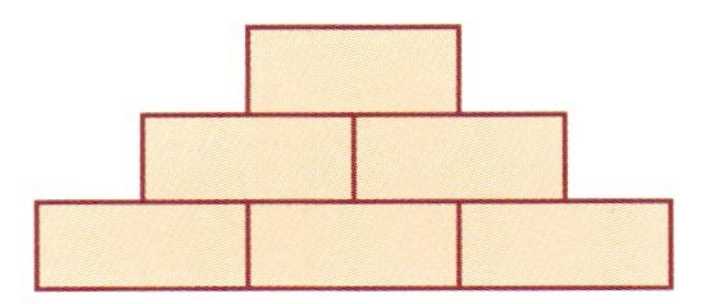

Each row has one brick fewer than the row beneath it.

a Show that, if the structure has n rows, the total number of bricks used is given by the formuls $S_n = \frac{n}{2}(n + 1)$.

b If the structure has 78 bricks, how many rows has it?

c If the child has 200 bricks, what is the maximum number of rows his structure can have?

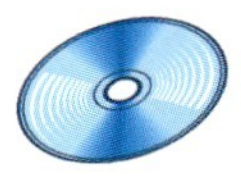

2.6 Geometric sequences and series

So far we have looked at sequences where there is a common difference between successive terms. There are, however, other types of sequence, e.g. 2, 4, 8, 16, 32. There is clearly a pattern to the way the numbers are generated as each term is double the previous term, but there is no common difference.

A sequence where there is a **common ratio** (r) between successive terms is known as a **geometric sequence**. For example:

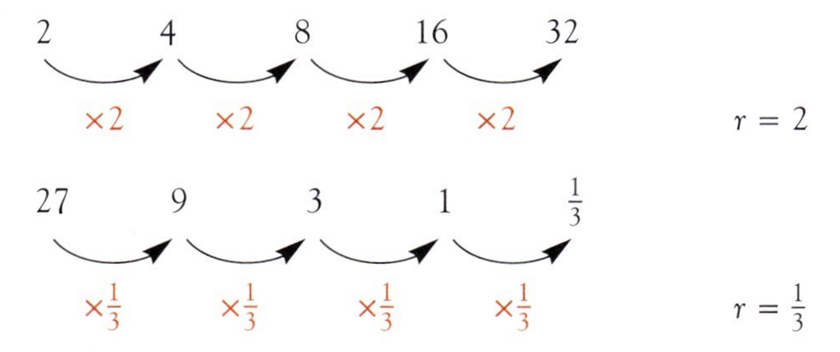

As with an arithmetic sequence, there are two main ways of describing a geometric sequence.

1 The term-to-term rule.

For example, for the following sequence,

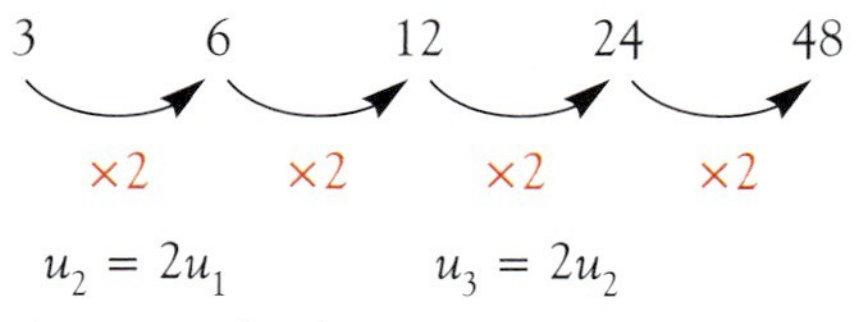

$u_2 = 2u_1$ $\quad u_3 = 2u_2$

the general rule is $u_{n+1} = 2u_n$, $u_1 = 3$.

2 The formula for the nth term of a geometric sequence.
As with an arithmetic sequence, this rule links each term to its position in the sequence. For example, for the following sequence,

Position	1	2	3	4	5	n
Term	3	6	12	24	48	

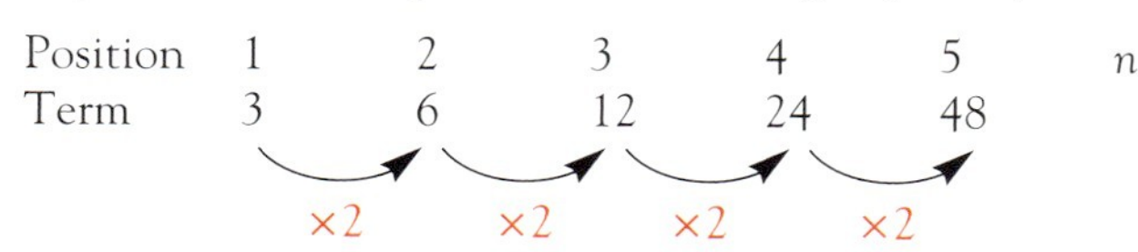

to reach the second term, the calculation is 3×2 or 3×2^1
to reach the third term, the calculation is $3 \times 2 \times 2$ or 3×2^2
to reach the fourth term, the calculation is $3 \times 2 \times 2 \times 2$ or 3×2^3

In general therefore

$$u_n = u_1 r^{n-1}$$

where u_1 is the first term and r is the common ratio.

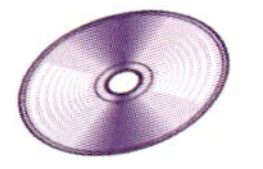

Compound interest

When money is invested in a bank account it may earn interest. When the interest paid is simple interest, the amount of money in an account forms an arithmetic sequence with the amount of interest being the common difference. When the interest paid is compound interest, the amount of money in an account forms a geometric sequence with the interest rate determining the common ratio.

For example, £100 is deposited in a bank account and left untouched for five years. If it earns compound interest at 10% per year, the amount in the account increases as shown below.

Number of years	0	1	2	3	4	5
Amount	100.00	110.00	121.00	133.10	146.41	161.05

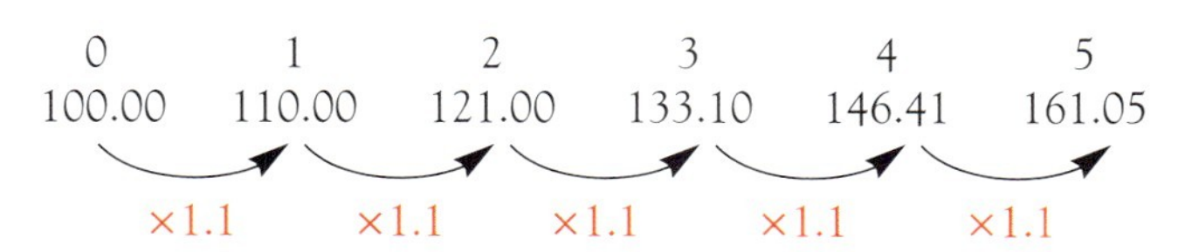

Simple interest and compound interest are dealt with in more detail in Sections 8.2 and 8.3.

Exercise 2.6.1

1 Identify which of the following are geometric sequences and which are not.

a 2, 6, 18, 54 **b** 25, 5, 1, $\frac{1}{5}$ **c** 1, 4, 9, 16,

d $-3, 9, -27, 81$ **e** $\frac{1}{2}, \frac{2}{3}, \frac{3}{4}, \frac{4}{5}$ **f** $\frac{1}{2}, \frac{2}{4}, \frac{4}{8}, \frac{8}{16}$

2 For the sequences in question 1 that are geometric, calculate:

i) the common ratio r
ii) the next two terms
iii) a formula for the nth term.

3 The nth term of a geometric sequence is given by the formula $u_n = -6 \times 2^{n-1}$.
  a Calculate u_1, u_2 and u_3.
  b What is the value of n, if $u_n = -768$?

4 Part of a geometric sequence is given below.

  __, −1, __, __, 64, __ where $u_2 = -1$ and $u_5 = 64$. Calculate:

  a the common ratio r
  b the value of u_1
  c the value of u_{10}.

5 Archie invests €1000 in an savings account paying 6% compound interest per year. Calculate the amount in the account after 5 years.

Geometric series

A geometric series is one where the terms of a geometric sequence are added together. For example, for the following geometric sequence,

Position	1	2	3	4	5	6
Term	5	10	20	40	80	160

the formula for the nth term is $u_n = 5 \times 2^{n-1}$.

Therefore the sum of the series above is given by:

$$5\sum_{1}^{6} 2^{n-1} = 5(1 + 2 + 4 + 8 + 16 + 32) = 315$$

A formula for the sum of S_n a geometric series can be derived as follows:

$$S_n = u_1 + u_1 r + u_1 r^2 + u_1 r^3 + \dots + u_1 r^{n-1}$$

Multiplying by r: $$rS_n = u_1 r + u_1 r^2 + u_1 r^3 + \dots + u_1 r^{n-1} + u_1 r^n$$

Subtracting the first equation from the second gives:

$$rS_n - S_n = u_1 r^n - u_1$$
$$S_n(r - 1) = u_1(r^n - 1)$$

Therefore $$S_n = \frac{u_1(r^n - 1)}{r - 1} \quad r \neq 1$$

where u_1 is the first term, r is the common ratio and n is the number of terms.
This form of the formula is particularly useful when $r > 1$ or $r < -1$.

If we multiply $S_n \times \frac{-1}{-1}$ a variation of the formula is obtained:

$$S_n = \frac{u_1(1 - r^n)}{1 - r} \quad r \neq 1$$

This form is more useful when $-1 < r < 1$ as it avoids working with negative values.

Worked example

A series is as follows:

$4 + 12 + 36 + 108 + \ldots + 26\,244$

a Calculate the number of terms in the series.

b Calculate the sum of the series.

a $u_1 = 4, r = 3$

Using $u_n = u_1 r^{n-1}$

$4 \times 3^{n-1} = 26\,244$

$3^{n-1} = 6561$

Using a calculator we can find that $3^8 = 6561$

Therefore $3^8 = 3^{n-1}$, which implies $8 = n - 1$, i.e. $n = 9$.

There are nine terms in the series.

b $S_n = \dfrac{u_1(r^n - 1)}{r - 1}$ $\quad u_1 = 4, r = 3$ and $n = 9$

$S_9 = \dfrac{4(3^9 - 1)}{3 - 1} = 39\,364$

This can be checked using your GDC as shown on pages 41–2.

Exercise 2.6.2

1 For each of the following geometric series, calculate:

i) the value of the common ratio r

ii) the sum of the first 10 terms.

a $\frac{1}{8} + \frac{1}{4} + \frac{1}{2} + 1 + 2$

b $-\frac{1}{9} + \frac{1}{3} - 1 + 3 - 9$

c $5 + 7.5 + 11.25 + 16.875$

d $10 + 1 + 0.1 + 0.01 + 0.001$

2 For each of the following geometric series, the first three terms and the last term are given.

i) Find the number of terms n.

ii) Calculate the sum of the series.

a $1 + 3 + 9 + \ldots + 2187$

b $\frac{1}{5} + \frac{2}{5} + \frac{4}{5} + \ldots + 12\frac{4}{5}$

c $8 - 4 + 2 - \ldots + \frac{1}{32}$

d $a + ar + ar^2 + \ldots + ar^{n-1}$

3 Evaluate each of the following.

a $\sum_{1}^{5} 4^n$

b $\sum_{1}^{7} 2(3)^{n-2}$

c $\sum_{4}^{8} \frac{2^{n-1}}{4}$

4 In a geometric series, $u_2 = \frac{1}{3}$ and $u_5 = 72$. Calculate:

a the common ratio r

b the first term u_1

c the value of S_6.

5 Three consecutive terms of a geometric series are $(p - 2)$, $(-p + 1)$ and $(2p - 2)$.

a Calculate the possible values of p.

b Calculate the term before $(p - 2)$ assuming p is the larger of the values in part **a**.

c If $u_3 = (p - 2)$, calculate S_8.

6 A new strain of a virus is detected in one patient in a hospital. After 3 days, 375 new cases of the virus are detected. Assuming the number of new cases continues to grow as a geometrical sequence, calculate:

   a the number of new cases after 7 days

   b the total number of infected people after 7 days.

Infinite geometric series

For all the series considered so far, if the series was continued, the sum would diverge (continue to grow). However, this isn't always the case: for some series, the sum converges to a value. For example, in the following series,

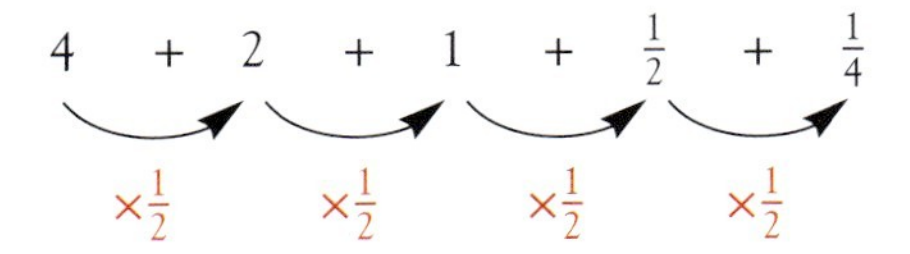

the amount being added each time decreases. This is represented visually in the diagram below.

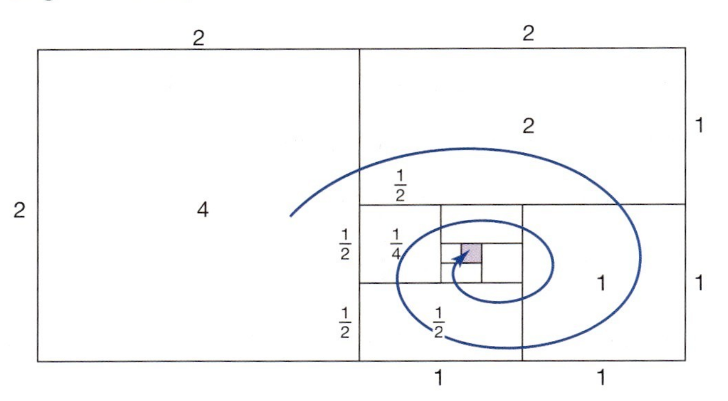

The area of each shape represents a term in the series. The total area of the large rectangle is 8. This suggests that the sum of the series, if continued infinitely, would also be 8, i.e. $S_\infty = 8$.

This can be proved algebraically using the formula $S_n = \dfrac{u_1(1 - r^n)}{1 - r}$.

(Note: The alternative version of the formula is used since $-1 < r < 1$.)

$u_1 = 4,\ r = \frac{1}{2}$

$$S_\infty = \frac{4\left(1 - \frac{1}{2}^n\right)}{1 - \frac{1}{2}} = \frac{4\left(1 - \frac{1}{2}^n\right)}{\frac{1}{2}} = 8\left(1 - \frac{1}{2}^n\right)$$

As $n \to \infty$, $\left(\frac{1}{2}\right)^n \to 0$. Therefore $S_\infty = 8(1 - 0) = 8$.

In general $S_n = \dfrac{u_1(1 - r^n)}{1 - r}$ becomes $S_\infty = \dfrac{u_1}{1 - r}$ for an infinite geometric series, as $(1 - r^n) \to 1$ as $n \to \infty$.

Therefore if $-1 < r < 1$, the sum to infinity of a series will have a value.

Worked example

A sequence is defined by $u_n = 3\left(\frac{1}{2}\right)^{n-1}$.

a Calculate u_1, u_2 and u_3.

b Write down the values of u_1 and r.

c Calculate the sum of the infinite series $\sum_{1}^{\infty} 3\left(\frac{1}{2}\right)^{n-1}$.

a $u_1 = 3 \times \left(\frac{1}{2}\right)^0 = 3$

$u_2 = 3 \times \left(\frac{1}{2}\right)^1 = \frac{3}{2}$

$u_3 = 3 \times \left(\frac{1}{2}\right)^2 = \frac{3}{4}$

b $u_1 = 3$, $r = \frac{1}{2}$

c $S_\infty = \dfrac{u_1}{1-r} = \dfrac{3}{1-\frac{1}{2}} = 6$

Exercise 2.6.3

1 Calculate the sum to infinity of the following series

a $18 + 6 + 2 + \frac{2}{3} + \ldots$

b $-8 + 4 - 2 + 1 - \ldots$

c $1 + \frac{1}{10} + \frac{1}{100} + \frac{1}{1000} + \ldots$

d $7 + 2 + \frac{4}{7} + \frac{8}{49} + \ldots$

2 Evaluate the following sums to infinity.

a $\sum_{1}^{\infty} \left(\frac{1}{4}\right)^n$

b $\sum_{1}^{\infty} \frac{2}{2^{n-1}}$

c $\sum_{5}^{\infty} \left(\frac{2}{3}\right)^n$

d $\sum_{10}^{\infty} \frac{5}{3^{n-1}}$

3 The second term of a geometric series is $\frac{3}{2}$. The sum to infinity of the same series is 6. Calculate:

a u_1

b the common ratio r.

4 In a geometric series $u_1 + u_2 = 12$. If $r = \frac{1}{3}$, find the sum of the infinite series.

2.7 Graphical solution of equations

A linear equation, when plotted, produces a straight line.

The following are all examples of linear equations:

$y = x + 1$ $\quad y = 2x - 1$ $\quad y = 3x$ $\quad y = -x - 2$ $\quad y = 4$

They all have a similar format, i.e. $y = mx + c$.

In the equation
$y = x + 1$, $m = 1$ and $c = 1$
$y = 2x - 1$, $m = 2$ and $c = -1$
$y = 3x$, $m = 3$ and $c = 0$
$y = -x - 2$, $m = -1$ and $c = -2$
$y = 4$, $m = 0$ and $c = 4$

Their graphs are shown below.

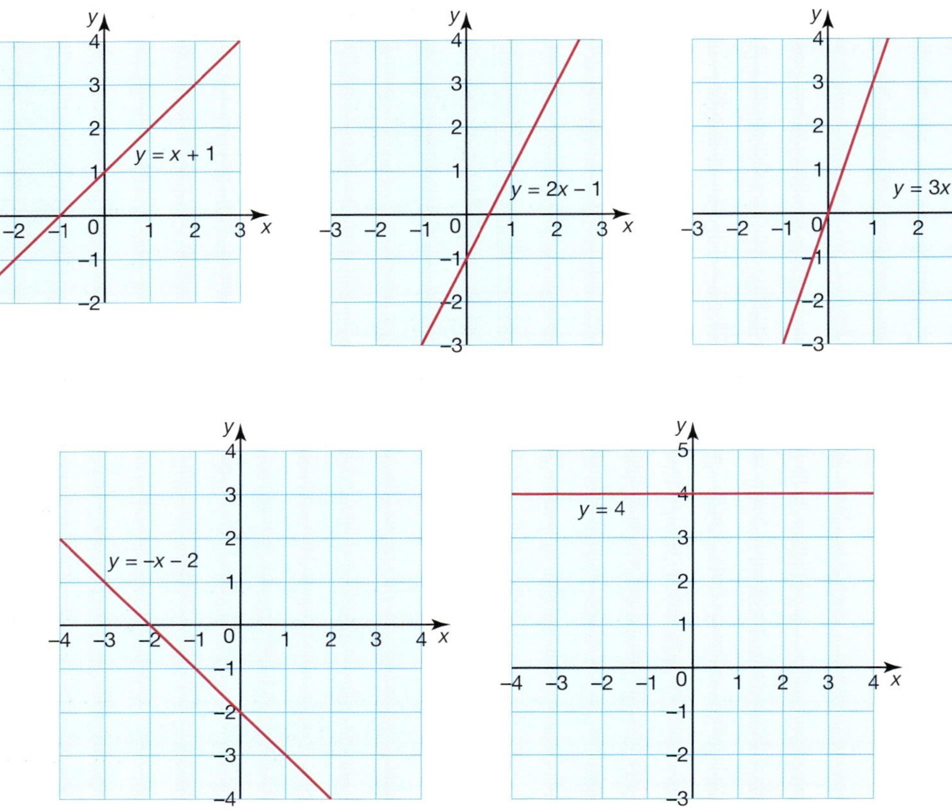

In Section 5.2 plotting and solving linear equations are studied in more detail. In this section we look exclusively at how to use your GDC and graphing software to draw linear graphs and to solve simple problems involving them.

Using a GDC or graphing software to plot a linear equation

In Topic 1 you saw how to plot a single linear equation using your GDC. For example, to graph the linear equation $y = 2x + 3$:

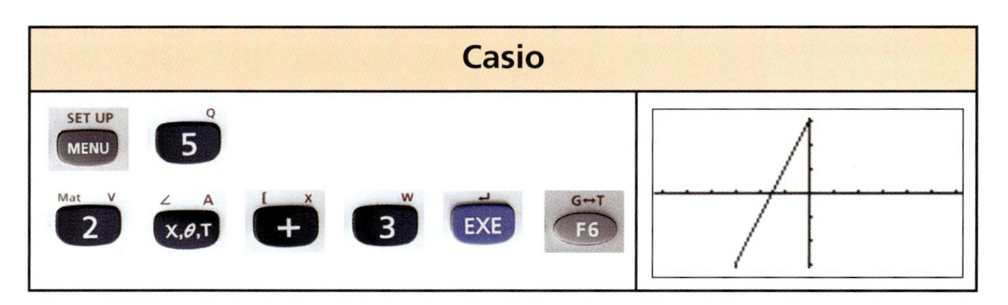

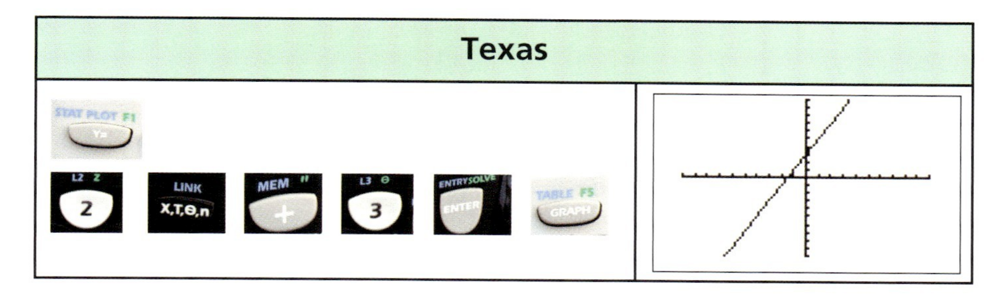

With graphing software too, the process is relatively straightforward:

Autograph

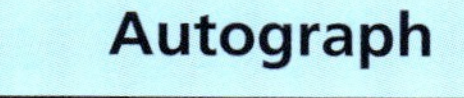

Select [icon] and enter the equation.

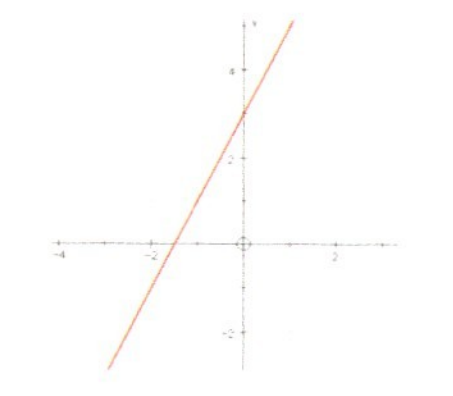

Note: To reposition the graph on the screen use [icon]. To change the scale on the axes use [icon].

GeoGebra

Type $f(x) = 2x + 3$ into the input box.

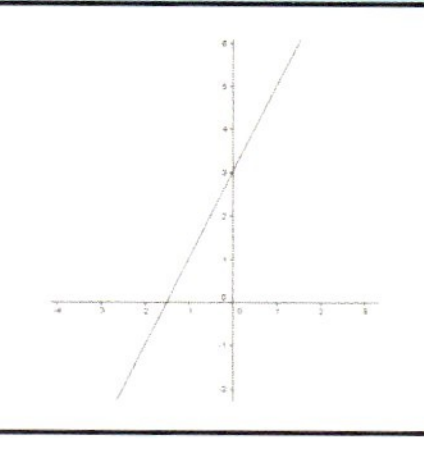

Note: To reposition the graph on the screen use [icon]. To change the scale on the axes select 'Options' followed by 'Drawing pad'.

Unless they are parallel to each other, when two linear graphs are plotted on the same axes, they will intersect at one point. Solving the equations simultaneously will give the coordinates of the point of intersection. Your GDC and graphing software will also be able to work out the coordinates of the point of intersection.

Worked example

Find the point of intersection of these linear equations.

$$y = 2x - 1 \text{ and } y = \tfrac{1}{2}x + 2$$

Using a GDC:

Casio

  and enter $y = 2x - 1$, 

Enter $y = \frac{1}{2}x + 2$, 

to graph the equations.

  followed by  to select 'intersect' in the 'graph solve' menu. The results are displayed at the bottom of the screen.

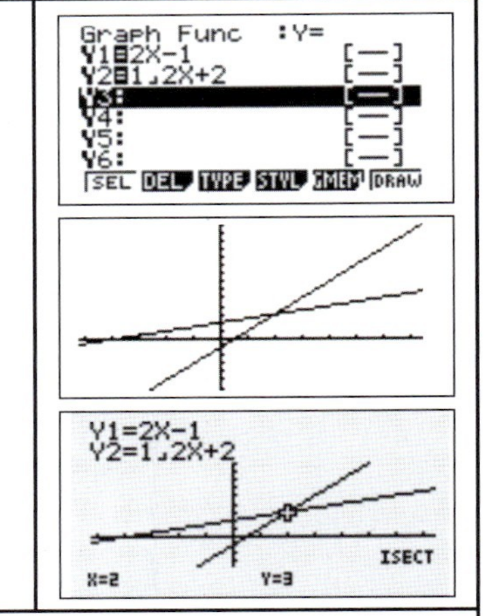

Note: Equations of lines have to be entered in the form $y = \ldots$, e.g. the equation $2x - 3y = 9$ would need to be rearranged to make y the subject, i.e. $y = \frac{2x - 9}{3}$ or $y = \frac{2}{3}x - 3$.

Texas

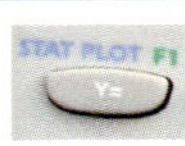 and enter $y = 2x - 1$, 

Then $y = \frac{1}{2}x + 2$, 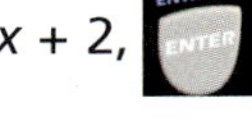

 to graph the equations.

 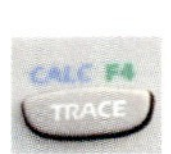 followed by  to select 'intersect' in the 'graph calc' menu.

Once the two lines are selected using , the results are displayed at the bottom of the screen.

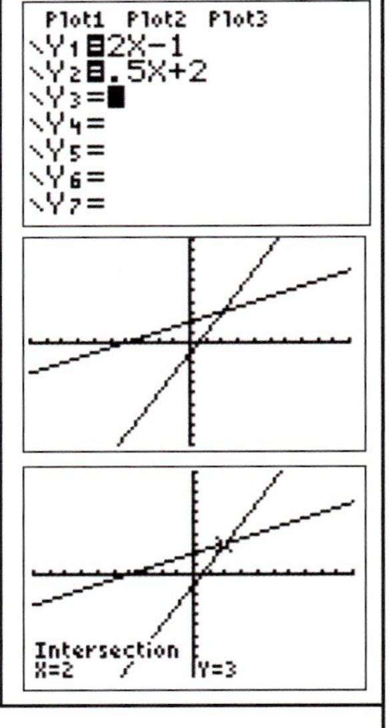

Note: See the note for the Casio above.

Using graphing software:

Autograph	
Select and enter each equation in turn. Use and select both lines. Click on 'Object' followed by the 'solve f(x) = g(x)' sub-menu. Click to display the 'results box'.	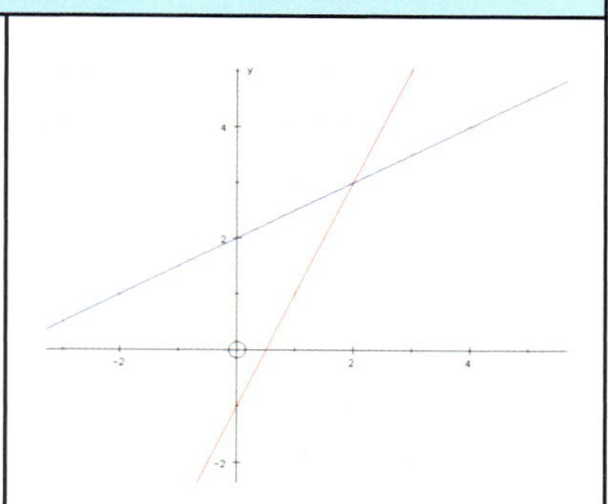
Note: To select the second line, keep the shift key pressed.	
GeoGebra	
Enter $f(x) = 2x - 1$ and $g(x) = \frac{1}{2}x + 2$ in the input field. Enter 'Intersect [$f(x)$, $g(x)$]' in the input field. The point of intersection is displayed and its coordinate written in the algebra window.	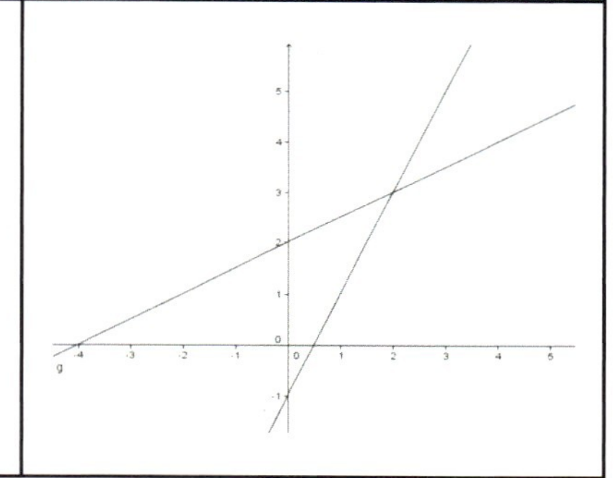

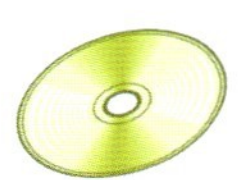

Exercise 2.7.1

1 Using either a GDC or graphing software, find the coordinates of the points of intersection of the following pairs of linear graphs.

- a $y = 5 - x$ and $y = x - 1$
- b $y = 7 - x$ and $y = x - 3$
- c $y = -2x + 5$ and $y = x - 1$
- d $x + 3y = -1$ and $y = \frac{1}{2}x + 3$
- e $x - y = 6$ and $x + y = 2$
- f $3x - 2y = 13$ and $2x + y = 4$
- g $4x - 5y = 1$ and $2x + y = -3$
- h $x = y$ and $x + y + 6 = 0$
- i $2x + y = 4$ and $4x + 2y = 8$
- j $y - 3x = 1$ and $y = 3x - 3$

2 By referring to the lines, explain your answers to parts **i** and **j** above.

Quadratic equations

An equation of the form $y = ax^2 + bx + c$, in which the highest power of the variable x is x^2, is known as a **quadratic equation**. The following are all examples of quadratic equations:

$y = x^2 + 2x - 4 \qquad y = -3x^2 + x + 2 \qquad y = x^2 \qquad y = \frac{1}{2}x^2 + 2$

When plotted, a quadratic graph has a specific shape known as a **parabola**. This will look like

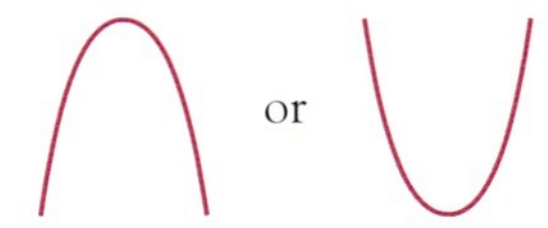

Depending on the values of a, b and c, the position and shape of the graph will vary, e.g.

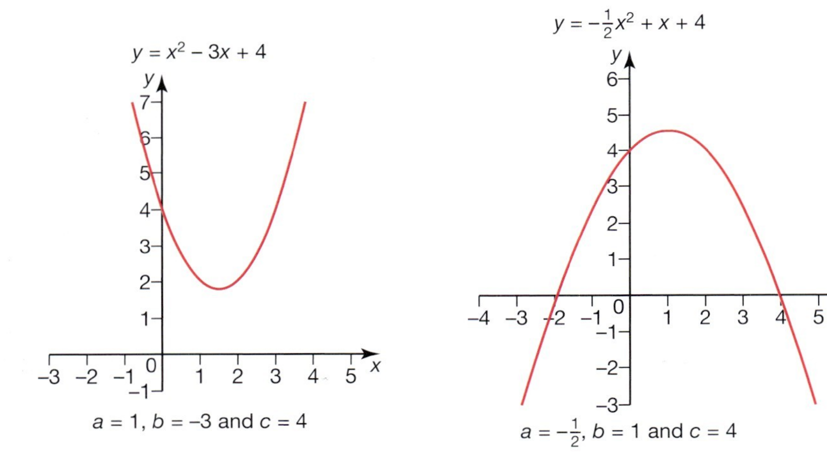

$a = 1$, $b = -3$ and $c = 4$

$a = -\frac{1}{2}$, $b = 1$ and $c = 4$

Solving a quadratic equation of the form $ax^2 + bx + c = 0$ implies finding where the graph crosses the x-axis, because $y = 0$ on the x-axis.

In the case of above $-\frac{1}{2}x^2 + x + 4 = 0$, we can see that the graph crosses the x-axis at $x = -2$ and 4. These are therefore the solutions to, or **roots** of, the equation. In the case of $x^2 + 3x + 4 = 0$, the graph does not cross the x-axis. Therefore the equation has no real solutions.

(Note: There are imaginary solutions, but these are not dealt with in this textbook.) A GDC can also be used to find the solution to quadratic equations:

Casio

and enter $y = -\frac{1}{2}x^2 + x + 4$,

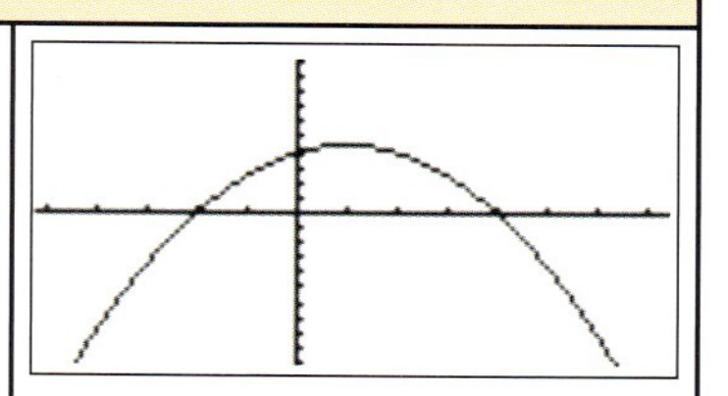

to graph the equation.

followed by

to select 'Root' in the 'graph solve' menu.

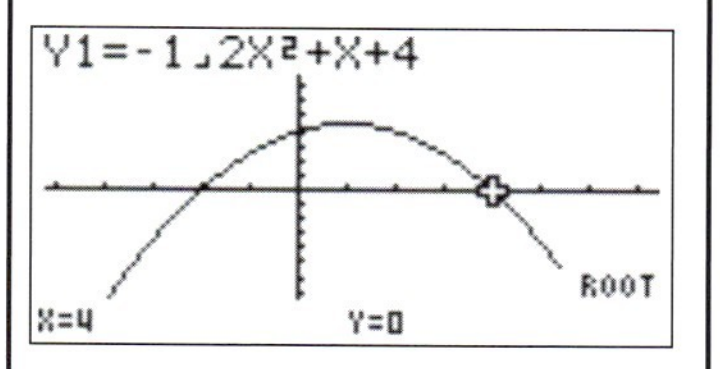

Use
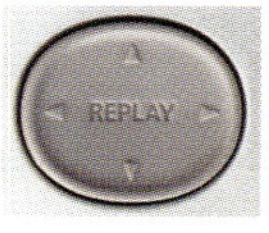

to find the second root.

The results are displayed at the bottom of the screen.

Texas

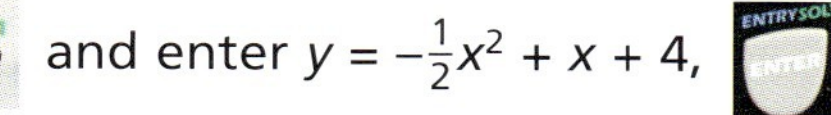
and enter $y = -\frac{1}{2}x^2 + x + 4$,

to graph the equation.

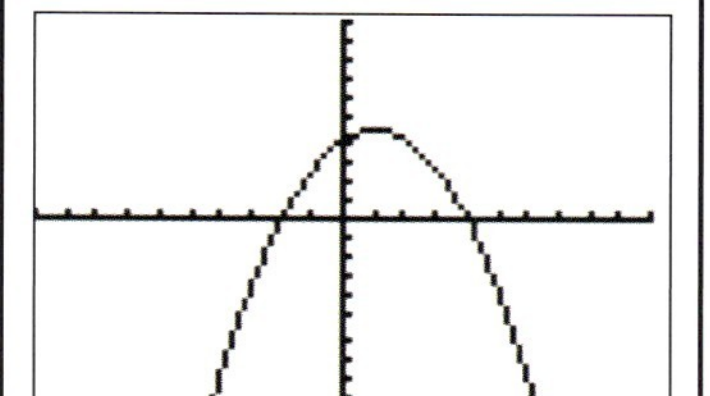

followed by

to select 'zero' in the 'graph calc' menu.

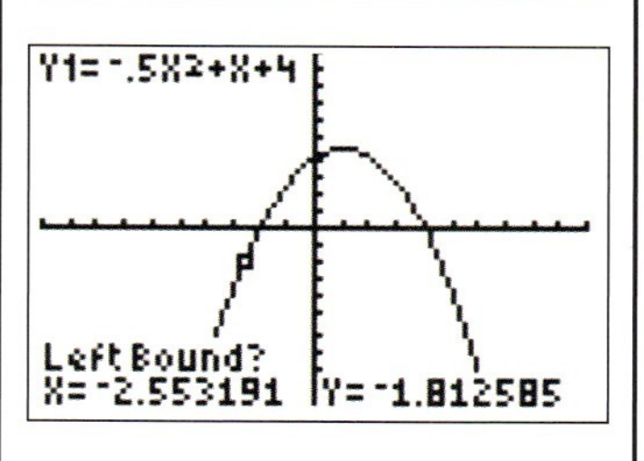

Use
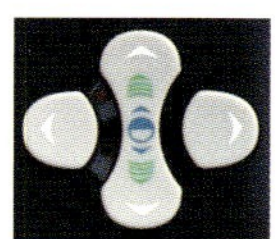
and follow the on-screen prompts to identify a point to the left and a point to the right of the root, in order for the calculator to give the root.

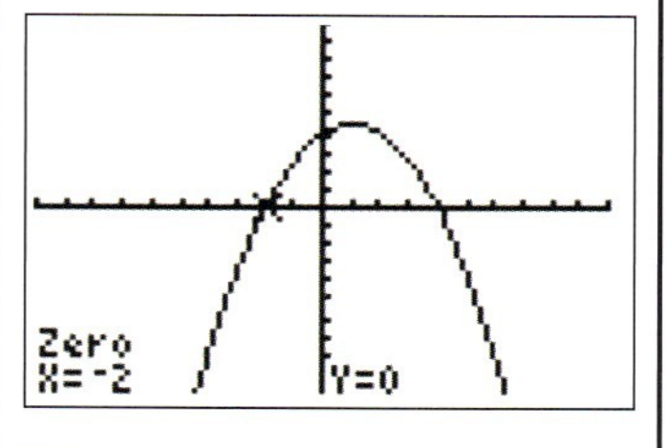

Alternatively, graphing software can be used:

Autograph

Select and enter the equation $y = -\frac{1}{2}x^2 + x + 4$.

Use and select the line.

Click on 'Object' followed by the 'solve f(x) = 0' sub-menu.

Click to display the 'results box'.

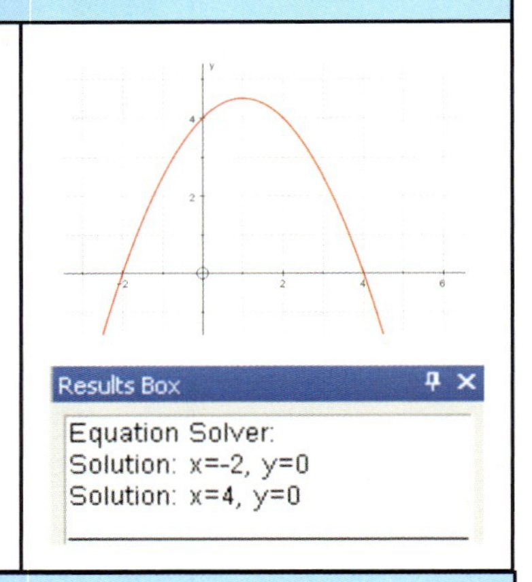

GeoGebra

Enter $f(x) = -\frac{1}{2}x^2 + x + 4$ in the input field.

Enter 'Root [$f(x$)]' in the input field.

The solutions are displayed and their coordinates written in the algebra window.

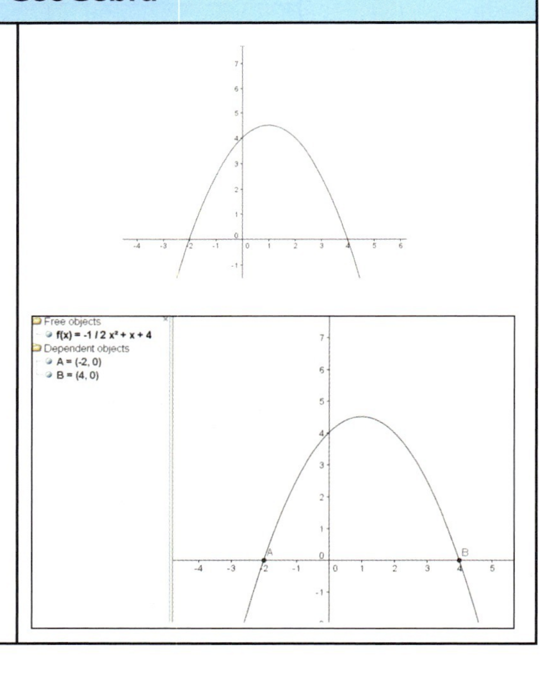

Exercise 2.7.2

Using either a GDC or graphing software:
i) graph the following quadratic equations
ii) find the coordinates of any roots.

1 a $y = x^2 - 3x + 2$
b $y = x^2 + 4x - 12$
c $y = -x^2 + 8x - 15$
d $y = x^2 + 2x + 6$
e $y = -x^2 + x + 4$

2 a $y = \frac{1}{2}x^2 - \frac{1}{2}x - 3$
b $4y = -x^2 + 6x + 16$
c $-2y = x^2 + 10x + 25$

Student assessment 1

Diagrams are not drawn to scale.

1 State whether each of the following numbers is rational or irrational.
a 1.6
b $\sqrt{3}$
c $0.\dot{7}$
d $0.\dot{7}\dot{3}$
e $\sqrt{121}$
f π

2 Round each of the following numbers to the degree of accuracy shown in brackets.
a 6472 (nearest 10)
b 88 465 (nearest 100)
c 64 785 (nearest 1000)
d 6.7 (nearest 10)

3 Round each of the following numbers to the number of decimal places shown in brackets.
a 3.84 (1 d.p.)
b 6.792 (1 d.p.)
c 0.8526 (2 d.p.)
d 1.5849 (2 d.p.)
e 9.954 (1 d.p.)
f 0.0077 (3 d.p.)

4 Round each of the following numbers to the number of significant figures shown in brackets.
a 42.6 (1 s.f.)
b 5.432 (2 s.f.)
c 0.0574 (1 s.f.)
d 48 572 (2 s.f.)
e 687 453 (1 s.f.)
f 687 453 (3 s.f.)

5 A cuboid's dimensions are given as 12.32 cm by 1.8 cm by 4.16 cm. Calculate its volume, giving your answer to three significant figures.

6 1 mile is 1760 yards. Estimate the number of yards in 11.5 miles.

7 Estimate the answers to the following. Do *not* work out an exact answer.
a $\dfrac{5.3 \times 11.2}{2.1}$
b $\dfrac{(9.8)^2}{(4.7)^2}$
c $\dfrac{18.8 \times (7.1)^2}{(3.1)^2 \times (4.9)^2}$

8 Estimate the shaded area of the figure below.

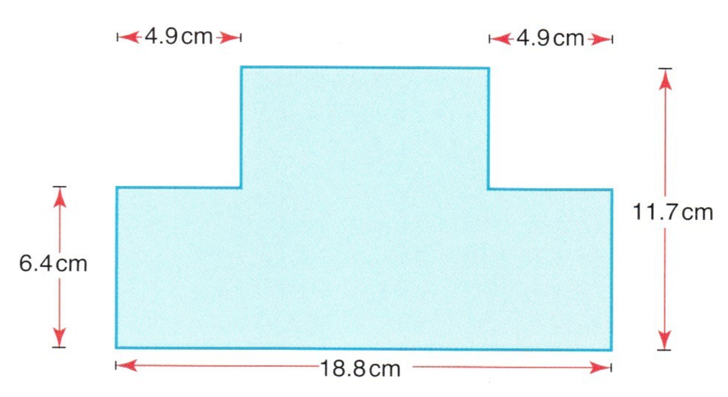

9 a Use a calculator to find the exact answer to question 8.
b Calculate your percentage error.

10 A boy estimates the weight of his bike to be 2.5 kg. It actually weighs 2.46 kg. What is his percentage error?

Student assessment 2

1 Write the following numbers in the form $a \times 10^k$, where $1 \le a < 10$ and $k \in \mathbb{Z}$.
   a 6 million
   b 0.0045
   c 3 800 000 000
   d 0.000 000 361
   e 460 million
   f 3

2 Write the following numbers in order of magnitude, starting with the smallest.

   6.2×10^7 5.5×10^{-3} 4.21×10^7

   4.9×10^8 3.6×10^{-5} 7.41×10^{-9}

3 Write the following numbers:

   6 million 820 000 0.0044 0.8 52 000
   a in the form $a \times 10^k$, where $1 \le a < 10$ and $k \in \mathbb{Z}$
   b in order of magnitude, starting with the largest.

4 Deduce the value of k in each of the following.
   a $4750 = 4.75 \times 10^k$
   b $6\,440\,000\,000 = 6.44 \times 10^k$
   c $0.0040 = 4.0 \times 10^k$
   d $1000^2 = 1 \times 10^k$
   e $0.9^3 = 7.29 \times 10^k$
   f $800^3 = 5.12 \times 10^k$

5 Write the answer to each of the following calculations in the form $a \times 10^k$, where $1 \le a < 10$ and $k \in \mathbb{Z}$.
   a $4000 \times 30\,000$
   b $(2.8 \times 10^5) \times (2.0 \times 10^3)$
   c $(3.2 \times 10^9) \div (1.6 \times 10^4)$
   d $(2.4 \times 10^8) \div (9.6 \times 10^2)$

6 The speed of light is $3 \times 10^8\,\text{m s}^{-1}$. Jupiter is 778 million kilometres from the Sun. Calculate the number of minutes it takes for sunlight to reach Jupiter.

7 A star is 500 light years away from Earth. If the speed of light is $3 \times 10^5\,\text{km s}^{-1}$, calculate the distance the star is from Earth. Give your answer in kilometres in the form $a \times 10^k$, where $1 \le a < 10$ and $k \in \mathbb{Z}$.

8 Convert 162 000 km to millimetres. Give your answer in the form $a \times 10^k$, where $1 \le a < 10$ and $k \in \mathbb{Z}$.

9 Convert 7 415 000 mg to kilograms. Give your answer correct to the nearest kilogram.

10 Using the formula $T_f = \frac{9}{5}T_c + 32$, where T_f and T_c are the temperatures in degrees Fahrenheit and Celsius respectively, convert the following to degrees Fahrenheit.
   a 40 °C
   b 500 ° C

Student assessment 3

1 For each of the following arithmetic sequences:
   i) write down a formula for the nth term
   ii) calculate the 10th term.
   a 1, 5, 9, 13, ...
   b 1, −2, −5 , −8, ...

2 For both of the following calculate u_5 and u_{100}.
   a $u_n = 6n - 3$
   b $u_n = -\frac{1}{2}n + 4$

3 Copy and complete both of the following tables of arithmetic sequences.

   a

Position	1	2	3	10		n
Term	17	14			−55	

   b

Position	2	6	10		n
Term	−4	−2		35	

4 A girl deposits $300 in a bank account. The bank offers simple interest 7% per year. Assuming the girl does not take any money out of the account, calculate the amount of money in the account after 5 years.

5 Part of a geometric sequence is given below.

   ___, ___, 27, ___, ___, −1

   where $u_3 = 27$ and $u_6 = -1$.

   Calculate:
   a the common ratio r
   b the value u_1
   c the value of n if $u_n = -\frac{1}{81}$.

6 Evaluate both of these series.

a $\sum_{1}^{10}(4n - 15)$

b $\sum_{5}^{18} -5n + 100$

7 The third and tenth terms of an arithmetic series are −6 and 15 respectively.

Calculate:

a the common difference

b the first term

c the value of S_{20}.

8 The third, fourth and fifth terms of an arithmetic series are $(2m + 2)$, $(3m + 1)$ and $(5m - 5)$.

Calculate:

a the common difference d

b the first term

c S_{10}.

9 In the following geometric series, the first three terms and the last term are given.

i) Find the number of terms, n.

ii) Calculate the sum of the series.

a $10 + 20 + 40 + \ldots + 10\,240$

b $128 - 64 + 32 + \ldots + \frac{1}{32}$

10 Evaluate these.

a $\sum_{1}^{5} 3^n$

b $\sum_{3}^{9} \frac{3^{n-2}}{5}$

11 Using either a GDC or graphing software, find the coordinates of the points of intersection of the following pairs of linear graphs.

a $y = -x + 3$ and $y = 2x - 3$

b $x + 5 = y$ and $2x + 3y - 5 = 0$

12 Using either a GDC or graphing software:

i) graph the following quadratic equations

ii) find the coordinates of any roots.

a $y = x^2 - 6x + 9$

b $y = -2x^2 + 20x - 48$

Topic 2 Discussion points, project ideas and theory of knowledge

1 Is there an underlying reason why the letters $\mathbb{N}$, $\mathbb{Z}$, $\mathbb{Q}$, $\mathbb{R}$ were chosen to represent sets of numbers?

2 Georg Cantor was a Russian mathematician educated in Berlin. He stated that $\mathbb{Q}$ is countable but $\mathbb{R}$ is not. Discuss this statement.

3 A possible topic for a project would be the Fibonacci sequence. But, be warned: whole lives have been spent on this study, and many books have been written on it.

4 What is the difference between brackets used in a sentence (like this), and brackets used in maths?

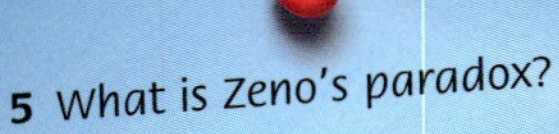

5 What is Zeno's paradox?

6 The sequence 1, 2, 3, 4, 5, … pairs with the sequence of square numbers, 1, 4, 9, 16, 25, …. Discuss whether there are more terms in the first sequence than in the second.

7 Is an error necessarily a mistake?

8 Is it possible to get an exact numerical answer to a problem? Discuss.

9 Some series are neither arithmetic nor geometric series. Investigate some different series.

10 Investigate the way that a knight moves on a chess board. This could form the basis of a project.

11 Investigate systems of measurement that preceded the SI system.

12 $n^2 - n + 41$ is always a prime number. Investigate this statement and try to prove or disprove it. Find other similar quadratic expressions as a basis for a project.

Topic 3

Sets, logic and probability

Syllabus content

3.1 Basic concepts of set theory: subsets; intersection; union; complement.

3.2 Venn diagrams and simple applications.

3.3 Sample space: event, A; complementary event, A'.

3.4 Basic concepts of symbolic logic: definition of a proposition; symbolic notation of propositions.

3.5 Compound statements: implication, $\Rightarrow$; equivalence, $\Leftrightarrow$; negation, $\neg$; conjunction, $\wedge$; disjunction, $\vee$; exclusive disjunction, $\veebar$.

Translation between verbal statements, symbolic form and Venn diagrams.

Knowledge and use of the 'exclusive disjunction' and the distinction between it and 'disjunction'.

3.6 Truth tables: the use of truth tables to provide proofs for the properties of connectives; concepts of logical contradiction and tautology.

3.7 Definition of implication: converse; inverse; contrapositive. Logical equivalence.

3.8 Equally likely events.

Probability of an event A given by $\mathrm{P}(A) = \dfrac{n(A)}{n(U)}$.

Probability of a complementary event, $\mathrm{P}(A') = 1 - \mathrm{P}(A)$.

3.9 Venn diagrams; tree diagrams; tables of outcomes. Solutions of problems using 'with replacement' and 'without replacement'.

3.10 Laws of probability.

Combined events:

$\mathrm{P}(A \cup B) = \mathrm{P}(A) + \mathrm{P}(B) - \mathrm{P}(A \cap B)$.

Mutually exclusive events:

$\mathrm{P}(A \cup B) = \mathrm{P}(A) + \mathrm{P}(B)$.

Independent events: $\mathrm{P}(A \cap B) = \mathrm{P}(A)\mathrm{P}(B)$.

Conditional probability: $\mathrm{P}(A \mid B) = \dfrac{\mathrm{P}(A \cap B)}{\mathrm{P}(B)}$.

Introduction

If the three areas of sets, logic and probability are looked at from a historical perspective, then logic came first. The study of logic developed in China, India and Greece, each independently of the other two, in the fourth century BC.

In the seventeenth century Pascal and others began to study probability. The study of sets did not truly begin until around 1900 when Georg Cantor and Richard Dedekind began work on the theory of sets.

Georg Cantor

3.1 Set theory

The modern study of set theory began with Georg Cantor and Richard Dedekind in an 1874 paper titled 'On a characteristic property of all real algebraic numbers'. It is most unusual to be able to put an exact date to the beginning of an area of mathematics.

The language of set theory is the most common foundation to all mathematics and is used in the definitions of nearly all mathematical objects.

A set is a well-defined group of objects or symbols. The objects or symbols are called the **elements** of the set. If an element e belongs to a set S, this is represented as $e \in S$. If e does not belong to set S this is represented as $e \notin S$.

Worked examples

1 A particular set consists of the following elements:
{South Africa, Namibia, Egypt, Angola, …}.
- a Describe the set.
- b Add another two elements to the set.
- c Is the set finite or infinite?

- a The elements of the set are countries of Africa.
- b e.g. Zimbabwe, Ghana
- c Finite. There is a finite number of countries in Africa.

2 Consider the set
$\{1, 4, 9, 16, 25, \ldots\}$.
- a Describe the set.
- b Write another two elements of the set.
- c Is the set finite or infinite?

- a The elements of the set are square numbers.
- b e.g. 36, 49
- c Infinite. There is an infinite number of square numbers.

Exercise 3.1.1

1 For each of the following sets:
 i) describe the set in words
 ii) write down another two elements of the set.
  a {Asia, Africa, Europe, ...}
  b {2, 4, 6, 8, ...}
  c {Sunday, Monday, Tuesday, ...}
  d {January, March, July, ...}
  e {1, 3, 6, 10, ...}
  f {Mehmet, Michael, Mustapha, Matthew, ...}
  g {11, 13, 17, 19, ...}
  h {a, e, i, ...}
  i {Earth, Mars, Venus, ...}
  j $A = \{x \mid 3 \leq x \leq 12\}$
  k $S = \{y \mid -5 \leq y \leq 5\}$

2 The number of elements in a set A is written as $n(A)$.
 Give the value of $n(A)$ for the finite sets in question 1 above.

Subsets

If all the elements of one set X are also elements of another set Y, then X is said to be a **subset** of Y.

This is written as $X \subseteq Y$.

If a set A is empty (i.e. it has no elements in it), then this is called the **empty set** and it is represented by the symbol $\varnothing$. Therefore $A = \varnothing$.

The empty set is a subset of all sets. For example, three girls, Winnie, Natalie and Emma, form a set A.

A = {Winnie, Natalie, Emma}
All the possible subsets of A are given below:
B = {Winnie, Natalie, Emma}
C = {Winnie, Natalie}
D = {Winnie, Emma}
E = {Natalie, Emma}
F = {Winnie}
G = {Natalie}
H = {Emma}
$I = \varnothing$

Note that the sets B and I above are considered as subsets of A,

i.e. $A \subseteq A$ and $\varnothing \subseteq A$.

However, sets C, D, E, F, G and H are considered **proper subsets** of A. This distinction in the type of subset is shown in the notation below. For proper subsets, we write:

$C \subset A$ and $D \subset A$ etc. instead of $C \subseteq A$ and $D \subseteq A$.

Similarly $G \nsubseteq H$ implies that G is not a subset of H
 $G \not\subset H$ implies that G is not a proper subset of H.

Worked example

$A = \{1, 2, 3, 4, 5, 6, 7, 8, 9, 10\}$

a List the subset B of even numbers.

b List the subset C of prime numbers.

a $B = \{2, 4, 6, 8, 10\}$

b $C = \{2, 3, 5, 7\}$

Exercise 3.1.2

1 P is the set of whole numbers less than 30.

  a List the subset Q of even numbers.
  b List the subset R of odd numbers.
  c List the subset S of prime numbers.
  d List the subset T of square numbers.
  e List the subset U of triangular numbers.

2 A is the set of whole numbers between 50 and 70.

  a List the subset B of multiples of 5.
  b List the subset C of multiples of 3.
  c List the subset D of square numbers.

3 $J = \{p, q, r\}$

  a List all the subsets of J.
  b List all the proper subsets of J.

4 State whether each of the following statements is true or false.

  a {Algeria, Mozambique} $\subseteq$ {countries in Africa}
  b {mango, banana} $\subseteq$ {fruit}
  c $\{1, 2, 3, 4\} \subseteq \{1, 2, 3, 4\}$
  d $\{1, 2, 3, 4\} \subset \{1, 2, 3, 4\}$
  e {volleyball, basketball} $\not\subseteq$ {team sport}
  f $\{4, 6, 8, 10\} \not\subset \{4, 6, 8, 10\}$
  g {potatoes, carrots} $\subseteq$ {vegetables}
  h {12, 13, 14, 15} $\not\subset$ {whole numbers}

The universal set

The **universal set** (U) for any particular problem is the set which contains all the possible elements for that problem.

The **complement** of a set A is the set of elements which are in U but not in A. The set is identified as A'. Notice that $U' = \varnothing$ and $\varnothing' = U$.

Worked examples

1 If $U = \{1, 2, 3, 4, 5, 6, 7, 8, 9, 10\}$ and $A = \{1, 2, 3, 4, 5\}$, what set is represented by A'?

  A' consists of those elements in U which are not in A.
  Therefore $A' = \{6, 7, 8, 9, 10\}$.

2 If U is the set of all three-dimensional shapes and P is the set of prisms, what set is represented by P'?

  P' is the set of all three-dimensional shapes except prisms.

Intersections and unions

The **intersection** of two sets is the set of all the elements that belong to both sets. The symbol $\cap$ is used to represent the intersection of two sets.

If $P = \{1, 2, 3, 4, 5, 6, 7, 8, 9, 10\}$ and $Q = \{2, 4, 6, 8, 10, 12, 14, 16, 18, 20\}$ then $P \cap Q = \{2, 4, 6, 8, 10\}$ as these are the numbers that belong to both sets.

The **union** of two sets is the set of all elements that belong to either or both sets and is represented by the symbol $\cup$.

Therefore in the example above,
$P \cup Q = \{1, 2, 3, 4, 5, 6, 7, 8, 9, 10, 12, 14, 16, 18, 20\}$.

Unions and intersections of sets can be shown diagrammatically using **Venn diagrams**.

3.2 Venn diagrams

Venn diagrams are the principal way of showing sets diagrammatically. They are named after the mathematician John Venn (1834–1923). The method consists primarily of entering the elements of a set into a circle or circles.

Some examples of the uses of Venn diagrams are shown below.

$A = \{2, 4, 6, 8, 10\}$ can be represented as:

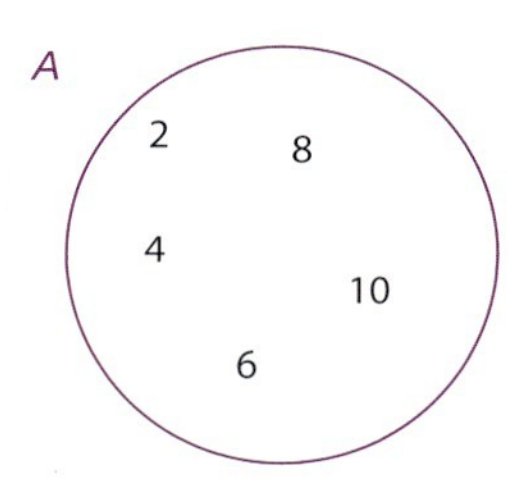

Elements which are in more than one set can also be represented using a Venn diagram.

$P = \{3, 6, 9, 12, 15, 18\}$ and $Q = \{2, 4, 6, 8, 10, 12\}$ can be represented as:

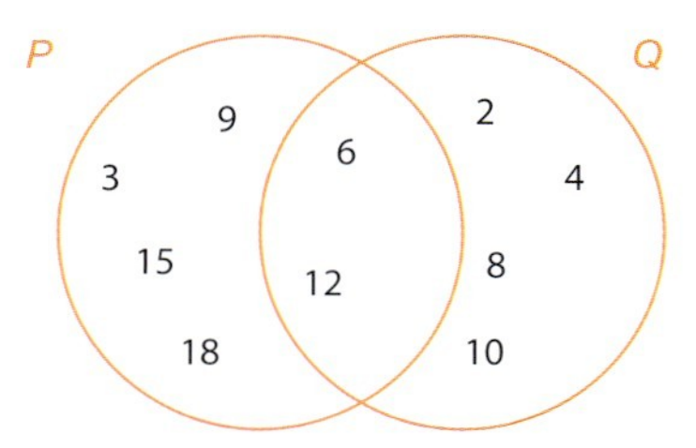

The elements which belong to both sets are placed in the region of overlap of the two circles.

As mentioned in the previous section, when two sets P and Q overlap as they do above, the notation $P \cap Q$ is used to denote the set of elements in the intersection, i.e. $P \cap Q = \{6, 12\}$. Note that $6 \in P \cap Q$; $8 \notin P \cap Q$.

$J = \{10, 20, 30, 40, 50, 60, 70, 80, 90, 100\}$ and $K = \{60, 70, 80\}$ can be represented as shown below; this is shown in symbols as $K \subset J$.

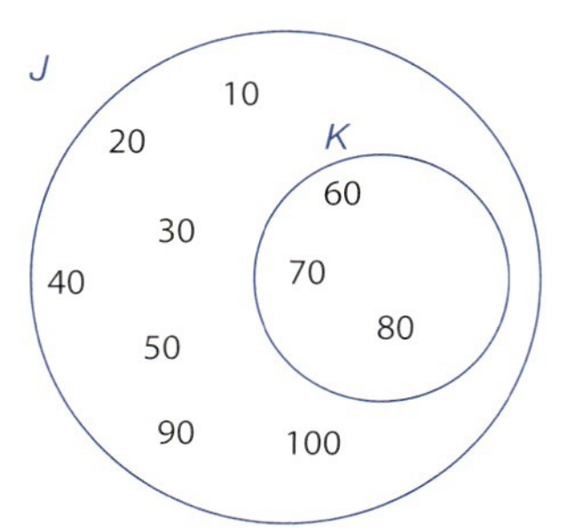

$X = \{1, 3, 6, 7, 14\}$ and $Y = \{3, 9, 13, 14, 18\}$ are represented as:

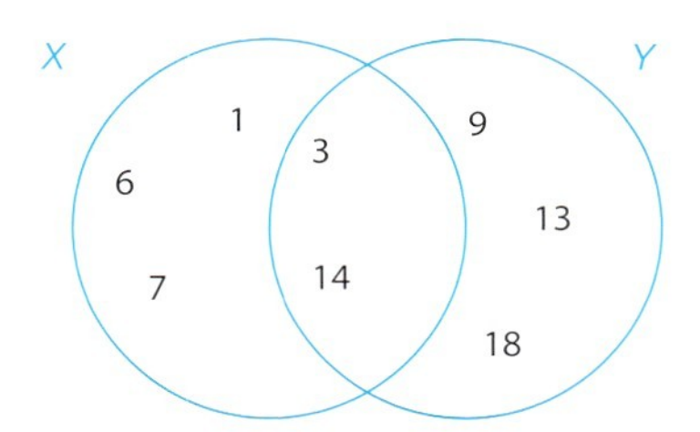

The union of two sets is everything which belongs to either or both sets and is represented by the symbol $\cup$. Therefore, in the example above, $X \cup Y = \{1, 3, 6, 7, 9, 13, 14, 18\}$.

Exercise 3.2.1

1 Using the Venn diagram, indicate whether the following statements are true or false. $\in$ means 'is an element of' and $\notin$ means 'is not an element of'.

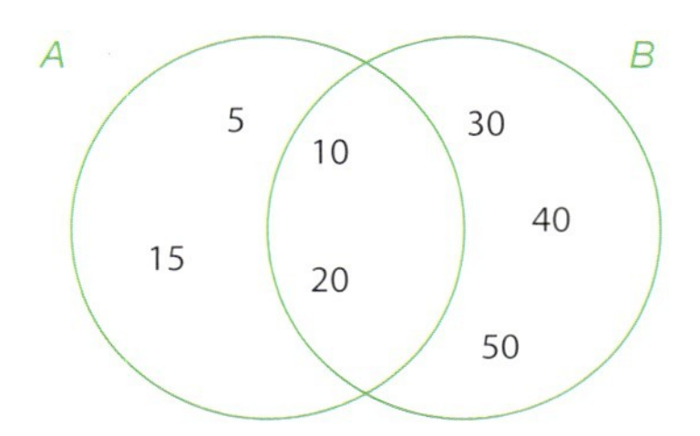

a $5 \in A$
b $20 \in B$
c $20 \notin A$
d $50 \in A$
e $50 \notin B$
f $A \cap B = \{10, 20\}$

2 Copy and complete the statement $A \cap B = \{...\}$ for each of the Venn diagrams below.

a *A* *B*
2, 10, 8, 4, 6, 9, 13, 3, 18

b *A* *B*
1, 16, 4, 9, 5, 7, 6, 8

c *A* *B*
Red, Orange, Blue, Indigo, Violet, Yellow, Green, Purple, Pink

3 Copy and complete the statement $A \cup B = \{...\}$ for each of the Venn diagrams in question 2 above.

4 Using the Venn diagram, copy and complete these statements.

a $U = \{...\}$

b $A' = \{...\}$

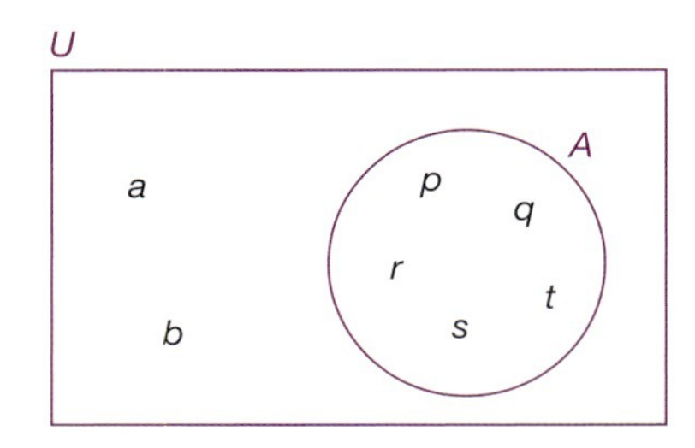

5 Using the Venn diagram, copy and complete the following statements.

a $U = \{...\}$

b $A' = \{...\}$

c $A \cap B = \{...\}$

d $A \cup B = \{...\}$

e $(A \cap B)' = \{...\}$

f $A \cap B' = \{...\}$

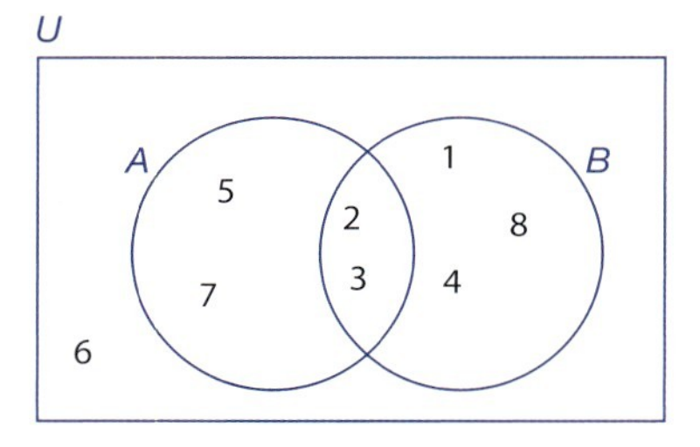

6 *A* *B* *C*
2, 10, 14, 6, 3, 9, 15, 12, 4, 8, 16, 20

a Using the Venn diagram, describe in words the elements of:

i) set A **ii)** set B **iii)** set C.

b Copy and complete the following statements.

i) $A \cap B = \{...\}$ **ii)** $A \cap C = \{...\}$ **iii)** $B \cap C = \{...\}$

iv) $A \cap B \cap C = \{...\}$ **v)** $A \cup B = \{...\}$ **vi)** $C \cup B = \{...\}$

7

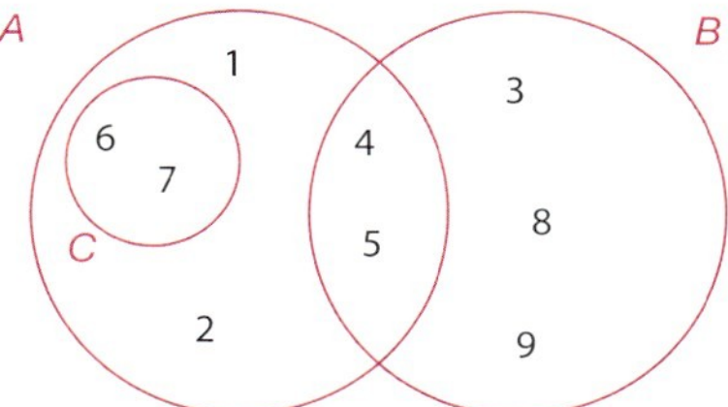

a Using the Venn diagram, copy and complete the following statements.

i) $A = \{...\}$ ii) $B = \{...\}$ iii) $C' = \{...\}$

iv) $A \cap B = \{...\}$ v) $A \cup B = \{...\}$ vi) $(A \cap B)' = \{...\}$

b State, using set notation, the relationship between C and A.

8

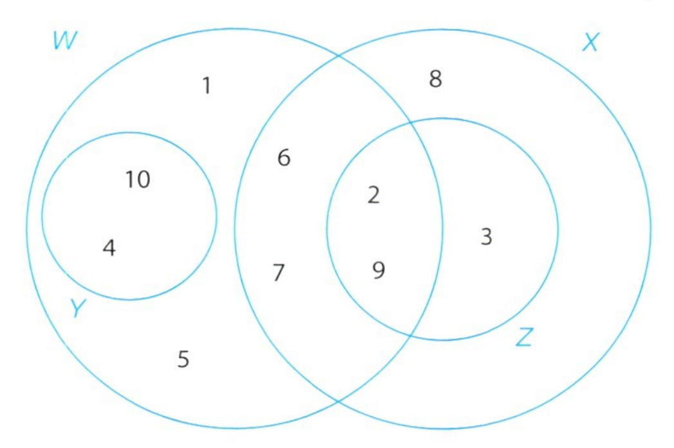

a Copy and complete the following statements.

i) $W = \{...\}$ ii) $X = \{...\}$ iii) $Z' = \{...\}$

iv) $W \cap Z = \{...\}$ v) $W \cap X = \{...\}$ vi) $Y \cap Z = \{...\}$

b Which of the named sets is a subset of X?

9 A = {Egypt, Libya, Morocco, Chad}
B = {Iran, Iraq, Turkey, Egypt}

a Draw a Venn diagram to illustrate the above information.

b Copy and complete the following statements.

i) $A \cap B = \{...\}$ ii) $A \cup B = \{...\}$

10 $P = \{2, 3, 5, 7, 11, 13, 17\}$
$Q = \{11, 13, 15, 17, 19\}$

a Draw a Venn diagram to illustrate the above information.

b Copy and complete the following statements.

i) $P \cap Q = \{...\}$ ii) $P \cup Q = \{...\}$

11 $B = \{2, 4, 6, 8, 10\}$
$A \cup B = \{1, 2, 3, 4, 6, 8, 10\}$
$A \cap B = \{2, 4\}$
Represent the above information on a Venn diagram.

12 $X = \{a, c, d, e, f, g, l\}$
$Y = \{b, c, d, e, h, i, k, l, m\}$
$Z = \{c, f, i, j, m\}$
Represent the above information on a Venn diagram.

13 $P = \{1, 4, 7, 9, 11, 15\}$
$Q = \{5, 10, 15\}$
$R = \{1, 4, 9\}$
Represent the above information on a Venn diagram.

Commutative, associative and distributive properties of sets

Set $A = \{2, 3, 4\}$, $B = \{1, 3, 5, 7\}$ and $C = \{3, 4, 5\}$.

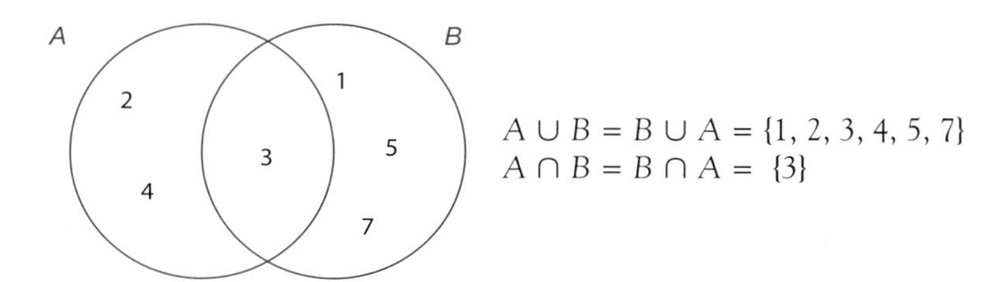

Therefore the union and intersection of sets are **commutative** (the same whichever way round the sets are ordered).

If C is added to the Venn diagram, we get:

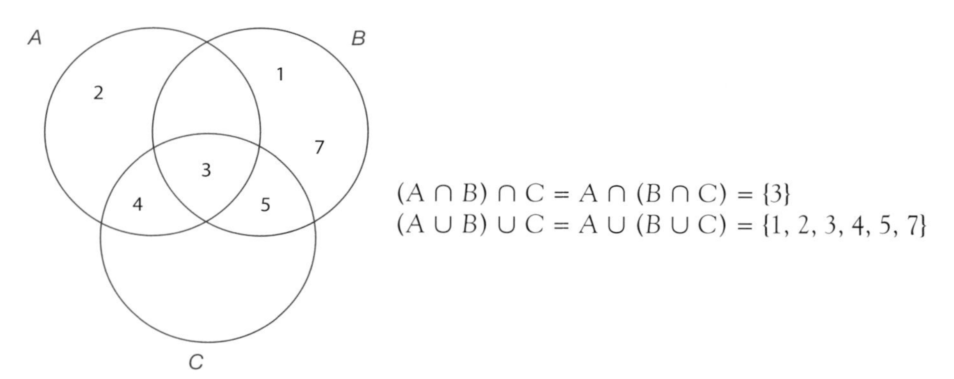

therefore the union and intersection of sets are **associative** (the order of the operations does not matter).

From the Venn diagram above it can also be seen that:

$A \cup (B \cap C) = (A \cup B) \cap (A \cup C) = \{2, 3, 4, 5\}$
$A \cap (B \cup C) = (A \cap B) \cup (A \cap C) = \{3, 4\}$.

Therefore the union is **distributive** over the intersection of sets.

Problems involving sets

Worked examples

1 In a class of 31 students, some study physics and some study chemistry. If 22 study physics, 20 study chemistry and 5 study neither, calculate the number of students who take both subjects.

The information given above can be entered in a Venn diagram in stages.
The students taking neither physics nor chemistry can be put in first (as shown).

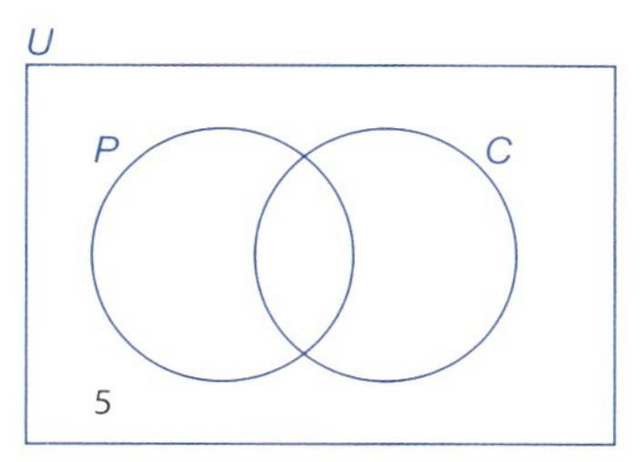

This leaves 26 students to be entered into the set circles.

If x students take both subjects then:

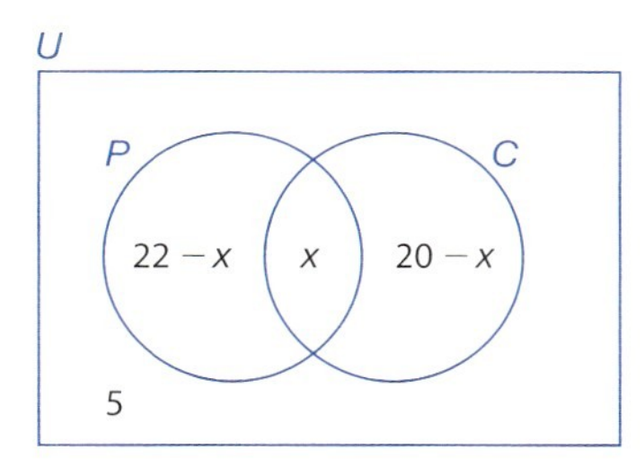

$n(P) = 22 - x + x$
$n(C) = 20 - x + x$
$P \cup C = 31 - 5 = 26$

Therefore $22 - x + x + 20 - x = 26$

$$42 - x = 26$$
$$x = 16$$

Substituting the value of x into the Venn diagram gives:

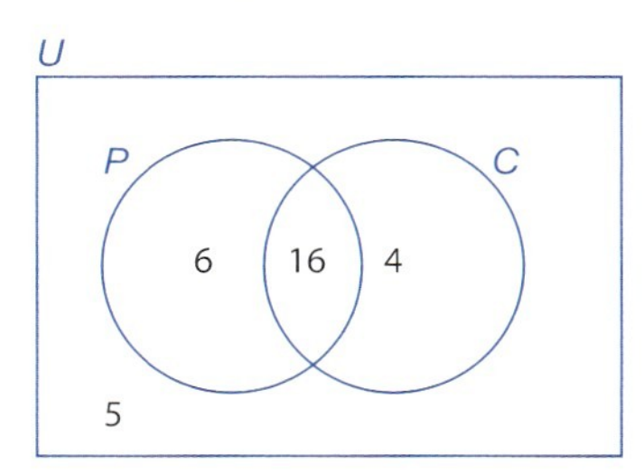

Therefore the number of students taking both physics and chemistry is 16.

2 In a region of mixed farming, farms keep goats, cattle or sheep. There are 77 farms altogether. 19 farms keep only goats, 8 keep only cattle and 13 keep only sheep. 13 keep both goats and cattle, 28 keep both cattle and sheep and 8 keep both goats and sheep.

a Draw a Venn diagram to show the above information.

b Calculate $n(G \cap C \cap S)$.

First of all draw a partly complete Venn diagram, filling in some of the information above.

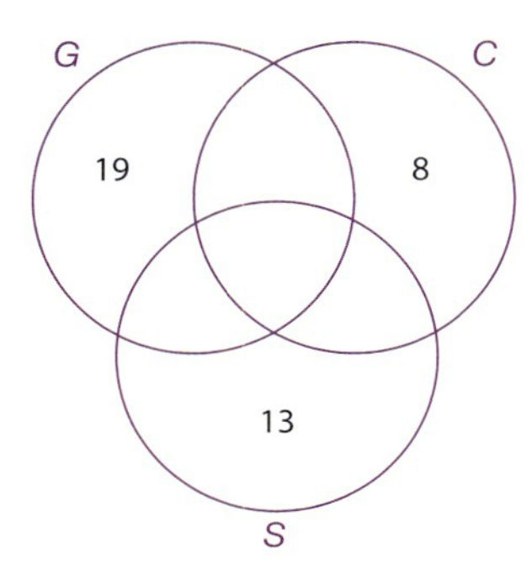

We know that:.

$n(\text{Only G}) + n(\text{Only C}) + n(\text{Only S}) = 19 + 8 + 13 = 40$

So the number of farms that keep two or more types of animal is $77 - 40 = 37$.
So, if $n(G \cap C \cap S) = x$ (i.e. x is the number of farms keeping cattle, sheep and goats), then

$$
\begin{aligned}
13 - x + 8 - x + 28 - x + x &= 37 \\
49 - 2x &= 37 \\
49 - 37 &= 2x \\
6 &= x
\end{aligned}
$$

It is then easy to complete the Venn diagram as shown:

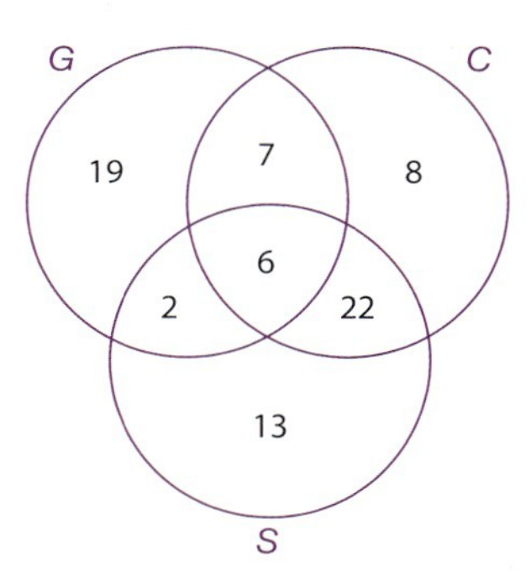

b As worked out in part **a** and shown in the diagram, $n(G \cap C \cap S) = 6$.

Exercise 3.2.2

1 In a class of 35 students, 19 take Spanish, 18 take French and 3 take neither. Calculate how many take:
  a both French and Spanish
  b just Spanish
  c just French.

2 In a year group of 108 students, 60 liked football, 53 liked tennis and 10 liked neither. Calculate the number of students who liked football but not tennis.

3 In a year group of 113 students, 60 liked hockey, 45 liked rugby and 18 liked neither. Calculate the number of students who:
   a liked both hockey and rugby
   b liked only hockey.

4 One year, 37 students sat an examination in physics, 48 sat an examination in chemistry and 45 sat an examination in biology. 15 students sat examinations in physics and chemistry, 13 sat examinations in chemistry and biology, 7 sat examinations in physics and biology and 5 students sat examinations in all three.
   a Draw a Venn diagram to represent this information.
   b Calculate $n(P \cup C \cup B)$.
   c Calculate $n(P \cap C)$.
   d Calculate $n(B \cap C)$.
   e How many students took an examination in only one subject?

5 On a cruise around the coast of Turkey, there are 100 passengers and crew. They speak Turkish, French and English.
   Out of the total of 100, 14 speak all three languages, 18 speak French and Turkish only, 16 speak English and French only, and 10 speak English and Turkish only.
   Of those speaking only one language, the number speaking only French or only English is the same and 6 more than the number that speak only Turkish.
   a How many speak only French?
   b How many speak only Turkish ?
   c In total, how many speak English?

6 In a group of 125 students who play tennis, volleyball or football, 10 play all three. Twice as many play tennis and football only. Three times as many play volleyball and football only, and 5 play tennis and volleyball only.
   If x play tennis only, $2x$ play volleyball only and $3x$ play football only, work out:
   a how many play tennis
   b how many play volleyball
   c how many play football.

3.3 Sample space

Set theory can be used to study probability.

A **sample space** is the set of all possible results of a trial or experiment. Each result or outcome is sometimes called an **event**.

Complementary events

A dropped drawing pin can land either pin up, U, or pin down, D. These are the only two possible outcomes and cannot both occur at the same time. The two events are therefore **mutually exclusive** (cannot happen at the same time) and **complementary** (the sum of their probabilities equal 1). The complement of an event A is written A′.

Therefore $P(A) + P(A') = 1$. In words this is read as 'the probability of event A happening added to the probability of event A not happening equals 1'.

Worked examples

1 A fair dice is rolled once. What is its sample space?

The sample space S is the set of possible outcomes or events. Therefore $S = \{1, 2, 3, 4, 5, 6\}$ and the number of outcomes or events is 6.

2 a What is the sample space, S, for two drawing pins dropped together.
b How many possible outcomes are there?

a $S = \{UU, UD, DU, DD\}$
b There are four possible outcomes.

3 The probability of an event B happening is $P(B) = \frac{3}{5}$. Calculate $P(B')$.

$P(B)$ and $P(B')$ are complementary events, so $P(B) + P(B') = 1$.

$P(B') = 1 - \frac{3}{5} = \frac{2}{5}$.

Exercise 3.3.1

1 What is the sample space and the number of events when three coins are tossed?

2 What is the sample space and number of events when a blue dice and a red dice are rolled? (Note: (1, 2) and (2, 1) are different events.)

3 What is the sample space and the number of events when an ordinary dice is rolled and a coin is tossed?

4 A mother gives birth to twins. What is the sample space and number of events for their sex?

5 What is the sample space if the twins in question 4 are identical?

6 Two women take a driving test.
a What are the possible outcomes?
b What is the sample space?

7 A tennis match is played as 'best of three sets'.
a What are the possible outcomes?
b What is the sample space?

8 If the tennis match in question 7 is played as 'best of five sets',
a what are the possible outcomes?
b what is the sample space?

3.4 Logic

Aristotle 384–322 BCE

In philosophy, traditional logic began with the Greek philosopher Aristotle. His six texts are collectively known as *The Organon*. Two of them, *Prior Analytics* and *De Interpretatione*, are the most important for the study of logic.

The fundamental assumption is that reasoning (logic) is built from **propositions**. A proposition is a statement that can be true or false. It consists of two terms: one term (the **predicate**) is affirmed (true) or denied (false) by the other term (the subject): for example

'All men [subject] are mortal [predicate].'

There are just four kinds of proposition in Aristotle's theory of logic.

A type: Universal and affirmative – 'All men are mortal.'
I type: Particular and affirmative – 'Some men are philosophers.'
E type: Universal and negative – 'No men are immortal.'
O type: Particular and negative – 'Some men are not philosophers.'

This is the fourfold scheme of propositions. The theory is a formal theory explaining which combinations of true **premises** give true conclusions.

A century later, in China, a contemporary of Confucius, Mozi 'Father Mo' (430 BC), is credited with founding the Mohist school of philosophy, which studied ideas of **valid inference** and correct conclusions.

What is logic?

Logic is a way to describe situations or knowledge that enables us to reason from existing knowledge to new conclusions. It is useful in computers and artificial intelligence where we need to represent the problems we wish to solve using a symbolic language.

Logic, unlike natural language, is precise and exact. (It is not always easy to understand logic, but it is necessary in a computer program.) An example of a logical argument is:

All students are poor.
I am a student.
By using logic, it follows that I am poor.

Note: if the original statement is false, the conclusion is still logical, even though it is false, e.g.

All students are rich. (is not true)
I am a student.
By using logic it follows that I am rich!

It is not the case that all students are rich but, if it were, I would be rich because I am a student. This is why computer programmers talk of 'Garbage in, garbage out'.

Logic systems are already in use for such things as the wiring systems of aircraft. The Japanese are using logic experiments with robots.

Exercise 3.4.1

1 You have four letters. A letter can be sent sealed or open. Stamps are either 10 cents or
15 cents.

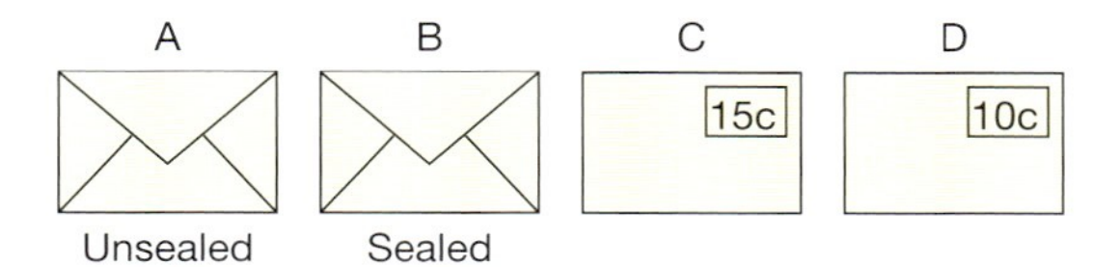

Which envelope must be turned over to test the rule 'If a letter is sealed, it must have a 15 cent stamp'?

2 You have four cards.
A letter A–Z is on one side.
A number 0–9 is on the other side.
You have these cards:

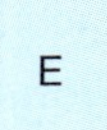

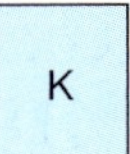

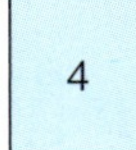

Which card do you turn over to test the rule 'If a card has a vowel on one side, it must have an even number on the other side'?

3.5 Sets and logical reasoning

Proposition: A proposition is a stated fact. It may also be called a **statement**. It can be true or false. For example,

'Nigeria is in Africa' is a true proposition.
'Japan is in Europe' is a false proposition.
These are examples of simple propositions.

Compound statement: Two or more simple propositions can be combined to form a compound proposition or compound statement.

Conjunction: Two simple propositions are combined with the word **and**, e.g.

p: Japan is in Asia.
q: The capital of Japan is Tokyo.

These can be combined to form: Japan is in Asia **and** the capital of Japan is Tokyo.

This is written $p \wedge q$, where $\wedge$ represents the word **and**.

Negation: The negation of any simple proposition can be formed by putting 'not' into the statement, e.g.

p: Ghana is in Africa.
q: Ghana is not in Africa.

Therefore $q = \neg p$ (i.e. p is the negation of q).

If p is true then q cannot also be true.

Implication: For two simple propositions p and q, $p \Rightarrow q$ means **if** p is true **then** q is also true, e.g.

p: It is raining.
q: I am carrying an umbrella.

Then $p \Rightarrow q$ states: If it is raining then I am carrying an umbrella.

Converse: This is the reverse of a proposition. In the example above the converse of $p \Rightarrow q$ is $q \Rightarrow p$. Note, however, although $p \Rightarrow q$ is true, i.e. If it is raining then I must be carrying an umbrella, its converse $q \Rightarrow p$ is not necessarily true, i.e. it is not necessarily the case that: If I am carrying an umbrella then it is raining.

Equivalent propositions: If two propositions are true and converse, then they are said to be equivalent. For, example if we have two propositions

p: Pedro lives in Madrid.
q: Pedro lives in the capital city of Spain.

these propositions can be combined as a compound statement:

If Pedro lives in Madrid, then Pedro lives in the capital city of Spain.

i.e. p implies q $(p \Rightarrow q)$

This statement can be manipulated to form its converse:

If Pedro lives in the capital of Spain, then Pedro lives in Madrid.

i.e. q implies p $(q \Rightarrow p)$

The two combined statements are both true and converse so they are said to be logically equivalent $(q \Leftrightarrow p)$. Logical equivalence will be discussed further, later in this section.

Disjunction: For two propositions, p and q, $p \vee q$ means **either** p **or** q is true **or both** are true, e.g.

p: It is sunny.
q: I am wearing flip-flops.

Then $p \vee q$ states either it is sunny or I am wearing flip-flops or it is both sunny and I am wearing flip-flops.

Exclusive disjunction: For two propositions, p and q, $p \veebar q$ means **either** p **or** q is true but **not both** are true, e.g.

p: It is sunny.
q: I am wearing flip-flops.

Then $p \veebar q$ states either it is sunny or I am wearing flip-flops only.

Valid arguments: An argument is valid if the conclusion follows from the premises (the statements). A premise is always assumed to be true, even though it might not be, e.g.

London is in France.	the first premise
France is in Africa.	the second premise
Therefore London is in Africa.	the conclusion

In this case, although both premises and the conclusion are false, the argument is logically valid.

The validity of an argument can be tested using Venn diagrams.

If p, q and r are three statements and if $p \Rightarrow q$ and $q \Rightarrow r$, then it follows that $p \Rightarrow r$.
In terms of sets, if A, B and C are all proper subsets ($\subset$) of the universal set U and if $A \subset B$ and $B \subset C$ then $A \subset C$.

Diagrammatically this can be represented as shown in the Venn diagram opposite:

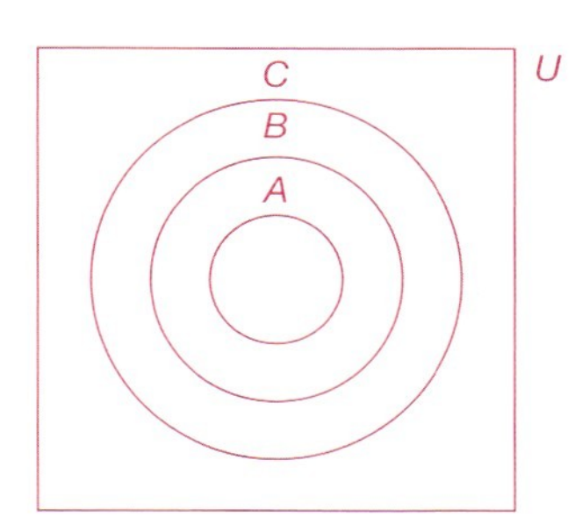

Exercise 3.5.1

1 Which of the following are propositions?

a Are you from Portugal?
b Capetown is in South Africa.
c Catalan is a Spanish language.
d Be careful with that.
e $x = 3$
f $x \neq 3$
g I play football.
h Go outside and play.
i Apples are good to eat.
j J is a letter of the alphabet.

2 Form compound statements using the word 'and' from the two propositions given and say whether the compound statement is true or false.

a t: Teresa is a girl. a: Abena is a girl.
b p: $x < 8$ q: $x > -1$
c a: A pentagon has 5 sides. b: A triangle has 4 sides.
d l: London is in England. e: England is in Europe.
e k: $x < y$ l: $y < z$
f m: $5 \in$ {prime number} n: $4 \in$ {even numbers}
g s: A square is a rectangle. t: A triangle is a rectangle.
h p: Paris is the capital of France. g: Ghana is in Asia.
i a: $37 \in$ {prime number} b: $51 \in$ {prime numbers}
j p: parallelograms $\in$ {rectangles} t: trapeziums $\in$ {rectangles}

The analogy of logic and set theory

The use of **No** or **Never** or **All ... do not** in statements (e.g. **No** French people are British people) means the sets are **disjoint**, i.e. they do not overlap.

The use of **All** or **If ... then** or **No ... not** in statements (e.g. There is **no** nurse who does **not** wear a uniform) means that one set is a subset of another.

The use of **Some** or **Most** or **Not all** in statements (e.g. **Some** televisions are very expensive) means that the sets intersect.

Worked examples

1 P is the set of French people and Q is the set of British people. Draw a Venn diagram to represent the sets.

The Venn diagram is as shown,
i.e. $P \cap Q = \varnothing$
In logic this can be written $p \veebar q$,
i.e. p or q but not both.

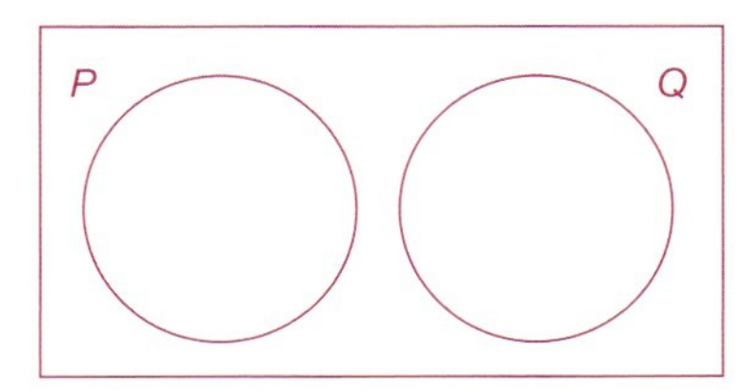

2 P is the set of nurses and Q is the set of people who wear uniform. Draw a Venn diagram to represent the sets.

P is a subset of Q as there are other people who wear uniforms apart from nurses, i.e. $P \subset Q$
In logic this can be written $p \Rightarrow q$.

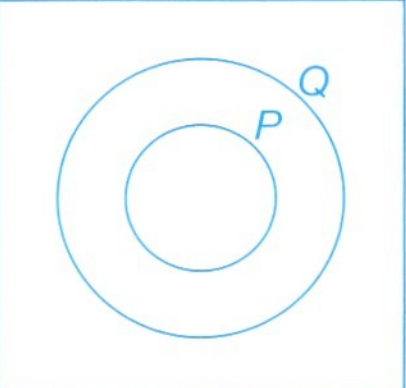

3 P is the set of televisions and Q is the set of expensive electrical goods. Draw a Venn diagram to represent the sets.

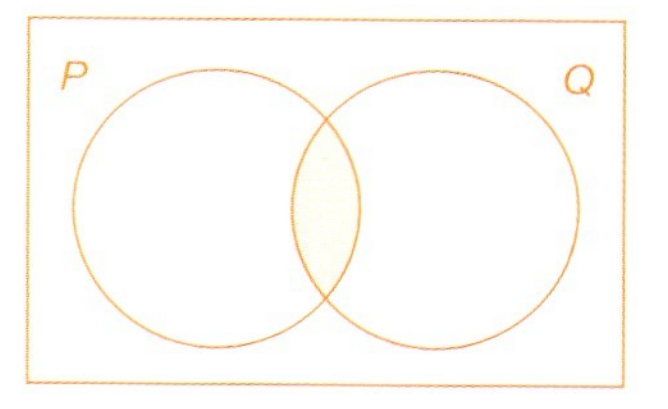

P intersects Q as there are expensive electrical goods that are not televisions and there are televisions that are not expensive.
In logic the intersection can be written $p \wedge q$.

Exercise 3.5.2

1 Illustrate the following sets using a Venn diagram.
   Q: students wearing football shirts
   P: professional footballers wearing football shirts
Shade the region that represents the statement 'Kofi is a professional footballer and a student'. How would you write this using logic symbols?

2 Illustrate the following sets using a Venn diagram.
   Q: students wearing football shirts
   P: professional footballers wearing shirts
Shade the region that satisfies the statement 'Maanu is either a student or a professional footballer but he is not both'. How would you write this using logic symbols?

3 Illustrate the following sets using a Venn diagram.
   P: maths students U: all students.
Shade the region that satisfies the statement 'Boamah is not a maths student'. How would you write this using logic symbols?

4 Illustrate the following sets using a Venn diagram.
   P: five-sided shapes U: all shapes
Shade the region that satisfies the statement 'A regular pentagon is a five-sided shape'. How would you write this using logic symbols?

5 Illustrate the following sets using a Venn diagram.
   Q: multiples of 5 U: integers
Shade the region where you would place 17. How would you write this using logic symbols?

6 Illustrate the following sets using a Venn diagram.
   P: people who have studied medicine
   Q: people who are doctors
Shade the region that satisfies the statement 'All doctors have studied medicine'. How would you write this using logic symbols?

7 Illustrate the statement people with too much money are never happy using a Venn diagram with these sets.
   P: people who have too much money
   Q: people who are happy
Shade the region that satisfies the statement 'People with too much money are never happy'. How would you write this using logic symbols?

8 Illustrate the following sets using a Venn diagram.
P: music lessons Q: lessons that are expensive
Shade the region that satisfies the statement 'Some music lessons are expensive'.
How would you write this using logic symbols?

3.6 Truth tables

In probability experiments, a coin when tossed can land on heads or tails. These are complementary events, i.e. $P(H) + P(T) = 1$.

In logic, if a statement is not uncertain, then it is either true (T) or false (F). If there are two statements, then either both are true, both are false or one is true and one is false.

A truth table is a clear way of showing the possibilities of statements.

Let proposition p be 'Coin A lands heads' and proposition q be 'Coin B lands heads'. The truth table below shows the different possibilities when the two coins are tossed. Alongside is a two-way table also showing the different outcomes. Note the similarity between the two tables.

p	**q**
T	T
T	F
F	T
F	F

Truth table

Coin A	**Coin B**
H	H
H	T
T	H
T	T

Two-way table

Symbols used in logic

There are some symbols that you will need to become familiar with when we study logic in more detail.

The following symbols refer to the relationship between two propositions p and q.

Symbol	**Meaning**
$\wedge$	p and q (conjunction)
$\vee$	p or q or both (inclusive disjunction)
$\underline{\vee}$	p or q but not both (exclusive disjunction)
$\Rightarrow$	If p then q (implication)
$\Leftrightarrow$	If $p \Rightarrow q$ and $q \Rightarrow p$ the statements are equivalent, i.e. $p \Leftrightarrow q$ (equivalence)
$\neg$	If p is true, q cannot be true. $p \neg q$ (negation)

Conjunction, disjunction and negation

Extra columns can be added to a truth table.

$p \wedge q$ (conjunction) means that both p and q must be true for the statement to be true.

p	q	$p \wedge q$
T	T	T
T	F	F
F	T	F
F	F	F

$p \vee q$ (inclusive disjunction) means that either p or q, or both, must be true for the statement to be true.

p	q	$p \vee q$
T	T	T
T	F	T
F	T	T
F	F	F

$p \veebar q$ (exclusive disjunction) means that either p or q, but not both, must be true for the statement to be true.

p	q	$p \veebar q$
T	T	F
T	F	T
F	T	T
F	F	F

$\neg p$ represents a negation, i.e. p must not be true for the statement to be true.

p	q	$\neg p$
T	T	F
T	F	F
F	T	T
F	F	T

Exercise 3.6.1

1 Copy and complete the truth table for three propositions p, q and r. It may help to think of spinning three coins and drawing a table of possible outcomes.

p	q	r
T	T	T
F	F	F

2 Copy and complete the truth table below for the three statements p, q and r.

p	q	r	$\neg p$	$p \vee q$	$\neg p \vee r$	$(p \vee q) \wedge (\neg p \vee r)$
T	T	T	F	T	T	T
T	T	F				
T						
T						
F						
F						
F						
F						

Logical contradiction and tautology

Logical contradiction

A contradiction or contradictory proposition is never true. For example, let p be the proposition that Rome is in Italy.

p: Rome is in Italy.

Therefore $\neg p$, the negation of p, is the proposition: Rome is not in Italy.

If we write $p \wedge \neg p$ we are saying Rome is in Italy and Rome is not in Italy. This cannot be true at the same time. This is an example of a logical contradiction.

A truth table is shown below for the above statement.

p	$\neg p$	$p \wedge \neg p$
T	F	F
F	T	F

Both entries in the final column are F. In other words a logical contradiction *must* be false.

Worked example

Show that the compound proposition below is a contradiction.

$(p \vee q) \wedge [(\neg p) \wedge (\neg q)]$

Construct a truth table:

p	q	$\neg p$	$\neg q$	$p \vee q$	$(\neg p) \wedge (\neg q)$	$(p \vee q) \wedge [(\neg p) \wedge (\neg q)]$
T	T	F	F	T	F	F
T	F	F	T	T	F	F
F	T	T	T	T	F	F
F	F	T	T	F	T	F

Because the entries in the last column are all false, the statement is a logical contradiction.

Tautology

The manager of the band *Muse* said to me recently: 'If *Muse's* album "Resistance" is a success, they will be a bigger band than U2.' He paused 'Or they will not'.

This is an example of a tautology: 'either it does or it doesn't'. It is always true.

A compound proposition is a tautology if it always true regardless of the truth values of its variables.

Consider the proposition: All students study maths or all students do not study maths. This is a tautology, as can be shown in a truth table by considering the result of $p \vee \neg p$.

p	$\neg p$	$p \vee \neg p$
T	F	T
F	T	T

Since the entries in the final column $p \vee \neg p$ are all true, this is a tautology.

Worked example

Show that $(p \vee q) \vee [(\neg p) \wedge (\neg q)]$ is a tautology by copying and completing the truth table below.

p	q	$\neg p$	$\neg q$	$p \vee q$	$(\neg p) \wedge (\neg q)$	$(p \vee q) \vee [(\neg p) \wedge (\neg q)]$
T	T	F	F	T	F	T
T	F	F	T	T	F	T
F	T	T	F	T	F	T
F	F	T	T	F	T	T

As the entries in the final column $(p \vee q) \vee [(\neg p) \wedge (\neg q)]$ are all true, the statement is a tautology.

Exercise 3.6.2

1 Describe each of the following as a tautology, a contradiction or neither. Use a truth table if necessary.

a $p \wedge \neg q$

b $q \wedge \neg q$

c $p \vee \neg q$

d $q \vee \neg q$

e $[p \vee (\neg q)] \wedge [q \vee (\neg q)]$

2 By drawing a truth table in each case, decide whether each of the following propositions is a tautology, contradiction or neither.

a $\neg p \wedge \neg q$

b $\neg(\neg p) \vee p$

c $q \wedge \neg r$

d $(p \wedge q) \wedge r$

e $(p \wedge q) \vee r$

3.7 Implication; converse; inverse; contrapositive and logical equivalence

Implication

'If' is a word introducing a conditional clause.

Later in your life someone might say to you, 'If you get a degree, then I will buy you a car'.

Let us look at this in a truth table.

p: You get a degree.
q: I will buy you a car.

p	q	$p \Rightarrow q$
T	T	T
T	F	F
F	T	T
F	F	T

The first row is simple:

You get a degree, I buy you a car, and therefore I have kept my promise.

The second row too is straightforward:

You get a degree, I don't buy you a car, and therefore I have broken my promise.

The last two rows seem more complicated, but think of them like this. If you do not get a degree, then I have kept my side of the bargain whether I buy you a car or not.

Therefore, the only way that this type of statement is false is if a 'promise' is broken.

Logically $p \Rightarrow q$ is true if:

p is false
or q is true
or p is false *and* q is true

Similarly $p \Rightarrow q$ is only false if p is true and q is false.

Worked examples

In the following statements, assume that the first phrase is p and the second phrase q.

Determine whether the statement $p \Rightarrow q$ is logically true or false.

1 'If $5 \times 4 = 20$, then the Earth moves round the Sun.'
As both p and q are true, then $p \Rightarrow q$ is true, i.e. the statement $p \Rightarrow q$ is logically correct.

2 'If the Sun goes round the Earth, then I am an alien.'
Since p is false, then $p \Rightarrow q$ is true whether I am an alien or not. Therefore the statement is logically true.

This means that witty replies like:

'If I could run faster, I could be a professional footballer'
'Yes and if you had wheels you'd be a professional skater' are logically true, since the premise p, 'if you had wheels', is false and therefore what follows is irrelevant.

Exercise 3.7.1

1 In the following statements, assume that the first phrase is p and the second phrase is q. Determine whether the statement $p \Rightarrow q$ is logically true or false.
 a If $2 + 2 = 5$ then $2 + 3 = 5$.
 b If the moon is round, then the Earth is flat.
 c If the Earth is flat, then the moon is flat.
 d If the Earth is round, then the moon is round.
 e If the Earth is round, then I am the man on the moon.

2 Descartes' phrase 'Cogito, ergo sum' translates as 'I think, therefore I am'.
 a Rewrite the sentence using one or more of the following: 'if', 'whenever' , 'it follows that', 'it is necessary' , 'unless' , 'only'.
 b Copy and complete the following sentence: 'Cogito ergo sum' only breaks down logically if Descartes thinks, but . . .'

Logical equivalence

There are many different ways that we can form compound statements from the propositions p and q using connectives. Some of the different compound propositions have the same truth values. These propositions are said to be equivalent. The symbol for equivalence is $\Leftrightarrow$.

Two propositions are logically equivalent when they have identical truth values.

Worked example

Use a truth table to show that $\neg(p \vee q)$ and $\neg p \vee \neg q$ are logically equivalent.

p	q	$p \wedge q$	$\neg(p \wedge q)$	$\neg p$	$\neg q$	$\neg p \vee \neg q$
T	T	T	F	F	F	F
T	F	F	T	F	T	T
F	T	F	T	T	F	T
F	F	F	T	T	T	T

Since the truth values for $\neg(p \wedge q)$ and $\neg p \vee \neg q$ (columns 4 and 7) are identical, the two statements are logically equivalent.

Converse

The statement 'All squares are rectangles' can be rewritten using the word 'if' as:

'If an object is a square, then it is a rectangle'. $p \Rightarrow q$. (true in this case)

The converse is:

$q \Rightarrow p$. 'If an object is a rectangle, then it is a square.' (false in this case)

Inverse

The inverse of the statement 'If an object is a square, then it is a rectangle' ($p \Rightarrow q$) is:

$\neg p \Rightarrow \neg q$. 'If an object is not a square, then it is not a rectangle.' (false in this case)

Contrapositive

The contrapositive of the statement 'If an object is a square, then it is a rectangle' ($p \Rightarrow q$) is:

$\neg q \Rightarrow \neg p$. 'If an object is not a rectangle, then it is not a square.' (true in this case)

Note:
A statement is logically equivalent to its contrapositive.
A statement is not logically equivalent to its converse or inverse.
The converse of a statement is logically equivalent to the inverse.

So if a statement is true, then its contrapositive is also true.
If a statement is false, then its contrapositive is also false.

And if the converse of a statement is true, then the inverse is also true.
If the converse of a statement is false, then the inverse is also false.

To summarize:

given a conditional statement:	$p \Rightarrow q$
the converse is:	$q \Rightarrow p$
the inverse is:	$\neg p \Rightarrow \neg q$
the contrapositive is:	$\neg q \Rightarrow \neg p$

Worked example

Statement: 'All even numbers are divisible by 2.'

a Rewrite the statement as a conditional statement.
b State the converse, inverse and contrapositive of the conditional statement. State whether each new statement is true or false.

a Conditional: 'If a number is even, then it is divisible by 2.' (true)
b Converse: 'If a number is divisible by 2, then it is an even number.' (true)
Inverse: 'If a number is not even, then it is not divisible by 2.' (true)
Contrapositive: 'If a number is not divisible by 2, then it is not an even number. (true)

Note: The contrapositive both switches the order and negates. It combines the converse and the inverse.

On a truth table it can be shown that a conditional statement and its contrapositive are logically equivalent

p	q	$\neg p$	$\neg q$	Implication $p \Rightarrow q$	Contrapositive $\neg q \Rightarrow \neg p$
T	T	F	F	T	T
T	F	F	T	F	F
F	T	T	F	T	T
F	F	T	T	T	T

Note: If we have a tautology, we must have logical equivalence. For example, 'If you cannot find the keys you have lost, then you are looking in the wrong place.' Obviously if you are looking in the right place then you can find your keys. (So the contrapositive is equivalent to the proposition.)

Exercise 3.7.2

1 Write each of the following as a conditional statement and then write its converse, inverse or contrapositive, as indicated in brackets.

Example Being interested in the Romans means that you will enjoy Italy. (converse)

Solution: Conditional statement. If you are interested in the Romans, then you will enjoy Italy.

Converse. If you enjoy Italy, then you are interested in the Romans.

- a You do not have your mobile phone, so you cannot send a text. (inverse)
- b A small car will go a long way on 20 euros worth of petrol. (contrapositive)
- c Speaking in French means that you will enjoy France more. (converse)
- d When it rains I do not play tennis. (inverse)
- e We stop playing golf when there is a threat of lightning. (inverse)
- f The tennis serve is easy if you practise it. (contrapositive)
- g A six-sided polygon is a hexagon. (contrapositive)
- h You are less than 160 cm tall, so you are smaller than me. (inverse)
- i The bus was full, so I was late. (contrapositive)
- j The road was greasy, so the car skidded. (converse)

2 Rewrite these statements using the conditional 'if'. Then state the converse, inverse and contrapositive. State whether each new statement is true or false.

- a Any odd number is a prime number.
- b A polygon with six sides is called an octagon.
- c An acute-angled triangle has three acute angles.
- d Similar triangles are congruent.
- e Congruent triangles are similar.
- f A cuboid has six faces.
- g A solid with eight faces is a regular octahedron.
- h All prime numbers are even numbers.

3.8 Probability

Although Newton and Galileo had some thoughts about chance, it is accepted that the study of what we now call probability began when Blaise Pascal (1623–1662) and Pierre de Fermat (of Fermat's last theorem fame) corresponded about problems connected with games of chance. Later Christiaan Huygens wrote the first book on the subject, *The Value of all Chances in Games of Fortune*, in 1657. This included a chapter entitled 'Gambler's Ruin'.

Pierre de Fermat

In 1821 Carl Friedrich Gauss (1777–1855), one of the greatest mathematicians who ever lived, worked on the 'Normal distribution', a very important contribution to the study of probability.

Probability is the study of chance, or the likelihood of an event happening. In this section we will be looking at theoretical probability. But, because probability is based on chance, what theory predicts does not necessarily happen in practice.

Probability of an event

A favourable outcome refers to the event in question actually happening. The total number of possible outcomes refers to all the different types of outcome one can get in a particular situation. In general:

$$\text{Probability of an event} = \frac{\text{number of favourable outcomes}}{\text{total number of equally likely outcomes}}$$

This can also be written as: $P(A) = \dfrac{n(A)}{n(U)}$,

where $P(A)$ is the probability of event A, $n(A)$ is the number of ways event A can occur and $n(U)$ is the total number of equally likely outcomes.

Therefore

if the probability = 0, it implies the event is impossible
if the probability = 1, it implies the event is certain to happen

Worked example

An ordinary, fair dice is rolled.

a Calculate the probability of getting a 6.
b Calculate the probability of not getting a 6.

a Number of favourable outcomes = 1 (i.e. getting a 6)

Total number of possible outcomes = 6 (i.e. getting a 1, 2, 3, 4, 5 or 6)
Probability of getting a 6, $P(6) = \frac{1}{6}$

b Number of favourable outcomes = 5 (i.e. getting a 1, 2, 3, 4, 5)
Total number of possible outcomes = 6 (i.e. getting a 1, 2, 3, 4, 5 or 6)
Probability of not getting a six, $P(6') = \frac{5}{6}$

From this it can be seen that the probability of not getting a 6 is equal to 1 minus the probability of getting a 6, i.e. $P(6) = 1 - P(6')$.

These are known as **complementary events**.

In general, for an event A, $P(A) = 1 - P(A')$.

Exercise 3.8.1

1 Calculate the theoretical probability, when rolling an ordinary, fair dice, of getting each of the following.

a a score of 1
b a score of 5
c an odd number
d a score less than 6
e a score of 7
f a score less than 7

2 **a** Calculate the probability of:
i) being born on a Wednesday
ii) not being born on a Wednesday.
b Explain the result of adding the answers to **a i)** and **ii)** together.

3 250 tickets are sold for a raffle. What is the probability of winning if you buy:

a 1 ticket
b 5 tickets
c 250 tickets
d 0 tickets?

4 In a class there are 25 girls and 15 boys. The teacher takes in all of their books in a random order. Calculate the probability that the teacher will:

a mark a book belonging to a girl first
b mark a book belonging to a boy first.

5 Tiles, each lettered with one different letter of the alphabet, are put into a bag. If one tile is drawn out at random, calculate the probability that it is:

a an A or P
b a vowel
c a consonant
d an X, Y or Z.
e a letter in your first name.

6 A boy was late for school 5 times in the previous 30 school days. If tomorrow is a school day, calculate the probability that he will arrive late.

7 3 red, 10 white, 5 blue and 2 green counters are put into a bag.

a If one is picked at random, calculate the probability that it is:
 i) a green counter
 ii) a blue counter.
b If the first counter taken out is green and it is not put back into the bag, calculate the probability that the second counter picked is:
 i) a green counter
 ii) a red counter.

8 A roulette wheel has the numbers 0 to 36 equally spaced around its edge. Assuming that it is unbiased, calculate the probability on spinning it of getting:

a the number 5
b an even number
c an odd number
d zero
e a number greater than 15
f a multiple of 3
g a multiple of 3 or 5
h a prime number.

9 The letters R, C and A can be combined in several different ways.

a Write the letters in as many different combinations as possible.
b If a computer writes these three letters at random, calculate the probability that:
 i) the letters will be written in alphabetical order
 ii) the letter R is written before both the letters A and C
 iii) the letter C is written after the letter A
 iv) the computer will spell the word CART if the letter T is added.

10 A normal pack of playing cards contains 52 cards. These are made up of four suits (hearts, diamonds, clubs and spades). Each suit consists of 13 cards. These are labelled ace, 2, 3, 4, 5, 6, 7, 8, 9, 10, Jack, Queen and King. The hearts and diamonds are red; the clubs and spades are black.

If a card is picked at random from a normal pack of cards calculate the probability of picking:

a a heart
b a black card
c a four
d a red King
e a Jack, Queen or King
f the ace of spades
g an even numbered card
h a seven or a club.

3.9 Combined events

In this section we look at the probability of two or more events happening: combined events. If only two events are involved, then two-way tables can be used to show the outcomes.

Two-way tables of outcomes

Worked example

a Two coins are tossed. Show all the possible outcomes in a two-way table.
b Calculate the probability of getting two heads.
c Calculate the probability of getting a head and a tail in any order.

a

Coin 2 \ Coin 1	Head	Tail
Head	HH	TH
Tail	HT	TT

b All four outcomes are equally likely, therefore the probability of getting HH is $\frac{1}{4}$.

c The probability of getting a head and a tail in any order, i.e. HT or TH, is $\frac{2}{4} = \frac{1}{2}$.

Exercise 3.9.1

1 a Two fair tetrahedral dice are rolled. If each is numbered 1–4, draw a two-way table to show all the possible outcomes.
  b What is the probability that both dice show the same number?
  c What is the probability that the number on one dice is double the number on the other?
  d What is the probability that the sum of both numbers is prime?

2 Two fair dice are rolled. Copy and complete the diagram to show all the possible combinations.

What is the probability of getting:
  a a double 3
  b any double
  c a total score of 11
  d a total score of 7
  e an even number on both dice
  f an even number on at least one dice
  g a 6 or a double
  h scores which differ by 3
  i a total which is either a multiple of 2 or 5?

Dice 2 \ Dice 1	1	2	3	4	5	6
6			3,6			
5			3,5			
4			3,4			
3			3,3			
2			3,2		5,2	6,2
1	1,1	2,1	3,1	4,1		

Tree diagrams

When more than two combined events are being considered, two-way tables cannot be used and therefore another method of representing information diagrammatically is needed. Tree diagrams are a good way of doing this.

Worked example

a If a coin is tossed three times, show all the possible outcomes on a tree diagram, writing each of the probabilities at the side of the branches.

b What is the probability of getting three heads?

c What is the probability of getting two heads and one tail in any order?

d What is the probability of getting at least one head?

e What is the probability of getting no heads?

a

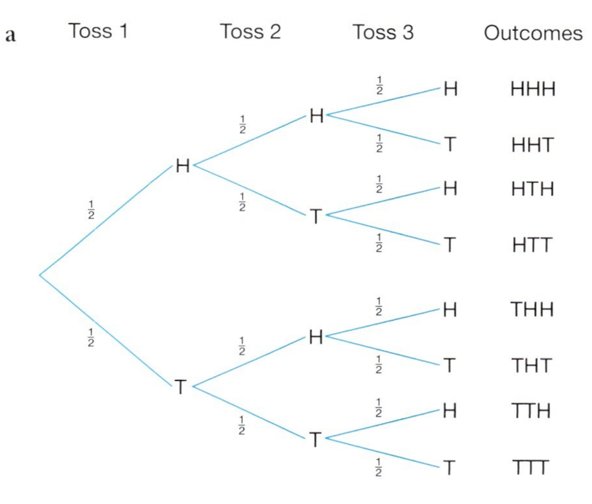

b To calculate the probability of getting three heads, multiply along the branches:

$P(HHH) = \frac{1}{2} \times \frac{1}{2} \times \frac{1}{2} = \frac{1}{8}$

c The successful outcomes are HHT, HTH, THH.

P(two heads, one tail any order)

$= P(HHT) + P(HTH) + P(THH)$

$= \left(\frac{1}{2} \times \frac{1}{2} \times \frac{1}{2}\right) + \left(\frac{1}{2} \times \frac{1}{2} \times \frac{1}{2}\right) + \left(\frac{1}{2} \times \frac{1}{2} \times \frac{1}{2}\right) = \frac{1}{8} + \frac{1}{8} + \frac{1}{8} = \frac{3}{8}$

Therefore the probability is $\frac{3}{8}$.

d This refers to any outcome with either one, two or three heads, i.e. all of them except TTT.

$P(TTT) = \frac{1}{2} \times \frac{1}{2} \times \frac{1}{2} = \frac{1}{8}$

$P(\text{at least one head}) = 1 - P(TTT) = 1 - \frac{1}{8} = \frac{7}{8}$

Therefore the probability is $\frac{7}{8}$.

e The only successful outcome for this event is TTT.

Therefore the probability is $\frac{1}{8}$, as shown in part **d**.

Exercise 3.9.2

1 a A computer uses the numbers 1, 2 and 3 at random to make three-digit numbers. Assuming that a number can be repeated, show on a tree diagram all the possible combinations that the computer can print.

b Calculate the probability of getting:
- i) the number 131
- ii) an even number
- iii) a multiple of 11
- iv) a multiple of 3
- v) a multiple of 2 or 3
- vi) a palindromic number.

2 a A family has four children. Draw a tree diagram to show all the possible combinations of boys and girls. [Assume P(girl) = P(boy).]

b Calculate the probability of getting:
- i) all girls
- ii) two girls and two boys
- iii) at least one girl
- iv) more girls than boys.

3 a A netball team plays three matches. In each match the team is equally likely to win, lose or draw. Draw a tree diagram to show all the possible outcomes over the three matches.

b Calculate the probability that the team:
- i) wins all three matches
- ii) wins more times than it loses
- iii) loses at least one match
- iv) doesn't win any of the three matches.

c Explain why it is not very realistic to assume that the outcomes are equally likely in this case.

4 A spinner is split into quarters.

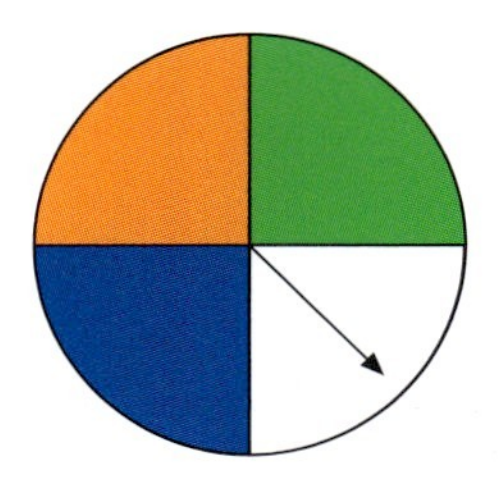

a If it is spun twice, draw a probability tree showing all the possible outcomes.

b Calculate the probability of getting:
- i) two greens
- ii) a green and a blue in any order
- iii) a blue and a white in any order.

Tree diagrams for unequal probabilities

In each of the cases considered so far, all of the outcomes have been assumed to be equally likely. However, this need not be the case.

Worked example

In winter, the probability that it rains on any one day is $\frac{5}{7}$.

a Using a tree diagram, show all the possible combinations for two consecutive days. Write each of the probabilities by the sides of the branches.
b Calculate the probability that it will rain on both days.
c Calculate the probability that it will rain on the first day but not the second day.
d Calculate the probability that it will rain on at least one day.

a

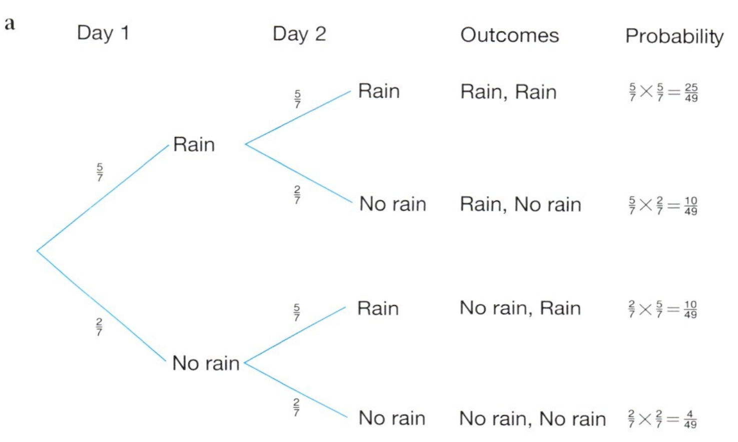

Note how the probability of each outcome is found by multiplying the probabilities for each of the branches.

b $P(R, R) = \frac{5}{7} \times \frac{5}{7} = \frac{25}{49}$

c $P(R, NR) = \frac{5}{7} \times \frac{2}{7} = \frac{10}{49}$

d The outcomes which satisfy this event are (R, R), (R, NR) and (NR, R).

Therefore the probability is $\frac{25}{49} + \frac{10}{49} + \frac{10}{49} = \frac{45}{49}$

Exercise 3.9.3

1 A particular board game involves players rolling a dice. However, before a player can start, he or she needs to roll a 6.

a Copy and complete the tree diagram below showing all the possible combinations for the first two rolls of the dice.

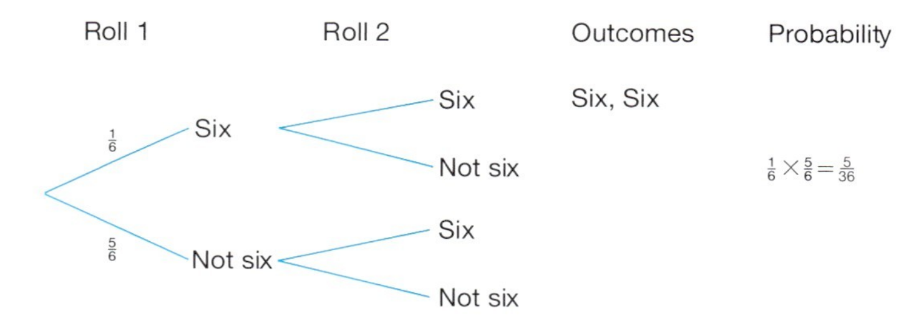

b Calculate the probability of each of the following.
- **i)** Getting a six on the first throw
- **ii)** Starting within the first two throws
- **iii)** Starting on the second throw
- **iv)** Not starting within the first three throws
- **v)** Starting within the first three throws

c If you add the answers to **b iv)** and **v)** what do you notice? Explain.

2 In Italy $\frac{3}{5}$ of the cars are foreign made. By drawing a tree diagram and writing the probabilities next to each of the branches, calculate each of these probabilities.

a The next two cars to pass a particular spot are both Italian.
b Two of the next three cars are foreign.
c At least one of the next three cars is Italian.

3 The probability that a morning bus arrives on time is 65%.

a Draw a tree diagram showing all the possible outcomes for three consecutive mornings.
b Label your tree diagram and use it to calculate the probability of each of the following.
- **i)** The bus is on time on all three mornings.
- **ii)** The bus is late the first two mornings.
- **iii)** The bus is on time two out of the three mornings.
- **iv)** The bus is on time at least twice.

4 Light bulbs are packaged in cartons of three; 10% of the bulbs are found to be faulty. Calculate the probability of finding two faulty bulbs in a single carton.

5 A volleyball team has a 0.25 chance of losing a game. Calculate the probability of the team achieving:

a two consecutive wins **b** three consecutive wins
c 10 consecutive wins.

Tree diagrams for probability problems with and without 'replacement'

In the examples considered so far, the probability for each outcome remained the same throughout the problem. However, this need not always be the case.

Worked examples

1 A bag contains three red balls and seven black balls. If the balls are put back after being picked, what is the probability of picking:

a two red balls
b a red ball and a black ball in any order.

This is selection with replacement. Draw a tree diagram to help visualise the problem.

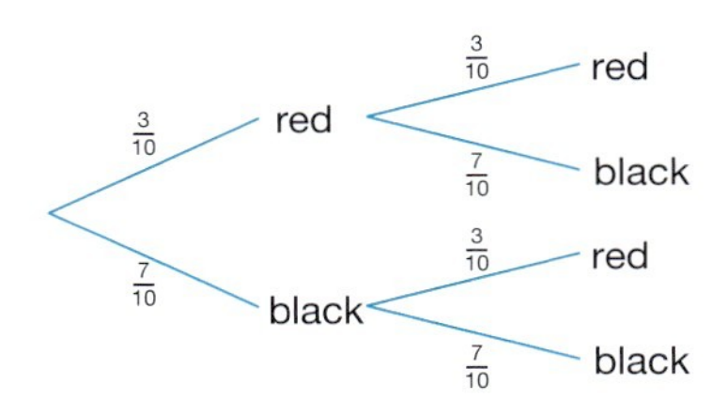

a The probability of a red followed by a red, P(RR) $= \frac{3}{10} \times \frac{3}{10} = \frac{9}{100}$.

b The probability of a red followed by a black or a black followed by a red is

$$P(RB) + P(BR) = \left(\frac{3}{10} \times \frac{7}{10}\right) + \left(\frac{7}{10} \times \frac{3}{10}\right) = \frac{21}{100} + \frac{21}{100} = \frac{42}{100}.$$

2 Repeat question 1, but this time each ball that is picked is not put back in the bag.

This is selection without replacement. The tree diagram is now as shown.

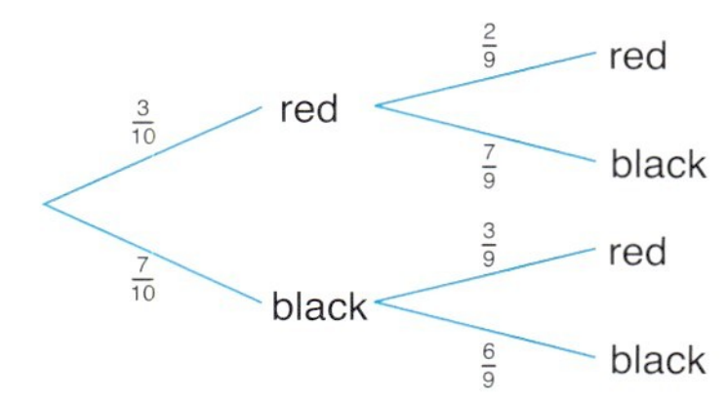

a $P(RR) = \frac{3}{10} \times \frac{2}{9} = \frac{6}{90}$.

b $P(RB) + P(BR) = \left(\frac{3}{10} \times \frac{7}{9}\right) + \left(\frac{7}{10} \times \frac{3}{9}\right) = \frac{21}{90} + \frac{21}{90} = \frac{42}{90}$.

Exercise 3.9.4

1 A bag contains five red balls and four black balls. If a ball is picked out at random, its colour is recorded and it is then put back in the bag, what is the probability of choosing:

a two red balls
b two black balls
c a red ball and a black ball in this order
d a red ball and a black ball in any order?

2 Repeat question 1 but, in this case, after a ball is picked at random, it is not put back in the bag.

3 A bag contains two black, three white and five red balls. If a ball is picked, its colour recorded and then put back in the bag, what is the probability of picking:

a two black balls
b a red and a white ball in any order?

4 Repeat question 3 but, in this case, after a ball is picked at random, it is not put back in the bag.

5 You buy five tickets for a raffle. 100 tickets are sold altogether. Tickets are picked at random. You have not won a prize after the first three tickets have been drawn.

a What is the probability that you win a prize with either of the next two draws?
b What is the probability that you do not win a prize with either of the next two draws?

6 A bowl of fruit contains one apple, one banana, two oranges and two pears. Two pieces of fruit are chosen at random and eaten.
   a Draw a probability tree showing all the possible combinations of the two pieces of fruit.
   b Use your tree diagram to calculate the probability that:
      i) both the pieces of fruit eaten are oranges
      ii) an apple and a banana are eaten
      iii) at least one pear is eaten.

Use of Venn diagrams in probability

You have seen earlier in this unit how Venn diagrams can be used to represent sets. They can also be used to solve problems involving probability.

Probability of event, A, $P(A) = \frac{n(A)}{n(U)}$

Worked examples

1 In a survey carried out in a college, students were asked for their favourite subject.

15 chose English
8 chose Science
12 chose Mathematics
5 chose Art

If a student is chosen at random, what is the probability that he or she likes Science best?

This can be represented on a Venn diagram as:

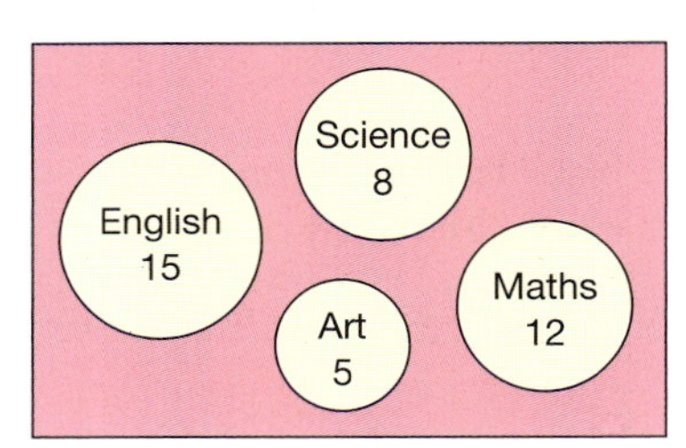

There are 40 students, so the probability is $\frac{8}{40} = \frac{1}{5}$.

2 A group of 21 friends decide to go out for the day to the local town; 9 of them decide to see a film at the cinema, 15 of them get together for lunch.
   a Draw a Venn diagram to show this information if set A represents those who see a film and set B those who have lunch.
   b Determine the probability that a person picked at random only went to the cinema.

a $9 + 15 = 24$; as there are only 21 people, this implies that 3 people see the film and have lunch. This means that $9 - 3 = 6$ only went to see a film and $15 - 3 = 12$ only had lunch.

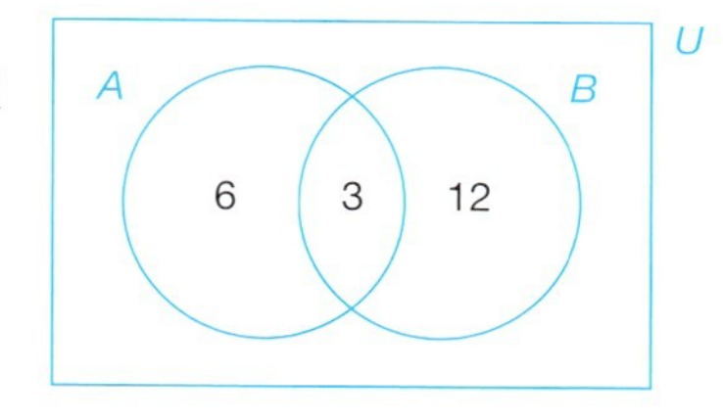

b The number who only went to the cinema is 6, as the other 3 who saw a film also went out for lunch. Therefore the probability is $\frac{6}{21} = \frac{2}{7}$.

Exercise 3.9.5

1 In a class of 30 students, 20 study French, 18 study Spanish and 5 study neither.
   a Draw a Venn diagram to show this information.
   b What is the probability that a student chosen at random studies both French and Spanish?

2 In a group of 35 students, 19 take Physics, 18 take Chemistry and 3 take neither. What is the probability that a student chosen at random takes:
   a both Physics and Chemistry
   b Physics only
   c Chemistry only?

3 108 people visited an art gallery; 60 liked the pictures, 53 liked the sculpture, 10 liked neither.
   What is the probability that a person chosen at random liked the pictures but not the sculpture?

4 In a series of examinations in a school:
   37 students took English
   48 students took French
   45 students took Spanish
   15 students took English and French
   13 students took French and Spanish
   7 students took English and Spanish
   5 students took all three.
   a Draw a Venn diagram to represent this information.
   b What is the probability that a student picked at random took:
      i) all three
      ii) English only
      iii) French only?

3.10 Laws of probability

Mutually exclusive events

Events that cannot happen at the same time are known as mutually exclusive events. For example, if a sweet bag contains 12 red sweets and 8 yellow sweets, let picking a red sweet be event *A*, and picking a yellow sweet be event *B*. If one sweet is picked, it is not possible to pick a sweet which is both red and yellow. Therefore these events are mutually exclusive.

This can be shown in a Venn diagram:

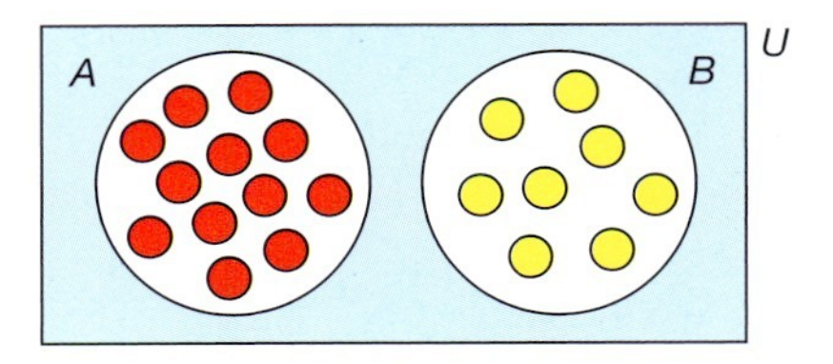

$P(A) = \frac{12}{20}$ whilst $P(B) = \frac{8}{20}$.

As there is no overlap, $P(A \cup B) = P(A) + P(B) = \frac{12}{20} + \frac{8}{20} = \frac{20}{20} = 1$.

i.e. the probability of mutually exclusive event A or event B happening is equal to the sum of the probabilities of event A and event B and the sum of the probabilities of all possible mutually exclusive events is 1.

Worked example

In a 50 m swim, the world record holder has a probability of 0.72 of winning. The probability of her finishing second is 0.25.

What is the probability that she either wins or comes second?

Since she cannot finish both first and second, the events are mutually exclusive. Therefore $P(1\text{st} \cup 2\text{nd}) = 0.72 + 0.25 = 0.97$.

Combined events

If events are not mutually exclusive then they may occur at the same time.

These are known as combined events.

For example, a pack of 52 cards contains four suits: clubs (♣), spades (♠), hearts (♥) and diamonds (♦). Clubs and spades are black; hearts and diamonds are red. Each suit contains 13 cards. These are ace, 2, 3, 4, 5, 6, 7, 8, 9,10, Jack, Queen and King.

A card is picked at random. Event A represents picking a black card; event B represents picking a King.

In a Venn diagram this can be shown as:

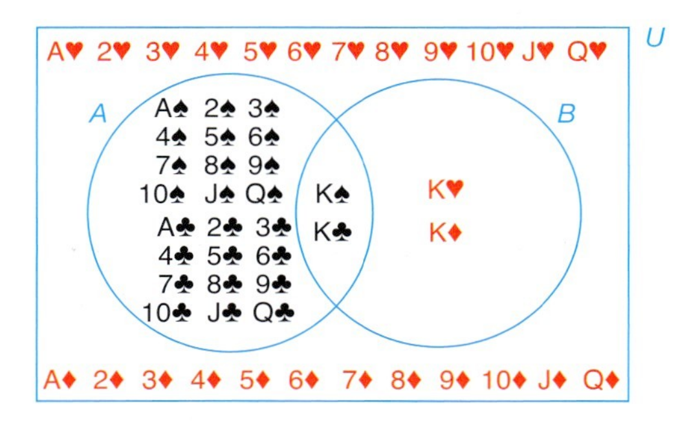

$P(A) = \frac{26}{52} = \frac{1}{4}$ and $P(B) = \frac{4}{52} = \frac{1}{13}$

However $P(A \cup B) \neq \frac{26}{52} + \frac{4}{52}$ because K♠ and K♣ belong to both events A and B and have therefore been counted twice. This is shown in the overlap of the Venn diagram.

Therefore, for combined events, $P(A \cup B) = P(A) + P(B) - P(A \cap B)$

i.e. the probability of event A or B is equal to the sum of the probabilities of A and B minus the probability of A and B.

Worked example

In a holiday survey of 100 people:

72 people have had a beach holiday
16 have had a skiing holiday
12 have had both.

What is the probability that one person chosen at random from the survey has had either a beach holiday (B) or a ski holiday (S)?

$P(B) = \frac{72}{100}$ $P(S) = \frac{16}{100}$ $P(B \cap S) = \frac{12}{100}$

Therefore $P(B \cup S) = \frac{72}{100} + \frac{16}{100} - \frac{12}{100} = \frac{76}{100}$

Independent events

A student may be born on 1 June, another student in his class may also be born on 1 June. These events are independent of each other (assuming they are not twins).

If a dice is rolled and a coin spun, the outcomes of each are also independent, i.e. the outcome of one does not affect the outcome of another.

For independent events, the probability of both events occurring is the product of each occurring separately, i.e.

$P(A \cap B) = P(A) \times P(B)$

Worked examples

1 I spin a coin and roll a dice.

a What is the probability of getting a head on the coin and a five on the dice?

b What is the probability of getting either a head on the coin or a five on the dice, but not both?

a $P(H) = \frac{1}{2}$ $P(5) = \frac{1}{6}$

Both events are independent therefore

$$\begin{aligned} P(H \cap 5) &= P(H) \times P(5) \\ &= \frac{1}{2} \times \frac{1}{6} \\ &= \frac{1}{12} \end{aligned}$$

b $P(H \cup 5)$ is the probability of getting a head, a five or both.

Therefore $P(H \cup 5) - P(H \cap 5)$ removes the probability of both events occurring.

The solution is

$$\begin{aligned} P(H \cup 5) - P(H \cap 5) &= P(H) + P(5) - P(H \cap 5) \\ &= \frac{1}{2} + \frac{1}{6} - \frac{1}{12} \\ &= \frac{7}{12} \end{aligned}$$

2 The probabilities of two events X and Y are given by:

$P(X) = 0.5$, $P(Y) = 0.4$, and $P(X \cap Y) = 0.2$.

a Are events X and Y mutually exclusive?
b Calculate $P(X \cup Y)$.
c What kind of events are X and Y?

a No: if the events were mutually exclusive, then $P(X \cap Y)$ would be 0 as the events could not occur at the same time.

b
$$\begin{aligned} P(X \cup Y) &= P(X) + P(Y) - P(X \cap Y) \\ &= 0.5 + 0.4 - 0.2 \\ &= 0.7 \end{aligned}$$

c Since $P(X \cap Y) = P(X) \times P(Y)$, i.e.
$0.2 = 0.5 \times 0.4$, events X and Y must be independent.

Conditional probability

Conditional probability refers to the probability of an event (A) occurring, which is in turn dependent on another event (B).

For example, a group of ten children play two tennis matches each. The table below shows which matches the children won and lost.

Child	First match	Second match
1	Won	Won
2	Lost	Won
3	Lost	Won
4	Won	Lost
5	Lost	Lost
6	Won	Lost
7	Won	Won
8	Won	Won
9	Lost	Won
10	Lost	Won

Let winning the first match be event A and winning the second match be event B. An example of conditional probability would be as follows: calculate the probability that a boy picked at random won his first match, if it is known that he won his second match.

Because we are told that the boy won his second match, this will affect the final probability.

This is written as $P(A \mid B)$, i.e. the probability of event A given that event B has happened.

$$P(A \mid B) = \frac{P(A \cap B)}{P(B)}$$

or $P(A \mid B) = \dfrac{n(A \cap B)}{n(B)}$ where n is the number of times that event happens.

Worked example

Using the table on the previous page, for a child picked at random:

a Calculate the probability that the child lost both matches.
b Calculate the probability that the child won his first match.
c Calculate the probability that a child won his first match, if it is known that he won his second match.

a $P(A' \cap B') = \frac{n(A' \cap B')}{n(U)} = \frac{1}{10}$

b $P(A) = \frac{n(A)}{n(U)} = \frac{5}{10} = \frac{1}{2}$

c $P(A \mid B) = \frac{n(A \cap B)}{n(B)} = \frac{3}{7}$

Note: The answers to **b** and **c** are different although both relate to the probability of a child winning his first match.

Venn diagrams are very useful for conditional probability as they show $n(A \cap B)$ clearly.

Worked example

In a class of 25 students, 18 play football, 8 play tennis and 6 play neither sport.

a Show this information on a Venn diagram.
b What is the probability that a student chosen at random plays both sports?
c What is the probability that a student chosen at random plays football, given that he also plays tennis?

a 18 + 8 + 6 = 32. As there are only 25 students, 7 must play both sports.

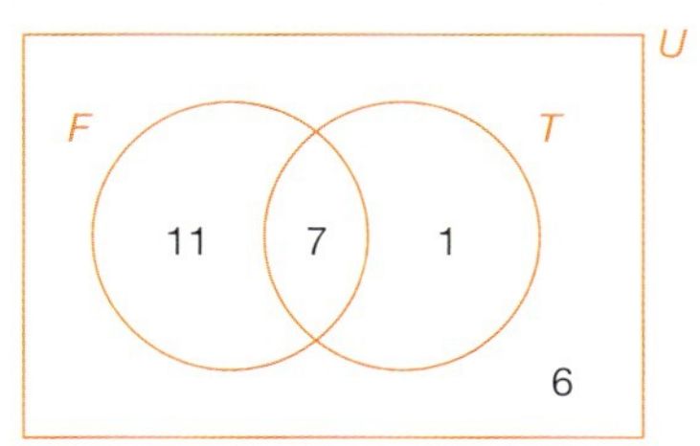

b $P(F \cap T) = \frac{n(F \cap T)}{n(U)} = \frac{7}{25}$

c $P(F \mid T) = \frac{n(F \cap T)}{n(T)} = \frac{7}{8}$

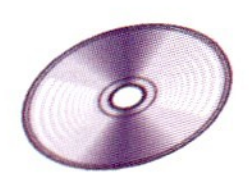

Exercise 3.10.1

1 The Jamaican 100 m women's relay team has a 0.5 chance of coming first in the final, 0.25 chance of coming second and 0.05 chance of coming third.
   a Are the events independent?
   b What is the team's chance of a medal?

2 I spin a coin and throw a dice.
  a Are the events independent?
  b What is the probability of getting:
    i) a head and a factor of 3
    ii) a head or a factor of 3
    iii) a head or a factor of 3, but not both?

3 What is the probability that two people picked at random both have a birthday in June?

4 Amelia takes two buses to work. On a particular day, the probability of her catching the first bus is 0.7 and the probability of her catching the second bus is 0.5. The probability of her catching neither is 0.1.
  a Are the events independent?
  b If A represents catching the first bus and B the second:
    i) State $P(A \cup B)'$.
    ii) Find $P(A \cup B)$.
    iii) Given that $P(A \cup B) = P(A) + P(B) - P(A \cap B)$, calculate $P(A \cap B)$.
    iv) Calculate the probability $P(A \mid B)$, i.e. the probability of Amelia having caught the first bus, given that she caught the second bus.

5 The probability of Marco having breakfast is 0.75. The probability that he gets a lift to work is 0.9 if he has had breakfast and 0.8 if he has not.
  a What is the probability of Marco having breakfast then getting a lift?
  b What is the probability of Marco not having breakfast then getting a lift?
  c What is the probability that Marco gets a lift?
  d If Marco gets a lift, what is the probability that he had breakfast?

6 Inês has a driving test on Monday and a Drama exam the next day. The probability of her passing the driving test is 0.73. The probability of her passing the Drama exam is 0.9. The probability of failing both is 0.05.
  Given that she has passed the driving test, what is the probability that she also passed her Drama exam?

7 An Olympic swimmer has a 0.6 chance of a gold medal in the 100 m freestyle, a 0.7 chance of a gold medal in the 200 m freestyle and a 0.1 chance of no gold medals. Given that she wins the 100 m race, what is the probability of her winning the 200 m race?

8 a How many pupils are in your class?
  b How likely do you think it is that two people in your class will share the same birthday? Very likely? Likely? Approx 50–50? Unlikely? Very unlikely?
  c Write down everybody's birthday. Did two people have the same birthday?
    Below is a way of calculating the probability that two people have the same birthday depending on how many people there are. To study this it is easiest to look at the probability of birthdays being different. When this probability is less than 50%, then the probability that two people will have the same birthday is greater than 50%.
    When the first person asks the second person, the probability of them *not* having the same birthday is $\frac{364}{365}$ (i.e. it is $\frac{1}{365}$ that they have the same birthday).

When the next person is asked, as the events are independent, the probability of all three having different birthdays is:

$$\left(\frac{364}{365}\right) \times \left(\frac{363}{365}\right) = 99.2\%$$

When the next person is asked, the probability of all four having different birthdays is:

$$\left(\frac{364}{365}\right) \times \left(\frac{363}{365}\right) \times \left(\frac{362}{365}\right) = 98.4\%$$

and so on....

d Copy and complete the table below until the probability is 50%.

Number of people	**Probability of them *not* having the same birthday**
2	$\frac{364}{365} = 99.7\%$
3	$\left(\frac{364}{365}\right) \times \left(\frac{363}{365}\right) = 99.2\%$
4	$\left(\frac{364}{365}\right) \times \left(\frac{363}{365}\right) \times \left(\frac{362}{365}\right) = 98.4\%$
5	
10	
15	
20	
etc.	

e Explain in words what your solution to part **d** means.

Student assessment 1

1 Describe the following sets in words.

a {1, 3, 5, 7}
b {1, 3, 5, 7, ...}
c {1, 4, 9, 16, 25, ...}
d {Arctic, Atlantic, Indian, Pacific}

2 Calculate the value of $n(A)$ for each of the sets shown below.

a A is the set of days of the week
b A is the set of prime numbers between 50 and 60
c $A = \{x \mid x \text{ is an integer and } -9 \leq x \leq -3\}$
d A is the set of students in your class

3 Copy this Venn diagram three times.

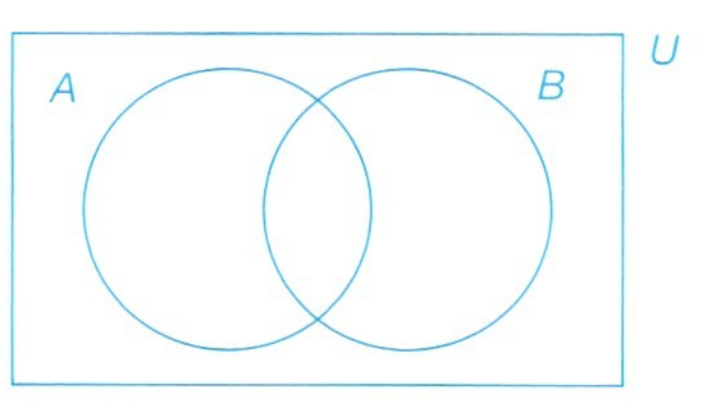

a On one copy shade and label the region which represents $A \cap B$.
b On another copy shade and label the region which represents $A \cup B$.
c On the third copy shade and label the region which represents $(A \cap B)'$.

4 If A = {w, o, r, k}, list all the subsets of A with at least three elements.

5 If $U = \{1, 2, 3, 4, 5, 6, 7, 8\}$ and $P = \{2, 4, 6, 8\}$, what set is represented by P'?

6 A hexagonal spinner is divided into equilateral triangles painted alternately red and black. What is the sample space when the spinner is spun three times?

7 If p is the proposition 'The Amazon river is in Africa', write the proposition $\neg p$ in words.

8 What is meant by $p \vee q$?

9 Calculate the theoretical probability of:
  a being born on a Saturday
  b being born on the 5th of a month in a non-leap year
  c being born on 20 June in a non-leap year
  d being born on 29 February.

10 A coin is tossed and an ordinary, fair dice is rolled.
  a Draw a two-way table showing all the possible combinations.
  b Calculate the probability of getting:
    i) a head and a 6
    ii) a tail and an odd number
    iii) a head and a prime number.

Student assessment 2

1 If $A = \{2, 4, 6, 8\}$, write all the proper subsets of A with two or more elements.

2 X = {lion, tiger, cheetah, leopard, puma, jaguar, cat}
Y = {elephant, lion, zebra, cheetah, gazelle}
Z = {anaconda, jaguar, tarantula, mosquito}
  a Draw a Venn diagram to represent the above information.
  b Copy and complete the statement $X \cap Y = \{...\}$.
  c Copy and complete the statement $Y \cap Z = \{...\}$.
  d Copy and complete the statement $X \cap Y \cap Z = \{...\}$.

3 U is the set of natural numbers, M is the set of even numbers and N is the set of multiples of 5.
  a Draw a Venn diagram and place the numbers 1, 2, 3, 4, 5, 6, 7, 8, 9, 10 in the appropriate places in it.
  b If $X = M \cap N$, describe set X in words.

4 A group of 40 people were asked whether they like tennis (T) and football (F). The number liking both tennis and football was three times the number liking only tennis. Adding 3 to the number liking only tennis and doubling the answer equals the number of people liking only football. Four said they did not like sport at all.
  a Draw a Venn diagram to represent this information.
  b Calculate $n(T \cap F)$.
  c Calculate $n(T \cap F')$.
  d Calculate $n(T' \cap F)$.

5 The Venn diagram below shows the number of elements in three sets P, Q and R.

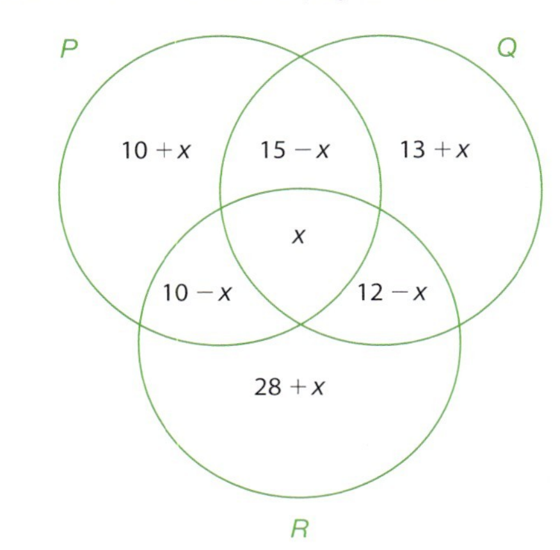

If $n(P \cup Q \cup R) = 93$ calculate:
  a x
  b $n(P)$
  c $n(Q)$
  d $n(R)$
  e $n(P \cap Q)$
  f $n(Q \cap R)$
  g $n(P \cap R)$
  h $n(R \cup Q)$
  i $n(P \cap Q)'$.

6 What is meant by $p \wedge q$?

7 Copy and complete the truth table below.

p	q	$p \wedge q$	$p \vee q$
T			
T			
F			
F			

8 What is a tautology? Give an example.

9 A goalkeeper expects to save one penalty out of every three. Calculate the probability that he:
- a saves one penalty out of the next three
- b fails to save any of the next three penalties
- c saves two out of the next three penalties.

Student assessment 3

1 The probability that a student takes English is 0.8. The probability that a student takes English and Spanish is 0.25.

What is the probability that a student takes Spanish, given that he takes English?

2 A card is drawn from a standard pack of cards.
- a Draw a Venn diagram to show the following:
  A is the set of aces
  B is the set of picture cards
  C is the set of clubs
- b From your Venn diagram find the following probabilities.
  - i) P(ace or picture card)
  - ii) P(not an ace or picture card)
  - iii) P(club or ace)
  - iv) P(club and ace)
  - v) P(ace and picture card)

3 Students in a school can choose to study one or more science subjects from physics, chemistry and biology.

In a year group of 120 students, 60 took physics, 60 took biology and 72 took chemistry; 34 took physics and chemistry, 32 took chemistry and biology and 24 took physics and biology; 18 took all three.
- a Draw a Venn diagram to represent this information.
- b If a student is chosen at random, what is the probability that:
  - i) the student chose to study only one subject
  - ii) the student chose physics or chemistry, and did not choose biology?

4 A class took an English test and a maths test. 40% passed both tests and 75% passed the English test.

What percentage of those who passed the English test also passed the maths test?

5 A jar contains blue and red counters. Two counters are chosen without replacement. The probability of choosing a blue then a red counter is 0.44. The probability of choosing a blue counter on the first draw is 0.5.

What is the probability of choosing a red counter on the second draw if the first counter chosen was blue?

6 In a group of children, the probability that a child has black hair is 0.7. The probability that a child has brown eyes is 0.55. The probability that a child has either black hair or brown eyes is 0.85.

What is the probability that a child chosen at random has both black hair and brown eyes?

7 A ball enters a chute at X.

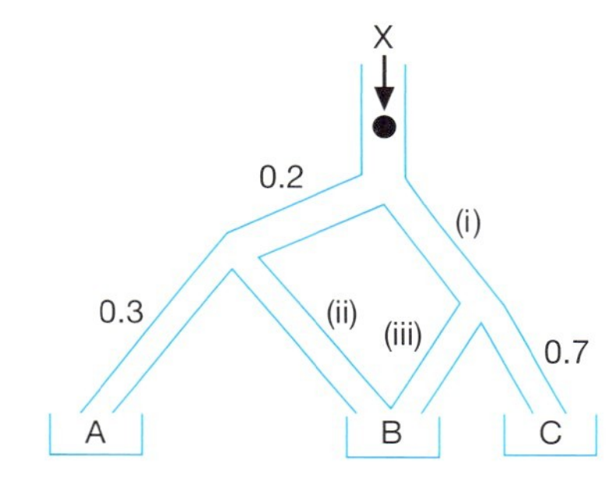

- a What are the probabilities of the ball going down each of the chutes labelled (i), (ii) and (iii)?
- b Calculate the probability of the ball landing in:
  - i) tray A
  - ii) tray C
  - iii) tray B.

Discussion points, project ideas and theory of knowledge

1 Research and discuss Russell's antinomy (not the element antimony).

2 Research and discuss 'Bertrand's Box Paradox'. This could be the starting point for a project on paradoxes.

3 Set theory is an area that could be studied as a project beyond the Mathematical Studies syllabus. Your teacher may suggest some areas of study.

4 Draw a Venn diagram to represent Belief, Truth and Knowledge. Discuss the statement 'Knowledge is found where belief and truth intersect.'

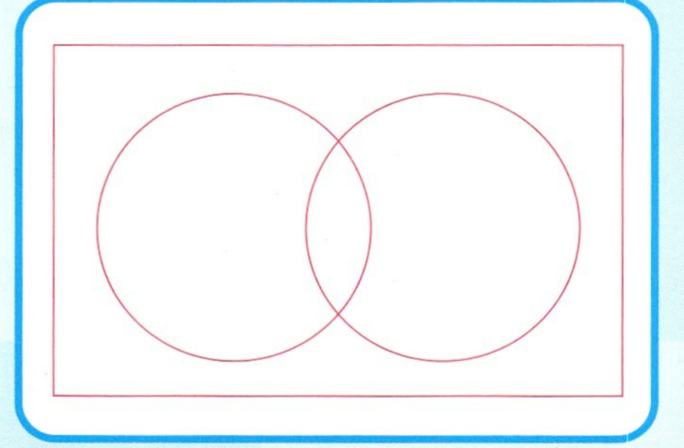

5 The set of whole numbers and the set of square numbers have an infinite number of elements. Does this mean that there are different values of infinity?

6 What is the difference between zero and an empty set? Is a vacuum an empty set?

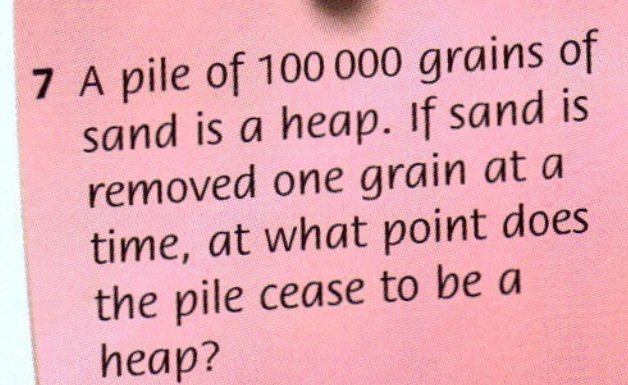

7 A pile of 100 000 grains of sand is a heap. If sand is removed one grain at a time, at what point does the pile cease to be a heap?

8 A teacher says that she will give a surprise test on one weekday of the following week. Why can the test not be a surprise if it is given on the Friday? By extending that reasoning, the test cannot be given on Thursday either, and so on. Discuss the fallacy of this reasoning.

9 Is it possible to draw a Venn diagram on a two-dimensional space (a piece of paper) to represent four sets which intersect at one place? (Think of U shapes). Is it possible to represent five sets in a similar way? This could make a possible project.

10 I put out three cards face down, one of which is an ace. I know the position of the ace. You pick a card, but do not turn it over. I then turn over a card which is not the ace. You are then offered the opportunity to change your pick to the remaining card. Design a probability experiment for many trials as a class activity. Why does the probability of you choosing the ace increase if you make the change? How much does the probability change by? This could be the starting point for a project.

11 What is a paradox? 'This statement is a lie.' Is that a paradox? Discuss.

Topic 4

Functions

Syllabus content

4.1 Concept of a function as a mapping.
Domain and range. Mapping diagrams.

4.2 Linear functions and their graphs, for example $f : x \mapsto mx + c$.

4.3 The graph of the quadratic function: $f(x) = ax^2 + bx + c$.
Properties of symmetry; vertex; intercepts.

4.4 The exponential expression: a^b; $b \in \mathbb{Q}$.
Graphs and properties of exponential functions.
$f(x) = a^x$; $f(x) = a^{\lambda x}$; $f(x) = ka^{\lambda x} + c$; $k, a, c, \lambda \in \mathbb{Q}$.
Growth and decay; basic concepts of asymptotic behaviour.

4.5 Graphs and properties of the sine and cosine functions:
$f(x) = a\sin bx + c$;
$f(x) = a\cos bx + c$; $a, b, c \in \mathbb{Q}$.
Amplitude and period.

4.6 Accurate graph drawing.

4.7 Use of a GDC to sketch and analyse some simple, unfamiliar functions.

4.8 Use of a GDC to solve equations involving simple combinations of some simple, unfamiliar functions.

Introduction

Leibniz was the first to use the term 'function', although not in the sense it is used in this topic. Euler used the term in work done around 1750. Both of these mathematicians were aware of the work of Girolamo Cardano, and the concept of 'function' may have developed from this. It is also possible that all three of these mathematicians were aware of the much earlier studies of Al-Karkhi.

Leonhard Euler (1707–1783)

Cardano was a famous Italian mathematician. In 1545 he published a book *Ars Magna* ('Great Art') in which he showed calculations involving solutions to cubic equations (equations of the form $ax^3 + bx^2 + cx + d = 0$) and quartic equations (equations of the form $ax^4 + bx^3 + cx^2 + dx + e = 0$).

Girolamo Cardano (1501–1576)

HIERONYMI CAR
DANI, PRÆSTANTISSIMI MATHE
MATICI, PHILOSOPHI, AC MEDICI,
ARTIS MAGNÆ,
SIVE DE REGVLIS ALGEBRAICIS,
Lib. unus. Qui & totius operis de Arithmetica, quod
OPVS PERFECTVM
inscripsit, est in ordine Decimus.

The title page of *Ars Magna*

His book (the title page is shown here) is one of the key historical texts on algebra. It was the first algebraic text written in Latin. In 1570, because of his interest in astrology, Cardano was arrested for heresy; no other work of his was ever published.

Al-Karkhi was one of the greatest Arab mathematicians. He lived in the eleventh century. He wrote many books on algebra and developed a theory of indices and a method of finding square roots.

4.1 A function as a mapping

A relationship between two sets can be shown as a **mapping**, which links elements in one set with elements in the other set.

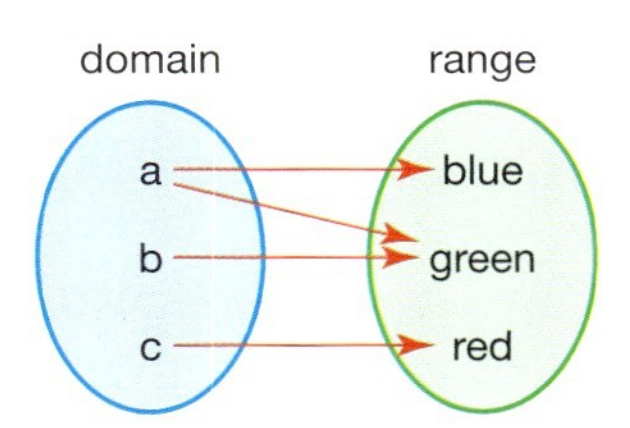

A **function** is a special type of relationship, which can be shown as a mapping with particular characteristics.

Consider the equation $y = 2x + 3$; $-1 \le x \le 3$. A table of results can be made and a mapping diagram drawn.

x	y
–1	1
0	3
1	5
2	7
3	9

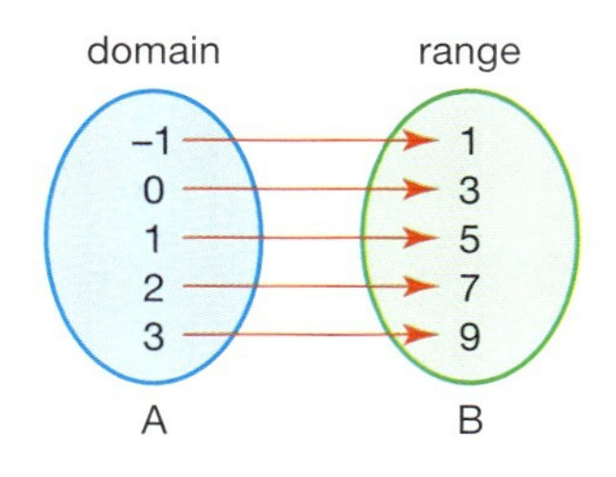

If a mapping is a function, each value in set B (the **range**: the values that the function can take) is produced from a value in set A (the **domain**: the values that x can take). The relationship above is therefore a function and, as a function, it can be written:

$$f(x) = 2x + 3; -1 \leq x \leq 3$$

or $$f: x \mapsto 2x + 3; -1 \leq x \leq 3$$

It is customary to write the domain after the function, because a different domain will produce a different range.

A mapping from A to B can be a **one-to-one mapping** or a **many-to-one mapping**.

The function above, $f(x) = 2x + 3; -1 \leq x \leq 3$, is a one-to-one function, as each value in the domain maps to a unique value in the range, i.e. no two values in the domain can map to the same value in the range. However, the function $f(x) = x^2; x \in \mathbb{Z}$, for example, is a many-to-one function, as some values in the range can be generated by more than one value in the domain, as shown.

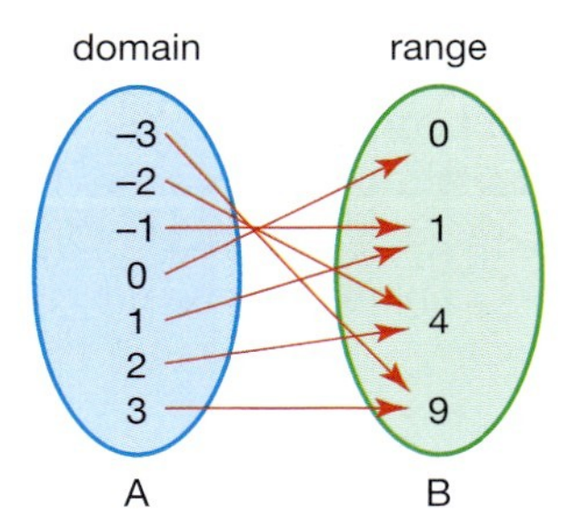

It is important to understand that, for a mapping to represent a function, one value in the domain (set A) must map to a single value in the range (set B). Therefore the mapping $f(x) = x^2; x \in \mathbb{Z}$ shown above represents a function.

Some mappings do not represent functions. Consider the relationship $y = \pm\sqrt{x}$.

The following table and mapping diagram can be produced.

x	y
1	± 1
4	± 2
9	± 3
16	± 4

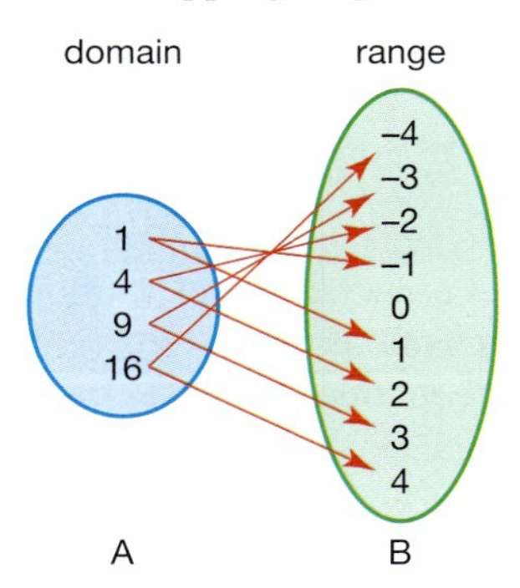

This relationship is not a function as one value in the domain produces more than one value in the range.

It is also important to remember the mathematical notation used to define different domains. The principal ones will be:

Notation	Meaning
$\mathbb{Z}$	The set of integers {0, ±1, ±2, ±3, …}
$\mathbb{Z}^+$	The set of positive integers {1, 2, 3, …}
$\mathbb{N}$	The set of natural numbers {0, 1, 2, 3, …}, i.e. positive integers and zero
$\mathbb{Q}$	The set of rational numbers, i.e. numbers that can be expressed as a fraction $\frac{a}{b}$
$\mathbb{R}$	The set of real numbers, i.e. numbers that exist

Calculating the range from the domain

The domain is the set of input values, whilst the range is the set of output values for a function. (Note: The range is not the difference between the greatest and least values, as in statistics.) The range is therefore dependent not only on the function itself, but also on the domain.

Worked example

Calculate the range for each of the following functions.

a $f(x) \mapsto x^3 - 3x;\ -2 \le x \le 3$
b $f(x) \mapsto x^3 - 3x;\ x \in \mathbb{R}$

a $f(x) \mapsto x^3 - 3x;\ -2 \le x \le 3$

The graph of the function is shown. As the domain is restricted to $-2 \le x \le 3$, the range is limited to values from -2 to 18.

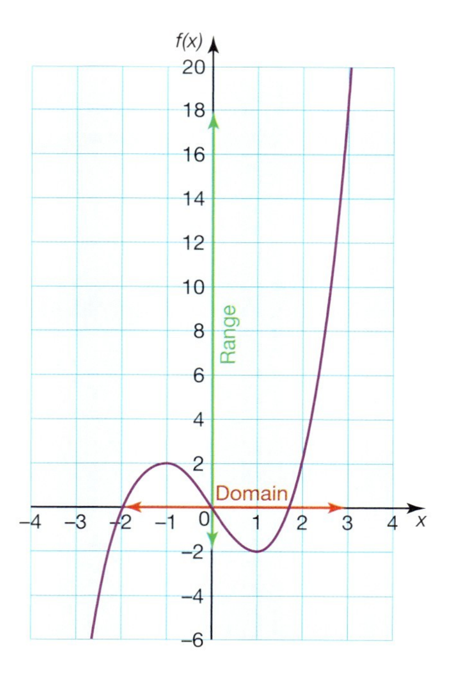

This is written as:
Range $-2 \le f(x) \le 18$.

b $f(x) \mapsto x^3 - 3x;\ x \in \mathbb{R}$

The graph will be similar to the one in part **a** except that the domain is not restricted. As the domain is for all real values of x, this implies that any real number can be an input value. As a result, in the range will also be all real values.

This is written as: Range $f(x) \in \mathbb{R}$.

Note: If the domain of a function is $x \in \mathbb{R}$, then it is often omitted.

Exercise 4.1.1

1 Which of the following mappings shows a function?

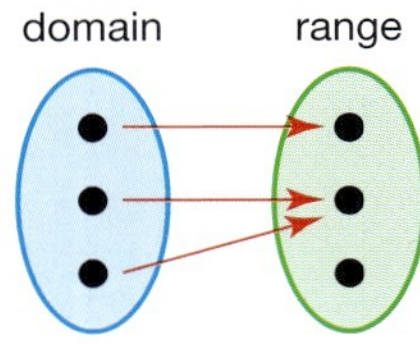

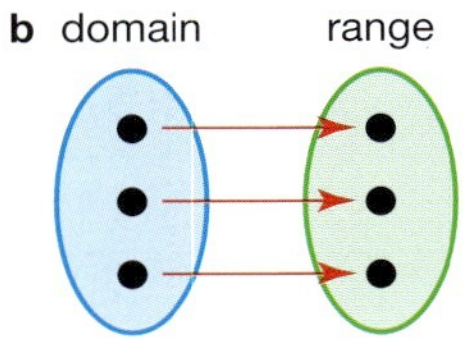

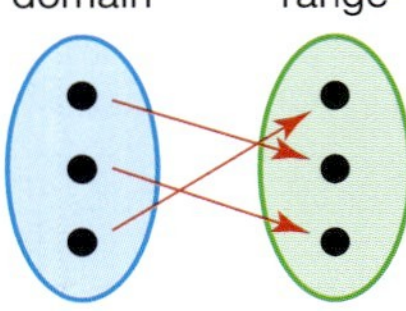

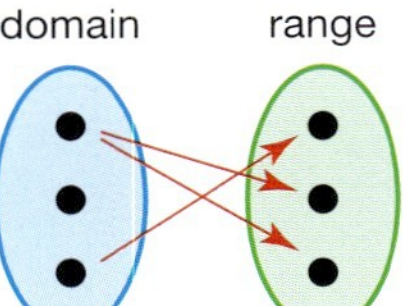

2 Give the domain and range of each of the following functions. You could plot the functions using your GDC.

a $f(x) = 2x - 1; -1 \leq x \leq 3$

b $f(x) = 3x + 2; -4 \leq x \leq 0$

c $f(x) = x^2 + 2; -3 \leq x \leq 3$

d $g(y) = \frac{1}{y}; y > 3$

e $h(t) = t + 3$

f $f(y) = 4$

g $f(n) = -n^2 + 2$

4.2 Linear functions and their graphs

(Note: This topic is covered further in Section 5.2.)

The function $g(x) = 2x + 3; -1 \leq x \leq 3$ gives the following graph.

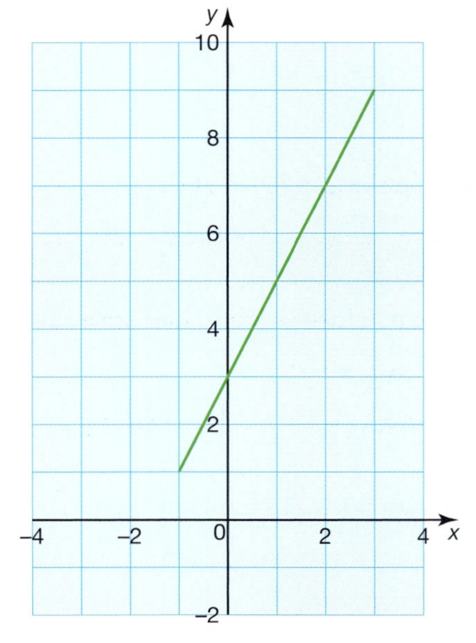

The domain of the function is $-1 \leq x \leq 3$, whilst its range is $1 \leq g(x) \leq 9$.

Evaluating linear functions

Worked examples

1 a For the linear function $h(t) = 2t - 5$, evaluate $h(7)$.
  b If $h(t) = 45$ evaluate t.

  a $h(7) = 2(7) - 5$
  $= 9$

  b $45 = 2t - 5 \Rightarrow 2t = 50 \Rightarrow t = 25$

2 Two linear functions are given as $f(x) = 2x + 1$ and $g(x) = 3x - 6$.

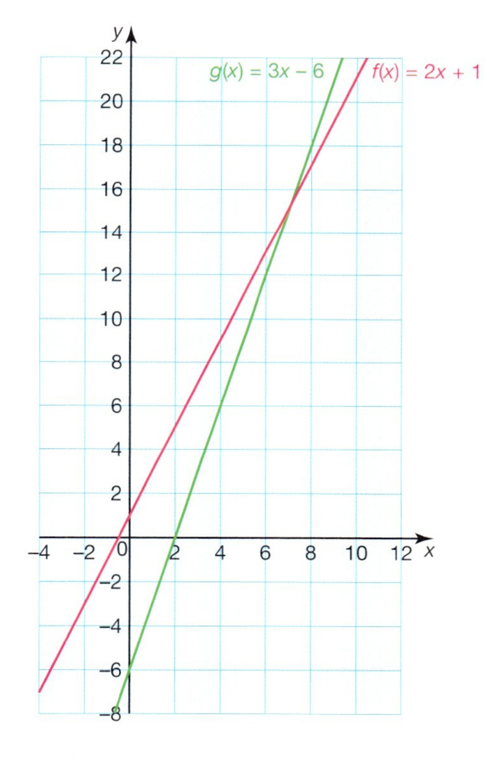

Find the values of x and $f(x)$ where $f(x) = g(x)$.

Graphically this can be interpreted as the coordinates of the point where the two functions intersect.

At the point of intersection the two functions are equal, i.e. $2x + 1 = 3x - 6$.

Solving the equation gives $x = 7$.

To find the y-coordinate, calculate either $f(7)$ or $g(7)$.

$f(7) = 2(7) + 1 = 15$.

Exercise 4.2.1

1 If $f(x) = \frac{3x + 2}{4}$, calculate each of the following.

a $f(2.5)$ b $f(0)$ c $f(-0.5)$ d $f(-6)$

2 If $g(x) = \frac{5x - 3}{3}$, calculate each of the following.

a $g(0)$ b $g(-3)$ c $g(-1.5)$ d $g(-9)$

3 If $h: x \mapsto \frac{-6x + 8}{4}$, calculate each of the following.

a $h(4)$ b $h(1.5)$ c $h(-2)$ d $h(-0.5)$

4 If $f(x) = \frac{-5x - 7}{-8}$, calculate each of the following.

a $f(3)$ b $f(-1)$ c $f(-7)$ d $f(-\frac{3}{5})$

5 A plumber charges a 50 euro callout charge and then 25 euros for each 15 minutes.

a Express the plumber's charges as a function $f(x)$, where x is the time in hours.
b Evaluate $f(2.5)$.

6 a Plot on the same graph the functions $f(x) = 4x + 2$ and $g(x) = 23 - 3x$.
b Evaluate x, where $f(x) = g(x)$.

Inverse functions

The inverse of a function is its reverse, i.e. it 'undoes' the function's effects. The inverse of the function $f(x)$ is written as $f^{-1}(x)$.

Worked examples

1 Find the inverse of both of the following functions.

a $f(x) = x + 2$ b $g(x) = 2x - 3$

a $f^{-1}(x) = x - 2$ b $g^{-1}(x) = \frac{(x + 3)}{2}$

2 If $f(x) = \frac{(x - 3)}{3}$, calculate both of the following.

a $f^{-1}(2)$ b $f^{-1}(-3)$

a $f^{-1}(x) = 3x + 3$
$f^{-1}(2) = 9$

b $f^{-1}(x) = 3x + 3$
$f^{-1}(-3) = -6$

Exercise 4.2.2

1 Find the inverse of each of the following functions.

a $f(x) = x + 3$ b $f(x) = x - 5$

c $f(x) = \frac{x}{3}$ d $f(x) = 4x$

e $f(x) = 2x + 5$ f $f(x) = 3x - 6$

g $f(x) = \frac{x + 4}{2}$ h $f(x) = \frac{1}{2}x + 3$

i $f(x) = 4(3x - 6)$ j $f(x) = -2(-3x + 2)$

2 If $f(x) = x - 4$, evaluate each of the following.

a $f^{-1}(2)$ b $f^{-1}(0)$ c $f^{-1}(-5)$

3 If $f(x) = 2x + 1$, evaluate each of the following.

a $f^{-1}(5)$ b $f^{-1}(0)$ c $f^{-1}(-11)$

4 If $g(x) = 6(x - 1)$, evaluate each of the following.

a $g^{-1}(12)$ b $g^{-1}(3)$ c $g^{-1}(6)$

5 If $g(x) = \frac{2x + 4}{3}$, evaluate each of the following.

a $g^{-1}(4)$ b $g^{-1}(0)$ c $g^{-1}(-6)$

4.3 Quadratic functions and their graphs

The general expression for a quadratic function takes the form $ax^2 + bx + c$, where a, b and c are constants. Some examples of quadratic functions are given below.

$y = 2x^2 + 3x - 12$ $y = x^2 - 5x + 6$ $y = 3x^2 + 2x - 3$

Plotting quadratic functions is also covered in Section 2.7. When the graph of a quadratic function is plotted, a smooth curve, called a **parabola**, is produced. For example:

$y = x^2$

x	−4	−3	−2	−1	0	1	2	3	4
y	16	9	4	1	0	1	4	9	16

$y = -x^2$

x	−4	−3	−2	−1	0	1	2	3	4
y	−16	−9	−4	−1	0	−1	−4	−9	−16

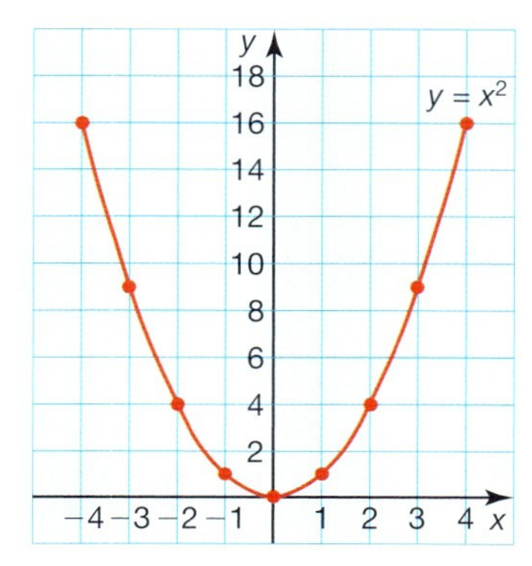

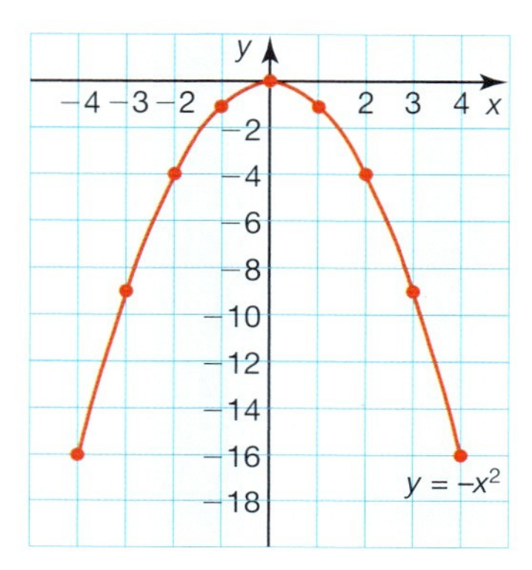

Both graphs above are symmetrical about the y-axis. The y-axis is therefore the axis of symmetry. For quadratic functions of the form $f(x) = ax^2 + bx + c$, the equation of the axis of symmetry is given by $x = -\frac{b}{2a}$. The vertex of the graph can also be found as it lies on the axis of symmetry.

Worked examples

1 a Plot a graph of the function $y = x^2 - 5x + 6$ for $0 \le x \le 5$.

b Deduce the equation of the axis of symmetry and the coordinates and nature of the vertex.

a First produce the table of values, then plot the points to draw the graph.

x	0	1	2	3	4	5
y	6	2	0	0	2	6

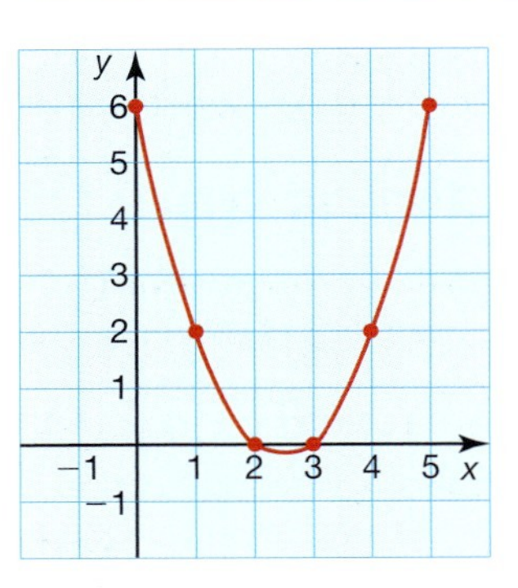

b It can be seen from the graph that the equation of the axis of symmetry is $x = \frac{5}{2}$.

It can also be calculated using the formula $x = -\frac{b}{2a}$, where $a = 1$ and $b = -5$.

$x = -\frac{-5}{2 \times 1} = \frac{5}{2}$,

When $x = \frac{5}{2}$, $y = \left(\frac{5}{2}\right)^2 - 5\left(\frac{5}{2}\right) + 6 = \frac{25}{4} - \frac{25}{2} + 6 = -\frac{1}{4}$.

So the vertex is at $\left(\frac{5}{2}, -\frac{1}{4}\right)$ and is a minimum.

2 a Plot a graph of the function $y = -x^2 + x + 2$ for $-3 \le x \le 4$.

b Deduce the equation of the axis of symmetry and the coordinates of the vertex.

a First produce the table of values, then plot the points to draw the graph.

x	−3	−2	−1	0	1	2	3	4
y	−10	-4	0	2	2	0	−4	−10

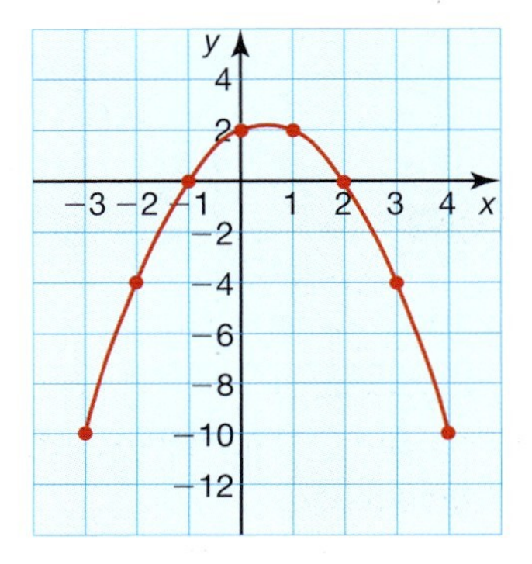

b It can be seen from the graph that the equation of the axis of symmetry is $x = \frac{1}{2}$.

It can also be calculated using the formula $x = -\frac{b}{2a}$, where $a = -1$ and $b = 1$.

$$x = -\frac{1}{2 \times (-1)} = \frac{1}{2}$$

When $x = \frac{1}{2}$, $y = -\left(\frac{1}{2}\right)^2 + \frac{1}{2} + 2 = -\frac{1}{4} + \frac{1}{2} + 2 = 2\frac{1}{4}$.

So the vertex is at $\left(\frac{1}{2}, 2\frac{1}{4}\right)$ and is a maximum.

Notice that when a is positive, the vertex is a minimum and when a is negative, the vertex is a maximum. Stationary points are dealt with further in Topic 7.

The above work can also be calculated and drawn using either your GDC or graphing software. Example 1 above is shown.

Casio

  to select the table mode. Enter the equation $y = x^2 - 5x + 6$.

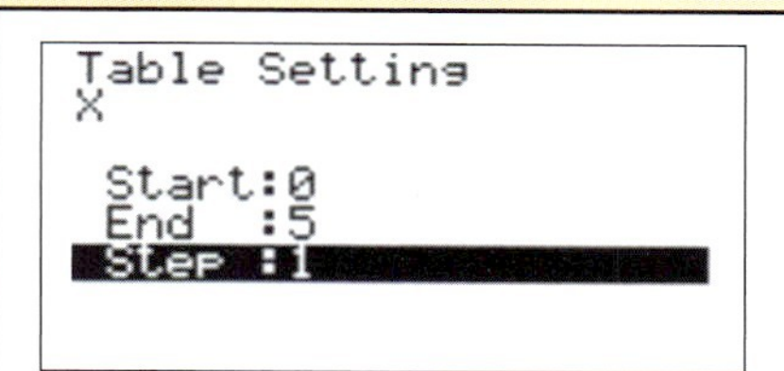

 to set the table parameters.

Enter Start:0, End:5 and Step:1 

 to tabulate the results. The first two columns are the x- and y-values. (Ignore the third column for the time being.)

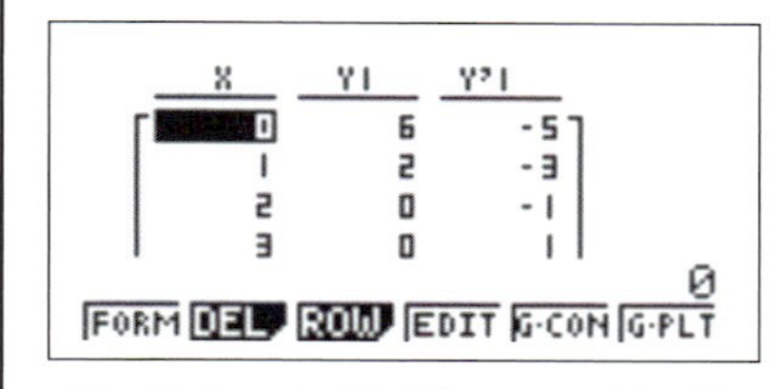

 to plot the points and draw a smooth curve through them.

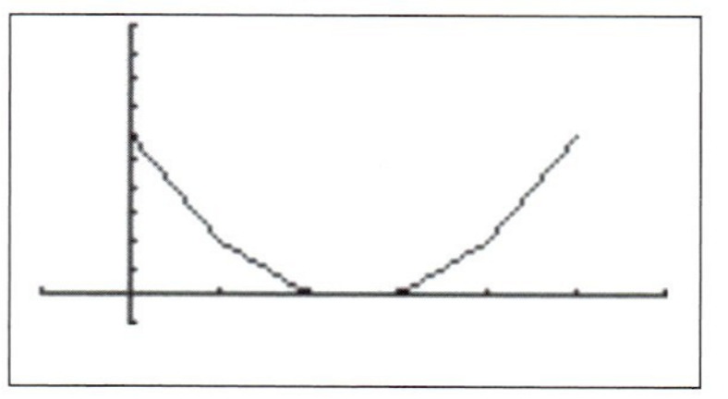

To find the minimum point:

to select the graphing mode. Enter the equation $y = x^2 - 5x + 6$.

 to graph the function.

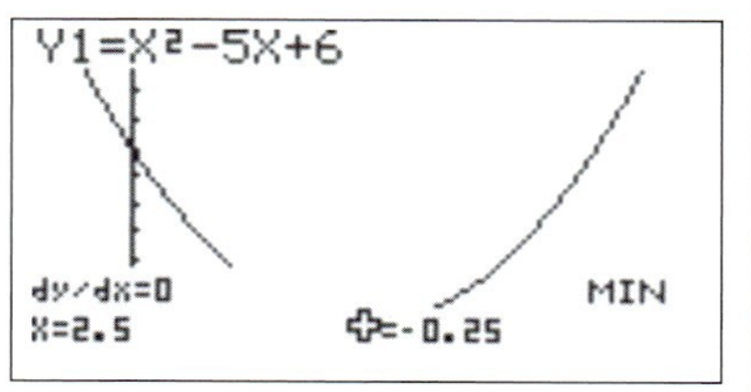

   to give the coordinate of the minimum point (2.5, −0.25).

Note: Once the minimum point is calculated, the x-value will give the equation of the axis of symmetry.

Texas

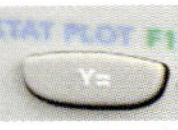

to select the function.

Enter the equation $y = x^2 - 5x + 6$.

to set the table parameters.

Enter Tblstart = 0, ΔTbl = 1, Indpnt: Auto, Depend:Auto.

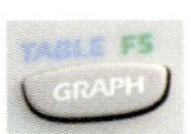

to tabulate the results.

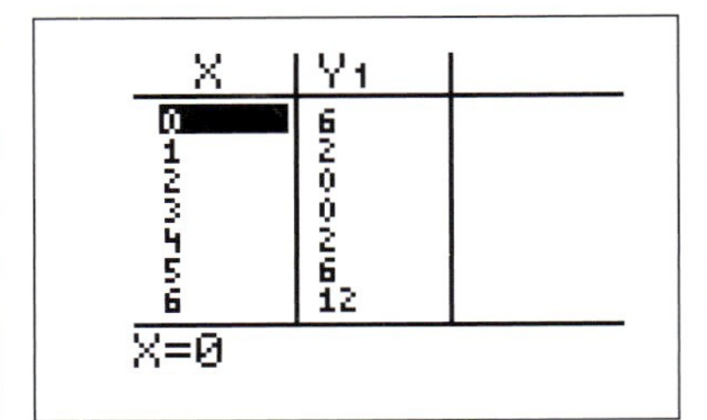

to graph the table of results.

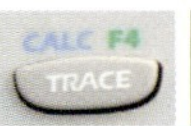

to calculate the coordinates of the minimum point.

Using the cursor select a point to the left of the minimum, then, when prompted, select a point to the right of the minimum. The calculator will then search for the minimum within this range.

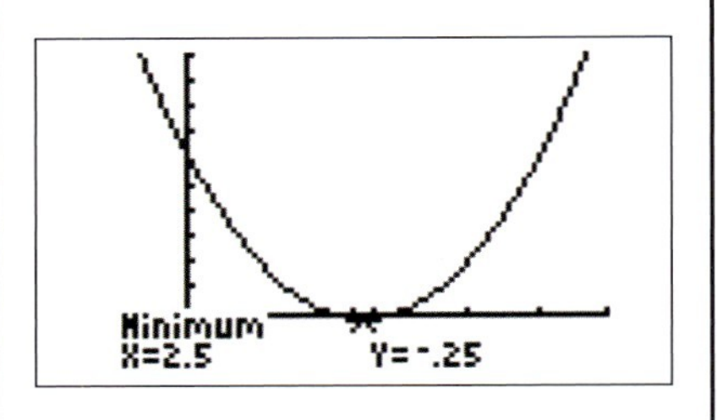

Note: Once the minimum point is calculated, the x-value will give the equation of the axis of symmetry.

Autograph

Select and enter the equation $y = x^2 - 5x + 6$.

To change the scale on the axes use .

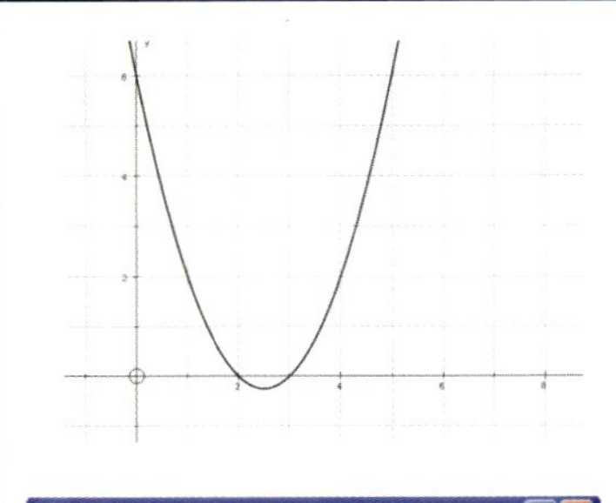

Select the curve then click 'Object' followed by 'Table of Values'. After entering the parameters, the results will appear in the results box.

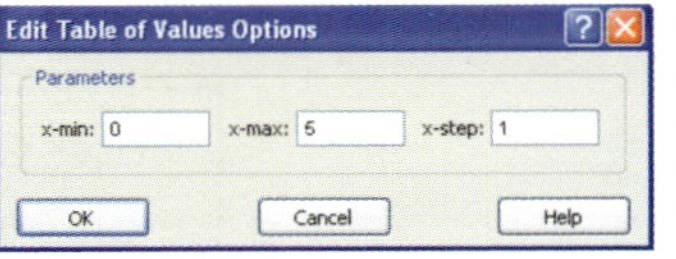

Autograph

To find the axis of symmetry, find the coordinate of the minimum point. Select 'Object' followed by 'Solve $f'(x) = 0$'.

This finds where the gradient of the graph is zero. The solution appears in the results box.

Results Box

Table of Values of Equation 1:

x	f(x)
0	6
1	2

Equation Solver:
Solution: x=2.5, y=-0.25

Note: The equation of the axis of symmetry is given by the x-coordinate of the minimum point.

GeoGebra

Type Function [x ^ 2 – 5x + 6,0,5] into the Input box. This produces the graph drawn within the limits $0 \leq x \leq 5$.

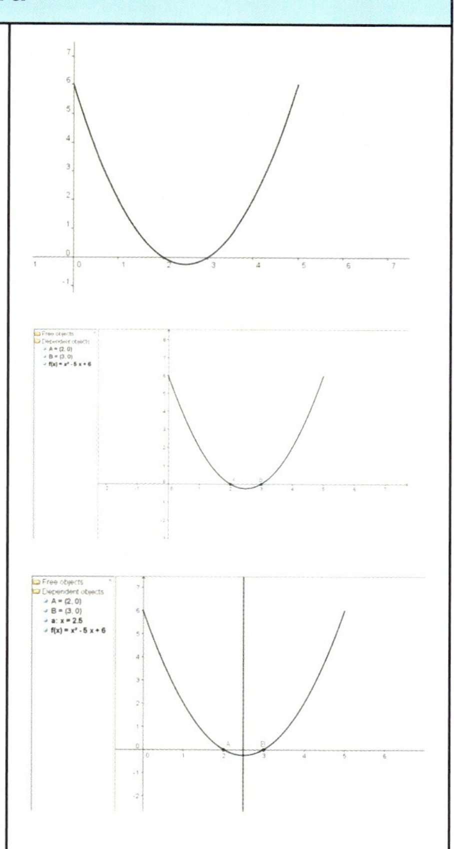

To change the scales of the axes select 'Options' → 'Drawing Pad' and enter the relevant information.
To find where the graph intersects the x-axis, type 'Root[f]'. The points are marked on the graph as A and B and their coordinates displayed in the algebra window.

Because of the symmetrical property of the graph, the axis of symmetry passes through the midpoint of A and B.
Type '$x = x(\text{A} + \text{B})/2$' in the input box.
The axis of symmetry is drawn and its equation displayed in the algebra window.

Accurate graph plotting

To plot an accurate graph, follow some simple rules.

- Label the axes accurately.
- Label the scale accurately.
- Use a table of results to obtain the coordinates of the points to plot on your graph.
- For linear functions, use a ruler to draw the straight lines.
- For curved functions, draw a smooth curve through the points rather than a succession of straight lines joining them.

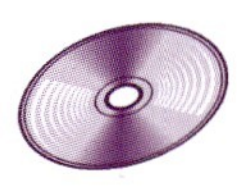

Exercise 4.3.1

For each of the following quadratic functions:

a construct a table of values and then plot the graph

b deduce or calculate the equation of the axis of symmetry and the coordinates and nature of the vertex.

1 $y = x^2 + x - 2,\ -4 \le x \le 3$

2 $y = -x^2 + 2x + 3,\ -3 \le x \le 5$

3 $y = x^2 - 4x + 4,\ -1 \le x \le 5$

4 $y = -x^2 - 2x - 1,\ -4 \le x \le 2$

5 $y = x^2 - 2x - 15,\ -4 \le x \le 6$

6 $y = 2x^2 - 2x - 3,\ -2 \le x \le 3$

7 $y = -2x^2 + x + 6,\ -3 \le x \le 3$

8 $y = 3x^2 - 3x - 6,\ -2 \le x \le 3$

9 $y = 4x^2 - 7x - 4,\ -1 \le x \le 3$

10 $y = -4x^2 + 4x - 1,\ -2 \le x \le 3$

Solving quadratic equations graphically

Worked example

a Draw a graph of $y = x^2 - 4x + 3$ for $-2 \le x \le 5$.

b Use the graph to solve the equation $x^2 - 4x + 3 = 0$.

a First produce the table of values, then plot the points to draw the graph.

x	−2	−1	0	1	2	3	4	5
y	15	8	3	0	−1	0	3	8

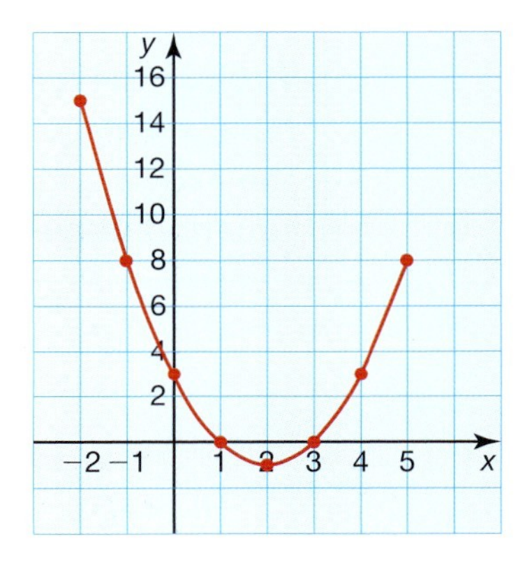

b To solve the equation, it is necessary to find the values of x when $y = 0$, i.e. where the graph crosses the x-axis. These points occur when $x = 1$ and $x = 3$ and are therefore the solutions.

The GDC and graphing software will also find the solution to quadratic equations graphically. In the explanations below, it is assumed that you can already plot the graph on your GDC or software.

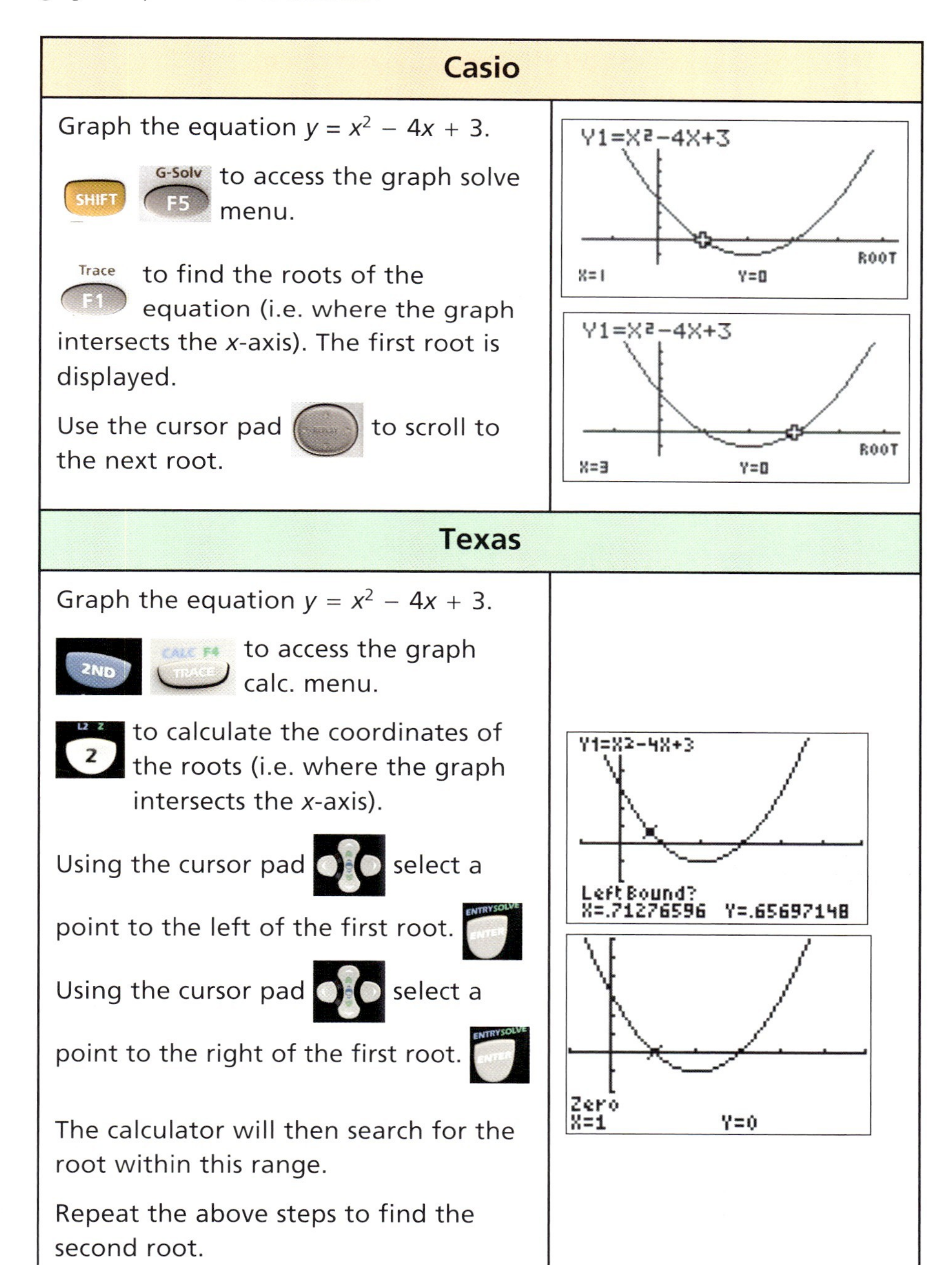

Casio

Graph the equation $y = x^2 - 4x + 3$.

SHIFT G-Solv F5 to access the graph solve menu.

Trace F1 to find the roots of the equation (i.e. where the graph intersects the x-axis). The first root is displayed.

Use the cursor pad to scroll to the next root.

Texas

Graph the equation $y = x^2 - 4x + 3$.

2ND CALC F4 TRACE to access the graph calc. menu.

2 to calculate the coordinates of the roots (i.e. where the graph intersects the x-axis).

Using the cursor pad select a point to the left of the first root. ENTER

Using the cursor pad select a point to the right of the first root. ENTER

The calculator will then search for the root within this range.

Repeat the above steps to find the second root.

Autograph	
Graph the equation $y = x^2 - 4x + 3$. Select the curve then click 'Object' followed by 'Solve $f(x) = 0$'. The points are marked on the graph and their coordinates can be displayed in the results box. This is accessed by selecting .	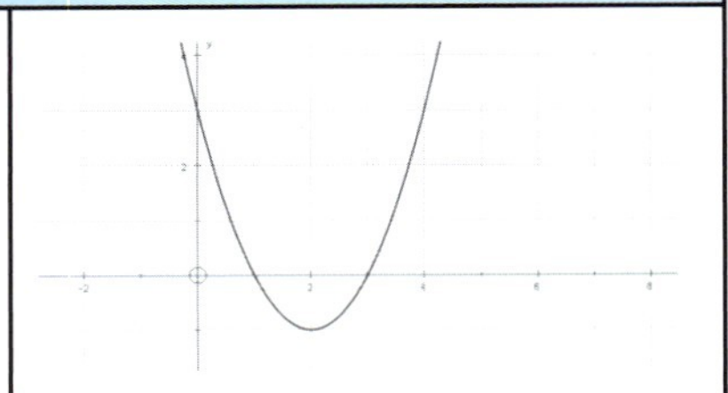
GeoGebra	
Graph the equation $y = x^2 - 4x + 3$. In the input box type 'Root[f]', this finds where the graph intersects the x-axis. The points are marked on the graph as A and B their coordinates are displayed in the algebra window.	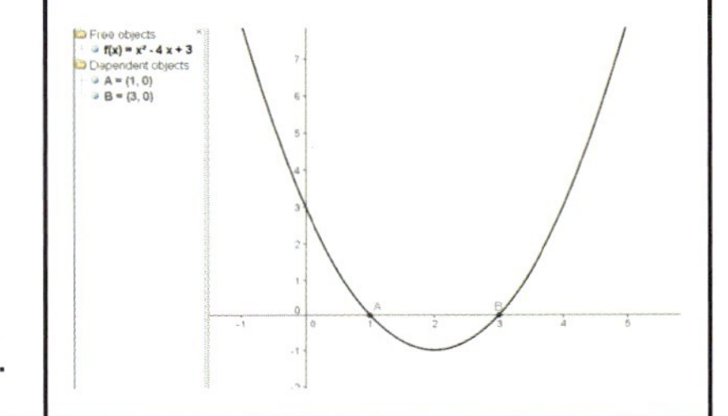

Exercise 4.3.2

Solve each of the quadratic equations below by plotting a graph of the function.

1. $x^2 - x - 6 = 0$
2. $-x^2 + 1 = 0$
3. $x^2 - 6x + 9 = 0$
4. $-x^2 - x + 12 = 0$
5. $x^2 - 4x + 4 = 0$
6. $2x^2 - 7x + 3 = 0$
7. $-2x^2 + 4x - 2 = 0$
8. $3x^2 - 5x - 2 = 0$

In the previous worked example, $y = x^2 - 4x + 3$, a solution could be found to the equation $x^2 - 4x + 3 = 0$ by reading off where the graph crossed the x-axis. This graph can, however, also be used to solve other quadratic equations.

Worked example

Use the graph of $y = x^2 - 4x + 3$ to solve the equation $x^2 - 4x + 1 = 0$.

$x^2 - 4x + 1 = 0$ can be rearranged to give:
$x^2 - 4x + 3 = 2$

Using the graph of $y = x^2 - 4x + 3$ and plotting the line $y = 2$ on the same graph gives the graph shown.

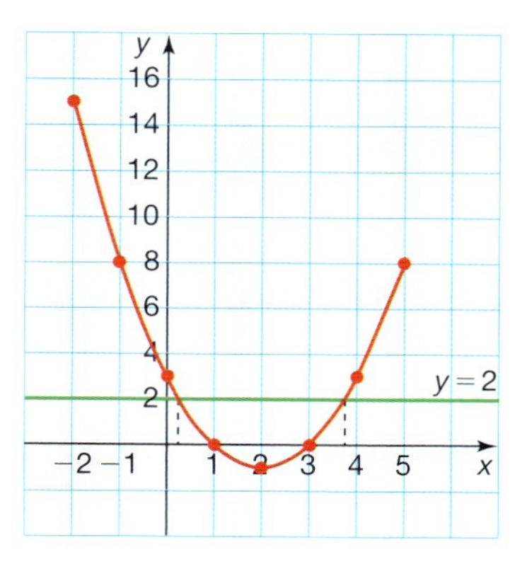

Where the curve and the line cross gives the solution to $x^2 - 4x + 3 = 2$ and hence the solution to $x^2 - 4x + 1 = 0$.

Therefore the solutions to $x^2 - 4x + 1 = 0$ are $x \approx 0.3$ and 3.7.

Exercise 4.3.3

Using the graphs that you drew for Exercise 4.3.2, solve the following quadratic equations. Show your method clearly.

1 $x^2 - x - 4 = 0$
2 $-x^2 - 1 = 0$
3 $x^2 - 6x + 8 = 0$
4 $-x^2 - x + 9 = 0$
5 $x^2 - 4x + 1 = 0$
6 $2x^2 - 7x = 0$
7 $-2x^2 + 4x = -1$
8 $3x^2 = 2 + 5x$

Factorizing quadratic expressions

In order to solve quadratic equations algebraically, it is necessary to know how to factorize quadratic expressions.

For example, the quadratic expression $x^2 + 5x + 6$ can be factorized by writing it as a product of two brackets: $(x + 3)(x + 2)$. A method for factorizing quadratics is shown below.

Worked examples

1 Factorize $x^2 + 5x + 6$.

On setting up a 2×2 grid, some of the information can immediately be entered. As there is only one term in x^2, this can be entered, as can the constant $+6$. The only two values which multiply to give x^2 are x and x. These too can be entered.

	x	
x	x^2	
		$+6$

We now need to find two values which multiply to give $+6$ and which add to give $+5$. The only two values which satisfy both these conditions are $+3$ and $+2$. The grid can then be completed.

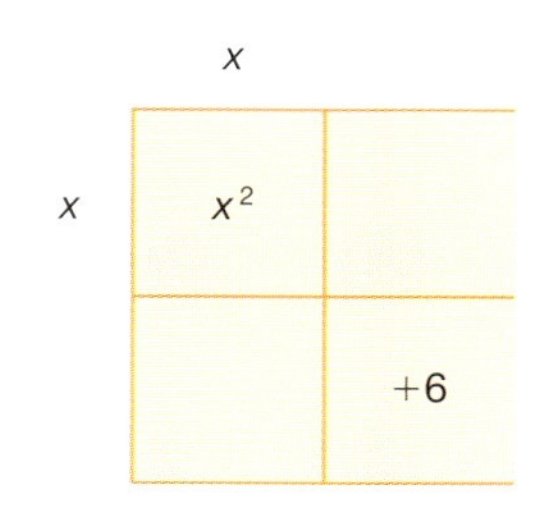

Therefore $x^2 + 5x + 6 = (x + 3)(x + 2)$.

2 Factorize $x^2 + 2x - 24$.

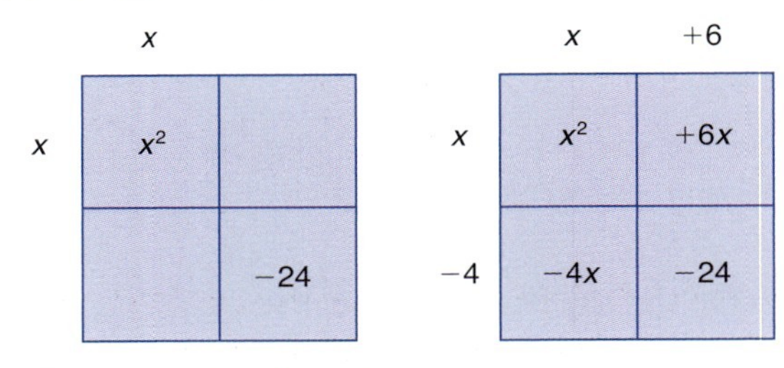

Therefore $x^2 + 2x - 24 = (x + 6)(x - 4)$.

3 Factorize $2x^2 + 11x + 12$.

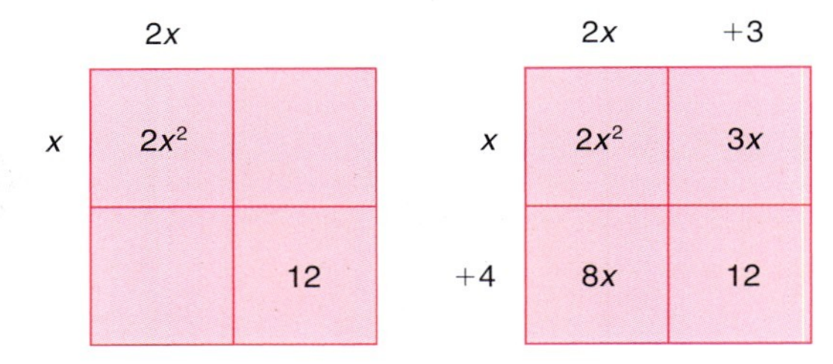

Therefore $2x^2 + 11x + 12 = (2x + 3)(x + 4)$.

4 Factorize $3x^2 + 7x - 6$.

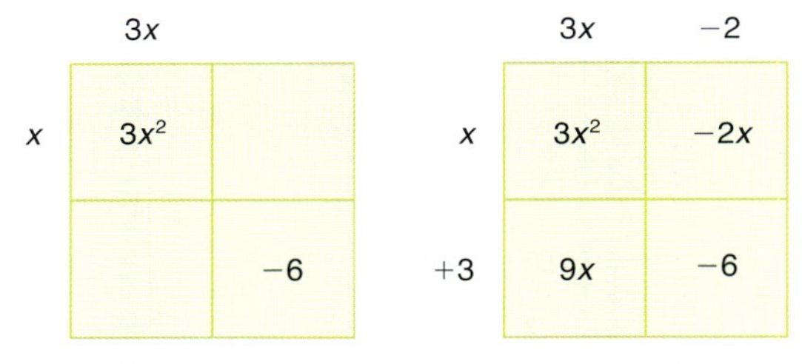

Therefore $3x^2 + 7x - 6 = (3x - 2)(x + 3)$.

Exercise 4.3.4

Factorize the following quadratic expressions.

1 a $x^2 + 7x + 12$ b $x^2 + 8x + 12$ c $x^2 + 13x + 12$
d $x^2 - 7x + 12$ e $x^2 - 8x + 12$ f $x^2 - 13x + 12$

2 a $x^2 + 6x + 5$ b $x^2 + 6x + 8$ c $x^2 + 6x + 9$
d $x^2 + 10x + 25$ e $x^2 + 22x + 121$ f $x^2 - 13x + 42$

3 a $x^2 + 14x + 24$ b $x^2 + 11x + 24$ c $x^2 - 10x + 24$
d $x^2 + 15x + 36$ e $x^2 + 20x + 36$ f $x^2 - 12x + 36$

4 a $x^2 + 2x - 15$ b $x^2 - 2x - 15$ c $x^2 + x - 12$
d $x^2 - x - 12$ e $x^2 + 4x - 12$ f $x^2 - 15x + 36$

5 a $x^2 - 2x - 8$ b $x^2 - x - 20$ c $x^2 + x - 30$
d $x^2 - x - 42$ e $x^2 - 2x - 63$ f $x^2 + 3x - 54$

6 a $2x^2 + 4x + 2$ b $2x^2 + 7x + 6$ c $2x^2 + x - 6$
d $2x^2 - 7x + 6$ e $3x^2 + 8x + 4$ f $3x^2 + 11x - 4$
g $4x^2 + 12x + 9$ h $9x^2 - 6x + 1$ i $6x^2 - x - 1$

Solving quadratic equations algebraically

$x^2 - 3x - 10 = 0$ is a quadratic equation which, when factorized, can be written as $(x - 5)(x + 2) = 0$.

Therefore either $(x - 5) = 0$ or $(x + 2) = 0$ since, if two things multiply to make zero, then one of them must be zero.

$x - 5 = 0$ or $x + 2 = 0$
$x = 5$ or $x = -2$

It is important to understand the relationship between a quadratic equation written in factorized form and the graph of the quadratic.

In the example above $x^2 - 3x - 10$ factorized to $(x - 5)(x + 2)$. The equation $x^2 - 3x - 10 = 0$ had solutions $x = 5$ and $x = -2$.

The graph of $y = x^2 - 3x - 10$ is as shown.

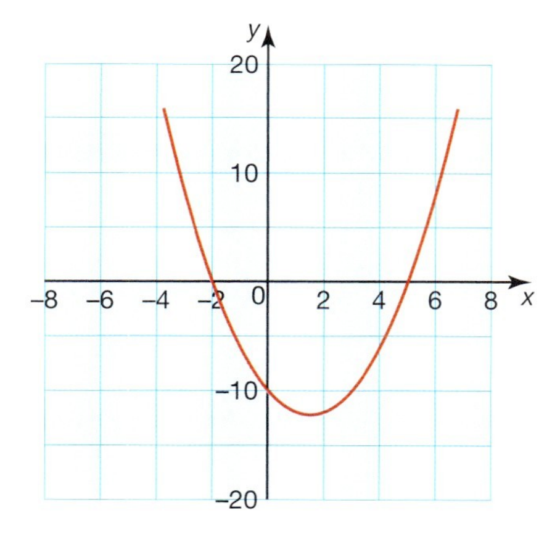

It crosses the x-axis at 5 and -2 because these are the points where the function has the value 0. These values are directly related to the factorized form.

Worked example

Solve each of the following equations to give two solutions for x.

a $x^2 - x - 12 = 0$
b $x^2 + 2x = 24$
c $x^2 - 6x = 0$
d $x^2 - 4 = 0$

a

$$\begin{aligned} x^2 - x - 12 &= 0 \\ (x-4)(x+3) &= 0 \end{aligned}$$

so either $x - 4 = 0$, $x = 4$ or $x + 3 = 0$, $x = -3$

b This becomes $x^2 + 2x - 24 = 0$

$$(x+6)(x-4) = 0$$

so either $x + 6 = 0$, $x = -6$ or $x - 4 = 0$, $x = 4$

c

$$\begin{aligned} x^2 - 6x &= 0 \\ x(x-6) &= 0 \end{aligned}$$

so either $x = 0$ or $x - 6 = 0$, $x = 6$

d

$$\begin{aligned} x^2 - 4 &= 0 \\ (x-2)(x+2) &= 0 \end{aligned}$$

so either $x - 2 = 0$, $x = 2$ or $x + 2 = 0$, $x = -2$

You can use your GDC to solve a quadratic equation. Although you are expected to be able to solve quadratic equations, your calculator is a useful tool for checking your answers.

Worked example

Solve the quadratic equation $x^2 - x - 12 = 0$ using your GDC.

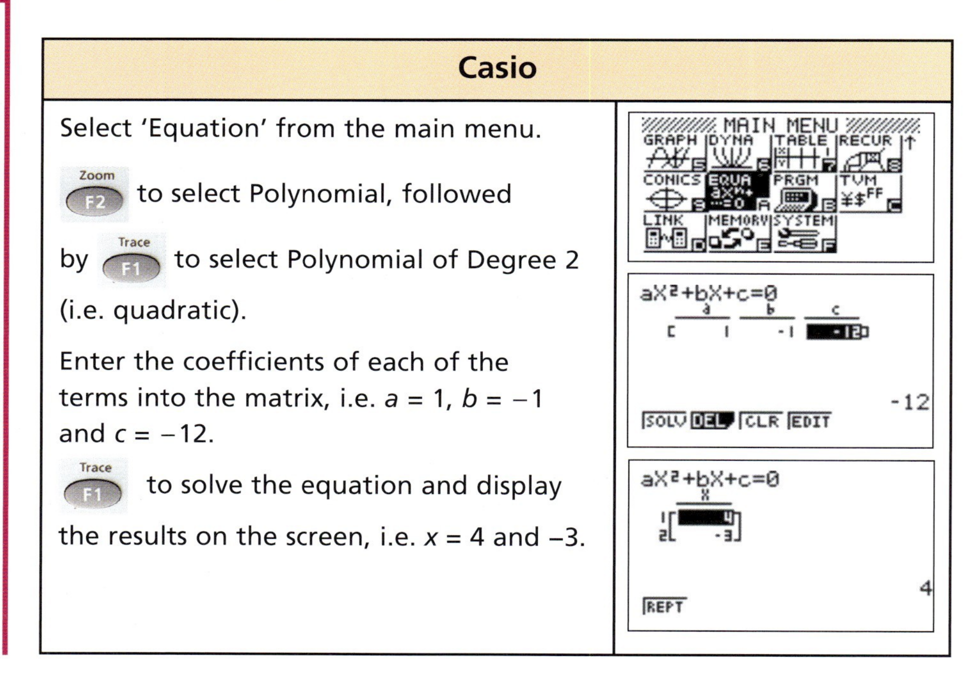

Casio

Select 'Equation' from the main menu.

[Zoom F2] to select Polynomial, followed by [Trace F1] to select Polynomial of Degree 2 (i.e. quadratic).

Enter the coefficients of each of the terms into the matrix, i.e. $a = 1$, $b = -1$ and $c = -12$.

[Trace F1] to solve the equation and display the results on the screen, i.e. $x = 4$ and -3.

Texas

 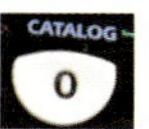 to access the equation solver and enter the equation $0 = x^2 - x - 12$.

EQUATION SOLVER
eqn:0=X²-X-12

Type an initial value for x, e.g. 0. Leave the bound at its default setting. The calculator will search for a solution in this range.

X²-X-12=0
X=■
bound={-1E99,1…

Highlight the initial value of x.

  to solve and display a solution to the equation, i.e. $x = -2.999...$

X²-X-12=0
▪X=-2.999999999…
bound={-1E99,1…
▪left-rt=0

To find the other solution restrict the bound to include the second solution, e.g. Bound = {0,5} and repeat the above steps.

X²-X-12=0
▪X=3.9999999999…
bound={0,5}
▪left-rt=0

Note: With this calculator, you need to know how many solutions there are and roughly where the solutions lie before using the equation solver.

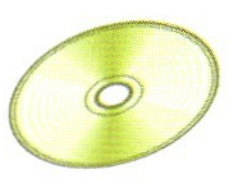

Exercise 4.3.5

1 Solve the following quadratic equations by factorizing.

a $x^2 + 7x + 12 = 0$
b $x^2 + 8x + 12 = 0$
c $x^2 + 3x - 10 = 0$
d $x^2 - 3x - 10 = 0$
e $x^2 + 5x = -6$
f $x^2 + 6x = -9$
g $x^2 - 2x = 8$
h $x^2 - x = 20$
i $x^2 + x = 30$
j $x^2 - x = 42$

2 Solve the following quadratic equations.

a $x^2 - 9 = 0$
b $x^2 = 25$
c $x^2 - 144 = 0$
d $4x^2 - 25 = 0$
e $9x^2 - 36 = 0$
f $x^2 - \frac{1}{9} = 0$
g $x^2 + 6x + 8 = 0$
h $x^2 - 6x + 8 = 0$
i $x^2 - 2x - 24 = 0$
j $x^2 - 2x - 48 = 0$

3 Solve the following quadratic equations.

a $x^2 + 5x = 36$
b $x^2 + 2x = -1$
c $x^2 - 8x = 0$
d $x^2 - 7x = 0$
e $2x^2 + 5x + 3 = 0$
f $2x^2 - 3x - 5 = 0$
g $x^2 + 12x = 0$
h $x^2 + 12x + 27 = 0$
i $2x^2 = 72$
j $3x^2 - 12 = 288$

In questions 4–10, construct equations from the information given and then solve them to find the unknown.

4 When a number x is added to its square, the total is 12. Find two possible values for x.

5 If the area of the rectangle below is $10\,\text{cm}^2$, calculate the only possible value for x.

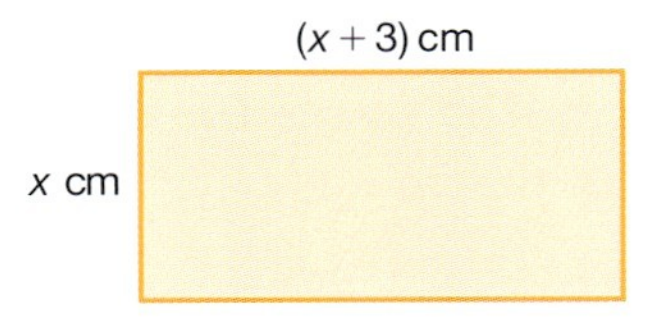

6 If the area of the rectangle below is $52\,\text{cm}^2$, calculate the only possible value for x.

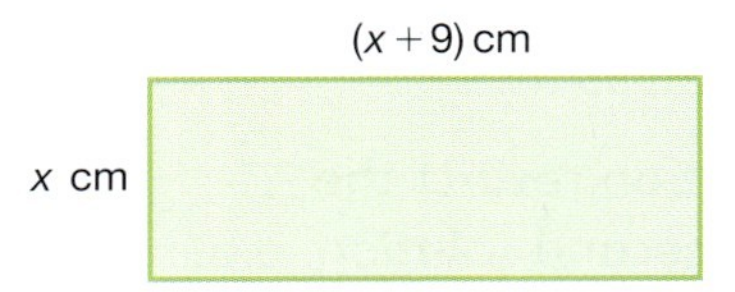

7 A triangle has a base length of $2x$ cm and a height of $(x - 3)$ cm. If its area is $18\,\text{cm}^2$, calculate its height and base length.

8 A triangle has a base length of $(x - 8)$ cm and a height of $2x$ cm. If its area is $20\,\text{cm}^2$, calculate its height and base length.

9 A right-angled triangle has a base length of x cm and a height of $(x - 1)$ cm. If its area is $15\,\text{cm}^2$, calculate its base length and height.

10 A rectangular garden has a square flowerbed of side length x m in one of its corners. The remainder of the garden consists of lawn and has dimensions as shown.

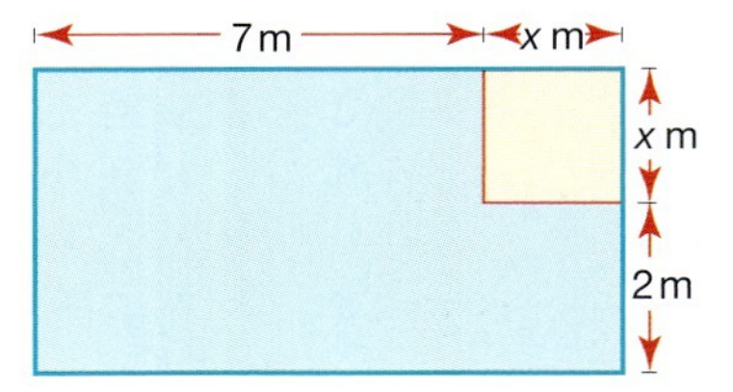

If the total area of the lawn is $50\,\text{m}^2$, calculate the length and width of the whole garden.

The quadratic formula

In general a quadratic equation takes the form $ax^2 + bx + c = 0$ where a, b and c are integers. Quadratic equations can be solved by the use of the quadratic formula, which states that:

$$x = \frac{-b \pm \sqrt{b^2 - 4ac}}{2a}$$

This is particularly useful when the quadratic equation has solutions but does not factorize neatly.

Worked examples

1 Solve the quadratic equation $x^2 + 7x + 3 = 0$.

$a = 1$, $b = 7$ and $c = 3$.

Substituting these values into the quadratic formula gives:

$$x = \frac{-7 \pm \sqrt{7^2 - 4 \times 1 \times 3}}{2 \times 1}$$

$$x = \frac{-7 \pm \sqrt{49 - 12}}{2}$$

$$x = \frac{-7 \pm \sqrt{37}}{2}$$

Therefore $x = \dfrac{-7 + 6.08}{2}$ or $x = \dfrac{-7 - 6.08}{2}$

$x = -0.46$ (2 d.p.) or $x = -6.54$ (2 d.p.)

2 Solve the quadratic equation $x^2 - 4x - 2 = 0$.

$a = 1$, $b = -4$ and $c = -2$.

Substituting these values into the quadratic formula gives:

$$x = \frac{-(-4) \pm \sqrt{(-4)^2 - (4 \times 1 \times -2)}}{2 \times 1}$$

$$x = \frac{4 \pm \sqrt{16 + 8}}{2}$$

$$x = \frac{4 \pm \sqrt{24}}{2}$$

Therefore $x = \dfrac{4 + 4.90}{2}$ or $x = \dfrac{4 - 4.90}{2}$

$x = 4.45$ (2 d.p.) or $x = -0.45$ (2 d.p.)

Exercise 4.3.6

Solve the following quadratic equations using the quadratic formula. Give your answers to two decimal places.

1 a $x^2 - x - 13 = 0$
  b $x^2 + 4x - 11 = 0$
  c $x^2 + 5x - 7 = 0$
  d $x^2 + 6x + 6 = 0$
  e $x^2 + 5x - 13 = 0$
  f $x^2 - 9x + 19 = 0$

2 a $x^2 + 7x + 9 = 0$
  b $x^2 - 35 = 0$
  c $x^2 + 3x - 3 = 0$
  d $x^2 - 5x - 7 = 0$
  e $x^2 + x - 18 = 0$
  f $x^2 - 8 = 0$

3 a $x^2 - 2x - 2 = 0$ b $x^2 - 4x - 11 = 0$
c $x^2 - x - 5 = 0$ d $x^2 + 2x - 7 = 0$
e $x^2 - 3x + 1 = 0$ f $x^2 - 8x + 3 = 0$

4 a $2x^2 - 3x - 4 = 0$ b $4x^2 + 2x - 5 = 0$
c $5x^2 - 8x + 1 = 0$ d $-2x^2 - 5x - 2 = 0$
e $3x^2 - 4x - 2 = 0$ f $-7x^2 - x + 15 = 0$

4.4 Exponential functions and their graphs

Note: You should be familiar with the laws of indices for this section.

Functions of the form $y = a^x$, where a is the base number and x the exponent (power), are known as **exponential functions**. Plotting an exponential function is done in the same way as for other functions.

Worked example

Plot the graph of the function $y = 2^x$ for $-3 \leq x \leq 3$.

x	−3	−2	−1	0	1	2	3
y	0.125	0.25	0.5	1	2	4	8

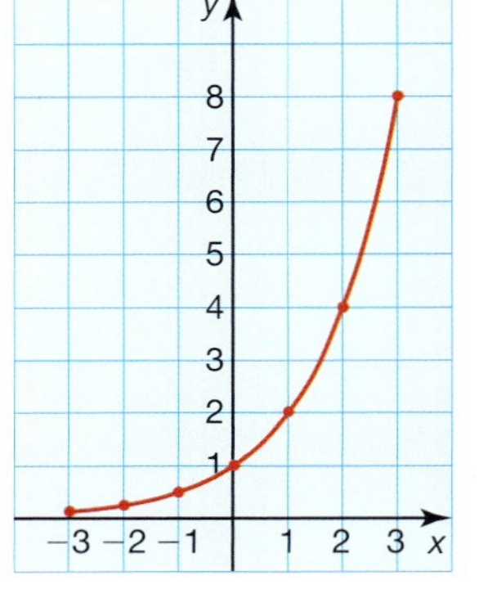

The graph above gets closer and closer to the x-axis as x gets smaller. For example, when $x = -3$, $y = 2^x$ is $y = 2^{-3} = \frac{1}{2^3} = \frac{1}{8}$; when $x = -4$, $y = 2^x$ is $y = 2^{-4} = \frac{1}{2^4} = \frac{1}{16}$.

Therefore the graph will get closer and closer to the x-axis but not cross it. The x-axis is therefore an **asymptote**.

Exercise 4.4.1

1 For each of the functions below:
i draw up a table of values of x and $f(x)$
ii plot a graph of the function
iii check your graph using either your GDC or graphing software.
a $f(x) = 3^x$, $-3 \leq x \leq 3$ b $f(x) = 1^x$, $-3 \leq x \leq 3$
c $f(x) = 2^x + 3$, $-3 \leq x \leq 3$ d $f(x) = 2^x + x$, $-3 \leq x \leq 3$
e $f(x) = 2^x - x$, $-3 \leq x \leq 3$ f $f(x) = 3^x - x^2$, $-3 \leq x \leq 3$

2 A tap is dripping at a constant rate into a container. The level (l cm) of the water in the container is given by the equation $l = 2^t - 1$ where t hours is the time taken.
a Calculate the level of the water after 3 hours.
b Calculate the level of the water in the container at the start.
c Calculate the time taken for the level of the water to reach 31 cm.
d Plot a graph showing the level of the water over the first 6 hours.
e From your graph, estimate the time taken for the water to reach 45 cm.

3 Draw a graph of $y = 4^x$ for values of x between -1 and 3. Use your graph to find approximate solutions to both of the following equations.
   a $4^x = 30$
   b $4^x = \frac{1}{2}$

4 Draw a graph of $y = 2^x$ for values of x between -2 and 5. Use your graph to find approximate solutions to the equation $2^x = 20$.

General equation of the exponential function

An exponential graph takes the form $y = ka^x + c$, $x \in \mathbb{R}$.

We can also write this as $y = ka^{\lambda x}$ where x is an integer ($x \in \mathbb{Z}$) and λ is a real number ($\lambda \in \mathbb{R}$).

The constants k, a and c affect the shape of the graph.

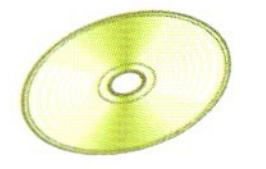

Exercise 4.4.2

Use the GeoGebra file '**4.4 Graphs of exponential functions**' on the CD. The file shows two exponential graphs: one of the form $y = ka^x + c$ and the other of the form $y = ka^{-x} + c$. By moving the sliders, each of the constants k, a and c can be changed and the resulting transformation observed.

1 By changing the value of c in the equation $y = ka^x + c$, describe mathematically the transformation that occurs.

2 By changing the value of k, describe mathematically the transformation that occurs.

3 What effect does the value of a have on the graph?

4 Describe the graphical relationship between $y = ka^x + c$ and $y = ka^{-x} + c$.

Exponential growth and decay

The story is told of the Chinese emperor who wanted to reward an advisor who had averted a famine. The advisor saw a chessboard and asked for one grain of rice to be placed on the first square, two on the next square, four on the next and so on. By the 64th square his reward would exceed the present world grain harvest! This is an example of **exponential growth**.

In England, the FA cup competition for football begins with 256 teams in round 1 playing a knockout system. This reduces to 128, 64, 32 and so on. This is an example of **exponential decay** (continuing until there is one winner).

Other examples of phenomena that experience exponential growth and decay are bacteria, viruses, population, electricity, air pressure, light passing into water, compound interest and radioactive decay.

The formulae for exponential growth and decay are:

Growth: $y = a(1 + r)^x$
Decay: $y = a(1 - r)^x$

where a is the initial amount
r is the growth/decay rate expressed as a fraction out of 100
x is the number of time intervals
y is the final amount.

Earlier we used the Chinese emperor's chessboard example to illustrate the effects of continued doubling, starting with 1. For this the general formula $y = a(1 + r)^x$ has $r = 1$ (as doubling is represented by 100% growth) and $a = 1$. Therefore we obtain the equation:

$$y = 1(1 + 1)^x \Rightarrow y = 2^x$$

In the example of the FA cup football competition, the rate of decay is 50% and the initial amount is 256. These values can be substituted into the formula $y = a(1 - r)^x$, where $a = 256$, $r = 0.5$. Therefore we obtain the equation:

$$y = 256(1 - 0.5)^x \Rightarrow y = 256 \times \left(\frac{1}{2}\right)^x$$

To find the number of rounds needed until there is one winner, we solve the following equation:

$$1 = 256 \times \left(\frac{1}{2}\right)^x$$

$$\frac{1}{256} = \frac{1}{2^x}$$

$$256 = 2^x$$

$$2^8 = 2^x$$

$$x = 8$$

Therefore the number of rounds is 8.

Notice that, as there is no constant term in these equations, the horizontal asymptote of the graph would be $y = 0$.

Another application of the formula is in compound interest and depreciation, briefly discussed in Section 2.6 in relation to geometric sequences and series. It will also be covered in more depth in Topic 8 on Financial mathematics.

The general formula $y = a(1 + r)^x$ is rewritten as:

$$A = C\left(1 + \frac{r}{100}\right)^n$$

where A is the final amount
C is the capital or starting amount
r is the interest rate, usually a percentage
n is the number of time periods, usually in years.

Worked examples

1 What is the final amount when 2500€ are invested at 6% per annum for 5 years at compound interest?

$A = C\left(1 + \frac{r}{100}\right)^n$

$A = 2500\left(1 + \frac{6}{100}\right)^5 = 2500 \times 1.06^5 = 3345.56$

The final amount is 3345.56€.

2 A new car cost US$20 000 and depreciates on average 20% each year for 8 years.

What is it worth after 8 years?

$A = C\left(1 + \frac{r}{100}\right)^n$

$A = 20\,000\left(1 - \frac{20}{100}\right)^8 = 20\,000 \times 0.8^8 = 3355.44$

The car is worth US$3355.44 after 8 years.

Exercise 4.4.3

1 A single virus can double every hour. Use your calculator to calculate the number of viruses there would be after 24 hours.

2 The half-life of plutonium 239 is 24 000 years. How long will 1 g of plutonium 239 take to decay to 1 mg?

3 In 1960 the population of China was 650 million. The population was expected to increase by 50% every 10 years. What would have been the projected population of China for 2010? Give your answer to two significant figures.

4 A pesticide has a half-life of 20 years. Its use was discontinued after it was found to be harmful. How much will remain of 100 g of the pesticide after 200 years?

5 A chess tournament has 512 entrants in a knockout event. How many remain after six rounds have been played?

6 A rain forest covers 1 000 000 km^2. It is thought that the rate at which it is being lost is 5% per year.

   a What area of the forest will be left in 20 years' time? Give your answer to three significant figures.

   b The rate of loss is in fact 10% per annum. How long will it take before the forest covers an area less than 500 000 km^2?

7 The number of bacteria in a Petri dish has reached 1 million from 1 bacterium in 20 hours. What is the rate of increase? Give your answer to two significant figures.

8 After seven years the population of a region was reduced to half of its original number. What is the average annual rate of reduction? Give your answer as a percentage to the nearest whole number.

4.5 Trigonometric functions and their graphs

Having studied the graphs and properties of linear, quadratic and exponential functions, this section looks at the graphs and properties of two of the trigonometric functions.

The sine curve

Plotting the function $y = \sin x$, $-180° \leq x < 540°$, produces the following graph.

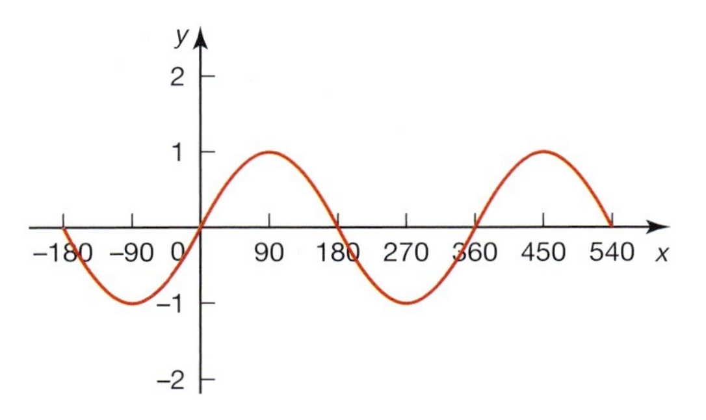

The graph can be plotted using a GDC.

Casio

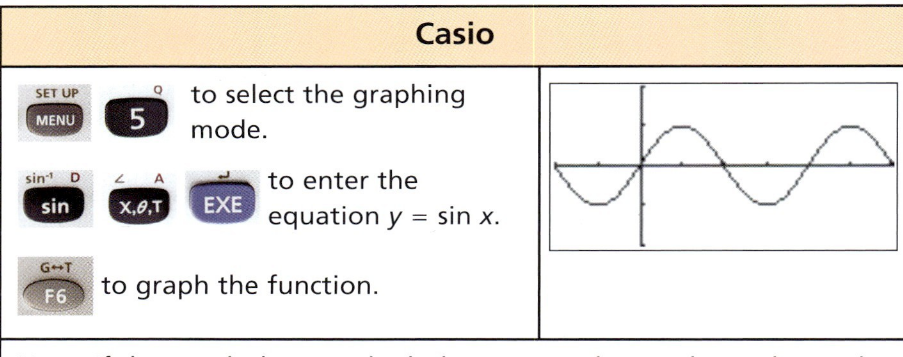

Note: If the graph does not look the same as the one here, the angle mode may not be set correctly. It can be changed by selecting

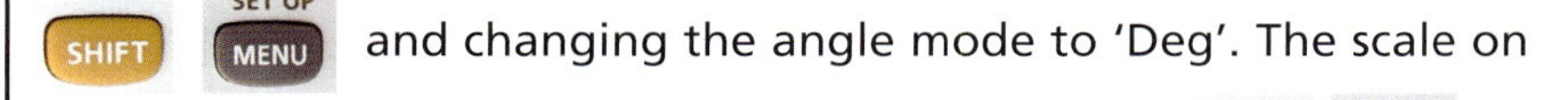

the axes may also need to be changed by selecting SHIFT V-Window F3 and entering the following: Xmin: −180, Xmax: 540, Xscale: 90, Ymin: −2, Ymax: 2, Yscale: 1.

Texas

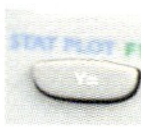 to enter the function.

  ENTER to enter the equation $y = \sin x$.

 to graph the function.

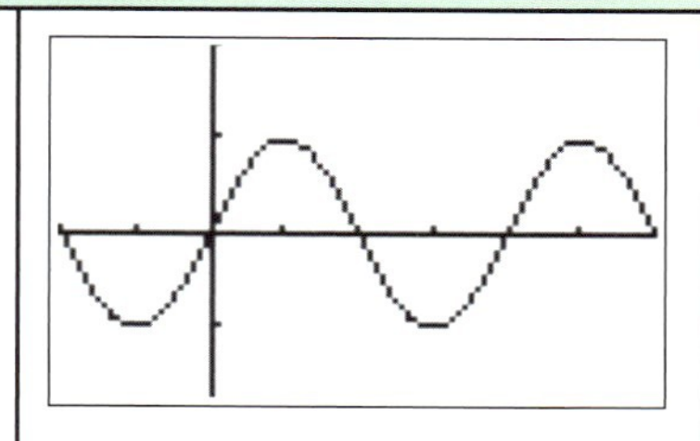

Note: If the graph does not look the same as the one here, the angle mode may not be set correctly. It can be changed by selecting MODE and changing the angle mode to 'Degree'.

The scale on the axes may also need to be changed by selecting  and entering the following: Xmin = −180, Xmax = 540, Xscl = 90, Ymin = −2, Ymax = 2, Yscl = 1.

Graphing software can also be used to visualize trigonometric functions.

Autograph

To use degree mode select .

Select and enter the equation $y = \sin x$.

To change the scale on the axes use .

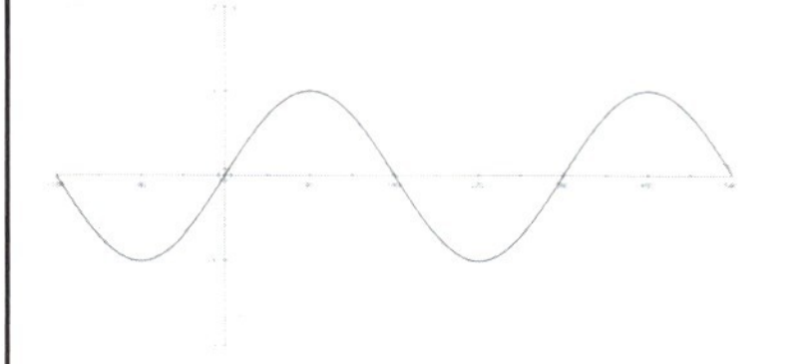

GeoGebra

Type $f(x) = \sin(x°)$ into the input box.

To change the scales of the axes select 'Options' → 'Drawing Pad' and enter the relevant information.

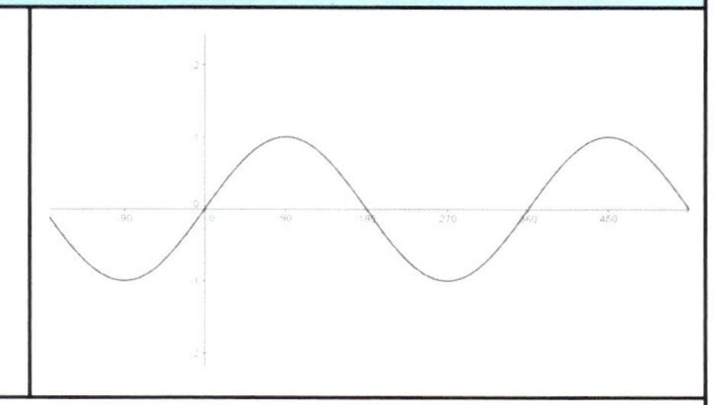

Note: It is important to enter the ° symbol, otherwise the default is to graph trigonometric functions in another unit, called radians.

As can be seen from the screens, the graph of $y = \sin x$ has certain properties.

- It oscillates between -1 and $+1$ and has an **amplitude** of 1.
- It has a **period** of 360° (i.e. it repeats itself every 360°).

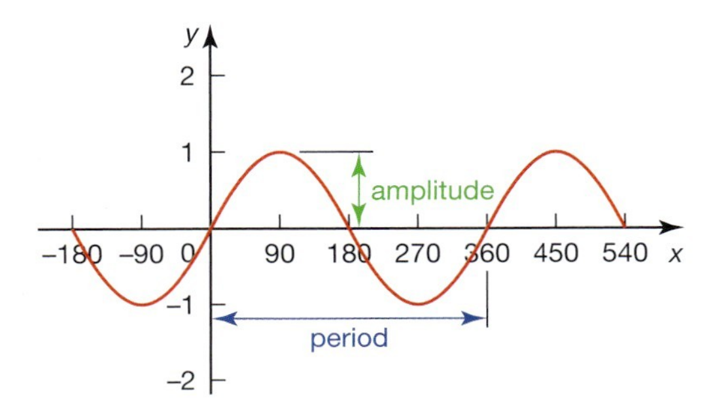

The cosine curve

Plotting the function $y = \cos x$, $-180° \leq x \leq 540°$ produces the following graph.

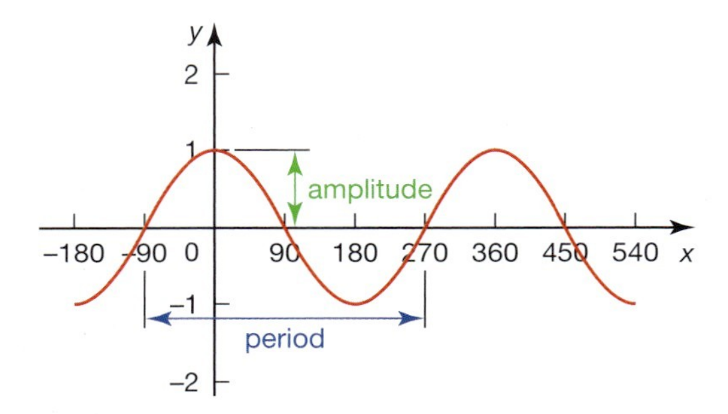

The graph of $y = \cos x$ has similar properties to that of $y = \sin x$

- It oscillates between -1 and $+1$ and has an amplitude of 1.
- It has a period of 360°.

The cosine curve is a translation of the sine curve of $\begin{pmatrix} -90° \\ 0 \end{pmatrix}$.

General equation of the sine and cosine functions

The sine and cosine curves earlier in this section are just one form of the sine and cosine functions. The general form of the sine function is $y = a \sin bx + c$ and of the cosine function is $y = a \cos bx + c$, where a, b and c can be any rational number, i.e. $a, b, c \in \mathbb{Q}$. So in the case of $y = \sin x$ and $y = \cos x$, $a = 1$, $b = 1$ and $c = 0$.

As the values of a, b and c change, then so do the amplitude, period and position of the graph.

Exercise 4.5.1

Using the GeoGebra file '**4.5 Transformations of the sine function**' on the CD, investigate the effect of changing the values of a, b and c on the graph of $y = a\sin bx + c$.

1 a By dragging the slider 'a', observe the effect on the graph when the value of a is changed.
  b Describe the effect on the graph of changing the value of a.

2 a By dragging the slider 'b', observe the effect on the graph when the value of b is changed.
  b Describe the effect on the graph of changing the value of b.

3 a By dragging the slider 'c', observe the effect on the graph when the value of c is changed.
  b Describe the effect on the graph of changing the value of c.

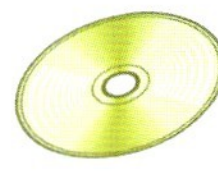

Using the GeoGebra file '**4.5 Transformations of the cosine function**' on the CD, investigate the effect of changing the values of a, b and c on the graph of $y = a\cos bx + c$.

4 a By dragging the slider 'a', observe the effect on the graph when the value of a is changed.
  b Describe the effect on the graph of changing the value of a.

5 a By dragging the slider 'b', observe the effect on the graph when the value of b is changed.
  b Describe the effect on the graph of changing the value of b.

6 a By dragging the slider 'c', observe the effect on the graph when the value of c is changed.
  b Describe the effect on the graph of changing the value of c.

Exercise 4.5.2

1 Without drawing the graphs, state the amplitude and period of each of these functions.

a $y = \frac{1}{2}\sin x$ b $y = 2\sin x + 2$
c $y = \sin 5x$ d $y = \sin \frac{1}{4}x$
e $y = 3\sin 2x$ f $y = -\sin 3x$
g $y = -2\sin \frac{1}{2}x$ h $y = -3\sin 2x - 1$

2 Without drawing the graphs, state the amplitude and period of each of these functions.

a $y = \cos x + 3$ b $y = 3\cos x$
c $y = 2\cos x - 2$ d $y = \cos \frac{1}{2}x$
e $y = \cos 2x - 1$ f $y = -\cos x + 1$
g $y = -\cos \frac{1}{3}x + 1$ h $y = -\cos(-2x) + 2$

Solving trigonometric equations graphically

Trigonometric equations can be solved both algebraically and graphically. This section looks at how to use your GDC or graphing software to solve equations graphically. (Unit 5 will look at solving simple trigonometric equations algebraically.)

Worked example

Solve the equation $2\cos x = 1$ for x in the range $0° \leq x \leq 360°$.

This can be solved in one of two ways:

1 Plot the graphs of $y = 2\cos x$ and $y = 1$ and find where they intersect.

2 Rearrange the equation as $2\cos x - 1 = 0$. Plot the graph of $y = 2\cos x - 1$ and find where it crosses the x-axis (i.e. where $y = 0$).

The explanations below demonstrate method 2. Given your increased familiarity with both the GDC and graphing software, some of the steps have been omitted.

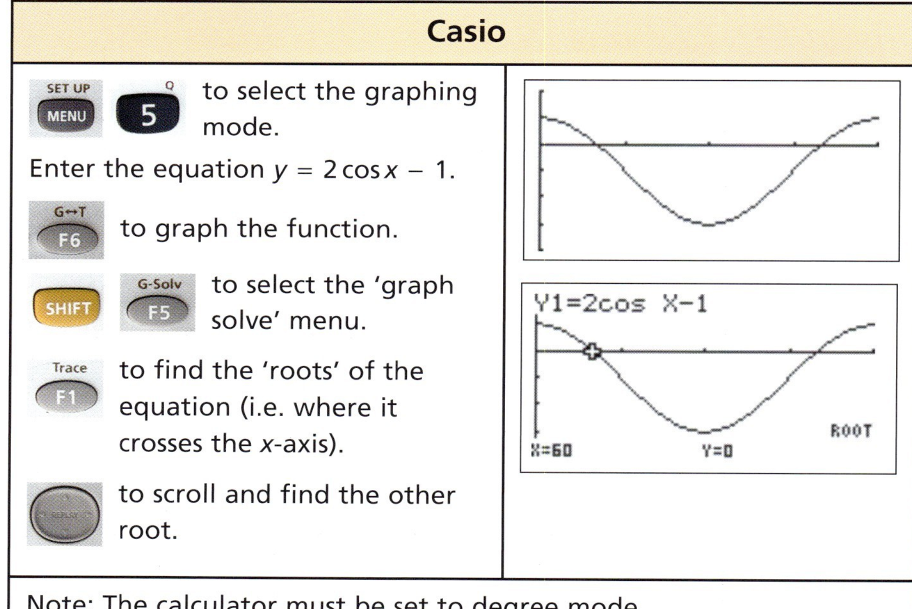

Note: The calculator must be set to degree mode.

The scale on the axes may need to be changed by selecting SHIFT V-Window F3 and entering the following: Xmin: 0, Xmax: 360, Xscale: 90, Ymin: −4, Ymax: 2, Yscale: 1.

Texas

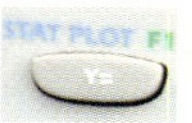 to enter the function.

Enter the equation $y = 2\cos x - 1$.

 to graph the function.

 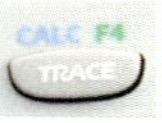 to select the graph calc. menu

 to find where the graph is zero (i.e. crosses the x-axis).

Use the cursor key to select a left bound, i.e. a point to the left of the first solution. Press enter and then use the cursor key to select a right bound. The calculator will search for a solution within this range. Press enter twice to display the answer.

The process is repeated for the second root.

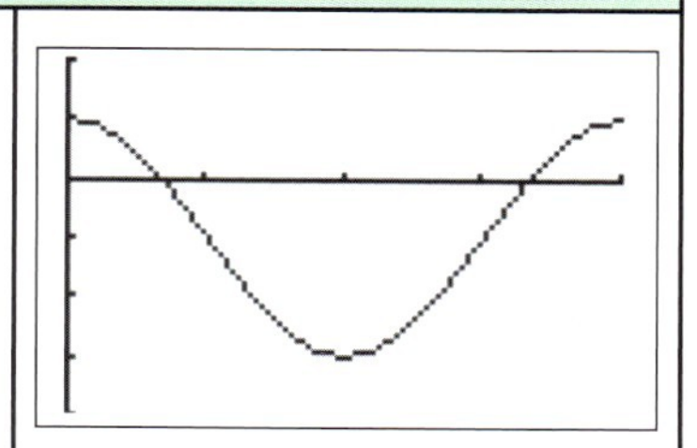

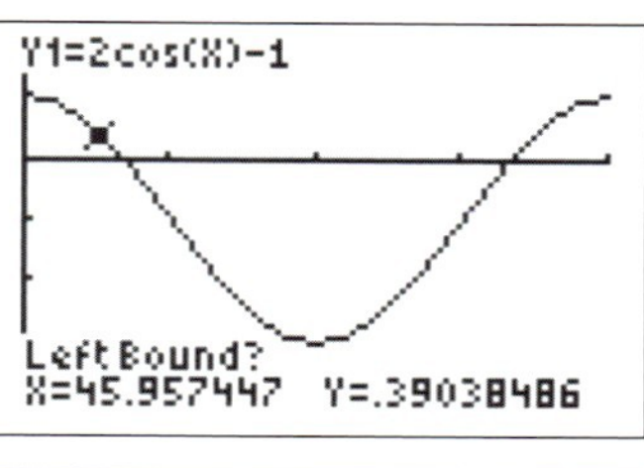

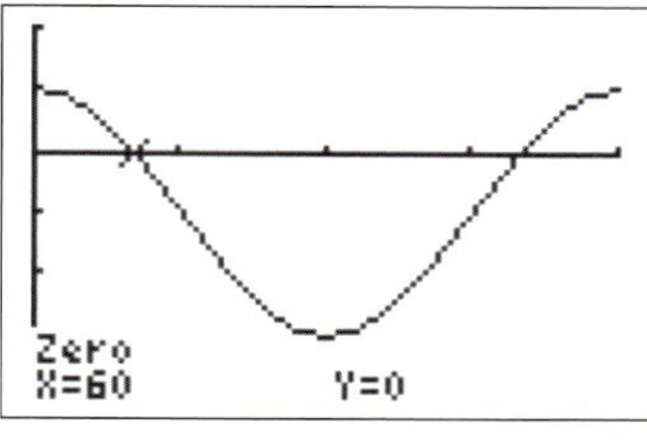

Note: The calculator must be set to degree mode.

The scale on the axes may also need to be changed by selecting  and entering the following: Xmin = 0, Xmax = 360, Xscl = 90, Ymin = −4, Ymax = 2, Yscl = 1.

Autograph

To use degree mode, select .

Select and enter the equation $y = 2\cos x - 1$.

To change the scale on the axes use .

Select the graph and choose 'Object' and 'Solve $f(x)=0$'.

The results are displayed at the bottom of the screen, or by accessing the 'results box' .

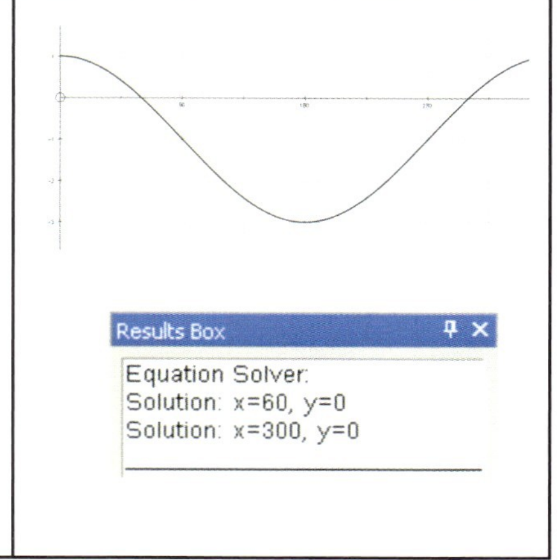

GeoGebra

Type $f(x) = 2\cos(x°) - 1$ into the input box.

To change the scales of the axes select 'Options' → 'Drawing Pad' and enter the relevant information.

To find the intercepts with the x-axis, select the 'intersect two objects' icon.

Click on the curve by the first solution and the x-axis. The solution appears in the algebra window. Repeat for the second intercept point.

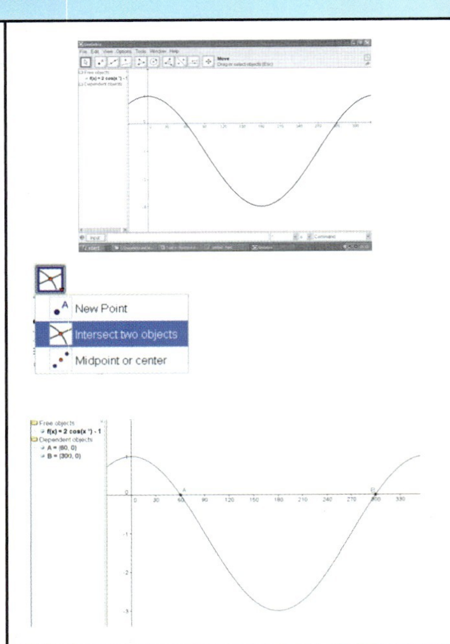

In the given range $0° \leq x \leq 360°$, there are two solutions: $x = 60°$ and $x = 300°$.

Exercise 4.5.3

Solve the following equations graphically, finding all the solutions in the range $0° \leq x \leq 360°$.

1 a $\sin x = 0.5$ b $2\sin\frac{1}{2}x = 0.5$

c $4\sin 3x = 2$ d $\sin x = \frac{x}{100}$

2 a $2\cos x - 3 = -2$ b $\cos x = \sin x$

c $\cos x = \sin 3x$ d $\cos 3x + 2 = \sin 3x + 1$

4.6 Accurate graph drawing

This topic has been covered within some of the earlier sections in relation to drawing graphs of different types of function (linear, quadratic, exponential and trigonometric). As a general reminder, here are the basic guidelines for drawing an accurate graph, as given earlier in this topic.

- Label the axes appropriately.
- Label the scale for each axis accurately.
- Use a table of results to obtain the coordinates of the points to plot.
- For linear functions, use a ruler to draw the straight lines.
- For curved functions, draw a smooth curve through the points rather than a succession of straight lines joining them.

4.7 Sketching unfamiliar functions

You should now know how to produce accurate graphs of linear, quadratic and exponential functions. This section will look at sketching some unfamiliar functions and describe some of their properties. A sketch should show the main features of the function but is not plotted using a table of values. It should show:

- the general shape of the function
- the position of relevant features such as:
  - intercepts
  - maxima, minima or points of inflexion
  - asymptotes.

The reciprocal function

The graph of $y = \frac{1}{x}$ is shown. This belongs to the family of curves whose equations take the form $y = \frac{a}{x + b}$, where a and b are rational.

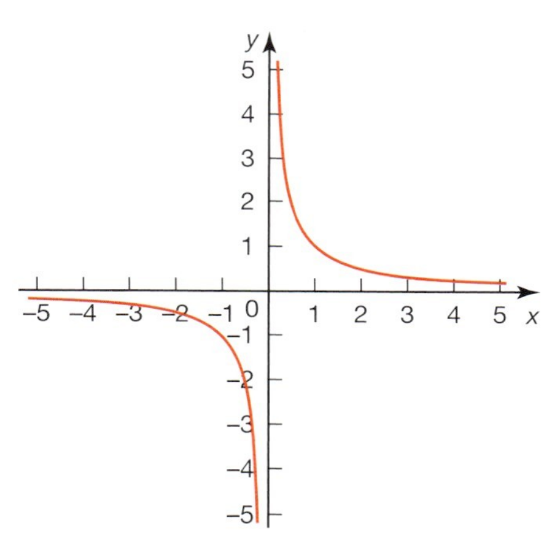

To investigate the effect that a has on the shape of the graph, let $b = 0$, i.e. investigate $y = \frac{a}{x}$.

This can be done using a GDC or graphing software as follows.

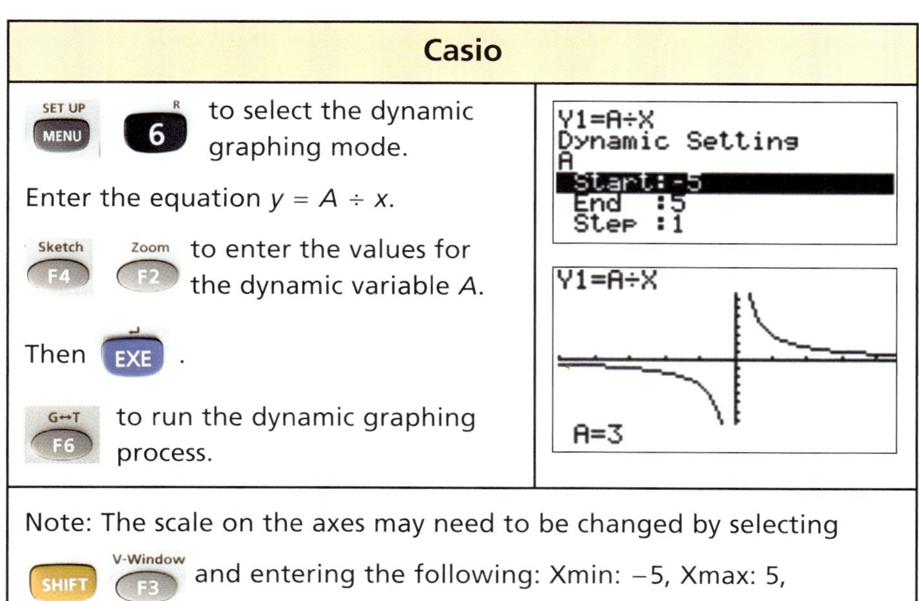

Texas

This calculator does not have a dynamic graphing facility. However, several of the functions can be graphed simultaneously to observe the family of curves, e.g. $y = \frac{1}{x}$, $y = \frac{2}{x}$, $y = \frac{3}{x}$ etc.

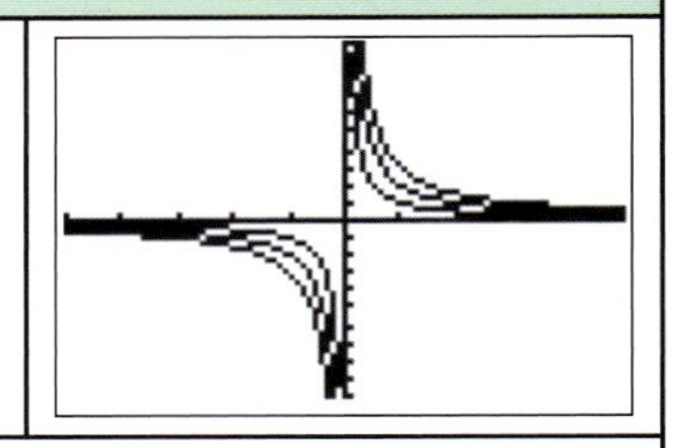

Note: The scale on the axes may need to be changed by selecting  and entering the following: Xmin = −5, Xmax = 5, Xscl = 1, Ymin = −10, Ymax = 10, Yscl = 1.

Autograph

Select and enter the equation $y = \frac{a}{x}$.

Select the constant controller.

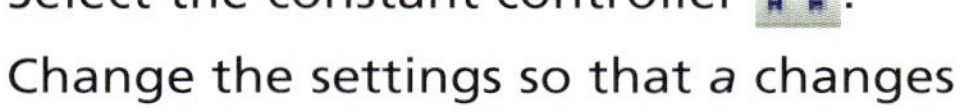

Change the settings so that a changes in increments of 1.

Use to change the value of a and see the graph move.

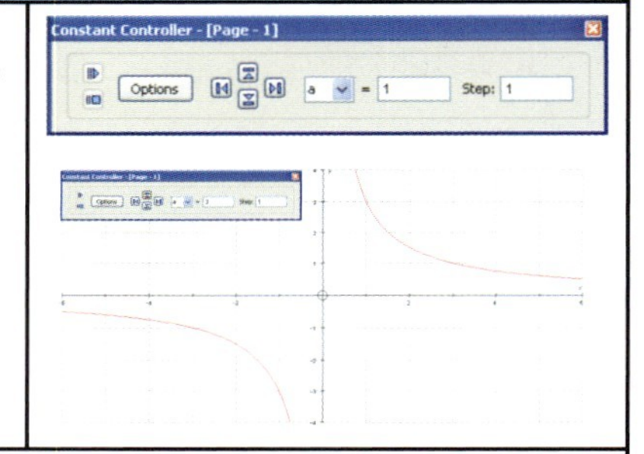

Note: Autograph can display the family of curves or animate the display, by selecting Options and selecting either 'family plot' or 'animation' and entering the parameters as required.

GeoGebra

Select the 'slider' tool and click on the drawing pad. This will enable you to enter the values for 'a' and the incremental change as shown.

Apply to place the slider on the drawing pad.

Type $f(x) = \frac{a}{x}$ into the input box.

The effect of 'a' can be observed by dragging the slider.

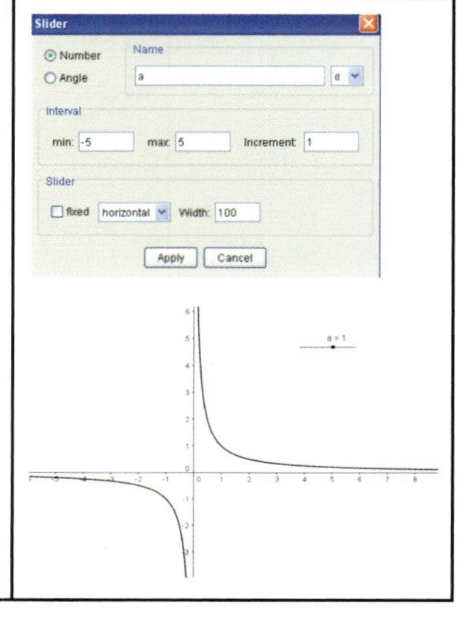

To see the effect of b on the shape of the graph $y = \frac{a}{x + b}$, let $a = 1$ and change the values of b in a similar way to that shown above.

Exercise 4.7.1

1 Describe the effect that changing a has on the shape of the graph $y = \frac{a}{x + b}$.

2 Describe the effect that changing b has on the shape of the graph $y = \frac{a}{x + b}$.

The graph $y = \frac{1}{x}$ has a particular property. It does not cross the y-axis (the line $x = 0$) or the x-axis (the line $y = 0$). The graph gets closer and closer to these lines but does not meet or cross them. These lines are known as **asymptotes** to the curve.

Therefore the graph of $y = \frac{1}{x}$ has a vertical asymptote at $x = 0$ and a horizontal asymptote at $y = 0$. Although clear from the graph, they can be calculated as follows.

Worked examples

1 Calculate the equations of any asymptotes for the graph of $y = \frac{1}{x}$.

Vertical asymptote: This occurs when the denominator $= 0$ as $\frac{1}{0}$ is undefined. Therefore the vertical asymptote is $x = 0$.

Horizontal asymptote: This can be deduced by looking at the value of y as x tends to infinity (written as $x \to \pm\infty$).

As $x \to +\infty$, $y \to 0$ as $\frac{1}{x} \to 0$ As $x \to -\infty$, $y \to 0$ as $\frac{1}{x} \to 0$

Therefore the horizontal asymptote is $y = 0$.

2 Calculate the equations of any asymptotes for the graph of $y = \frac{1}{x - 3} + 1$ and give the coordinates of any points where the graph crosses either axis.

Vertical asymptote: This occurs when the denominator is 0.

$x - 3 = 0$, so $x = 3$. Therefore the vertical asymptote is $x = 3$.

Horizontal asymptote: Look at the value of y as $x \to \pm\infty$.

As $x \to +\infty$, $y \to 1$ as $\frac{1}{x - 3} \to 0$ As $x \to -\infty$, $y \to 1$ as $\frac{1}{x - 3} \to 0$

Therefore the horizontal asymptote is $y = 1$.

To find where the graph intercepts the y-axis, let $x = 0$.

Substituting $x = 0$ into $\frac{1}{x - 3} + 1$ gives $y = -\frac{1}{3} + 1 = \frac{2}{3}$

To find where the graph intercepts the x-axis, let $y = 0$.

Substituting $y = 0$ into $\frac{1}{x - 3} + 1$ gives:

$$0 = \frac{1}{x - 3} + 1$$

$$\frac{1}{x - 3} = -1$$

$$x = 2$$

Therefore the intercepts with the axes occur at $\left(0, \frac{2}{3}\right)$ and $(2, 0)$.

The graph of $y = \dfrac{1}{x-3} + 1$ can now be sketched as shown.

Note: The asymptotes are shown with dashed lines.

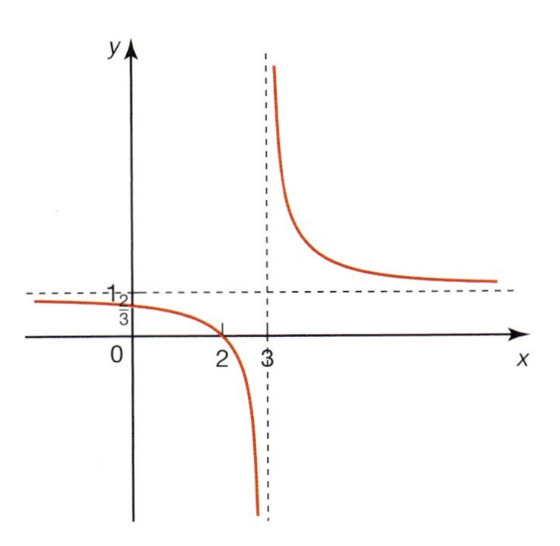

Exercise 4.7.2

For each of the equations in questions 1–3:

i) calculate the equations of the vertical and horizontal asymptotes
ii) calculate the coordinates of any points where the graph intercepts the axes
iii) sketch the graph
iv) check your graph using either a GDC or graphing software.

1 **a** $y = \dfrac{1}{x+1}$ **b** $y = \dfrac{1}{x+3}$

c $y = \dfrac{2}{x-4}$ **d** $y = \dfrac{-1}{x-3}$

2 **a** $y = \dfrac{1}{x} + 2$ **b** $y = \dfrac{1}{x} - 3$

c $y = \dfrac{1}{x-1} + 4$ **d** $y = \dfrac{1}{x+4} - 1$

3 **a** $y = \dfrac{1}{2x+1}$ **b** $y = \dfrac{1}{2x-1} + 1$

c $y = \dfrac{2}{3x-1}$ **d** $y = \dfrac{-1}{4x-1} + 2$

Worked examples

1 Use a GDC or graphing software to graph the function $y = \dfrac{1}{(x-2)(x+1)}$ and determine the equations of any asymptotes.

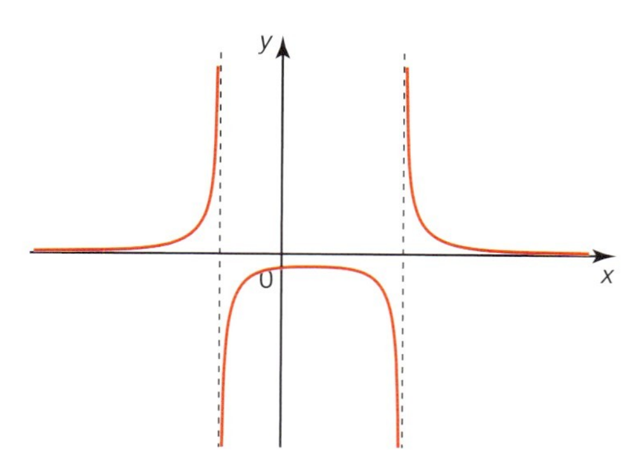

Graphing software will produce the following graph of the function:

The equations of any asymptotes can be determined as before.

Vertical asymptote: This occurs when the denominator is 0: $(x-2)(x+1) = 0$

Therefore the vertical asymptotes are $x = 2$ and $x = -1$.

Horizontal asymptote: Look at the value of y as $x \to \pm\infty$.

As $x \to \pm\infty$,

$y \to 0$ as $\dfrac{1}{(x-2)(x+1)} \to 0$

Therefore the horizontal asymptote is $y = 0$.

Therefore a more informative sketch is:

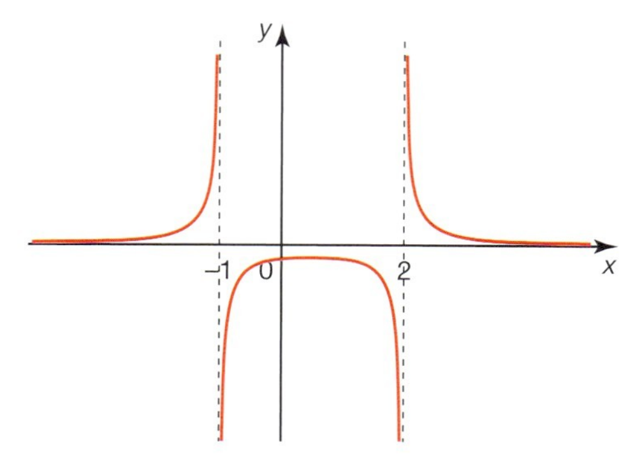

2 Use a GDC or graphing software to graph the function $y = \dfrac{1}{x^2 + x - 6} + 2$ and determine the equations of any asymptotes.

The graphing software will produce the following graph of the function:

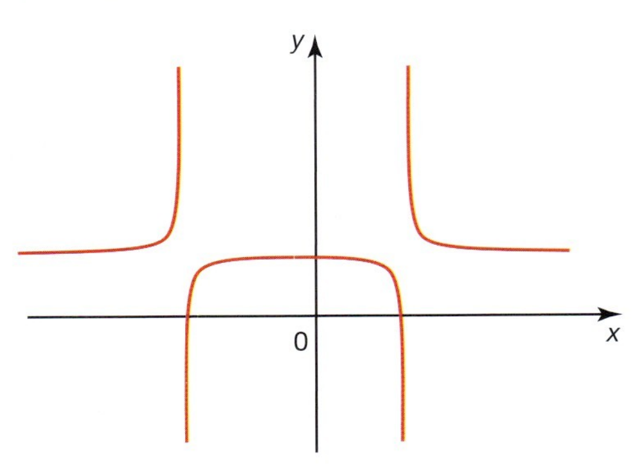

The equations of the asymptotes are determined as before.

Vertical asymptote: This occurs where the denominator is 0.

$x^2 + x - 6 = 0$

Factorising gives $(x - 2)(x + 3) = 0$

Therefore the vertical asymptotes are $x = 2$ and $x = -3$.

Horizontal asymptote: Look at the value of y as $x \to \pm\infty$.

As $x \to \pm\infty$, $y \to 2$ as $\dfrac{1}{x^2 + x - 6} \to 0$

Therefore the horizontal asymptote is $y = 2$.

Therefore a more informative sketch is:

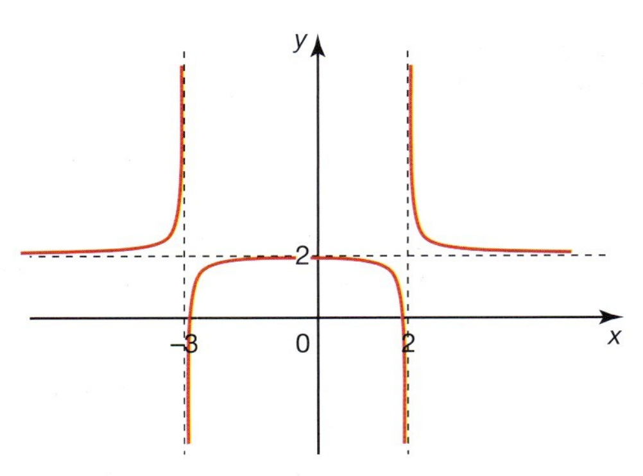

Exercise 4.7.3

For each of the equations in the following questions:

i) calculate the equations of any vertical or horizontal asymptotes

ii) calculate the coordinates of any points where the graph intercepts the axes

iii) sketch the graph of the function with the aid of a GDC or graphing software.

1 a $y = \dfrac{1}{(x - 1)(x - 2)}$

b $y = \dfrac{1}{(x - 4)(x + 3)}$

c $y = \dfrac{1}{x(x - 5)} + 1$

d $y = \dfrac{1}{(x + 2)^2} - 3$

2 a $y = \dfrac{1}{x^2 + 3x - 4}$ b $y = \dfrac{1}{x^2 + 3x - 10}$

c $y = \dfrac{1}{x^2 + 2x + 1} + 3$ d $y = \dfrac{1}{2x^2 - 7x - 4} - 1$

Higher-order polynomials

Earlier in this topic you saw how a quadratic equation can sometimes be factorized, e.g.

$$x^2 - 4x - 5 = (x - 5)(x + 1)$$

This can be used to find where the graph of $y = x^2 - 4x - 5$ intercepts the x-axis, i.e. when $y = 0$. $(x - 5)(x + 1) = 0$ gives $x = 5$ and $x = -1$ as shown.

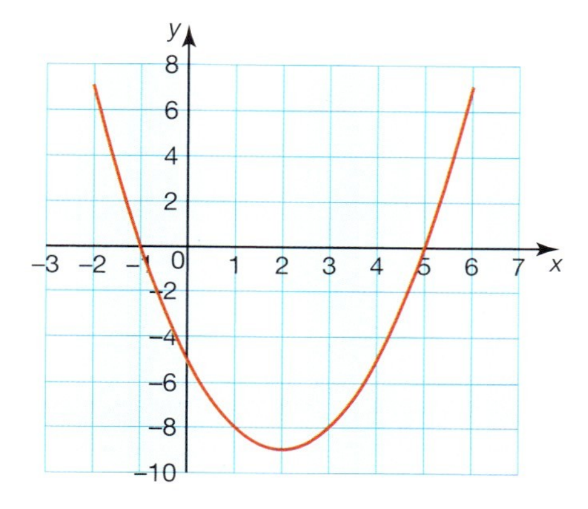

The reverse is therefore also true. A quadratic equation that intercepts the x-axis at $x = 3$ and -6 can be written as $y = (x - 3)(x + 6) = x^2 + 3x - 18$.

The same is true of higher-order polynomials.

Worked example

Using a GDC or graphing software, graph the cubic equation $y = x^3 - 2x^2 - 5x + 6$ and, by finding its roots, rewrite the equation in factorized form.

Casio

Select the graphing mode and enter the equation $y = x^3 - 2x^2 - 5x + 6$.

G↔T F6 to graph the function.

SHIFT G-Solv F5 to select the 'graph solve' menu.

Trace F1 to find the 'roots' of the graph.

 to scroll and find the other roots.

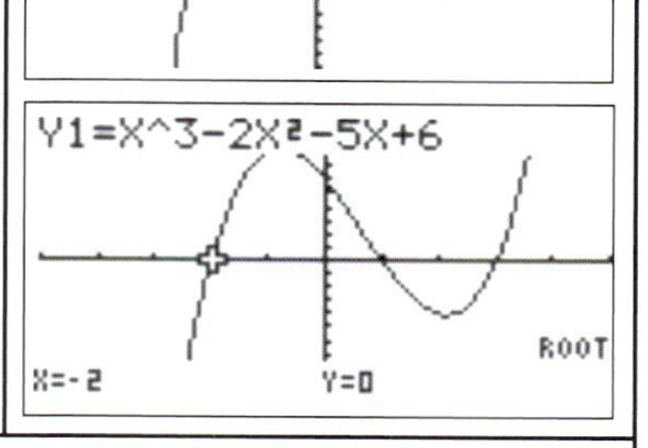

Note: Although the GDC will also find the y-intercept, it is quicker simply to substitute $x = 0$ into the equation, i.e. the y-intercept is $+6$.

Worked example

Texas

 and enter the equation $y = x^3 - 2x^2 - 5x + 6$.

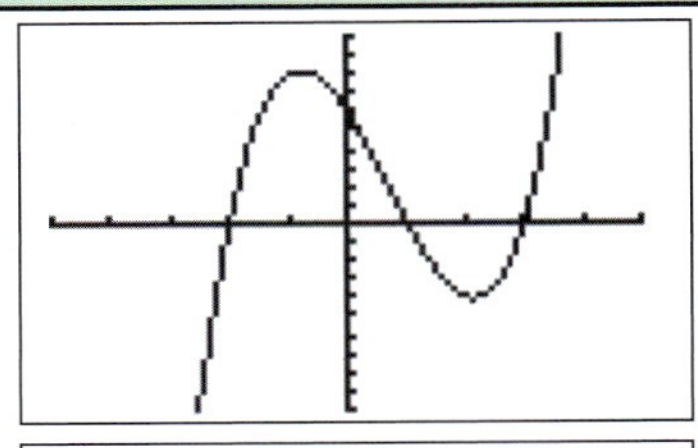

 to graph the function.

  to select the 'graph calc.' menu.

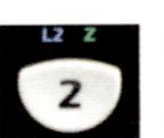 to find where the graph is zero (i.e. crosses the x-axis).

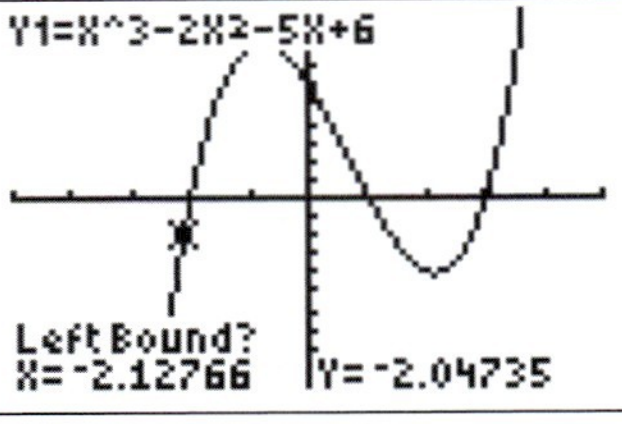

Use the cursor key to select a left bound. Press enter and then use the cursor key to select a right bound. The calculator will search for a solution within this range. Press enter twice to display the answer.

Repeat the process for the other roots.

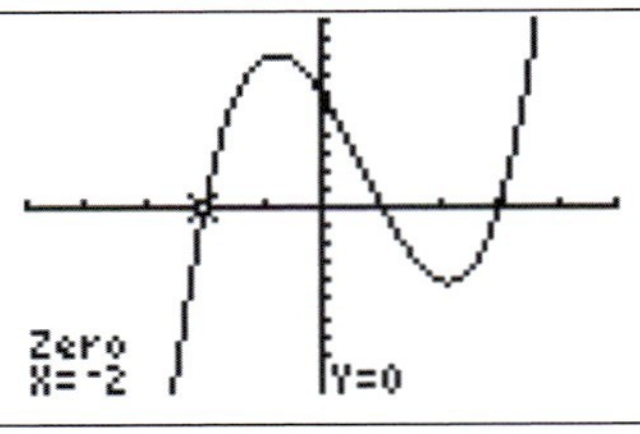

Note: Although the GDC will also find the y-intercept by entering a value of $x = 0$, it is quicker simply to substitute $x = 0$ into the equation, i.e. the y-intercept is $+6$.

Autograph

Select and enter the equation $y = x^3 - 2x^2 - 5x + 6$.

Select the graph and choose 'Object' and 'Solve $f(x) = 0$'.

The results are displayed at the bottom of the screen, or by accessing the 'results box' .

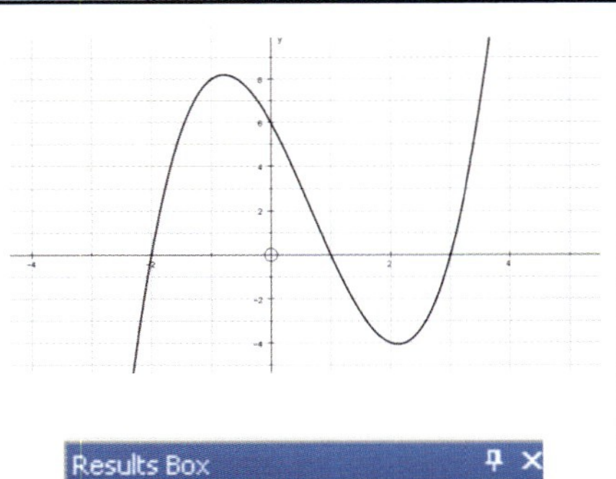

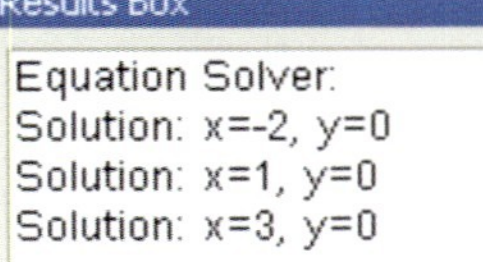

GeoGebra	
Type $f(x) = x \wedge 3 - 2x \wedge 2 - 5x + 6$ into the input box. Type: Root[f]. This finds the roots for the polynomial f. The results are displayed in the algebra window.	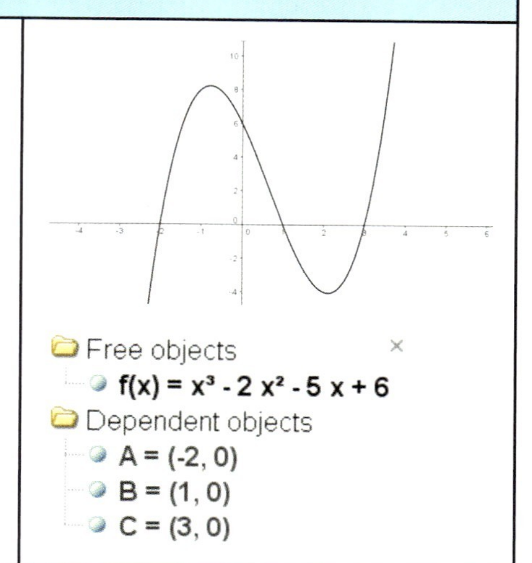

The roots of the equation are therefore $x = -2$, 1 and 3, so the equation can be written in factorized form as $y = (x + 2)(x - 1)(x - 3)$.

The y-intercept is found by substituting $x = 0$ into the equation, giving $y = 6$.

You can also use your GDC in a similar way to find the stationary points of the curve. The algebraic method for finding the stationary points of a curve is covered in Topic 7.

Exercise 4.7.4

For each of the equations in the following questions:

i) Use a GDC or graphing software to sketch the function.
ii) Determine the y-intercept by using $x = 0$.
iii) Rewrite the equation in factorized form by finding the roots of the equation.

1 a $y = x^3 + 6x^2 + 11x + 6$
  b $y = x^3 - 3x^2 + 3$
  c $y = x^3 - x^2 - 12x$
  d $y = x^3 - 4x$

2 a $y = x^3 - 3x + 2$
  b $y = x^3 - 4x^2 + 4x$
  c $y = -x^4 + 2x^3 + 3x^2$
  d $y = -x^3 + x^2 + 22x - 40$

4.8 Solving unfamiliar equations graphically

When an equation has to be solved, it can be done graphically or algebraically. The following example shows how equations can be solved graphically.

Worked example

Solve $x + 2 = \frac{3}{x}$.

Method 1: Rearrange the equation to form the quadratic equation $x^2 + 2x - 3 = 0$. This produces the graph shown. The roots can be calculated as shown earlier in the topic.

Method 2: Both sides of the original equation can be plotted separately and the x-coordinates of their points of intersection calculated.

Solution: $x = 1$ and $x = -3$.

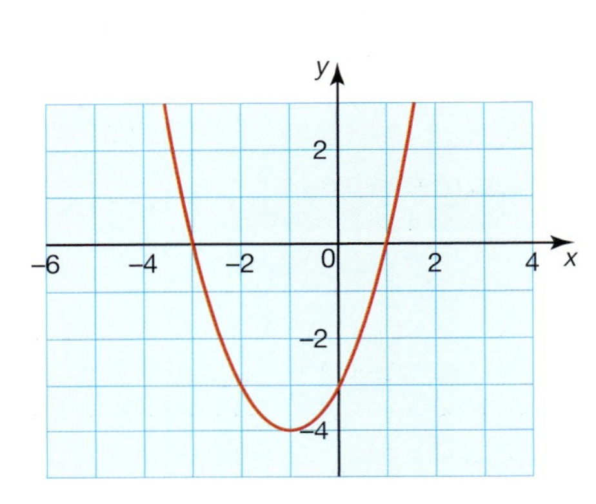

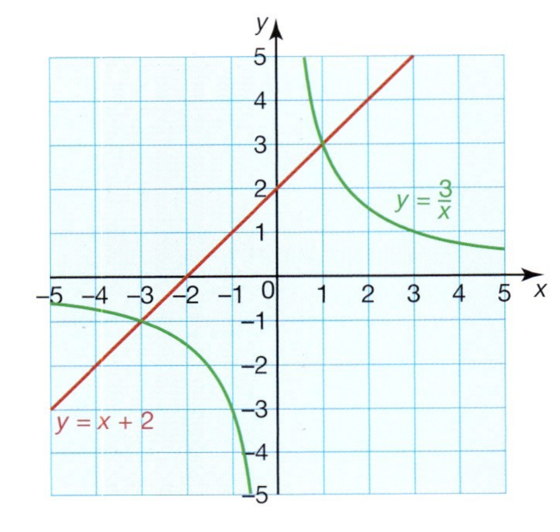

Casio

Select the graphing mode and enter the equations $y = x + 2$ and $y = \frac{3}{x}$.

 to graph the functions.

  to select the 'graph solve' menu.

 to find the intersection of the graphs.

 to scroll and find the other points of intersection.

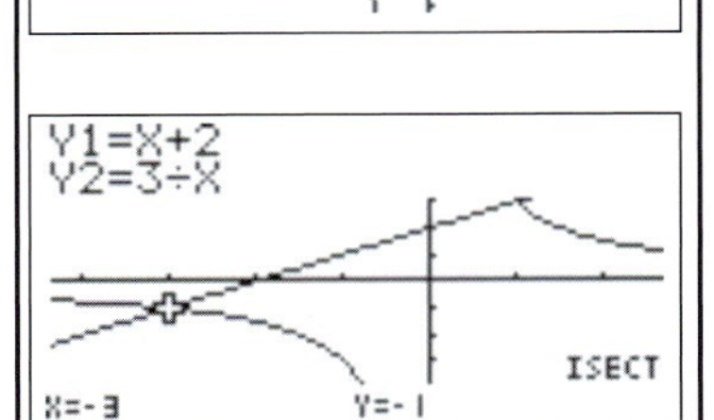

Texas

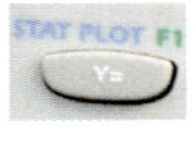 and enter the equations $y = x + 2$ and $y = \frac{3}{x}$.

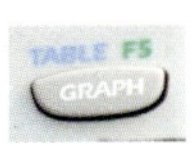 to graph the functions.

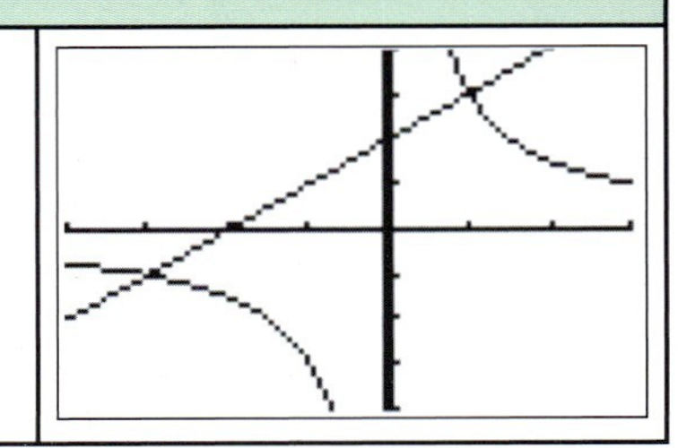

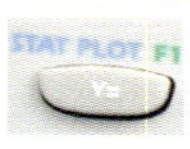 and enter the equations $y = x + 2$ and $y = \frac{3}{x}$.

 to graph the functions.

 to select the 'graph calc.' Menu.

 to find the intersection of the two graphs.

   to select both graphs and confirm the first intersection point.

Repeat the above steps for the second point of intersection. When prompted for the 'guess', move the cursor over the second point of intersection.

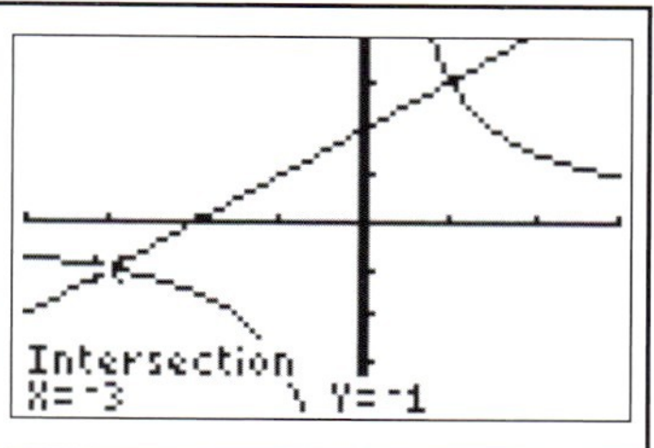

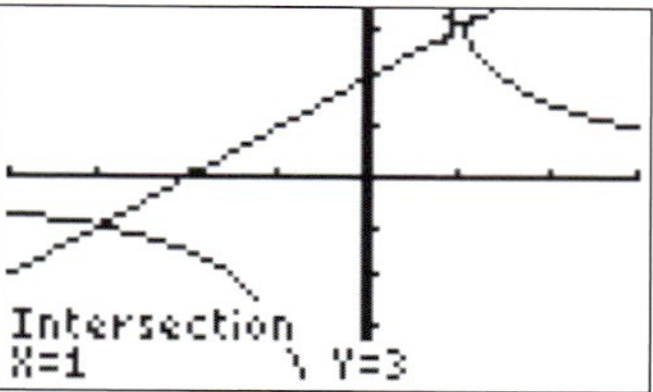

Autograph

Select and enter the equations $y = x + 2$ and $y = \frac{3}{x}$.

Select the graphs, choose 'Object' and 'Solve $f(x) = g(x)$'.

The results are displayed at the bottom of the screen, or by accessing the 'results box'.

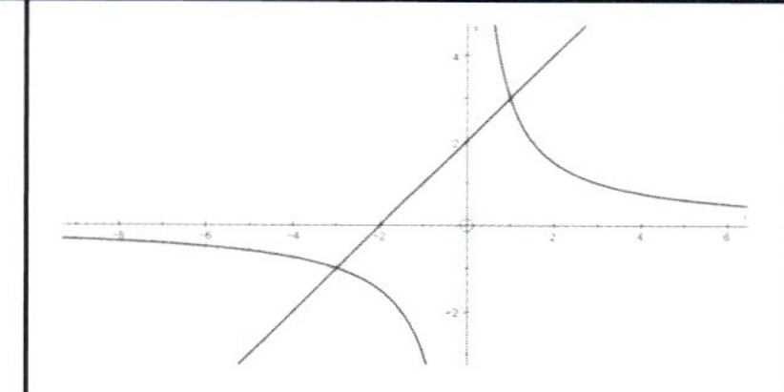

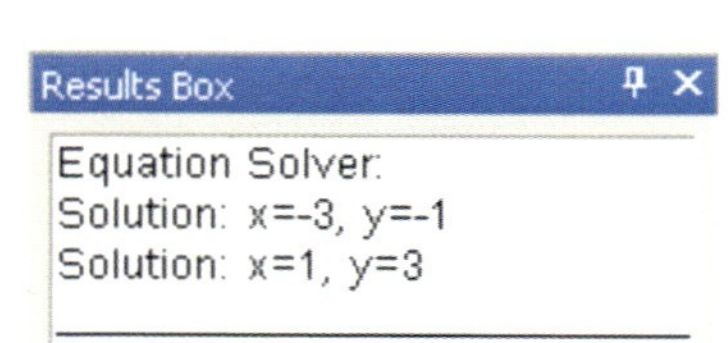

GeoGebra

Type $f(x) = x + 2$ and $g(x) = \frac{3}{x}$ into the input box.

To find the points of intersection, select the 'intersect two objects' icon.

Click on the curve and the straight line; the intersection point is marked. Its coordinates appear in the algebra window.

Repeat for the second point of intersection.

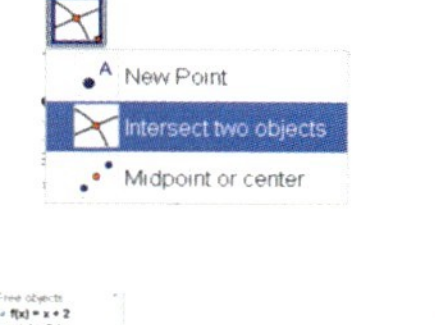

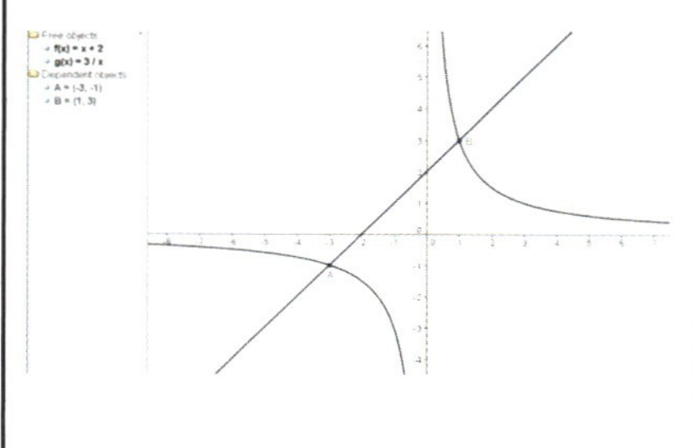

Exercise 4.8.1

1 Using either a GDC or graphing software, solve the following equations.

a $x + 1 = \frac{20}{x}$ b $\frac{3}{x} = 2x + 5$ c $\frac{6}{x+1} = \frac{1}{x^2}$

d $2x - \frac{35}{x} = 9$ e $\frac{4x^2}{5} = x - \frac{1}{5}$ f $x^2 - 3x - 6 = -\frac{8}{x}$

2 Solve the following equations using either a GDC or graphing software. Give your answers to one decimal place.

a $6x = 2^x$ b $2^x - 1 = 2 - x^2$ c $\left(\frac{1}{2}\right)^x - 3 = x^3 - 6x^2$

d $1 - \frac{1}{x} = \frac{1}{3^x}$ e $\frac{2}{x} - 5 = -\frac{1}{x-2}$ f $3 + \sin x = \frac{1}{\sin x}$

Student assessment 1

1 Which of the following is not a function? Give reasons for your answer.

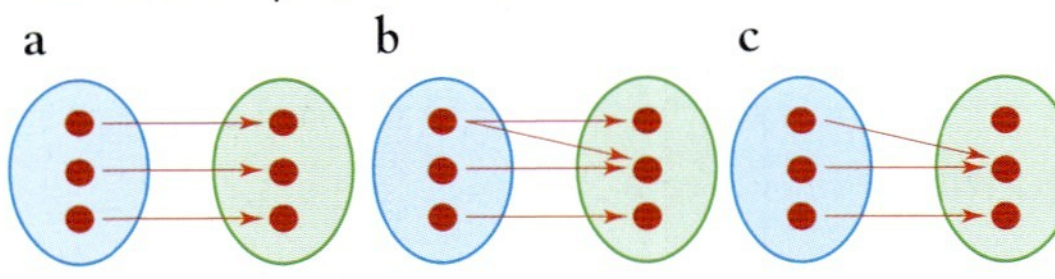

2 State the domain and range for these functions.
a $f(x) = 3x - 2; -3 \le x \le 1$
b $g(x) = x^2; x \in \mathbb{R}$

3 Calculate the range of $f: q \mapsto 4q + 7; -1 \le q \le 5$.

4 For the function $g: x \mapsto \frac{3x-2}{2}$, evaluate:
a $g(4)$ b $g(0)$ c $g(-3)$

5 Find the inverse of these functions.
a $f(x) = -3x + 9$ b $g(x) = \frac{3(x-6)}{2}$

6 If $h(x) = \frac{3}{2}(3 - x)$, evaluate:
a $h^{-1}(-3)$ b $h^{-1}\left(\frac{3}{2}\right)$

7 Calculate the coordinates of the point of intersection of each pair of linear functions.
a $f(p) = -2 - 2p$ and $f(q) = 2q + 10$
b $y = 2x - 1$ and $x + y = 8$
c $3x + 2y = 7$ and $5x + 2y = 7$

8 The graph of the temperature conversion from Celsius to Fahrenheit uses the formula $F = \frac{9}{5}C + 32$, where F and C are the temperatures in degrees Fahrenheit and Celsius respectively.
a Plot a graph of $F = \frac{9}{5}C + 32; 0 \le C \le 100$.
b Use the graph to estimate the following temperatures in degrees Fahrenheit.
i) 10 °C ii) 40 °C iii) 80 °C
c Use the formula to check your answers to part **b**.

9 The cost C in euro of producing a CD for a band is given by $C(n) = 210\,000 + 1.5n$ where n is the number of copies made. Each CD sells for 12 euro.
a Write down the function $R(n)$ for the revenue.
b Draw on the same axes graphs for cost and revenue.
c What value of n represents the break-even point (i.e. the point where costs = revenue)?
d What is the profit if the CD sells 250 000 copies?

Student assessment 2

1 a Copy and complete the table below for the function $f(x) = x^2 + 3x - 9$.

x	−3	−2	−1	0	1	2	3
f(x) = x² + 3x − 9							

b Plot a graph of $f(x)$ for the given domain.
c Deduce the range for $f(x)$.

2 Factorize each of the following quadratic functions.
   a $f(x) = x^2 - 9x + 18$ b $h(y) = 3y^2 + y - 2$
   c $f(x) = x^2 - 3x - 10$ d $h(x) = 2x^2 + 7x - 4$

3 Solve the following quadratic equations.
   a $x^2 + 6x + 8 = 0$ b $2x^2 + 10 = 12x$
   c $x^2 + 10x + 25 = 0$ d $3x^2 - 4 = 7x$

4 Using the quadratic formula $x = \frac{-b \pm \sqrt{b^2 - 4ac}}{2a}$ for solving quadratic equations of the form $ax^2 + bx + c = 0$, solve the following equations.
   a $4x^2 - 6x + 1 = 0$ b $5x^2 - 12x - 3 = 0$

5 Calculate the final amount saved if 4000€ is put into an account paying 7.5% interest per year, for 10 years.

6 The number of a type of bacteria increases by 25% every hour. If the initial number is x, calculate in terms of x the number of bacteria after 24 hours.

7 An oil-field is being depleted by 10% each year. It currently holds 10 million barrels of oil. How long will it be before it is reduced to 1 million barrels?

8 The amount of light which passes through water decreases off a coral reef by 12.5% per metre. At how many metres is there only 10% of the light at the surface? Give your answer to the nearest metre.

Student assessment 3

1 a Sketch the graph of $f(x) = \sin x$; $0° \leq x \leq 360°$.
   b On the same axes sketch the function $g(x) = 2 \sin x$.

2 a Sketch the graph of $f(x) = \cos x$; $-180° \leq x \leq 180°$.
   b On the same axes sketch the function $g(x) = \cos 2x + 1$.

3 a Sketch the graph of $f(x) = \sin x$; $0° \leq x \leq 360°$.
   b On the same axes sketch the function $g(x) = -\sin x + 2$.

4 a Sketch the graph of $f(x) = \cos x$; $-180° \leq x \leq 180°$.
   b On the same axes sketch the function $g(x) = 3\cos \frac{1}{2}x$.

5 Solve the equation $3 \sin 2x = 2$ graphically, finding all the solutions in the range $0° \leq x \leq 180°$.

6 Solve the equation $3 \cos x = -1$ graphically, finding all the solutions in the range $0° \leq x \leq 360°$.

7 For the equation $y = \frac{2}{x - 3} + 1$:
   a calculate the equations of any vertical and horizontal asymptotes
   b calculate the coordinates of any points where the graph intercepts the axes
   c sketch the graph.

8 For the equation $y = \frac{1}{2x - 4} - 1$:
   a calculate the equations of any vertical and horizontal asymptotes
   b calculate the coordinates of any points where the graph intercepts the axes
   c sketch the graph.

9 For the function $f(x) = \frac{1}{(x + 1)(x - 5)}$:
   a calculate the equations of any vertical and horizontal asymptotes
   b calculate the coordinates of any points where the graph intercepts the axes
   c sketch the graph.

10 For the function $f(x) = \frac{1}{x^2 - 5x + 6} + 2$:
   a calculate the equations of any vertical and horizontal asymptotes
   b calculate the coordinates of any points where the graph intercepts the axes
   c sketch the graph.

Discussion points, project ideas and theory of knowledge

1 The study of supply and demand, and graphs representing them, is fundamental to the study of economics. This type of graph work could form the basis of a project.

2 The terms mapping, function, domain and range have a specific meaning in this topic. How does each meaning differ from its more general meaning?

3 Some people consider the circle to be a perfect shape. Why do you think that is? What about the ellipse and the parabola? What do you think are perfect shapes?

4 What type of graph is obtained by a study of radioactive decay? Discuss the long-term problem of what to do with radioactive waste.

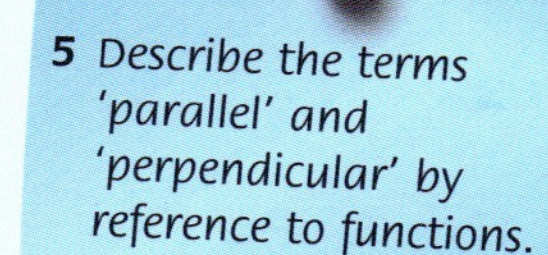

5 Describe the terms 'parallel' and 'perpendicular' by reference to functions.

6 Would reversing the position of the x- and y-axes have any effect on the study of graphs of functions?

7 Any business is concerned with cash flow, and cost and revenue functions. The study of these in a company could form the basis of a project.

8 Discuss the statement that 'to know about Descartes, Newton and Gauss has no relevance to learning mathematics'. Can mathematics be studied without any reference to its historical context?

9 There are a number of unusual and unfamiliar functions which produce interesting graphs. The study of these could form the basis of a project.

10 Leonhard Euler wrote seventy volumes of work on mathematics. Is a great mathematician one who is prolific like him, or one who makes only one major breakthrough?

Topic

5 Geometry and trigonometry

Syllabus content

5.1 Coordinates in two dimensions: points; lines; midpoints.

Distance between points.

5.2 Equation of a line in two dimensions: the forms $y = mx + c$ and $ax + by + d = 0$.

Gradient; intercepts.

Points of intersection of lines; parallel lines; perpendicular lines.

5.3 Right-angled trigonometry.

Use of the ratios of sine, cosine and tangent.

5.4 The sine rule: $\frac{a}{\sin A} = \frac{b}{\sin B} = \frac{c}{\sin C}$.

The cosine rule: $a^2 = b^2 + c^2 - 2bc \cos A$;

$\cos A = \frac{b^2 + c^2 - a^2}{2bc}$.

Area of a triangle: $\frac{1}{2} ab \sin C$.

Construction of labelled diagrams from verbal statements.

5.5 Geometry of three-dimensional shapes: cuboid; prism; pyramid; cylinder; sphere; hemisphere; cone.

Lengths of lines joining vertices with vertices, vertices with midpoints and midpoints with midpoints; sizes of angles between two lines and between lines and planes.

Introduction

On 22 October 1707 four English war ships, *The Association* (the flagship of Admiral Sir Clowdisley Shovell) and three others, struck the Gilstone Ledges off the Scilly Isles and more than two thousand men drowned. Why? Because the Admiral had no way of knowing exactly where he was. He needed two coordinates to place his position on the sea. He only had one, his latitude.

The story of how to solve the problem of fixing the second coordinate (longitude) is told in Dava Sobel's book *Longitude*. The British Government offered a prize of £20 000 (millions of pounds at today's prices) to anyone who could solve the problem of how to fix longitude at sea.

Cartesian coordinates, using (x, y) coordinates, are named after Descartes. René Descartes (1596–1650) was a French philosopher and mathematician. He is considered one of the most original thinkers of all time. His greatest work is *The Meditations* (published in 1641) which asked 'How and what do I know?'

His work in mathematics formed a link between algebra and geometry. He believed that mathematics was the supreme science in that the whole phenomenal world could be interpreted in terms of mathematical laws.

In this topic we will look at fixing positions in one and two dimensions, the properties of triangles and the geometrical properties of common three-dimensional shapes.

5.1 Coordinates

To fix a point in two dimensions (2D), its position is given in relation to a point called the **origin**. Through the origin, axes are drawn perpendicular to each other. The horizontal axis is known as the ***x*-axis**, and the vertical axis is known as the ***y*-axis**.

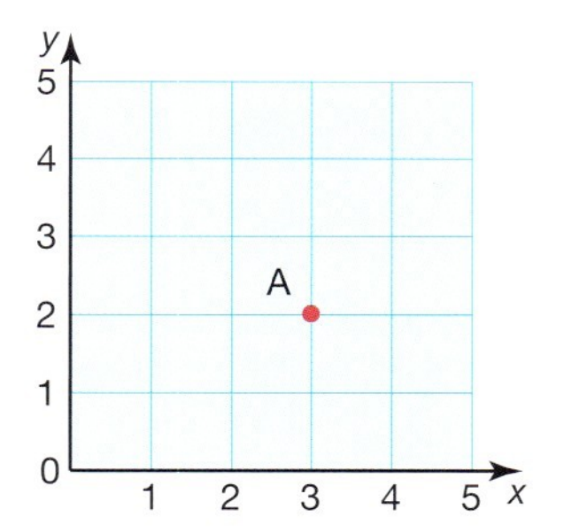

The x-axis is numbered from left to right. The y-axis is numbered from bottom to top.

The position of point A is given by two coordinates: the x-coordinate first, followed by the y-coordinate. So the coordinates of point A are (3, 2).

A number line can extend in both directions by extending the x- and y-axes below zero, as shown in the grid below.

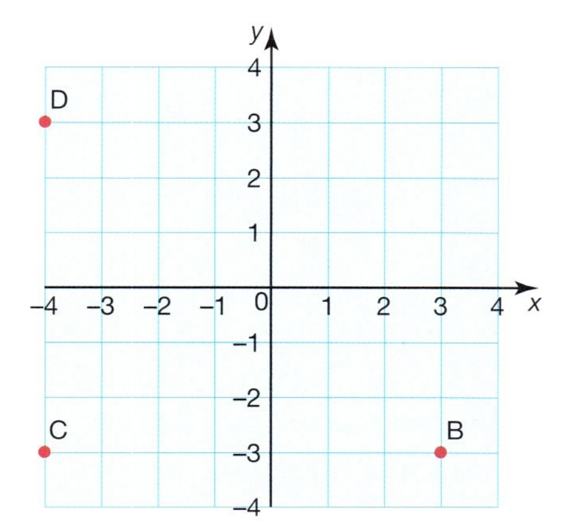

Points B, C, and D can be described by their coordinates:

Point B is at (3, −3)
Point C is at (−4, −3)
Point D is at (−4, 3)

Calculating the distance between two points

To calculate the distance between two points, the coordinates need to be given. Once these are known, Pythagoras' theorem can be used to calculate the distance.

Worked example

The coordinates of two points are (1, 3) and (5, 6). Draw a pair of axes, plot the given points and calculate the distance between them.

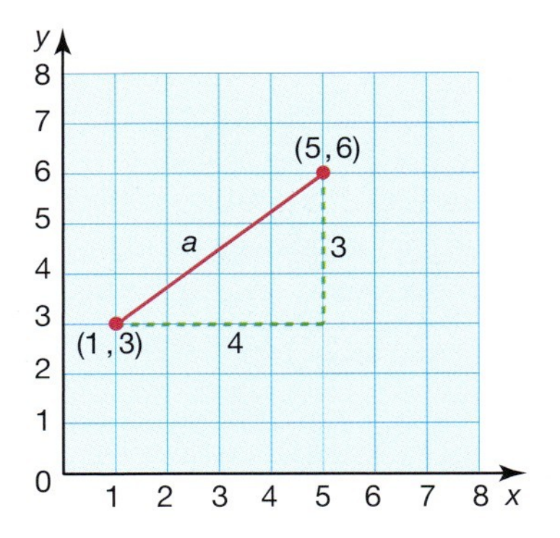

By dropping a perpendicular from the point (5, 6) and drawing a line across from (1, 3), a right-angled triangle is formed. The length of the hypotenuse of the triangle is the length we wish to find.

Using Pythagoras' theorem, we have:

$$a^2 = 3^2 + 4^2$$
$$a^2 = 25$$
$$a = \sqrt{25}$$
$$a = 5$$

The distance between the two points is 5 units.

To find the distance between the points directly from the coordinates, use the formula:

$$d = \sqrt{(x_1 - x_2)^2 + (y_1 - y_2)^2}$$

Worked example

Without plotting the points, calculate the distance between the points (1, 3) and (5, 6).

$$\begin{aligned} d &= \sqrt{(1 - 5)^2 + (3 - 6)^2} \\ &= \sqrt{(-4)^2 + (-3)^2} \\ &= \sqrt{25} \\ &= 5 \end{aligned}$$

The distance between the two points is 5 units.

The midpoint of a line segment

To find the midpoint of a line segment, use the coordinates of its end points. To find the x-coordinate of the midpoint, find the mean of the x-coordinates of the end points. Similarly, to find the y-coordinate of the midpoint, find the mean of the y-coordinates of the end points.

Worked examples

1 Find the coordinates of the midpoint of the line segment AB where A is (1, 3) and B is (5, 6).

The x-coordinate at the midpoint will be $\frac{1 + 5}{2} = 3$

The y-coordinate at the midpoint will be $\frac{3 + 6}{2} = 4.5$

So the coordinates of the midpoint are (3, 4.5).

2 Find the coordinates of the midpoint of a line segment PQ, where P is (−2, −5) and Q is (4, 7).

The x-coordinate at the midpoint will be $\frac{-2 + 4}{2} = 1$

The y-coordinate at the midpoint will be $\frac{-5 + 7}{2} = 1$

So the coordinates of the midpoint are (1, 1).

Exercise 5.1.1

1 **i)** Plot each of the following pairs of points.
ii) Calculate the distance between each pair of points.
iii) Find the coordinates of the midpoint of the line segment.

a (5, 6) (1, 2)
b (6, 4) (3, 1)
c (1, 4) (5, 8)
d (0, 0) (4, 8)
e (2, 1) (4, 7)
f (0, 7) (−3, 1)
g (−3, −3) (−1, 5)
h (4, 2) (−4, −2)
i (−3, 5) (4, 5)
j (2, 0) (2, 6)
k (−4, 3) (4, 5)
l (3, 6) (−3, −3)

2 Without plotting the points:
i) calculate the distance between each of the following pairs of points
ii) find the coordinates of the midpoint of the line segment.

a (1, 4) (4, 1)
b (3, 6) (7, 2)
c (2, 6) (6, −2)
d (1, 2) (9, −2)
e (0, 3) (−3, 6)
f (−3, −5) (−5, −1)
g (−2, 6) (2, 0)
h (2, −3) (8, 1)
i (6, 1) (−6, 4)
j (−2, 2) (4, −4)
k (−5, −3) (6, −3)
l (3, 6) (5, −2)

5.2 Straight lines

Lines are made of an infinite number of points. This section deals with those points which form a straight line. Each point on a straight line, if plotted on a pair of axes, will have particular coordinates. The relationship between the coordinates of the points on a straight line indicates the equation of that straight line.

Gradient

The **gradient** of a straight line refers to its 'steepness' or 'slope'. The gradient of a straight line is constant, i.e. it does not change. The gradient can be calculated by considering the coordinates of any two points (x_1, y_1), (x_2, y_2) on the line. It is calculated using the following formula:

$$\text{Gradient} = \frac{\text{vertical distance between the two points}}{\text{horizontal distance between the two points}}$$

By considering the x- and y-coordinates of the two points this can be rewritten as:

$$\text{Gradient} = \frac{y_2 - y_1}{x_2 - x_1}$$

Worked examples

1 The coordinates of two points on a straight line are (1, 3) and (5, 7). Plot the two points on a pair of axes and calculate the gradient of the line joining them.

$$\text{Gradient} = \frac{7-3}{5-1} = \frac{4}{4} = 1$$

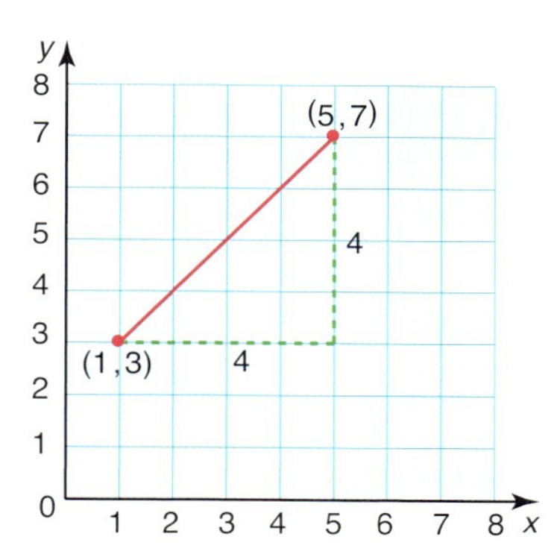

Note: It does not matter which point we choose to be (x_1, y_1) or (x_2, y_2) as the gradient will be the same. In the example above, reversing the points:

$$\text{Gradient} = \frac{3-7}{1-5} = \frac{-4}{-4} = 1$$

2 The coordinates of two points on a straight line are (2, 6) and (4, 2). Plot the two points on a pair of axes and calculate the gradient of the line joining them.

$$\text{Gradient} = \frac{2-6}{4-2} = \frac{-4}{2} = -2$$

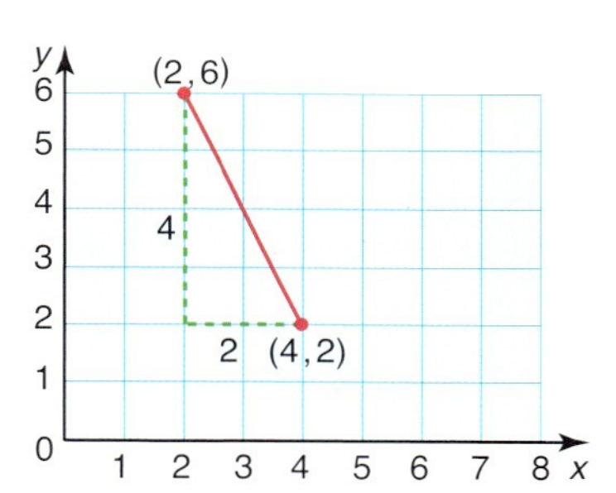

To check whether or not the sign of the gradient is correct, the following guideline is useful:

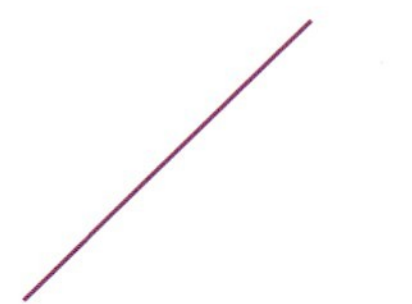

A line sloping this way has a positive gradient

A line sloping this way has a negative gradient

Parallel lines have the same gradient. Conversely, lines which have the same gradient are parallel. If two lines are parallel to each other, their gradients m_1 and m_2 are equal: i.e. $m_1 = m_2$.

L_1, L_2, L_3 all have the same gradient so are parallel.

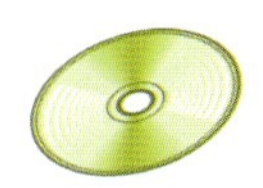

The x-axis and the y-axis on a graph intersect at right angles. They are perpendicular to each other. In the graph below, L_1 and L_2 are perpendicular to each other.

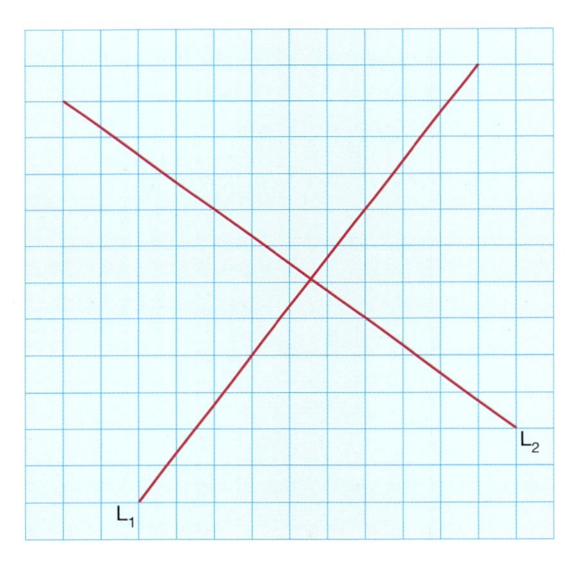

The gradient m_1 of line L_1 is $\frac{4}{3}$ and the gradient m_2 of line L_2 is $-\frac{3}{4}$.

The product of m_1m_2 gives the result -1, i.e. $\frac{4}{3} \times \left(-\frac{3}{4}\right) = -1$.

If two lines are perpendicular to each other, the product of their gradients is -1, i.e. $m_1m_2 = -1$.

Therefore the gradient of one line is the negative reciprocal of the other line, i.e.: $m_1 = \dfrac{-1}{m_2}$.

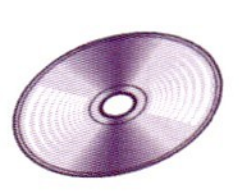

Exercise 5.2.1

1 With the aid of axes if necessary, calculate:
 i) the gradient of the line joining the following pairs of points
 ii) the gradient of a line perpendicular to this line.

a (5, 6) (1, 2) b (6, 4) (3, 1)
c (1, 4) (5, 8) d (0, 0) (4, 8)
e (2, 1) (4, 7) f (0, 7) (−3, 1)
g (−3, −3) (−1, 5) h (4, 2) (−4, −2)
i (−3, 5) (4, 5) j (2, 0) (2, 6)
k (−4, 3) (4, 5) l (3, 6) (−3, −3)

2 With the aid of axes if necessary, calculate:
 i) the gradient of the line joining the following pairs of points
 ii) the gradient of a line perpendicular to this line.

a (1, 4) (4, 1) b (3, 6) (7, 2)
c (2, 6) (6, −2) d (1, 2) (9, −2)
e (0, 3) (−3, 6) f (−3, −5) (−5, −1)
g (−2, 6) (2, 0) h (2, −3) (8, 1)
i (6, 1) (−6, 4) j (−2, 2) (4, −4)
k (−5, −3) (6, −3) l (3, 6) (5, −2)

Equation of a straight line

The coordinates of every point on a straight line all have a common relationship. This relationship, when expressed algebraically as an equation in terms of x and/or y, is known as the equation of the straight line.

Worked examples

1 By looking at the coordinates of some of the points on the line below, establish the equation of the straight line.

x	y
1	4
2	4
3	4
4	4
5	4
6	4

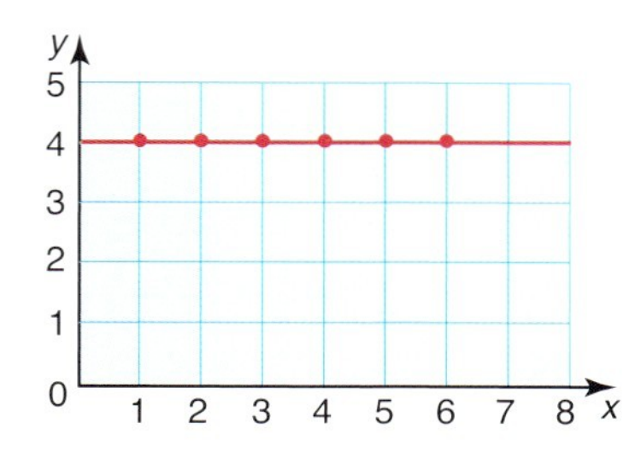

Some of the points on the line have been identified and their coordinates entered in the table. By looking at the table it can be seen that the only rule that all the points have in common is that $y = 4$.

Hence the equation of the straight line is $y = 4$.

2 By looking at the coordinates of some of the points on the line below, establish the equation of the straight line.

x	y
1	2
2	4
3	6
4	8

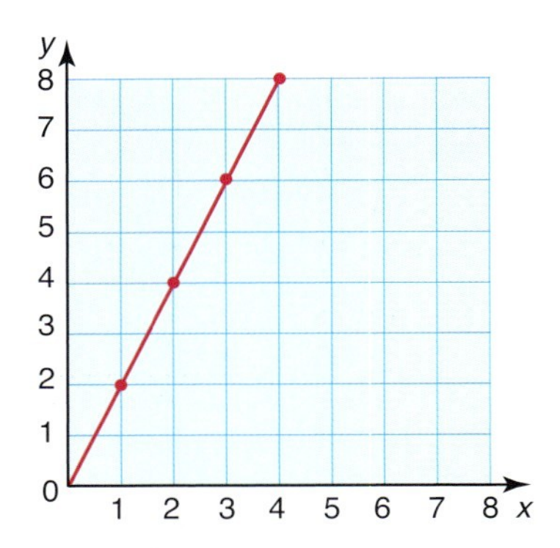

Once again, by looking at the table it can be seen that the relationship between the x- and y-coordinates is that each y-coordinate is twice the corresponding x-coordinate.

Hence the equation of the straight line is $y = 2x$.

Exercise 5.2.2

1 For each of the following, identify the coordinates of some of the points on the line and use these to find the equation of the straight line.

a 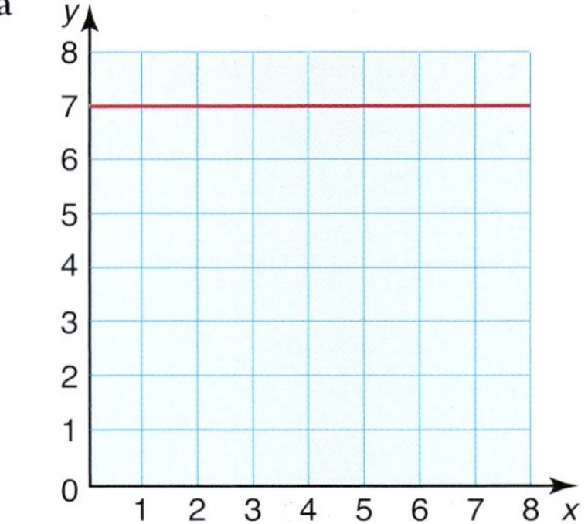

b 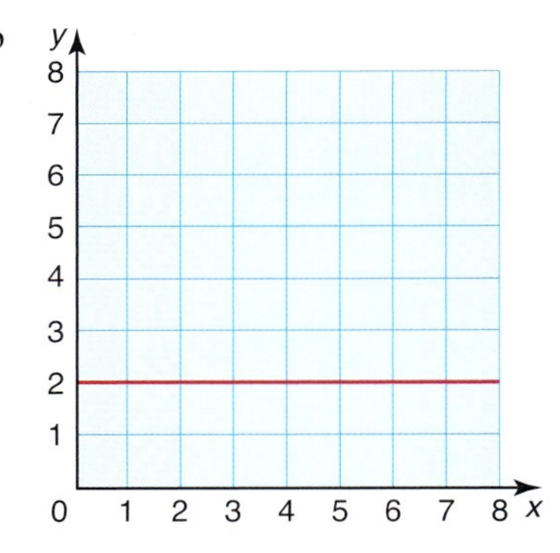

c 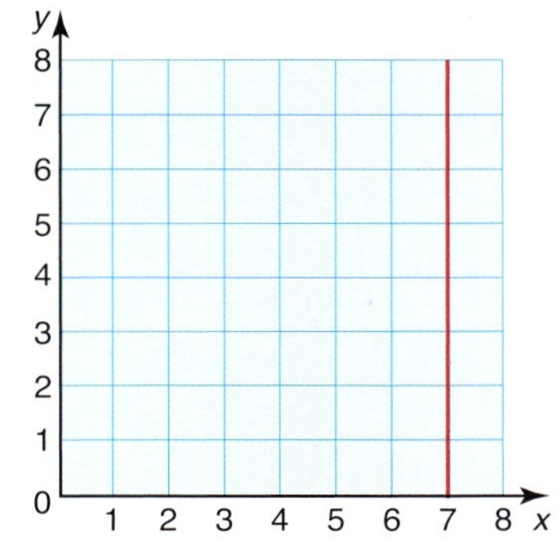

d 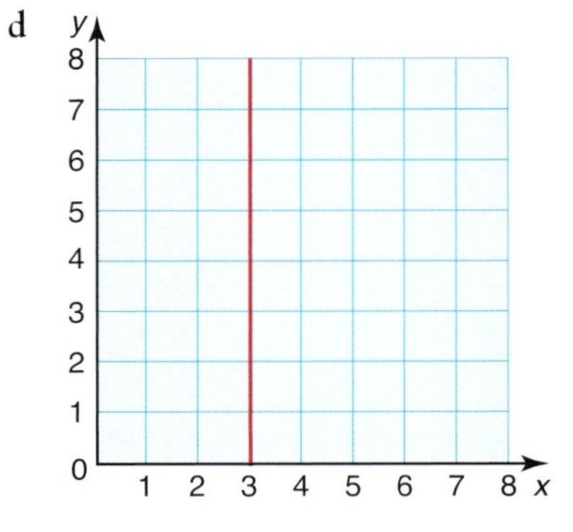

e

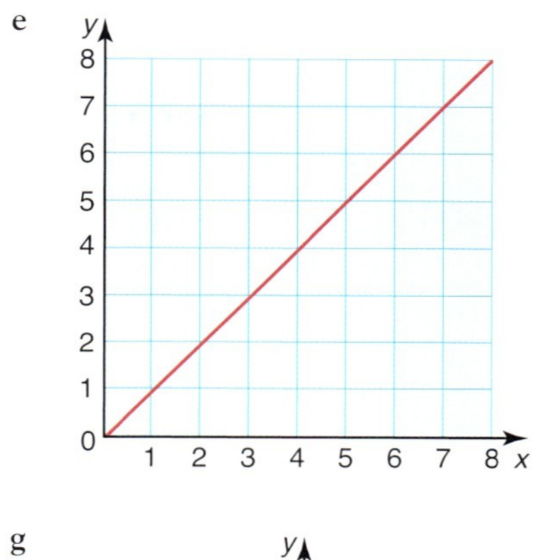

f

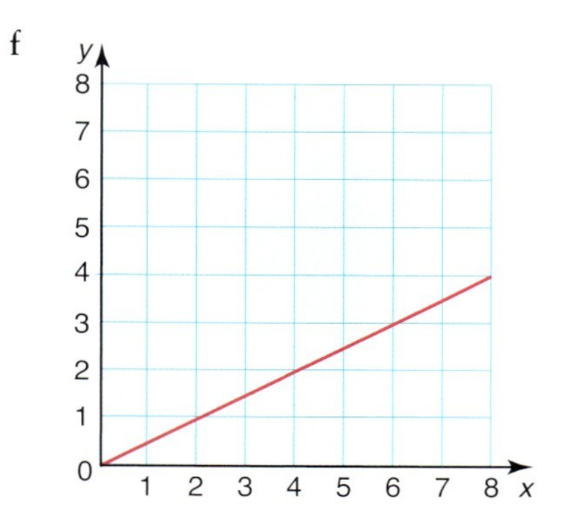

g

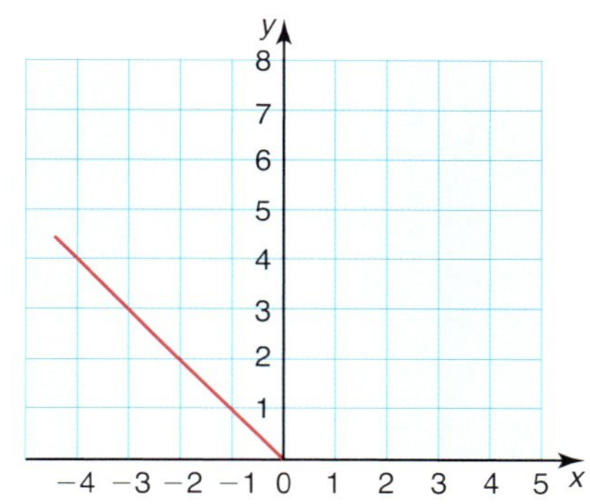

h

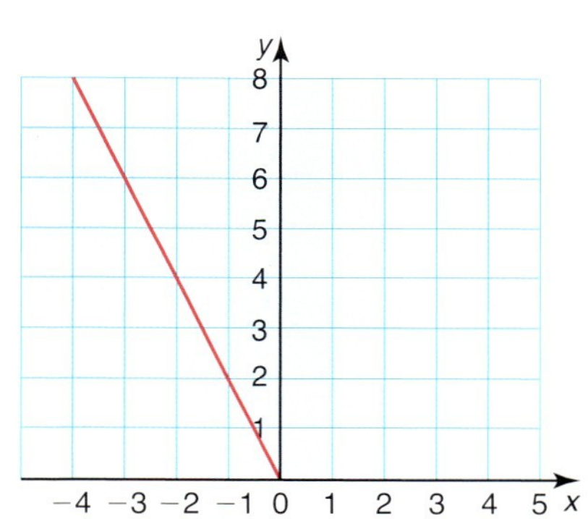

2 For each of the following, identify the coordinates of some of the points on the line and use these to find the equation of the straight line.

a

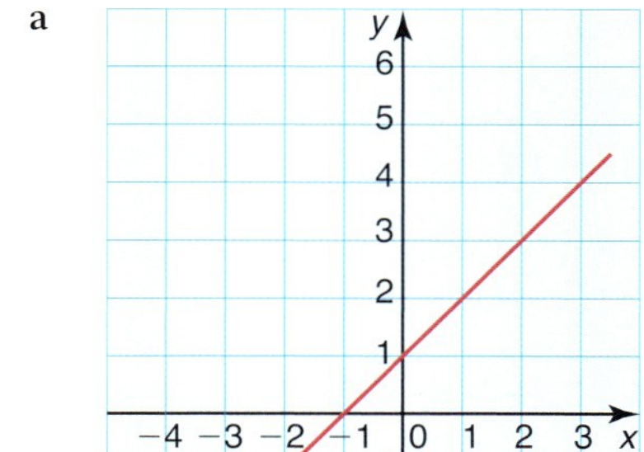

b

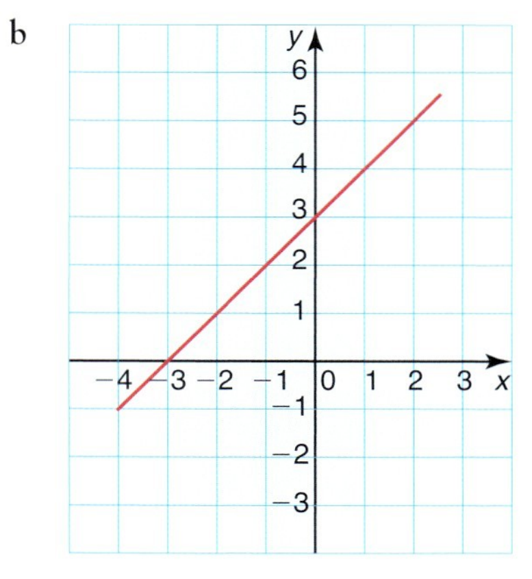

c

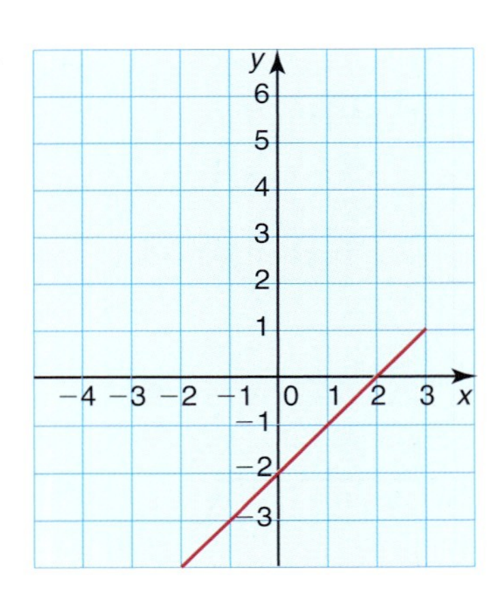

d

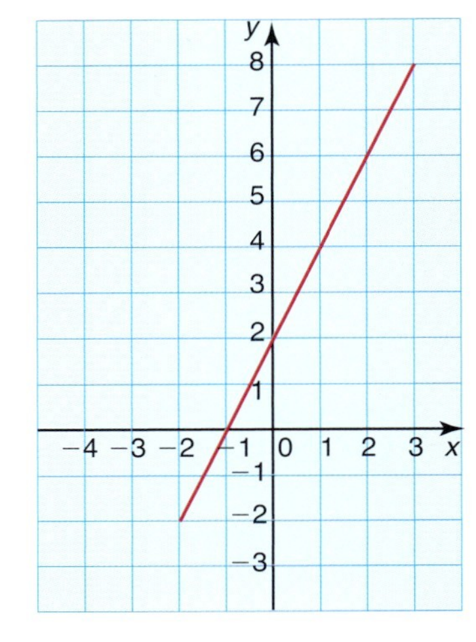

e

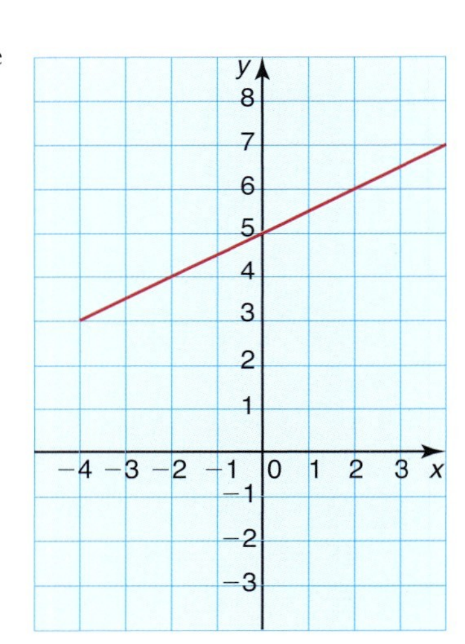

f

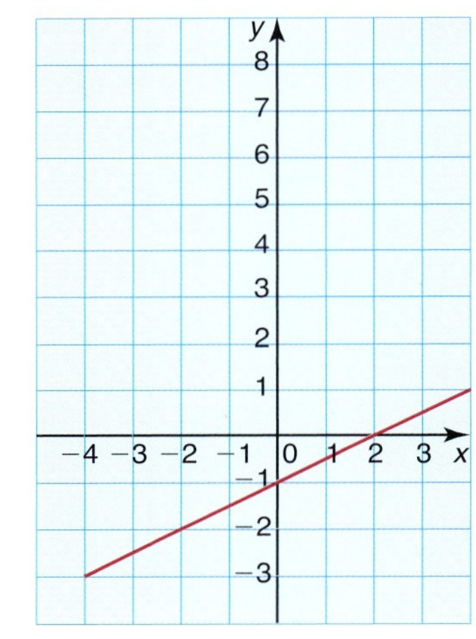

3 For each of the following, identify the coordinates of some of the points on the line and use these to find the equation of the straight line.

a

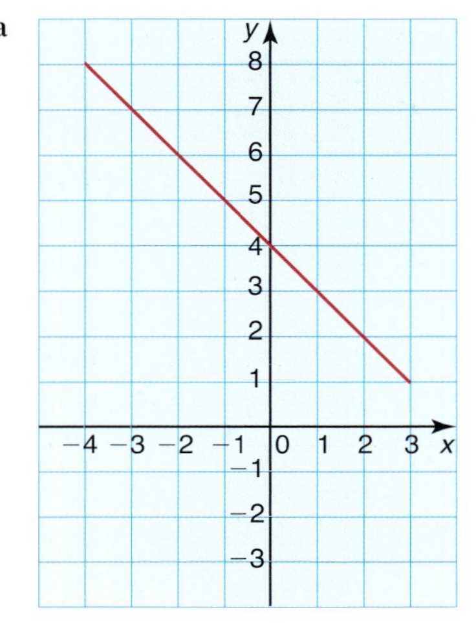

b

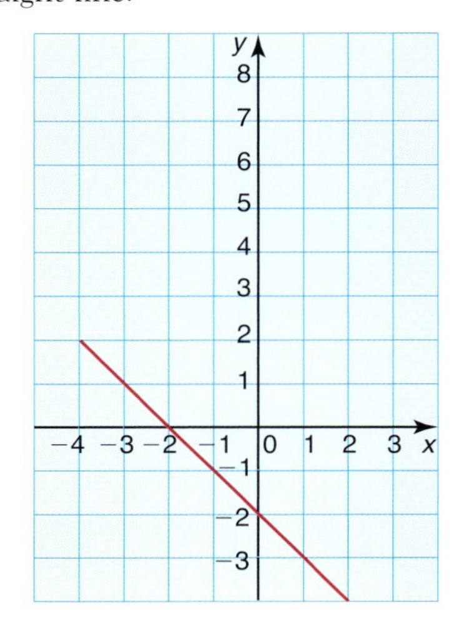

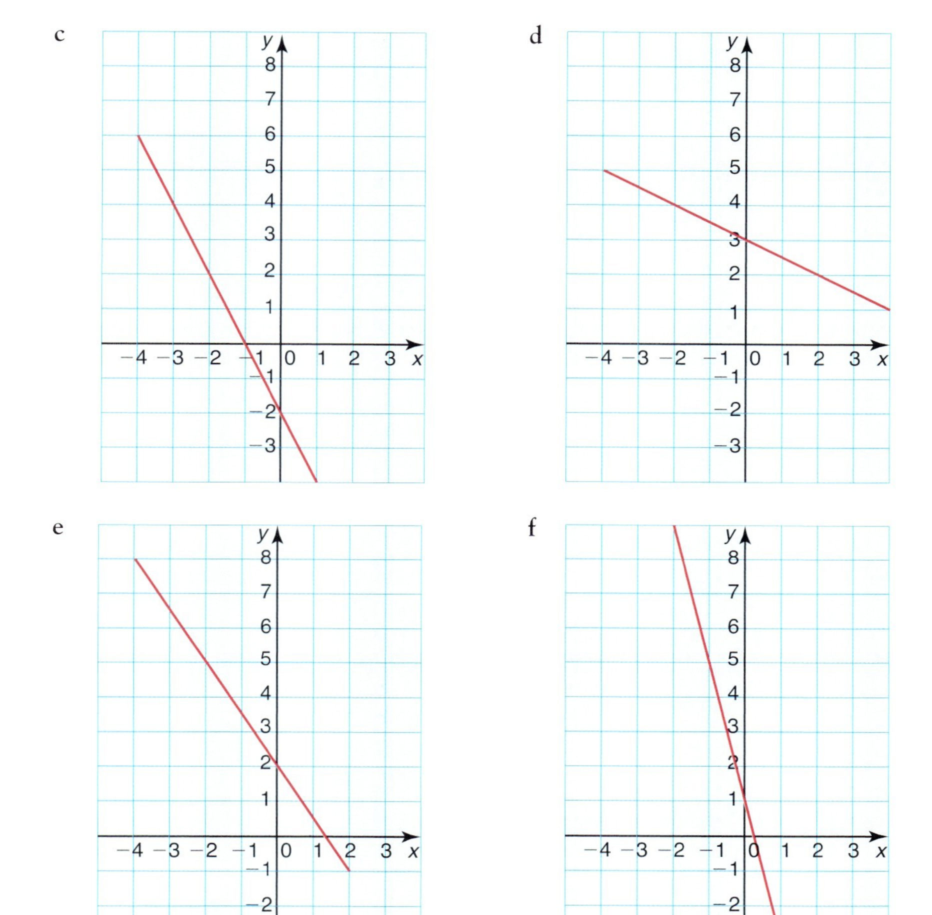

4 a For each of the graphs in questions 2 and 3, calculate the gradient of the straight line.

b What do you notice about the gradient of each line and its equation?

c What do you notice about the equation of the straight line and where the line intersects the y-axis?

5 Copy the diagrams in question 2. Draw two lines on the diagram parallel to the given line.

a Write the equation of these new lines in the form $y = mx + c$.

b What do you notice about the equations of these new parallel lines?

6 In question 3 you identified the equation of the lines shown in the form $y = mx + c$. Change the value of the intercept c and then draw the new line. What do you notice about this new line and the first line?

In general the equation of any straight line can be written in the form:

$y = mx + c$

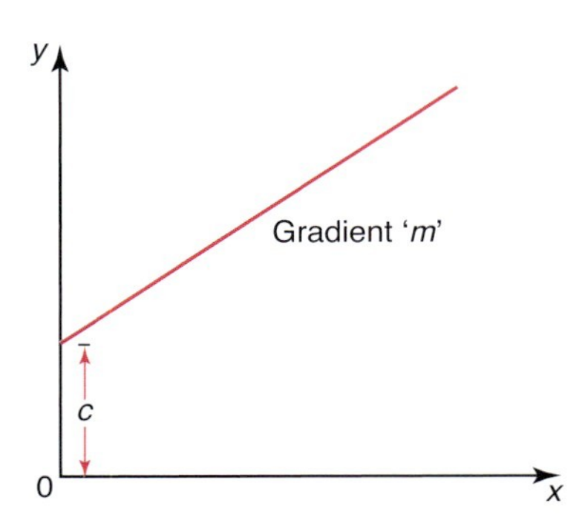

where m represents the gradient of the straight line and c the intercept with the y-axis. This is shown in the diagram.

By looking at the equation of a straight line written in the form $y = mx + c$, it is therefore possible to deduce the line's gradient and intercept with the y-axis without having to draw it.

Worked examples

1 Find the gradient and y-intercept of these straight lines.

a $y = 3x - 2$ **b** $y = -2x + 6$

a Gradient $= 3$
y-intercept $= -2$

b Gradient $= -2$
y-intercept $= 6$

2 Find the gradient and y-intercept of these straight lines.

a $2y = 4x + 2$ **b** $y - 2x = -4$

c $-4y + 2x = 4$ **d** $\dfrac{y + 3}{4} = -x + 2$

a $2y = 4x + 2$ needs to be rearranged into gradient–intercept form (i.e. $y = mx + c$):

$y = 2x + 1$

Gradient $= 2$
y-intercept $= 1$

b Rearranging $y - 2x = -4$ into gradient–intercept form, gives:

$y = 2x - 4$

Gradient $= 2$
y-intercept $= -4$

c Rearranging $-4y + 2x = 4$ into gradient–intercept form, gives:

$y = \frac{1}{2}x - 1$

Gradient $= \frac{1}{2}$
y-intercept $= -1$

d Rearranging $\dfrac{y + 3}{4} = -x + 2$ into gradient–intercept form, gives:

$y + 3 = -4x + 8$
$y = -4x + 5$

Gradient $= -4$
y-intercept $= 5$

Exercise 5.2.3

For the following linear equations, find both the gradient and y-intercept in each case.

1 a $y = 2x + 1$ b $y = 3x + 5$ c $y = x - 2$
d $y = \frac{1}{2}x + 4$ e $y = -3x + 6$ f $y = -\frac{2}{3}x + 1$
g $y = -x$ h $y = -x - 2$ i $y = -(2x - 2)$

2 a $y - 3x = 1$ b $y + \frac{1}{2}x - 2 = 0$ c $y + 3 = -2x$
d $y + 2x + 4 = 0$ e $y - \frac{1}{4}x - 6 = 0$ f $-3x + y = 2$
g $2 + y = x$ h $8x - 6 + y = 0$ i $-(3x + 1) + y = 0$

3 a $2y = 4x - 6$ b $2y = x + 8$ c $\frac{1}{2}y = x - 2$
d $\frac{1}{4}y = -2x + 3$ e $3y - 6x = 0$ f $\frac{1}{3}y + x = 1$
g $6y - 6 = 12x$ h $4y - 8 + 2x = 0$ i $2y - (4x - 1) = 0$

4 a $2x - y = 4$ b $x - y + 6 = 0$ c $-2y = 6x + 2$
d $12 - 3y = 3x$ e $5x - \frac{1}{2}y = 1$ f $-\frac{2}{3}y + 1 = 2x$
g $9x - 2 = -y$ h $-3x + 7 = -\frac{1}{2}y$ i $-(4x - 3) = -2y$

5 a $\dfrac{y + 2}{4} = \dfrac{1}{2}x$ b $\dfrac{y - 3}{x} = 2$ c $\dfrac{y - x}{8} = 0$
d $\dfrac{2y - 3x}{2} = 6$ e $\dfrac{3y - 2}{x} = -3$ f $\dfrac{\frac{1}{2}y - 1}{x} = -2$
g $\dfrac{3x - y}{2} = 6$ h $\dfrac{6 - 2y}{3} = 2$ i $\dfrac{-(x + 2y)}{5x} = 1$

6 a $\dfrac{3x - y}{y} = 2$ b $\dfrac{-x + 2y}{4} = y + 1$
c $\dfrac{y - x}{x + y} = 2$ d $\dfrac{1}{y} = \dfrac{1}{x}$
e $\dfrac{-(6x + y)}{2} = y + 1$ f $\dfrac{2x - 3y + 4}{4} = 4$
g $\dfrac{y + 1}{x} + \dfrac{3y - 2}{2x} = -1$ h $\dfrac{x}{y + 1} + \dfrac{1}{2y + 2} = -3$
i $\dfrac{-(-y + 3x)}{-(6x - 2y)} = 1$ j $\dfrac{-(x - 3y) - (-x - 2y)}{4 + x - y} = -2$

The equation of a line through two points

The equation of a straight line can be deduced once the coordinates of two points on the line are known.

Worked examples

Calculate the equation of the straight line passing through the points (−3, 3) and (5, 5).

Plotting the two points gives:

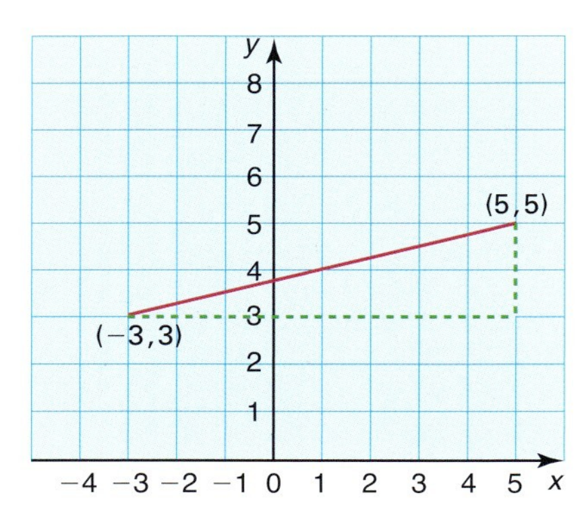

The equation of any straight line can be written in the general form $y = mx + c$. Here we have:

$$\text{Gradient} = \frac{5-3}{5-(-3)} = \frac{2}{8} = \frac{1}{4}$$

The equation of the line now takes the form $y = \frac{1}{4}x + c$.

Since the line passes through the two given points, their coordinates must satisfy the equation. So to calculate the value of c, the x- and y-coordinates of one of the points are substituted into the equation.

Substituting (5, 5) into the equation gives:

$$5 = \tfrac{1}{4} \times 5 + c$$

$$5 = 1\tfrac{1}{4} + c$$

Therefore $c = 5 - 1\frac{1}{4} = 3\frac{3}{4}$.

The equation of the straight line passing through (−3, 3) and (5, 5) is:

$y = \frac{1}{4}x + 3\frac{3}{4}$.

We have seen that the equation of a straight line takes the form $y = mx + c$. It can, however, also take the form $ax + by + d = 0$. It is possible to write the equation $y = mx + c$ in the form $ax + by + d = 0$ by rearranging the equation.

In the example above, $y = \frac{1}{4}x + 3\frac{3}{4}$ can be rewritten as $y = \frac{x}{4} + \frac{15}{4}$.

Multiplying both sides of the equation by 4 produces the equation $4y = x + 15$. This can be rearranged to $-x + 4y - 15 = 0$, which is the required form with $a = -1$, $b = 4$ and $c = -15$.

Exercise 5.2.4

Find the equation of the straight line that passes through each of the following pairs of points. Express your answers in the form:

i) $y = mx + c$

ii) $ax + by + d = 0$.

1 **a** (1, 1) (4, 7) **b** (1, 4) (3, 10) **c** (1, 5) (2, 7)
d (0, −4) (3, −1) **e** (1, 6) (2, 10) **f** (0, 4) (1, 3)
g (3, −4) (10, −18) **h** (0, −1) (1, −4) **i** (0, 0) (10, 5)

2 **a** (−5, 3) (2, 4) **b** (−3, −2) (4, 4) **c** (−7, −3) (−1, 6)
d (2, 5) (1, −4) **e** (−3, 4) (5, 0) **f** (6, 4) (−7, 7)
g (−5, 2) (6, 2) **h** (1, −3) (−2, 6) **i** (6, −4) (6, 6)

Drawing straight-line graphs

To draw a straight-line graph only two points need to be known. Once these have been plotted the line can be drawn between them and extended if necessary at both ends. It is important to check your line is correct by taking a point from your graph and ensuring it satisfies the original equation.

Worked examples

1 Plot the line $y = x + 3$.

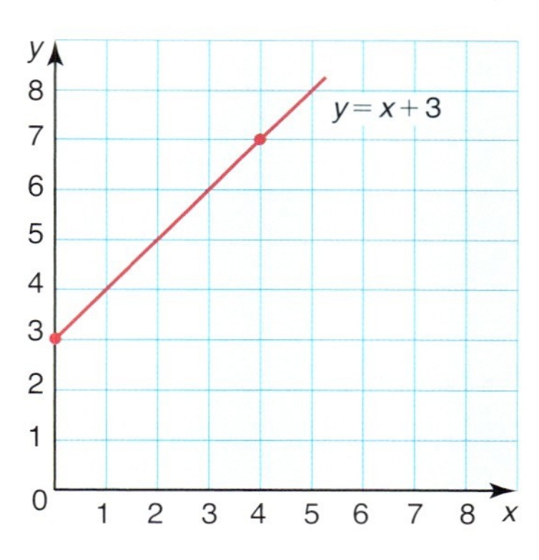

To identify two points, simply choose two values of x. Substitute these into the equation and calculate their corresponding y-values.

When $x = 0$, $y = 3$.
When $x = 4$, $y = 7$.

Therefore two of the points on the line are (0, 3) and (4, 7).
The straight line $y = x + 3$ is plotted as shown.
Check using the point (1, 4).

When $x = 1$, $y = x + 3 = 4$, so (1, 4) satisfies the equation of the required line.

2 Plot the line $y = -2x + 4$.

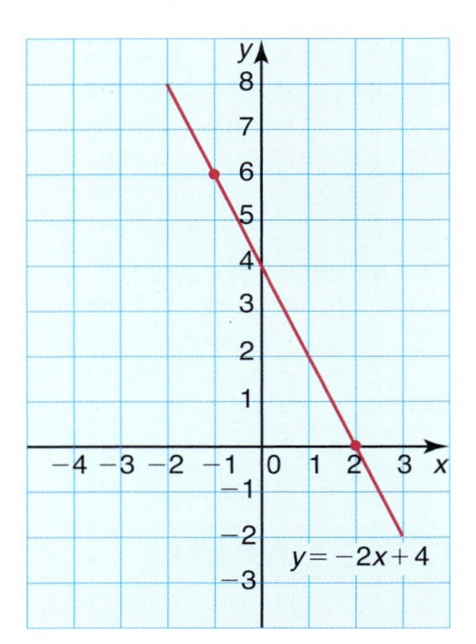

When $x = 2$, $y = 0$.
When $x = -1$, $y = 6$.

The coordinates of two points on the line are $(2, 0)$ and $(-1, 6)$, and the line is plotted as shown.

Check using the point $(0, 4)$.

When $x = 0$, $y = -2x + 4 = 4$, so $(0, 4)$ satisfies the equation of the required line.

Note that, in questions of this sort, it is often easier to rearrange the equation into gradient–intercept form first.

Exercise 5.2.5

1 Plot the following straight lines.

a $y = 2x + 3$
b $y = x - 4$
c $y = 3x - 2$
d $y = -2x$
e $y = -x - 1$
f $-y = \frac{1}{2}x + 4$
g $-y = 3x - 3$
h $2y = 4x - 2$
i $y - 4 = 3x$

2 Plot the following straight lines.

a $-2x + y = 4$
b $-4x + 2y = 12$
c $3y = 6x - 3$
d $2x = x + 1$
e $3y - 6x = 9$
f $2y + x = 8$
g $x + y + 2 = 0$
h $3x + 2y - 4 = 0$
i $4 = 4y - 2x$

3 Plot the following straight lines.

a $\frac{x + y}{2} = 1$
b $x + \frac{y}{2} = 1$
c $\frac{x}{3} + \frac{y}{2} = 1$
d $y + \frac{x}{2} = 3$
e $\frac{y}{5} + \frac{x}{3} = 0$
f $\frac{-(2x + y)}{4} = 1$
g $\frac{y - (x - y)}{3x} = -1$
h $\frac{y}{2x + 3} - \frac{1}{2} = 0$
i $-2(x + y) + 4 = -y$

Solving simultaneous equations

The process of solving two equations and finding a common solution is known as solving equations simultaneously. Simultaneous equations can be solved algebraically by elimination or by substitution.

By elimination

The aim of this method is to eliminate one of the unknowns by either adding or subtracting the two equations.

Worked example

Solve the following simultaneous equations by finding the values of x and y which satisfy both equations.

a $3x + y = 9$
$5x - y = 7$

b $4x + y = 23$
$x + y = 8$

a $3x + y = 9$ (1)
$5x - y = 7$ (2)

By adding equations (1) + (2) the variable y is eliminated:

$8x = 16$
$x = 2$

To find the value of y, substitute $x = 2$ into either equation (1) or (2).
Substituting $x = 2$ into equation (1):

$3x + y = 9$
$6 + y = 9$
$y = 3$

To check that the solution is correct, substitute the values of x and y into equation (2). If it is correct then the left-hand side of the equation will equal the right-hand side.

$5x - y = 7$
$10 - 3 = 7$
$7 = 7$

b $4x + y = 23$ (1)
$x + y = 8$ (2)

By subtracting the equations i.e. (1) − (2), the variable y is eliminated:

$3x = 15$
$x = 5$

By substituting $x = 5$ into equation (2), y can be calculated:

$x + y = 8$
$5 + y = 8$
$y = 3$

Check by substituting both values into equation (1):

$$4x + y = 23$$
$$20 + 3 = 23$$
$$23 = 23$$

By substitution

The same equations can also be solved by the method known as substitution.

Worked example

Solve these simultaneous equations by finding the values of x and y which satisfy both equations.

a $3x + y = 9$
$5x - y = 7$

b $4x + y = 23$
$x + y = 8$

a $3x + y = 9$ (1)
$5x - y = 7$ (2)

Equation (2) can be rearranged to give: $y = 5x - 7$ This can now be substituted into equation (1):

$$3x + (5x - 7) = 9$$
$$3x + 5x - 7 = 9$$
$$8x - 7 = 9$$
$$8x = 16$$
$$x = 2$$

To find the value of y, $x = 2$ is substituted into either equation (1) or (2) as before giving $y = 3$.

b $4x + y = 23$ (1)
$x + y = 8$ (2)

Equation (2) can be rearranged to give $y = 8 - x$. This can be substituted into equation (1):

$$4x + (8 - x) = 23$$
$$4x + 8 - x = 23$$
$$3x + 8 = 23$$
$$3x = 15$$
$$x = 5$$

y can be found as before, giving a result of $y = 3$.

Graphically

Simultaneous equations can also be solved graphically by plotting the lines on the same pair of axes and finding the point of intersection. The Casio GDC can solve simultaneous equations algebraically. This gives the point of intersection if the graphs do not need to be drawn. Currently the Texas GDC cannot solve simultaneous equations algebraically. However, both GDCs can solve them graphically.

Worked example

Using a GDC or graphing software, find the point of intersection of the lines given by the equations below, either algebraically or a graphical method.

$3x + y = 9$ and $5x - y = 7$

Method 1: Solving algebraically using the equation solve facility

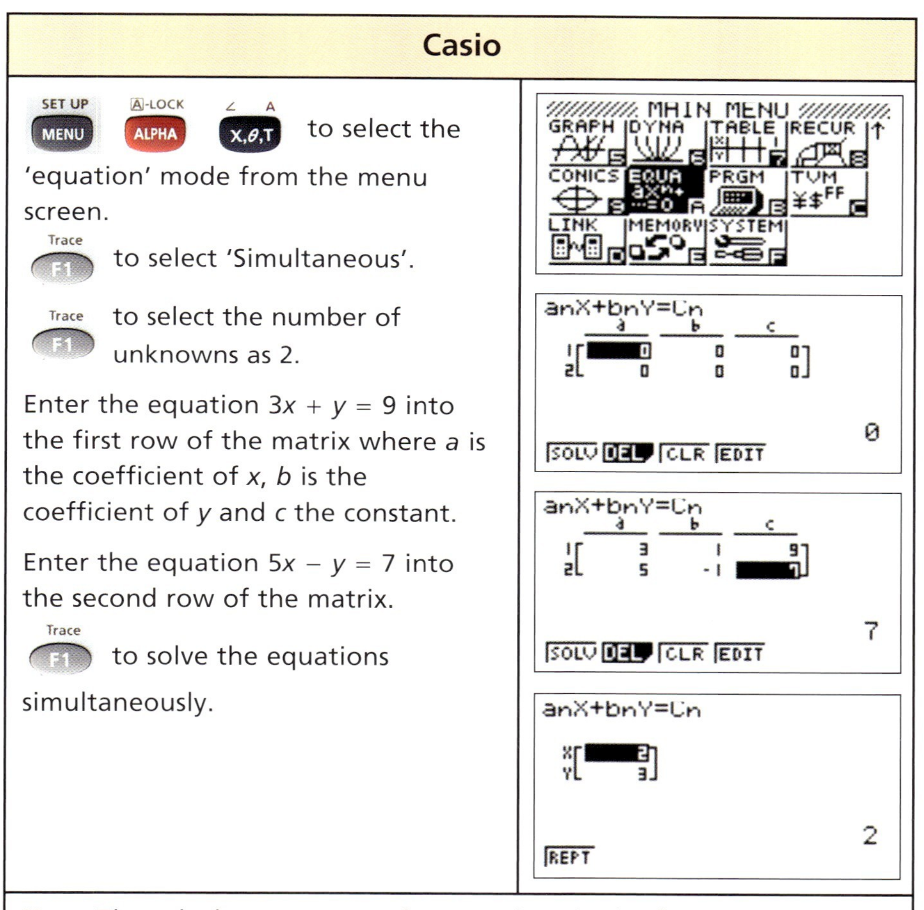

Note: The calculator requests the equations in the form $ax + by = c$ not $ax + by + d = 0$.

The point of intersection of the lines with equations $3x + y = 9$ and $5x - y = 7$ is $(2, 3)$.

Method 2: Solving graphically

Your GDC and graphing software can solve simultaneous equations graphically. Firstly, rearrange each equation into gradient–intercept form, i.e. $y = mx + c$:

$3x + y = 9 \rightarrow y = -3x + 9$
$5x - y = 7 \rightarrow y = 5x - 7$

Then use the GDC or graphing software to plot both lines and find the point of intersection.

Casio

  to select the graph mode.

Enter the equations $y = -3x + 9$ and $y = 5x - 7$

 to graph both equations.

   to select 'intersect' from the 'graph solve' menu.

The results are displayed on the screen.

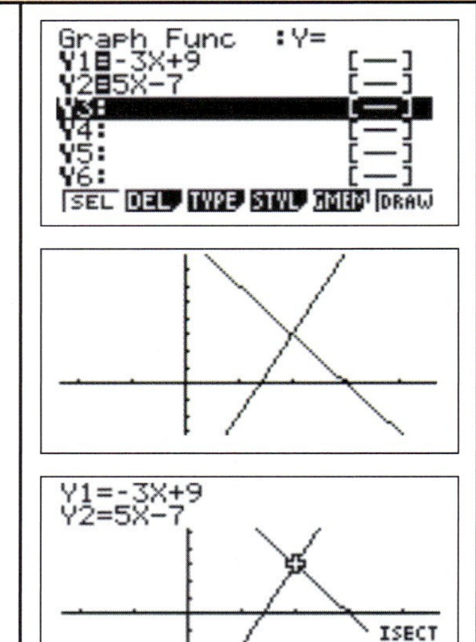

Texas

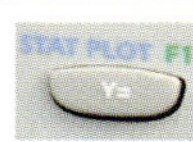 to enter the equations $y = -3x + 9$ and $y = 5x - 7$

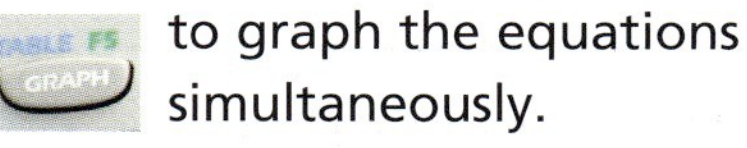 to graph the equations simultaneously.

   to calculate the coordinates of the point of intersection.

Using the cursor select the first 'curve', then when prompted select the second 'curve'. Finally move the cursor over the point of intersection.

 The calculator will give the coordinates of the point of intersection.

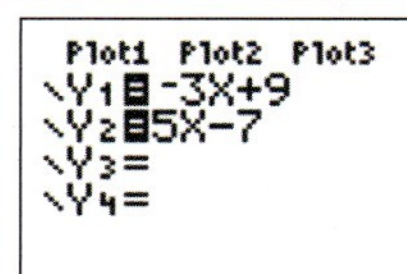

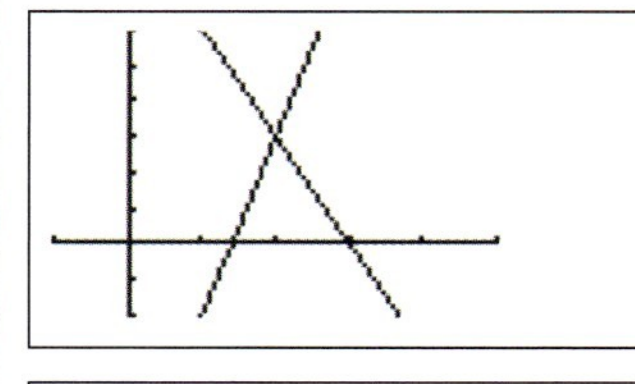

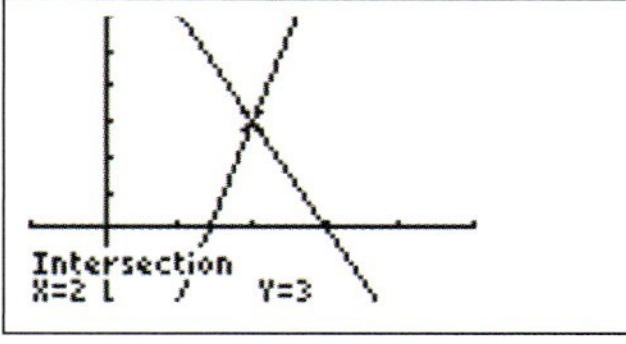

Autograph

Select and enter the equations $3x + y = 9$ and $5x - y = 7$.

Select both graphs and click on 'Object' followed by 'Solve $f(x) = g(x)$'.

The results are displayed in the results box by selecting .

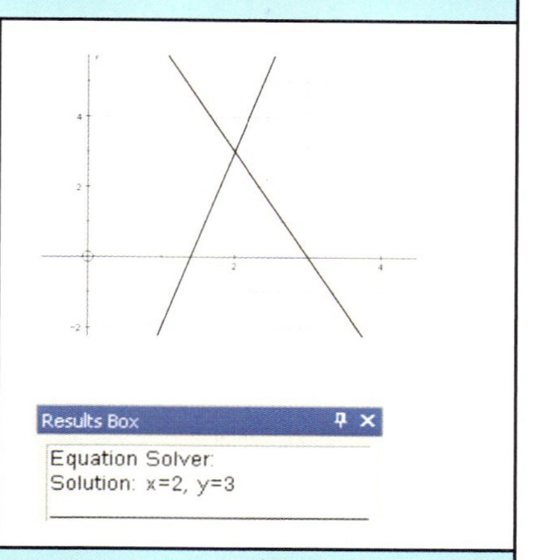

GeoGebra

Enter the equations $3x + y = 9$ and $5x - y = 7$ in turn into the input box.

Type 'Intersect [a, b]' in the input box.

The point of intersection is marked on the graph and its coordinates appear in the algebra window.

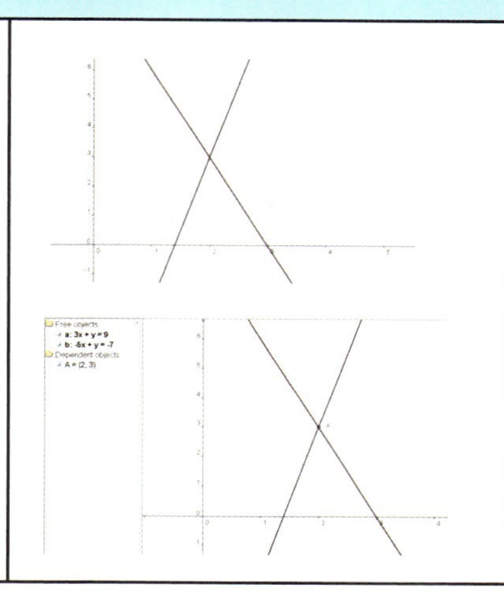

Note: The letters 'a' and 'b' are in the command 'Intersect [a, b]' as Geogebra automatically assigns the lines the labels 'a' and 'b'.

The point of intersection of the lines with equations $3x + y = 9$ and $5x - y = 7$ is $(2, 3)$.

Exercise 5.2.6

Solve these simultaneous equations either algebraically or graphically.

1 a $x + y = 6$
$x - y = 2$

b $x + y = 11$
$x - y - 1 = 0$

c $x + y = 5$
$x - y = 7$

d $2x + y = 12$
$2x - y = 8$

e $3x + y = 17$
$3x - y = 13$

f $5x + y = 29$
$5x - y = 11$

2 a $3x + 2y = 13$
$4x = 2y + 8$

b $6x + 5y = 62$
$4x - 5y = 8$

c $x + 2y = 3$
$8x - 2y = 6$

d $9x + 3y = 24$
$x - 3y = -14$

e $7x - y = -3$
$4x + y = 14$

f $3x = 5y + 14$
$6x + 5y = 58$

3 a $2x + y = 14$
$x + y = 9$

b $5x + 3y = 29$
$x + 3y = 13$

c $4x + 2y = 50$
$x + 2y = 20$

d $x + y = 10$
$3x = -y + 22$

e $2x + 5y = 28$
$4x + 5y = 36$

f $x + 6y = -2$
$3x + 6y = 18$

4 a $x - y = 1$
$2x - y = 6$

b $3x - 2y = 8$
$2x - 2y = 4$

c $7x - 3y = 26$
$2x - 3y = 1$

d $x = y + 7$
$3x - y = 17$

e $8x - 2y = -2$
$3x - 2y = -7$

f $4x - y = -9$
$7x - y = -18$

5 a $x + y = -7$
$x - y = -3$

b $2x + 3y = -18$
$2x = 3y + 6$

c $5x - 3y = 9$
$2x + 3y = 19$

d $7x + 4y = 42$
$9x - 4y = -10$

e $4x - 4y = 0$
$8x + 4y = 12$

f $x - 3y = -25$
$5x - 3y = -17$

6 a $2x + 3y = 13$
$2x - 4y + 8 = 0$

b $2x + 4y = 50$
$2x + y = 20$

c $x + y = 10$
$3y = 22 - x$

d $5x + 2y = 28$
$5x + 4y = 36$

7 a $-4x = 4y$
$4x - 8y = 12$

b $3x = 19 + 2y$
$-3x + 5y = 5$

c $3x + 2y = 12$
$-3x + 9y = -12$

d $3x + 5y = 29$
$3x + y = 13$

e $-5x + 3y = 14$
$5x + 6y = 58$

f $-2x + 8y = 6$
$2x = 3 - y$

Further simultaneous equations

If neither x nor y can be eliminated by simply adding or subtracting the two equations then it is necessary to multiply one or both of the equations. The equations are multiplied by a number in order to make the coefficients of x (or y) numerically equal.

Worked example

Solve these simultaneous equations

a $3x + 2y = 22$
$x + y = 9$

b $5x - 3y = 1$
$3x + 4y = 18$

a $3x + 2y = 22$ (1)
$x + y = 9$ (2)

To eliminate y, equation (2) is multiplied by 2:

$3x + 2y = 22$ (1)
$2x + 2y = 18$ (3)

By subtracting (3) from (1), the variable y is eliminated:

$x = 4$

Substituting $x = 4$ into equation (2), we have:

$x + y = 9$
$4 + y = 9$
$y = 5$

Check by substituting both values into equation (1):

$3x + 2y = 22$
$12 + 10 = 22$
$22 = 22$

b $5x - 3y = 1$ (1)
$3x + 4y = 18$ (2)

To eliminate the variable y, equation (1) is multiplied by 4 and equation (2) is multiplied by 3.

$20x - 12y = 4$ (3)
$9x + 12y = 54$ (4)

By adding equations (3) and (4) the variable y is eliminated:

$29x = 58$
$x = 2$

Substituting $x = 2$ into equation (2) gives:

$3x + 4y = 18$
$6 + 4y = 18$
$4y = 12$
$y = 3$

Check by substituting both values into equation (1):

$5x - 3y = 1$
$10 - 9 = 1$
$1 = 1$

Exercise 5.2.7

Solve these equations either algebraically or graphically.

1 a $2x + y = 7$
$3x + 2y = 12$

b $5x + 4y = 21$
$x + 2y = 9$

c $x + y = 7$
$3x + 4y = 23$

d $2x - 3y = -3$
$3x + 2y = 15$

e $4x = 4y + 8$
$x + 3y = 10$

f $x + 5y = 11$
$2x - 2y = 10$

2 a $x + y = 5$
$3x - 2y + 5 = 0$

b $2x - 2y = 6$
$x - 5y = -5$

c $2x + 3y = 15$
$2y = 15 - 3x$

d $x - 6y = 0$
$3x - 3y = 15$

e $2x - 5y = -11$
$3x + 4y = 18$

f $x + y = 5$
$2x - 2y = -2$

3 a $3y = 9 + 2x$
$3x + 2y = 6$

b $x + 4y = 13$
$3x - 3y = 9$

c $2x = 3y - 19$
$3x + 2y = 17$

d $2x - 5y = -8$
$-3x - 2y = -26$

e $5x - 2y = 0$
$2x + 5y = 29$

f $8y = 3 - x$
$3x - 2y = 9$

4 a $4x + 2y = 5$
$3x + 6y = 6$

b $4x + y = 14$
$6x - 3y = 3$

c $10x - y = -2$
$-15x + 3y = 9$

d $-2y = 0.5 - 2x$
$6x + 3y = 6$

e $x + 3y = 6$
$2x - 9y = 7$

f $5x - 3y = -0.5$
$3x + 2y = 3.5$

5.3 Right-angled trigonometry

Lord Nelson would have used trigonometry in navigation

Trigonometry and the trigonometric ratios developed from the ancient study of the stars. The study of right-angled triangles probably originated with the Egyptians and the Babylonians, who used them extensively in construction and engineering.

The ratios, which are introduced in this chapter, were set out by Hipparchus of Rhodes about 150 BC.

Trigonometry was used extensively in navigation at sea, particularly in the sailing ships of the eighteenth and nineteenth centuries, when it formed a major part of the examination to become a lieutenant in the Royal Navy.

The trigonometric ratios

There are three basic trigonometric ratios: sine, cosine and tangent and you should already be familiar with these.

Each of the trigonometric ratios relates an angle of a right-angled triangle to a ratio of the lengths of two of its sides.

The sides of the triangle have names, two of which are dependent on their position in relation to a specific angle.

The longest side (always opposite the right angle) is called the **hypotenuse**. The side opposite the angle is called the **opposite** side and the side next to the angle is called the **adjacent** side.

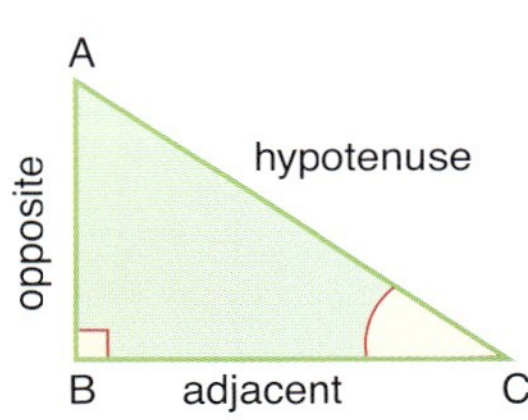

Note that, when the chosen angle is at A, the sides labelled opposite and adjacent change.

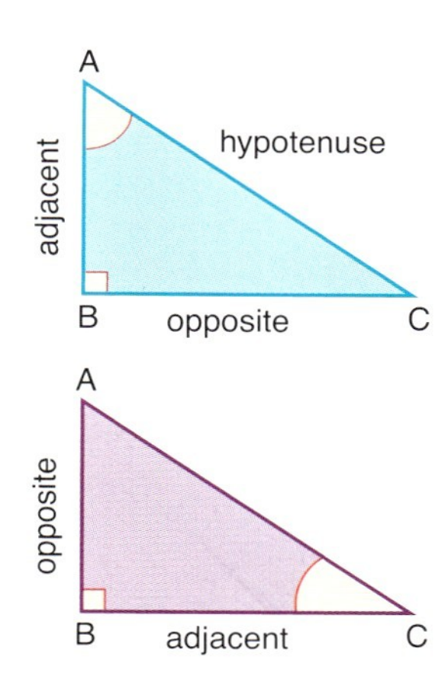

Tangent

The tangent ratio is:

$$\tan C = \frac{\text{length of opposite side}}{\text{length of adjacent side}}$$

Worked examples

1 Calculate the size of angle BAC.

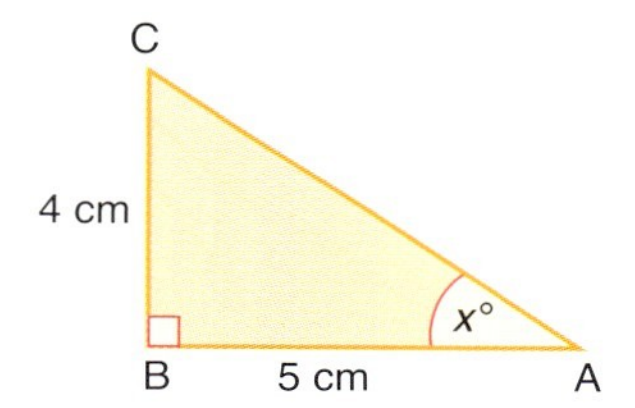

$$\tan x° = \frac{\text{opposite}}{\text{adjacent}} = \frac{4}{5}$$

$$x = \tan^{-1}\left(\frac{4}{5}\right)$$

$$x = 38.7 \text{ (3 s.f.)}$$

$$\angle BAC = 38.7° \text{ (3 s.f.)}$$

2 Calculate the length of QR.

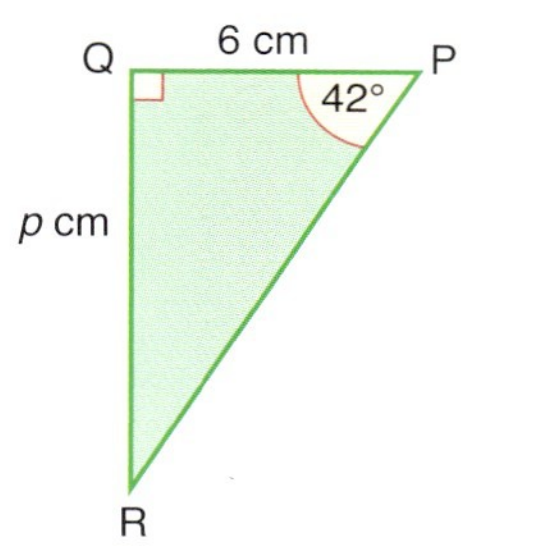

$$\tan 42° = \frac{p}{6}$$

$$6 \times \tan 42° = p$$

$$p = 5.40 \text{ (3 s.f.)}$$

$$QR = 5.40 \text{ cm (3 s.f.)}$$

Sine

The sine ratio is:

$$\sin N = \frac{\text{length of opposite side}}{\text{length of hypotenuse}}$$

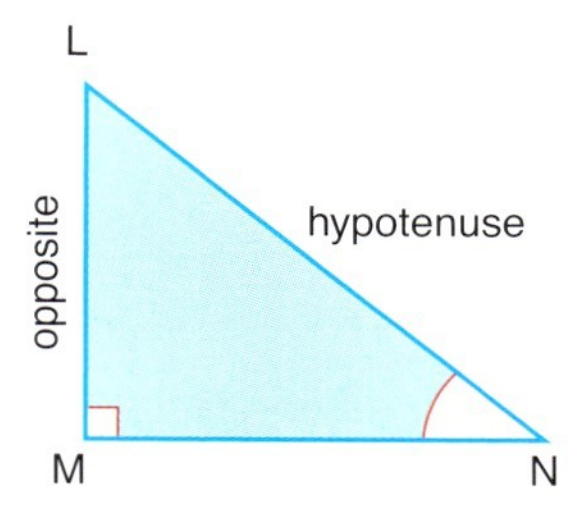

Worked examples

1 Calculate the size of angle BAC.

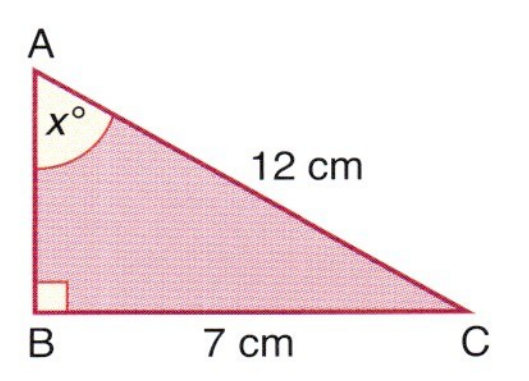

$$\sin x = \frac{\text{opposite}}{\text{hypotenuse}} = \frac{7}{12}$$

$$x = \sin^{-1}\left(\tfrac{7}{12}\right)$$

$$x = 35.7 \text{ (3 s.f)}$$

$$\angle BAC = 35.7° \text{ (3 s.f)}$$

2 Calculate the length of PR.

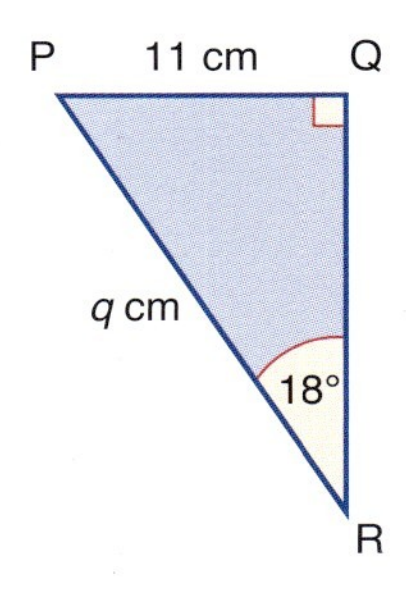

$$\sin 18° = \frac{11}{q}$$

$$q \times \sin 18° = 11$$

$$q = \frac{11}{\sin 18°}$$

$$q = 35.6 \text{ (3 s.f.)}$$

$$PR = 35.6\,\text{cm (3 s.f.)}$$

Cosine

The cosine ratio is:

$$\cos Z = \frac{\text{length of adjacent side}}{\text{length of hypotenuse}}$$

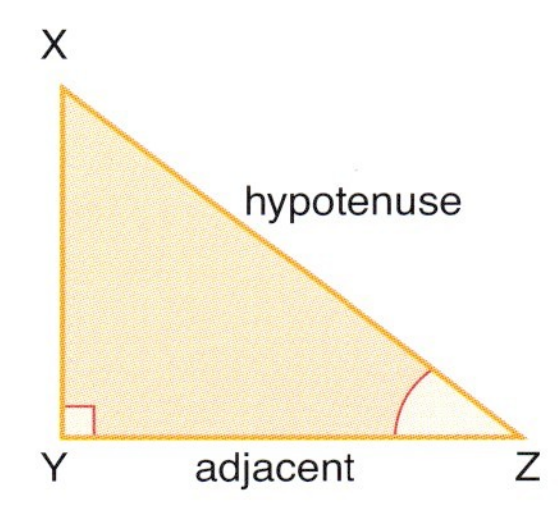

Worked examples

1 Calculate the length XY.

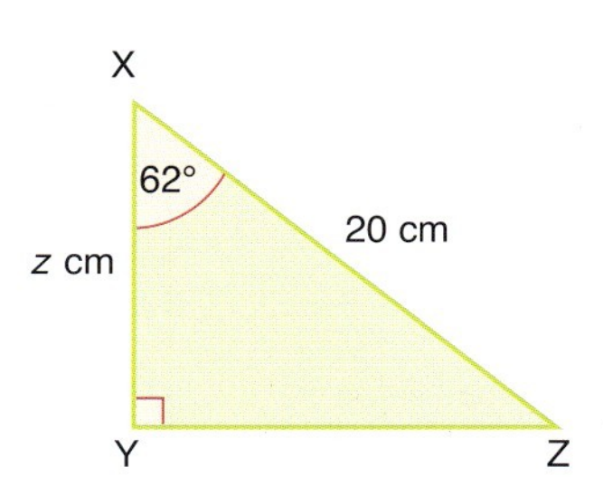

$$\cos 62° = \frac{\text{adjacent}}{\text{hypotenuse}} = \frac{z}{20}$$

$$z = 20 \times \cos 62°$$

$$z = 9.39 \text{ (3 s.f.)}$$

$$XY = 9.39\,\text{cm (3 s.f.)}$$

2 Calculate the size of angle ABC.

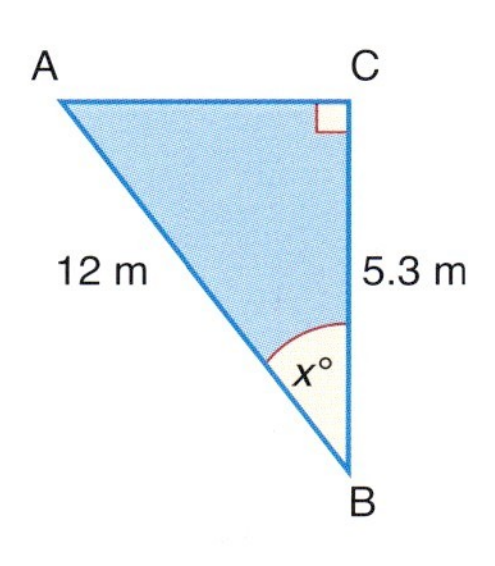

$$\cos x = \frac{\text{adjacent}}{\text{hypotenuse}}$$

$$\cos x = \frac{5.3}{12}$$

$$x = \cos^{-1}\left(\frac{5.3}{12}\right)$$

$$x = 63.8 \text{ (3 s.f.)}$$

$$\angle ABC = 63.8° \text{ (3 s.f.)}$$

Pythagoras' theorem

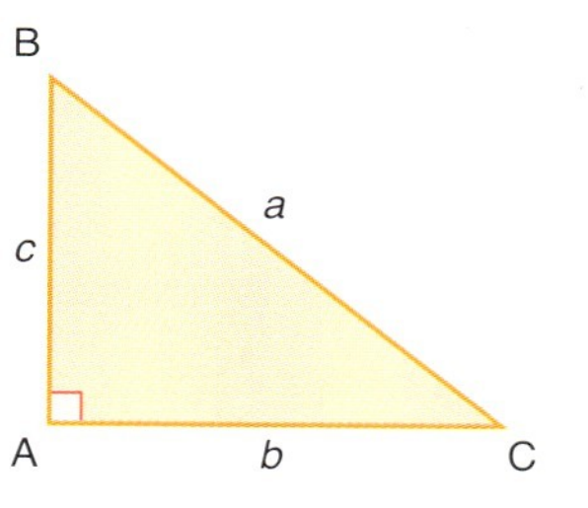

Pythagoras' theorem states the relationship between the lengths of the three sides of a right-angled triangle:

$a^2 = b^2 + c^2$

Worked examples

1 Calculate the length of the side BC.

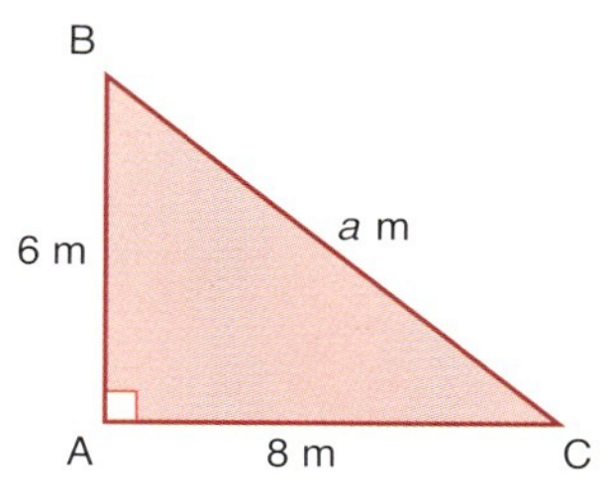

Using Pythagoras' theorem:

$$a^2 = b^2 + c^2$$
$$a^2 = 8^2 + 6^2$$
$$a^2 = 64 + 36 = 100$$
$$a = \sqrt{100}$$
$$a = 10$$

BC = 10 m

2 Calculate the length of the side AC.

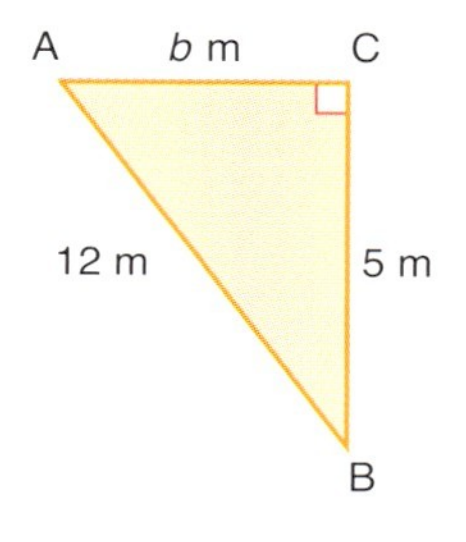

Using Pythagoras' theorem:

$$a^2 = b^2 + c^2$$
$$a^2 - c^2 = b^2$$
$$b^2 = 144 - 25 = 119$$
$$b = \sqrt{119}$$
$$b = 10.9 \text{ (3 s.f.)}$$

AC = 10.9 m (3 s.f.)

Angles of elevation and depression

The **angle of elevation** is the angle above the horizontal through which a line of view is raised. The **angle of depression** is the angle below the horizontal through which a line of view is lowered.

Worked examples

1 The base of a tower is 60 m away from a point X on the ground. If the angle of elevation of the top of the tower from X is 40° calculate the height of the tower. Give your answer to the nearest metre.

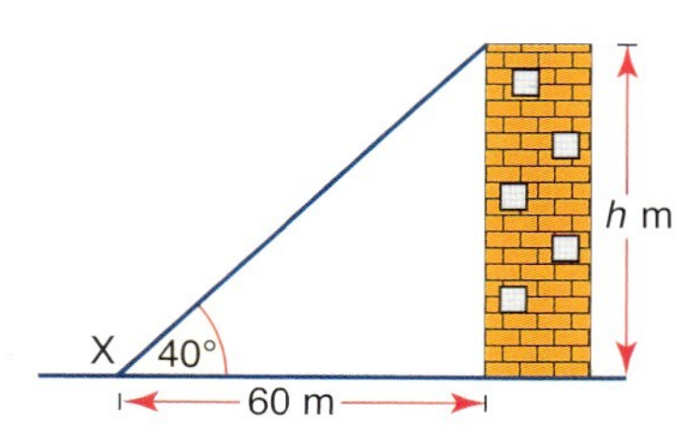

$$\tan 40° = \frac{h}{60}$$

$$h = 60 \times \tan 40° = 50$$

The height is 50 m.

2 An aeroplane receives a signal from a point X on the ground. If the angle of depression of point X from the aeroplane is 30°, calculate the height at which the plane is flying.
Give your answer to the nearest 0.1 kilometre.

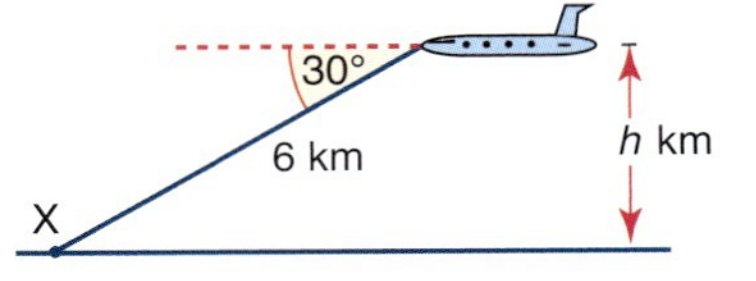

$$\sin 30° = \frac{h}{6}$$

$$h = 6 \times \sin 30° = 3.0$$

The height is 3.0 km.

Exercise 5.3.1

1 Calculate the unknown length in each of the diagrams.

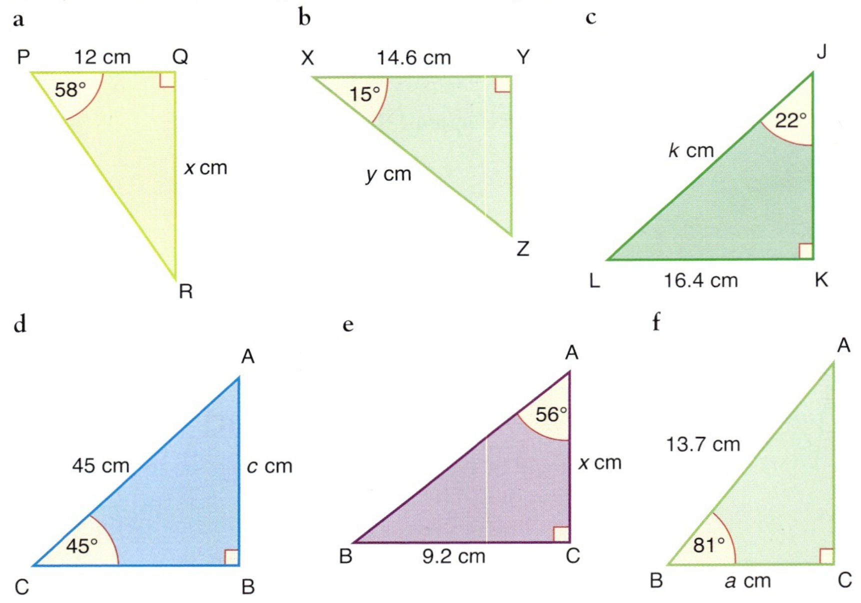

2 Calculate the size of the marked angle $x°$ in each of the following diagrams.

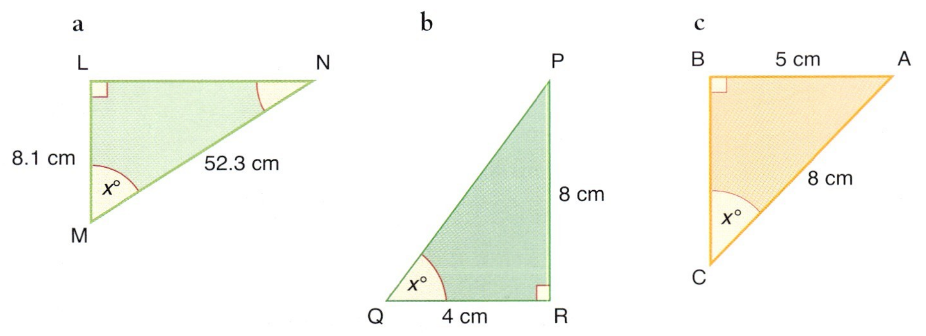

3 Calculate the unknown length or angle in each of the following diagrams

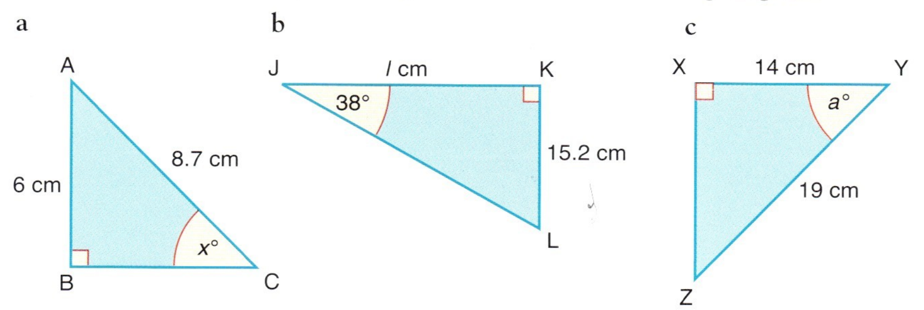

4 A sailing boat sets off from a point X and heads towards Y, a point 17 km north. At point Y it changes direction and heads towards point Z, a point 12 km away on a bearing of 090°. Once at Z the crew want to sail back to X. Calculate:

  a the distance ZX
  b the bearing of X from Z.

5 An aeroplane sets off from G on a bearing of 024° towards H, a point 250 km away. At H it changes course and heads towards J on a bearing of 055° and a distance of 180 km away.

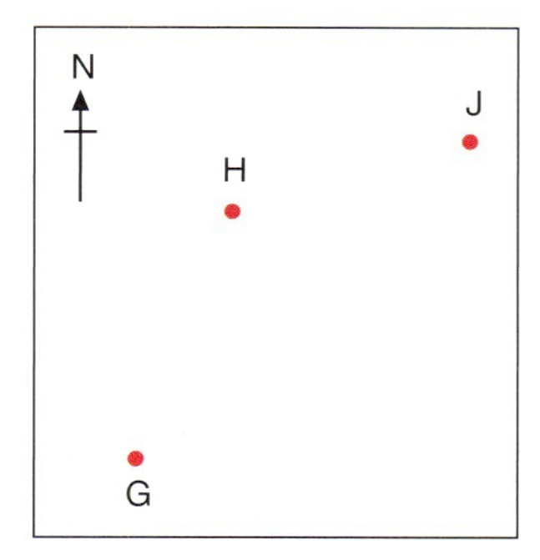

  a How far is H to the north of G?
  b How far is H to the east of G?
  c How far is J to the north of H?
  d How far is J to the east of H?
  e What is the shortest distance between G and J?
  f What is the bearing of G from J?

6 Two trees are standing on flat ground. The angle of elevation of their tops from a point X on the ground is 40°. If the horizontal distance between X and the small tree is 8 m and the distance between the tops of the two trees is 20 m, calculate:

  a the height of the small tree
  b the height of the tall tree
  c the horizontal distance between the trees.

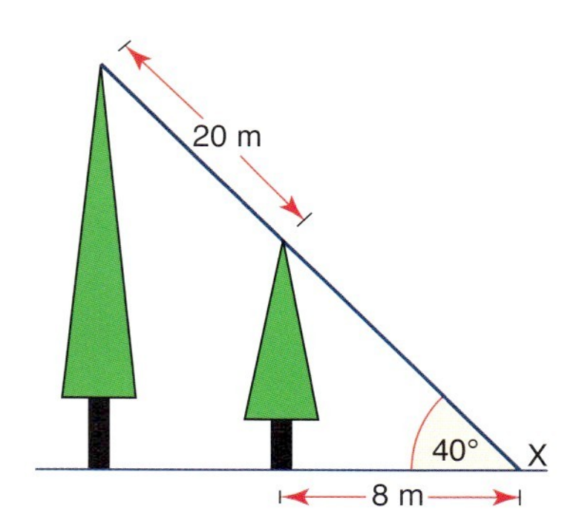

7 PQRS is a quadrilateral. The sides RS and QR are the same length. The sides QP and RS are parallel.

Calculate:
- a angle SQR
- b angle PSQ
- c length PQ
- d length PS
- e the area of PQRS.

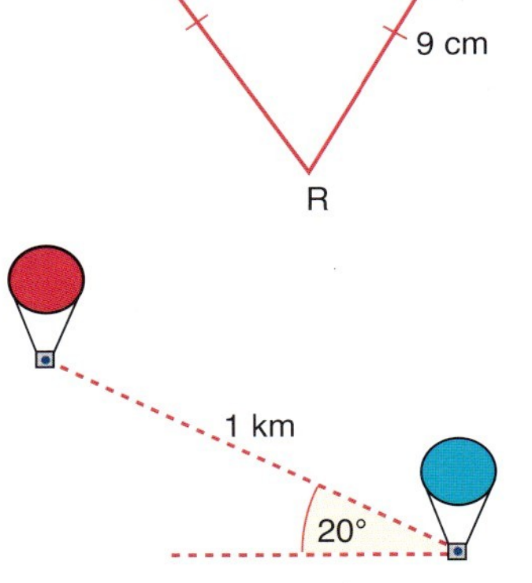

8 Two hot air balloons are 1 km apart in the air. If the angle of elevation of the higher from the lower balloon is 20°, calculate:
- a the vertical height between the two balloons
- b the horizontal distance between the two balloons.

Give your answers to the nearest metre.

9 A plane is flying at an altitude of 6 km directly over the line AB. It spots two boats A and B on the sea. If the angles of depression of A and B from the plane are 60° and 30° respectively, calculate the horizontal distance between A and B.

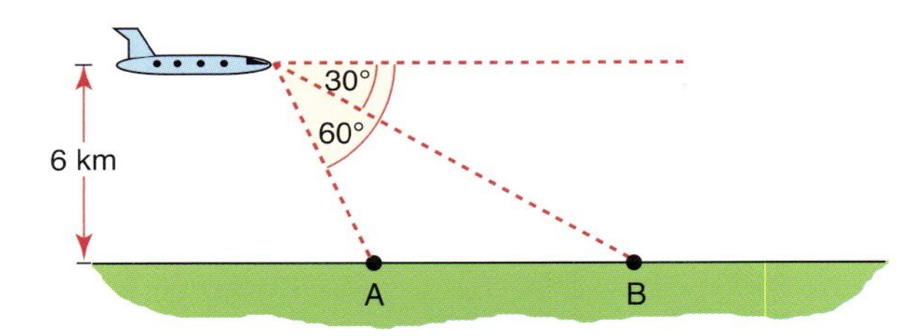

10 Two people A and B are standing either side of a transmission mast. A is 130 m away from the mast and B is 200 m away. If the angle of elevation of the top of the mast from A is 60°, calculate:
- a the height of the mast to the nearest metre
- b the angle of elevation of the top of the mast from B.

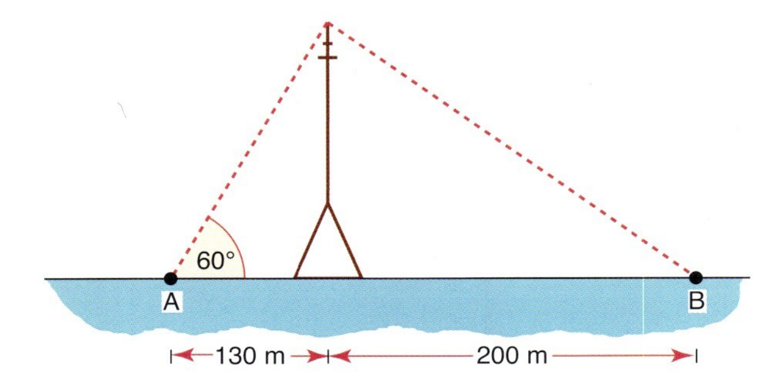

5.4 Trigonometry and non-right-angled triangles

Angles between 0° and 180°

When calculating the size of angles using trigonometry, there is often more than one possible solution. Most calculators, however, will give only the first solution. To be able to calculate the value of a second possible solution, we need to look at the shape of trigonometrical graphs in more detail.

Note: The sine and cosine functions are also covered in Section 4.5.

The sine curve

The graph of $y = \sin x$ is plotted below for x in the range $0 \leq x \leq 360°$, where x is the size of the angle in degrees.

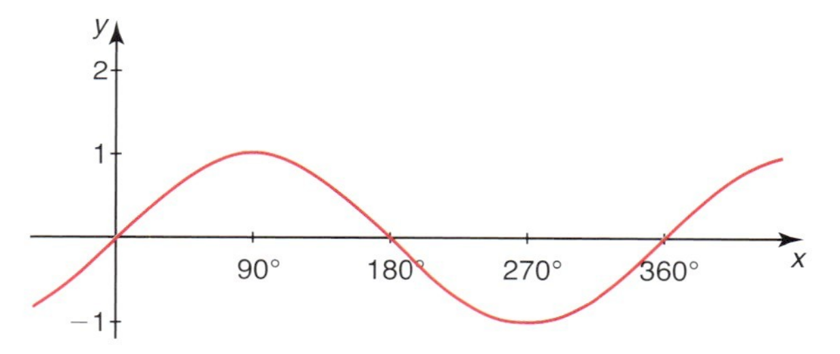

The graph of $y = \sin x$ has:

- a period of 360° (i.e. it repeats itself every 360°)
- a maximum value of 1 (at 90°)
- a minimum value of −1 (at 270°).

Worked example

$\sin 30° = 0.5$. Which other angle between 0° and 180° has a sine of 0.5?

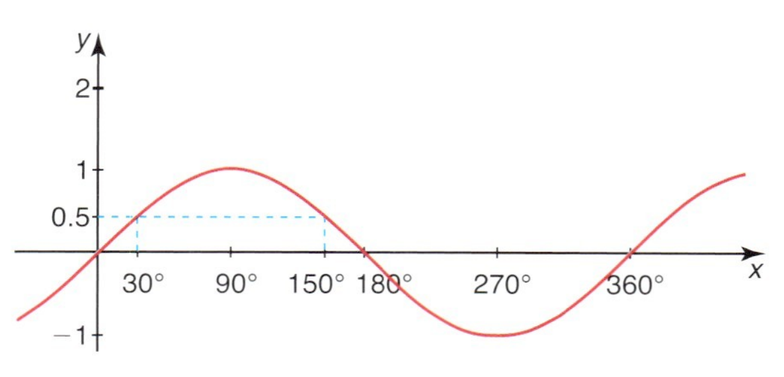

From the graph above it can be seen that $\sin 150° = 0.5$.
$\sin x = \sin(180° - x)$

Exercise 5.4.1

1 Express each of the following in terms of the sine of another angle between 0° and 180°.

a $\sin 60°$ b $\sin 80°$ c $\sin 115°$
d $\sin 140°$ e $\sin 128°$ f $\sin 167°$

2 Express each of the following in terms of the sine of another angle between 0° and 180°.

a $\sin 35°$ b $\sin 50°$ c $\sin 30°$
d $\sin 48°$ e $\sin 104°$ f $\sin 127°$

3 Solve these equations for $0° \leq x \leq 180°$. Give your answers to the nearest degree.

a $\sin x = 0.33$ b $\sin x = 0.99$ c $\sin x = 0.09$
d $\sin x = 0.95$ e $\sin x = 0.22$ f $\sin x = 0.47$

4 Solve these equations for $0° \leq x \leq 180°$. Give your answers to the nearest degree.

a $\sin x = 0.94$ b $\sin x = 0.16$ c $\sin x = 0.80$
d $\sin x = 0.56$ e $\sin x = 0.28$ f $\sin x = 0.33$

The cosine curve

The graph of $y = \cos x$ is plotted below for x in the range $0 \leq x \leq 360°$, where x is the size of the angle in degrees.

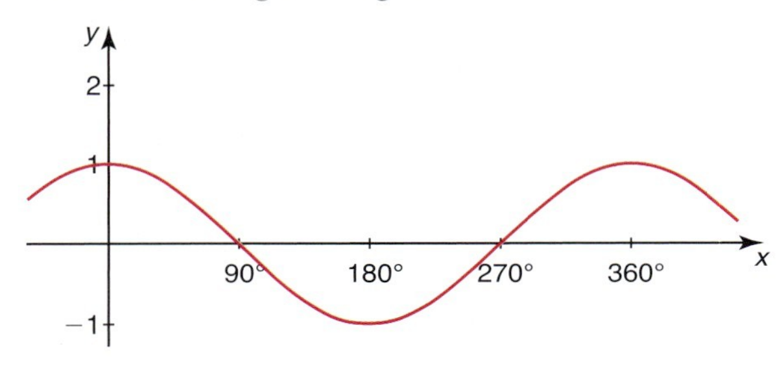

As with the sine curve, the graph of $y = \cos x$ has:

- a period of 360°
- a maximum value of 1 (at 0° and 360°)
- a minimum value of −1 (at 180°).

Note that $\cos x° = -\cos(180 - x)°$.

Worked examples

1 $\cos 60° = 0.5$. Which angle between 0° and 180° has a cosine of −0.5?

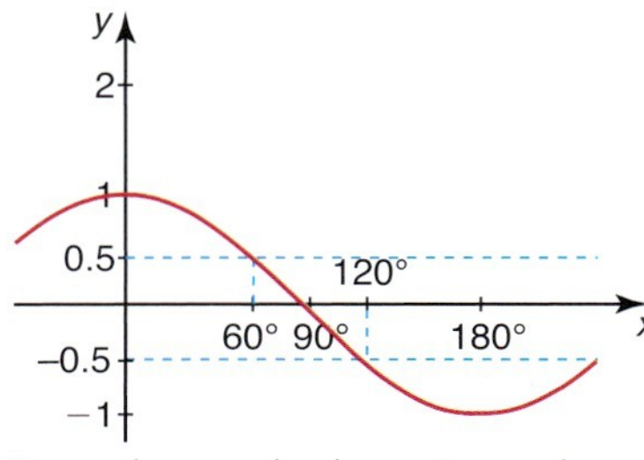

From the graph above it can be seen that $\cos 120° = -0.5$ as the curve is symmetrical.

2 The cosine of which angle between 0° and 180° is equal to the negative of cos 50°?

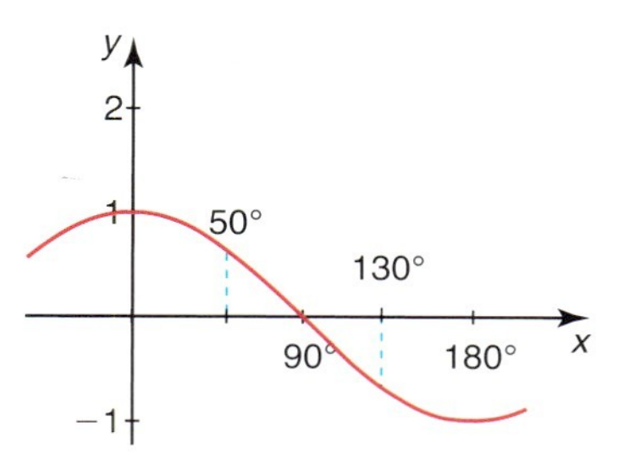

cos 130° = −cos 50°

Exercise 5.4.2

1 Express each of the following in terms of the cosine of another angle between 0° and 180°.

a cos 20° b cos 85° c cos 32°
d cos 95° e cos 147° f cos 106°

2 Express each of the following in terms of the cosine of another angle between 0° and 180°.

a cos 98° b cos 144° c cos 160°
d cos 143° e cos 171° f cos 123°

3 Express each of the following in terms of the cosine of another angle between 0° and 180°.

a −cos 100° b cos 90° c −cos 110°
d −cos 45° e −cos 122° f −cos 25°

4 The cosine of which acute angle has the same value as:

a cos 125° b cos 107° c −cos 120°
d −cos 98° e −cos 92° f −cos 110°?

The sine rule

With right-angled triangles we can use the basic trigonometric ratios of sine, cosine and tangent. The **sine rule** is a relationship which can be used with non-right-angled triangles.

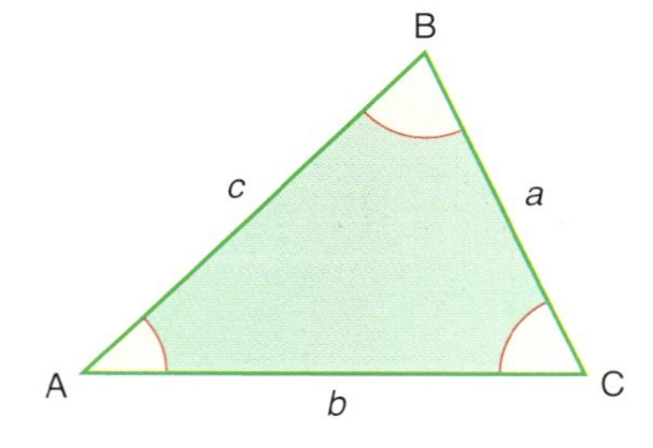

The sine rule states that:

$$\frac{a}{\sin A} = \frac{b}{\sin B} = \frac{c}{\sin C}$$

It can also be written as:

$$\frac{\sin A}{a} = \frac{\sin B}{b} = \frac{\sin C}{c}$$

Worked examples

1 Calculate the length of side BC.

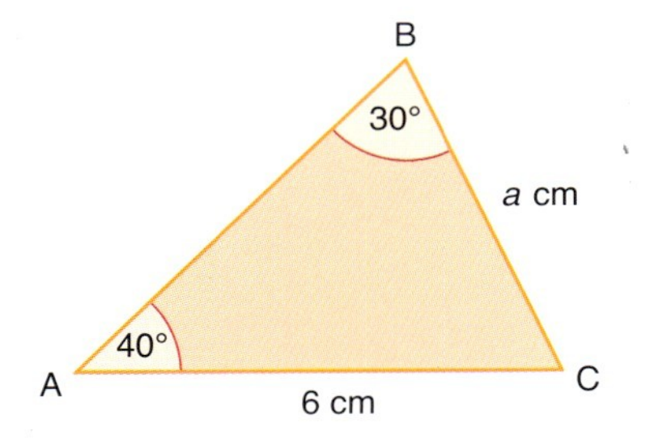

Using the sine rule:

$$\frac{a}{\sin A} = \frac{b}{\sin B}$$

$$\frac{a}{\sin 40°} = \frac{6}{\sin 30°}$$

$$a = \frac{6}{\sin 30°} \times \sin 40°$$

$$a = 7.71 \text{ (3 s.f.)}$$

BC = 7.71 cm (3 s.f.)

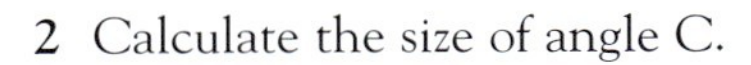

2 Calculate the size of angle C.

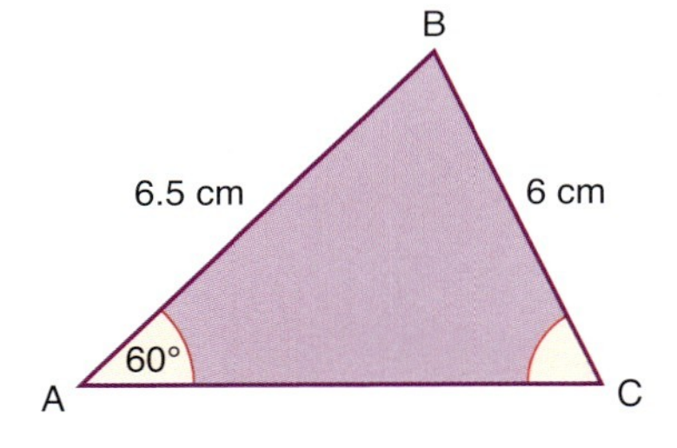

Using the sine rule:

$$\frac{\sin A}{a} = \frac{\sin C}{c}$$

$$\frac{\sin 60°}{6} = \frac{\sin C}{6.5}$$

$$\sin C = \frac{6.5 \times \sin 60°}{6}$$

$$C = \sin^{-1}(0.94)$$

$$C = 69.8° \text{ (3 s.f.)}$$

Exercise 5.4.3

1 Calculate the length of the side marked x in each of the following. Give your answers to one decimal place.

a

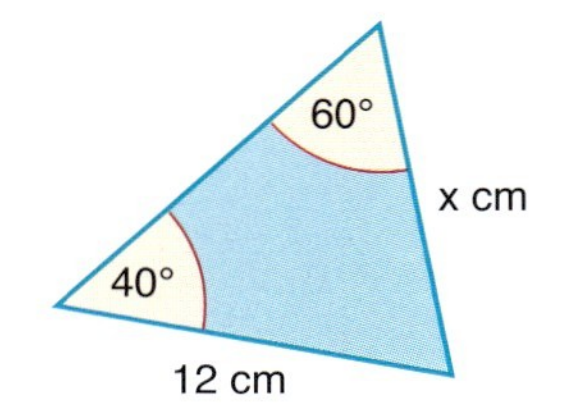

b

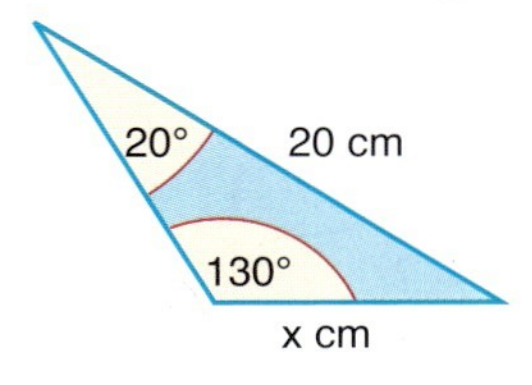

c

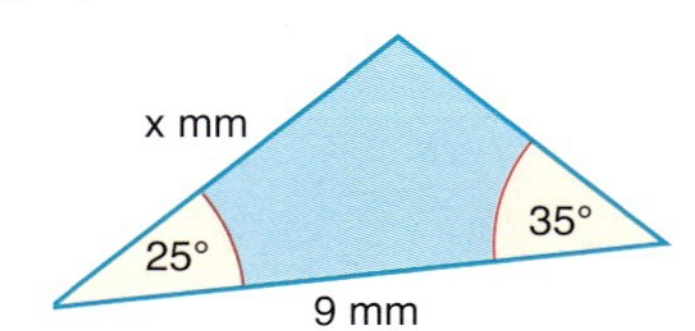

d

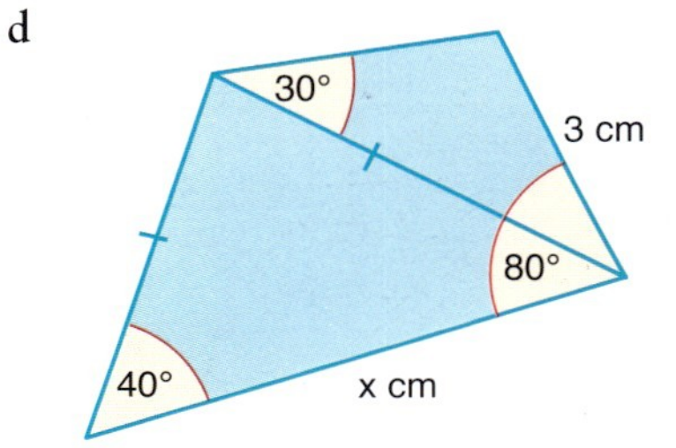

2 Calculate the size of the angle marked $\theta°$ in each of the following. Give your answers to one decimal place.

a

b

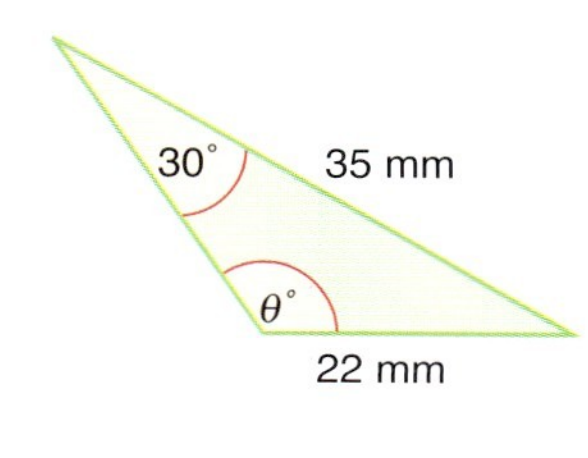

c

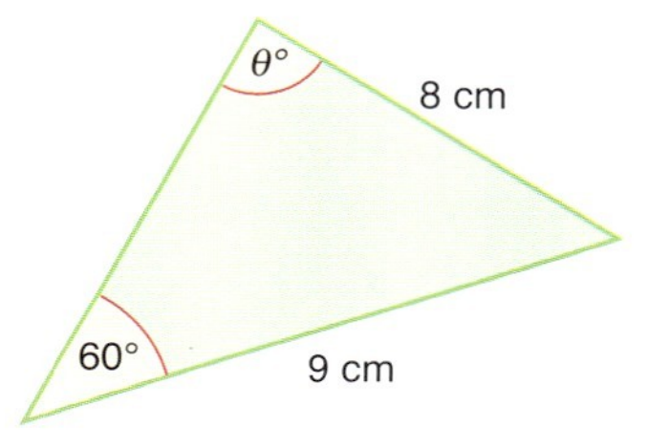

d

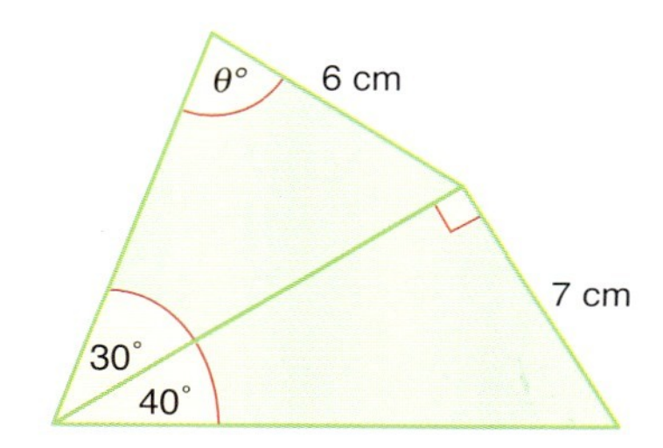

3 Triangle ABC has the following dimensions:

AC = 10 cm, AB = 8 cm and angle ACB = 20°.

a Calculate the two possible values for angle ABC.
b Sketch and label the two possible shapes for triangle ABC.

4 Triangle PQR has the following dimensions:

PQ = 6 cm, PR = 4 cm and angle PQR = 40°.

a Calculate the two possible values for angle QRP.
b Sketch and label the two possible shapes for triangle PQR.

The cosine rule

The **cosine rule** is another relationship which can be used with non-right-angled triangles.

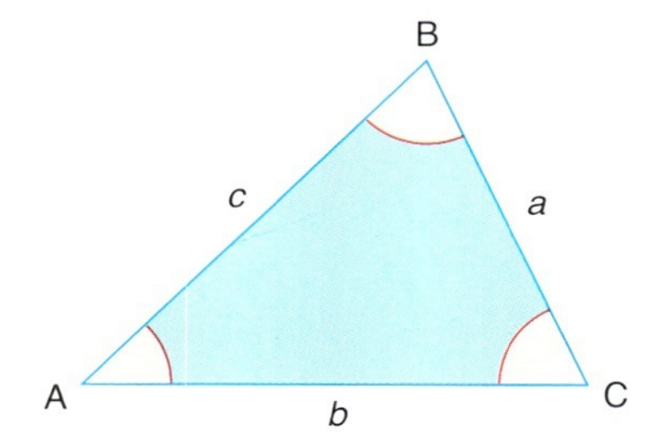

The cosine rule states that:

$a^2 = b^2 + c^2 - 2bc \cos A$

Worked examples

1 Calculate the length of the side BC.

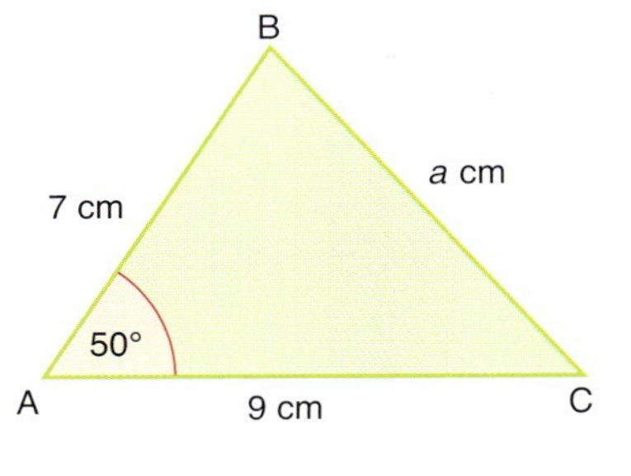

Using the cosine rule:

$$a^2 = b^2 + c^2 - 2bc\cos A$$
$$a^2 = 9^2 + 7^2 - (2 \times 9 \times 7 \times \cos 50°)$$
$$= 81 + 49 - (126 \times \cos 50°) = 49.0$$
$$a = \sqrt{49.0}$$
$$a = 7.00 \text{ (3 s.f.)}$$

BC = 7.00 cm (3 s.f.)

2 Calculate the size of angle A.

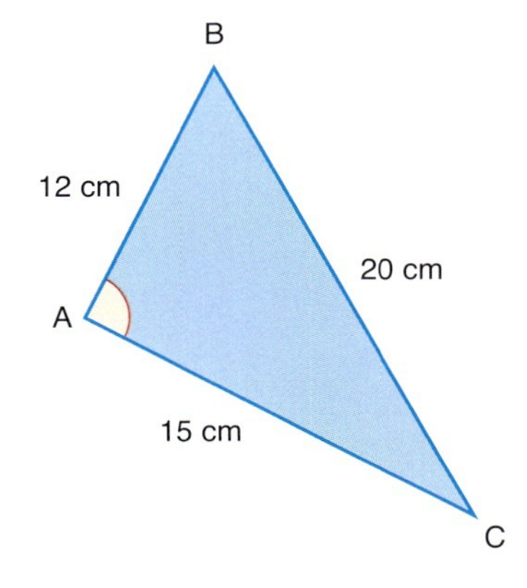

Using the cosine rule:

$$a^2 = b^2 + c^2 - 2bc\cos A$$

Rearranging the equation gives:

$$\cos A = \frac{b^2 + c^2 - a^2}{2bc}$$

$$\cos A = \frac{15^2 + 12^2 - 20^2}{2 \times 15 \times 12} = 0.086$$

$$A = \cos^{-1}(-0.086)$$
$$A = 94.9° \text{ (3 s.f.)}$$

Exercise 5.4.4

1 Calculate the length of the side marked x in each of the following. Give your answers to one decimal place.

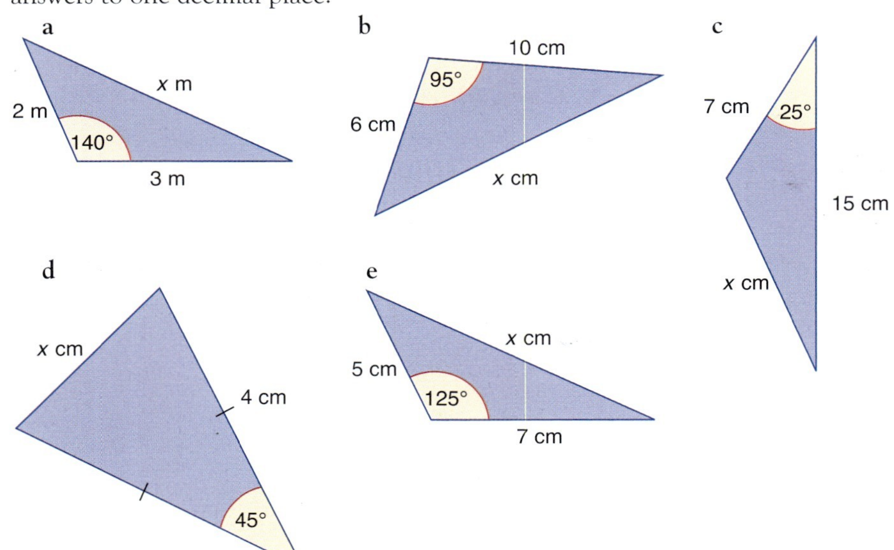

2 Calculate the angle marked $\theta°$ in each of the following. Give your answers to one decimal place.

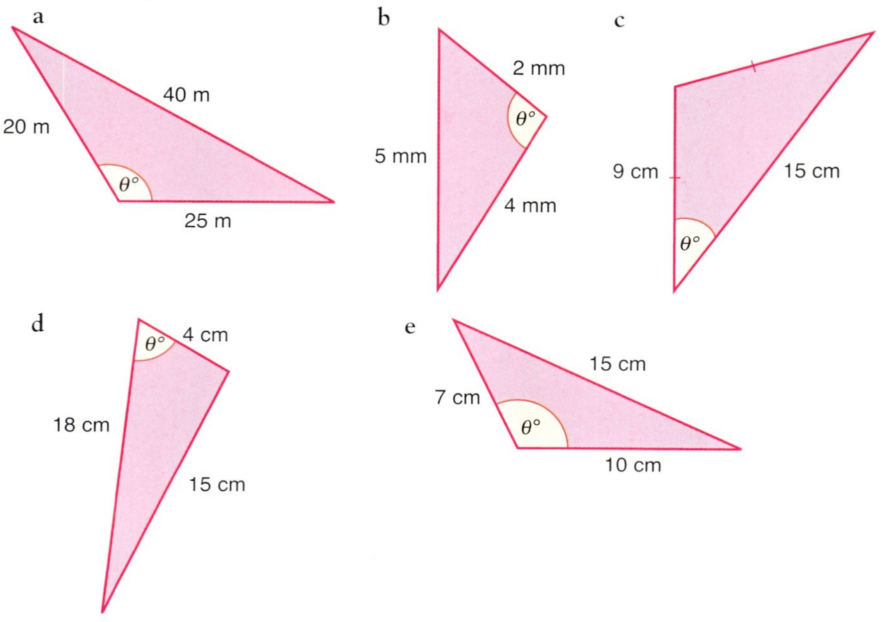

3 Four players W, X, Y and Z are on a rugby pitch. The diagram shows a plan view of their relative positions.

Calculate:

a the distance between players X and Z
b angle ZWX
c angle WZX
d angle YZX
e the distance between players W and Y.

X
20 m
W
24 m
30 m
80°
Y
40 m
Z

4 Three yachts A, B and C are racing off the 'Cape'. Their relative positions are shown in the diagram.

Calculate the distance between B and C to the nearest 10 m.

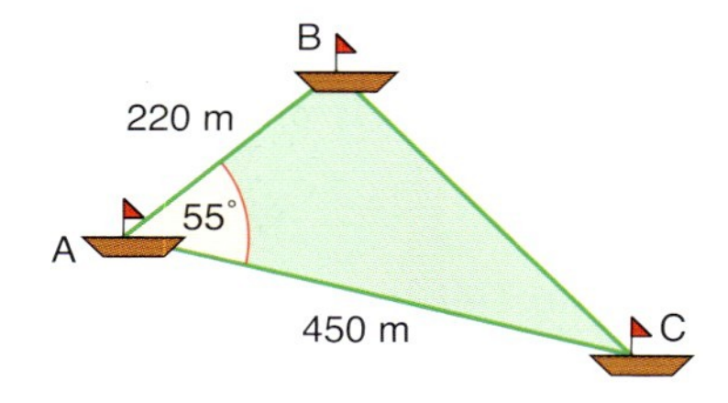

5 A girl standing on a cliff top at A can see two buoys X and Y, 200 m apart, floating on the sea. The angle of depression of Y from A is 45°, and the angle of depression of X from A is 60° (see diagram).

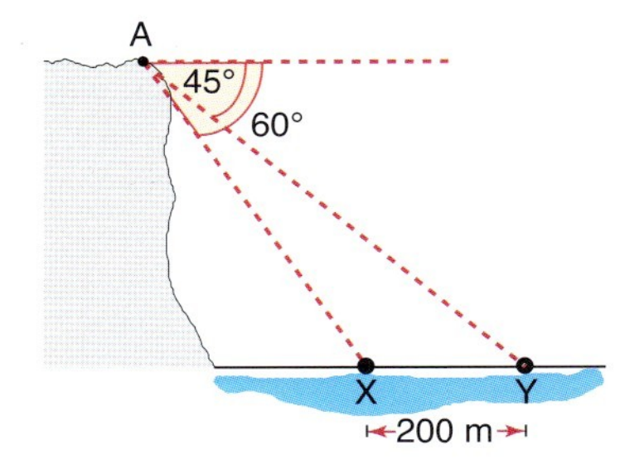

If A, X, Y are in the same vertical plane, calculate:

a the distance AY
b the distance AX
c the vertical height of the cliff.

6 There are two trees standing on one side of a river bank. On the opposite side is a boy standing at X.

Using the information given, calculate the distance between the two trees.

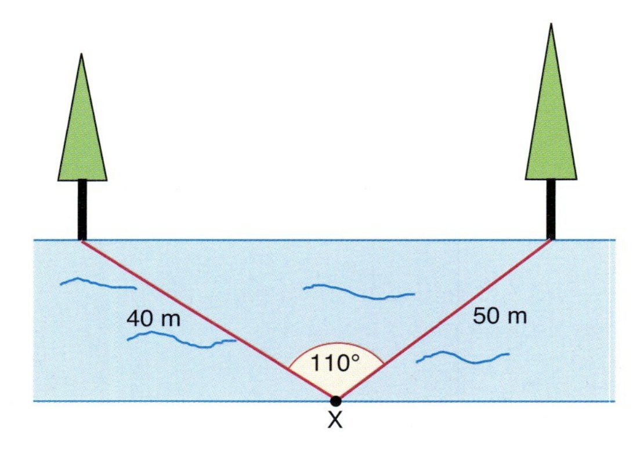

The area of a triangle

The area of a triangle is given by the formula:

$$\text{Area} = \tfrac{1}{2}bh$$

where b is the base and h is the vertical height of the triangle.

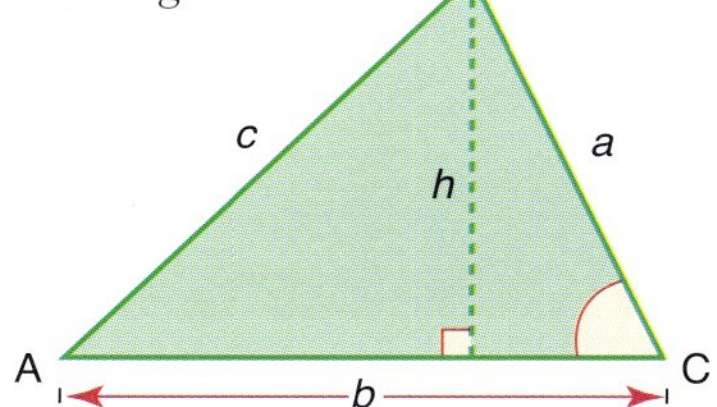

From trigonometric ratios we also know that:

$$\sin C = \frac{h}{a}$$

Rearranging, we have:

$$h = a \sin C$$

Substituting for h in the original formula gives another formula for the area of a triangle:

$$\text{Area} = \tfrac{1}{2}ab \sin C$$

Exercise 5.4.5

1 Calculate the area of the following triangles. Give your answers to one decimal place.

a

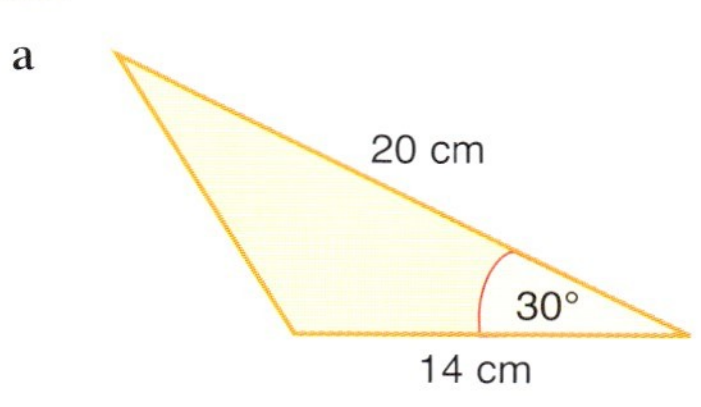

b

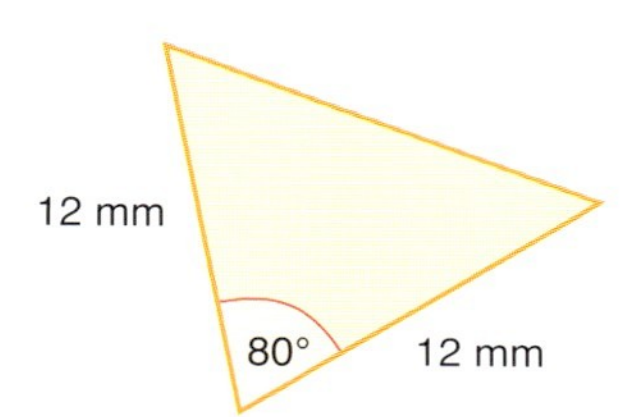

c

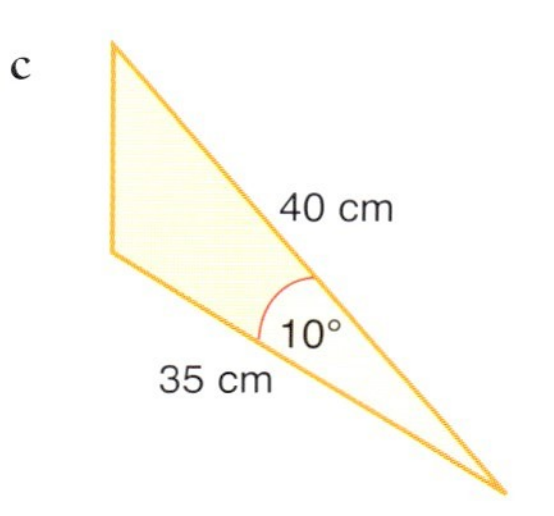

d

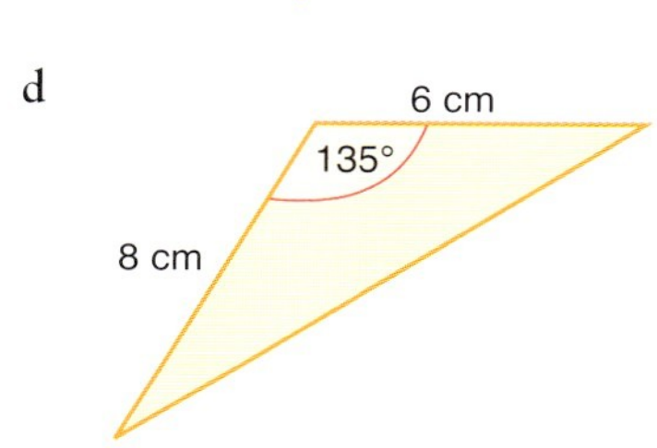

2 Calculate the value of x in each of the following. Give your answers correct to one decimal place.

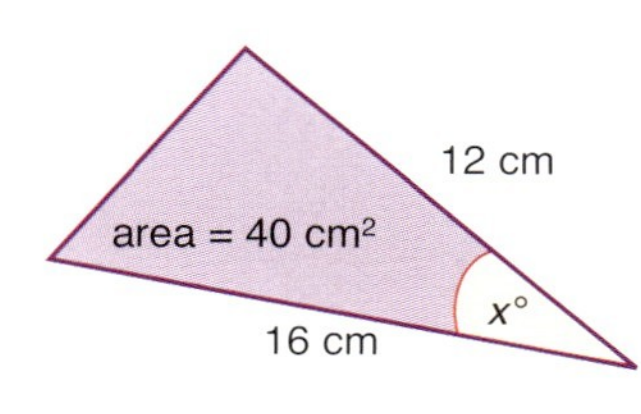

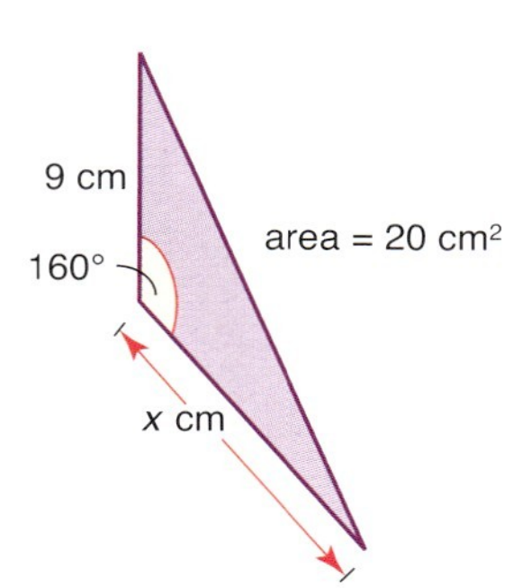

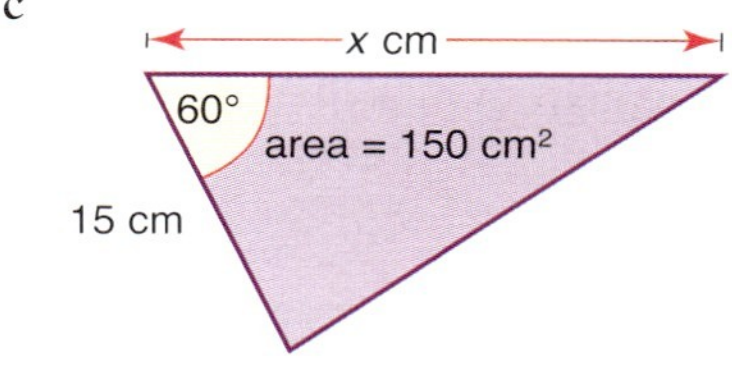

d

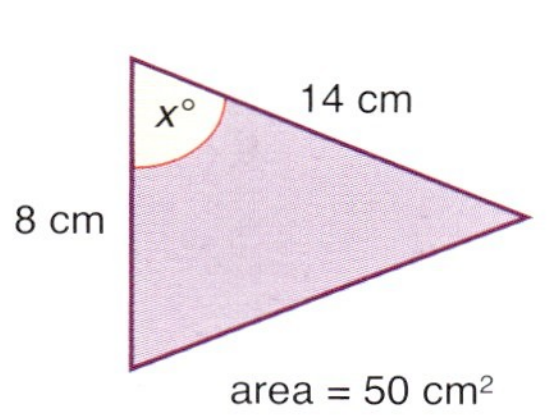

3 ABCD is a school playing field. The following lengths are known:

OA = 83 m, OB = 122 m, OC = 106 m, OD = 78 m

Calculate the area of the school playing field to the nearest 100 m^2.

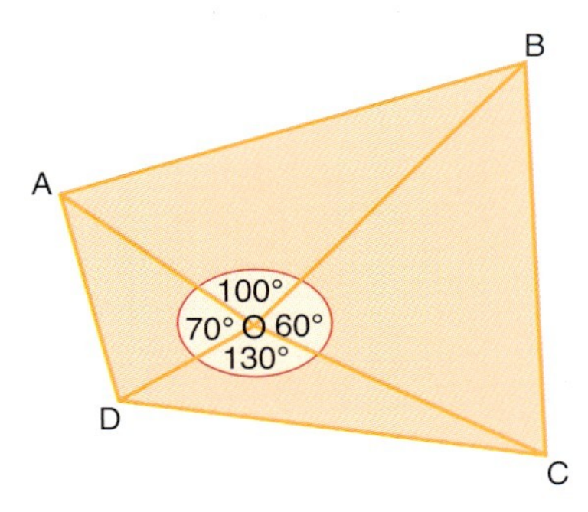

4 The roof of a garage has a slanting length of 3 m and makes an angle of 120° at its vertex. The height of the garage is 4 m and its depth is 9 m.

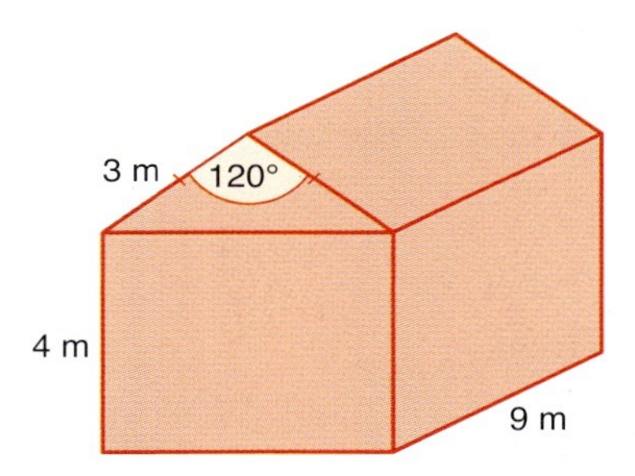

Calculate:

a the cross-sectional area of the roof

b the volume occupied by the whole garage.

5.5 Geometry of three-dimensional shapes

Egyptian society around 2000 BC was very advanced, particularly in its understanding and development of new mathematical ideas and concepts. One of the most important pieces of Egyptian work is called the 'Moscow Papyrus' – so called because it was taken to Moscow in the middle of the nineteenth century.

The Moscow Papyrus was written in about 1850 BC and is important because it contains 25 mathematical problems. One of the key problems is the solution to finding the volume of a truncated pyramid. Although the solution was not written in the way we write it today, it was mathematically correct and translates into the formula:

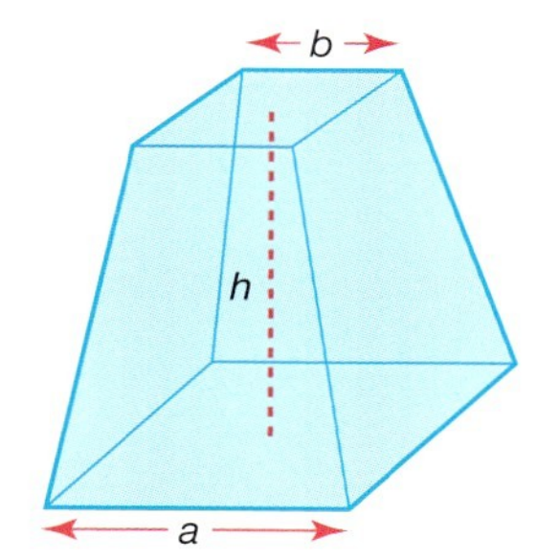

$$V = \frac{(a^2 + ab + b^2)h}{3}$$

The application of trigonometry first to triangles and then to other two-dimensional shapes led to the investigation of angles between a line and a plane, and then further to the application of trigonometry to three-dimensional solids such as the cuboid, pyramid, cylinder, cone and sphere.

Trigonometry in three dimensions

Worked examples

1 The diagram shows a cube of edge length 3 cm.

a Calculate the length EG.
b Calculate the length AG.
c Calculate the angle EGA.
d Calculate the distance from the midpoint X of AB to the midpoint Y of BC.

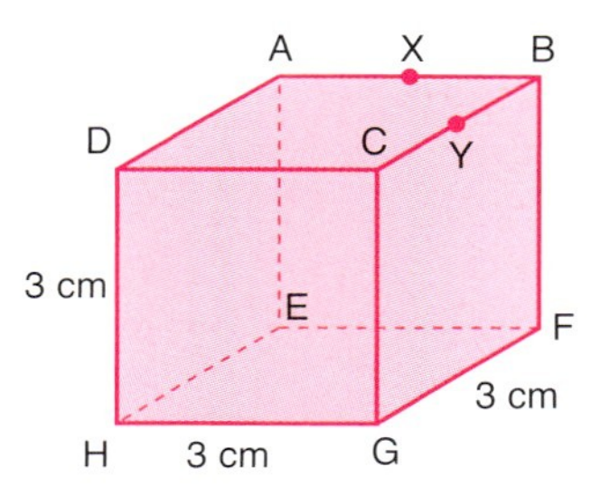

a Triangle EHG (below) is right angled. Use Pythagoras' theorem to calculate the length EG.

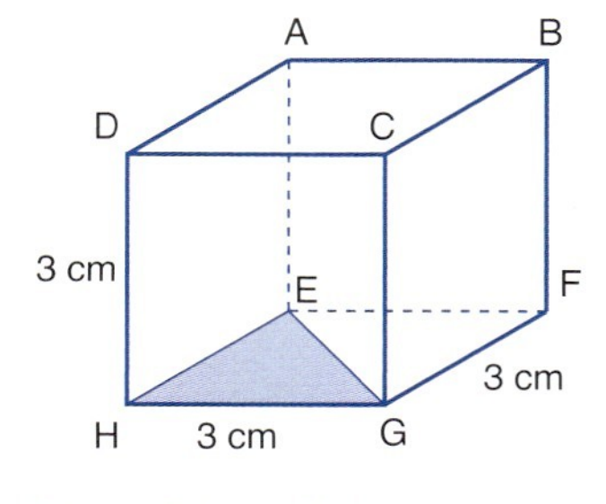

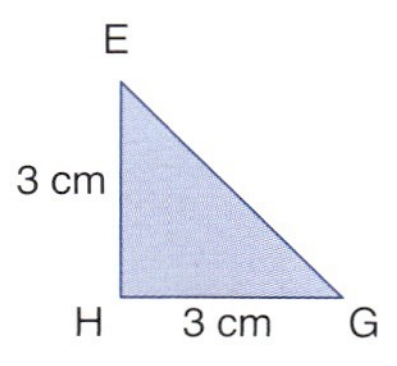

$EG^2 = EH^2 + HG^2$

$EG^2 = 3^2 + 3^2 = 18$

$EG = \sqrt{18}\,\text{cm}$

b Triangle AEG (below) is right angled. Use Pythagoras' theorem to calculate the length AG.

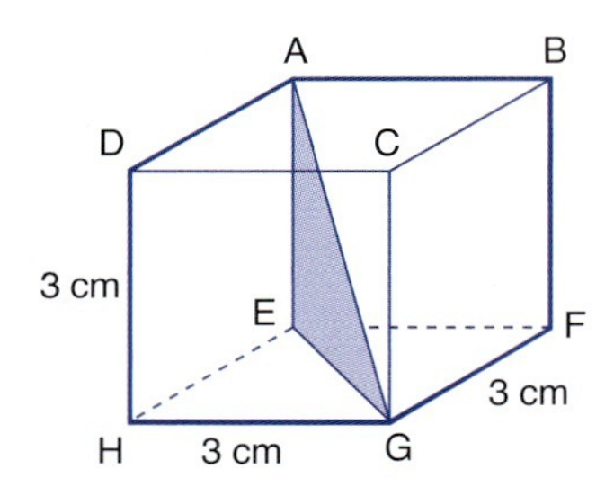

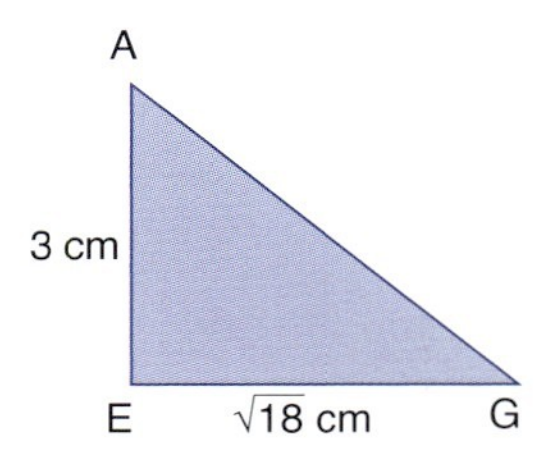

$AG^2 = AE^2 + EG^2$

$AG^2 = 3^2 + (\sqrt{18})^2$

$AG^2 = 9 + 18$

$AG = \sqrt{27}$ cm

c To calculate angle EGA, use the triangle EGA:

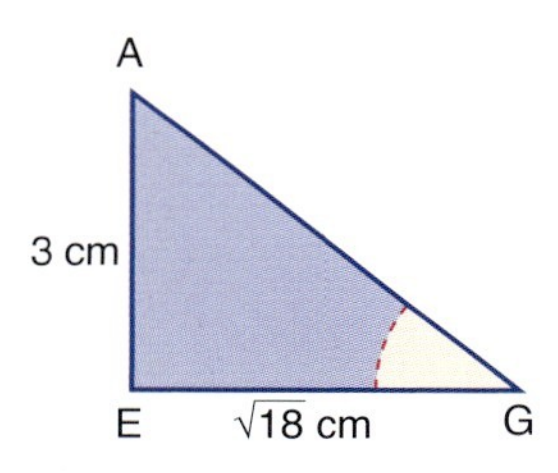

$\tan G = \dfrac{3}{\sqrt{18}}$

Angle G (EGA) = 35.3° (3 s.f.)

d To calculate the distance from X to Y, use Pythagoras' theorem:

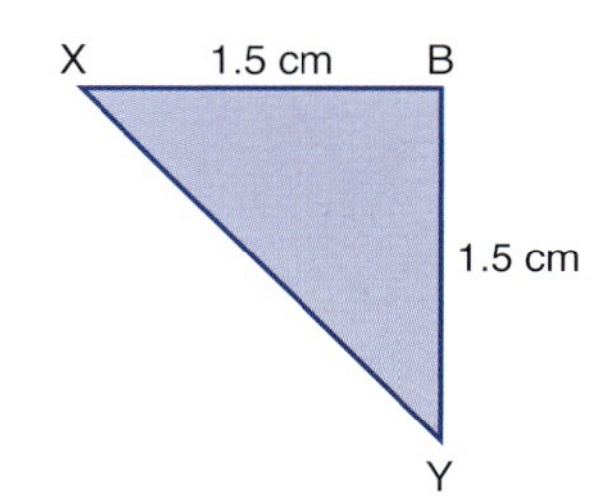

$XY^2 = XB^2 + BY^2$
$XY^2 = 1.5^2 + 1.5^2$
$XY^2 = 4.5$
$XY = 2.12$ cm

2 In the cuboid ABCDEFGH, AB is 12 cm, AD = 9 cm and BF = 5 cm.

a **i)** Calculate the length AC.

ii) Calculate the length BE.

iii) Calculate the length HB.

b X is the midpoint of CG and Y is the midpoint of GH.

i) Calculate the length XY.

ii) Calculate the length XA.

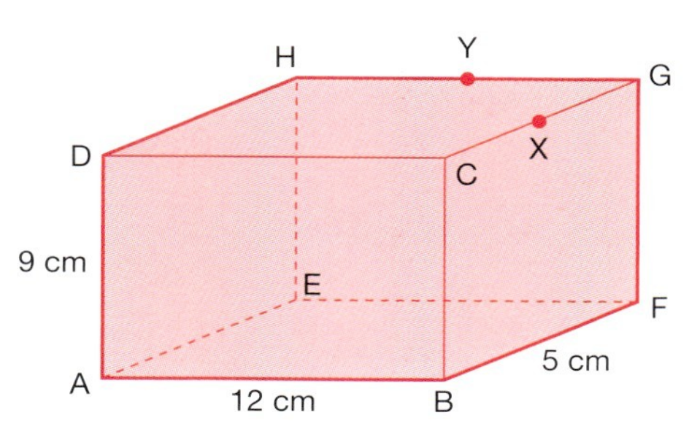

a **i)** To calculate AC, consider triangle ABC.
Triangle ABC is right angled, so use Pythagoras'theorem:

$AC^2 = AB^2 + AC^2$
$AC^2 = 12^2 + 9^2$
$AC^2 = 144 + 81$
$AC^2 = 225$
$AC = 15\text{ cm}$

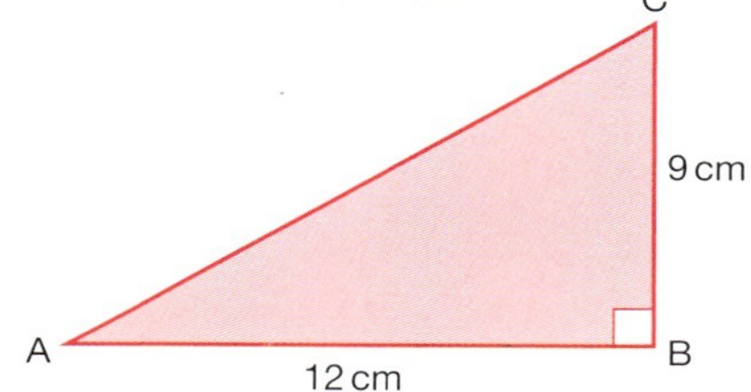

ii) To calculate BE, consider triangle ABE.
Triangle ABE is right angled, so use Pythagoras'theorem:

$BE^2 = AB^2 + BE^2$
$BE^2 = 12^2 + 5^2$
$BE^2 = 169$
$BE = 13\text{ cm}$

iii) To calculate HB, consider triangle EBH.
Triangle EBH is right angled, so use Pythagoras'theorem:

$HB^2 = EH^2 + BE^2$
$HB^2 = 9^2 + 13^2$
$HB^2 = 81 + 169$
$HB = 15.8\text{ cm}$

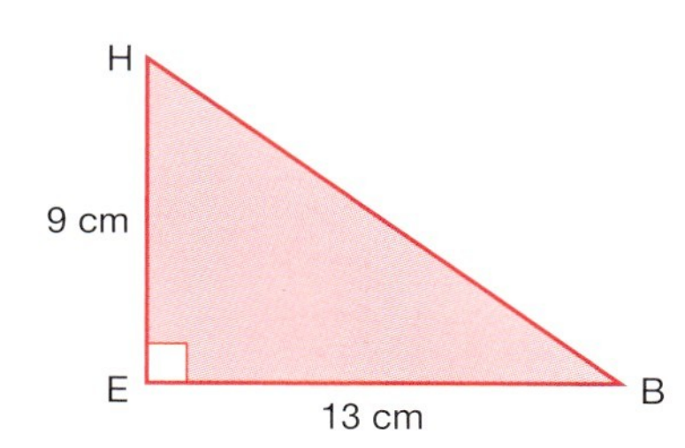

b i) To calculate XY, consider the triangle XYG with a right angle at G.

$XY^2 = XG^2 + YG^2$
$XY^2 = 2.5^2 + 6^2$
$XY^2 = 6.25 + 36$
$XY = \sqrt{42.25} = 6.5\,\text{cm}$

ii) To calculate XA consider the triangle AXZ.

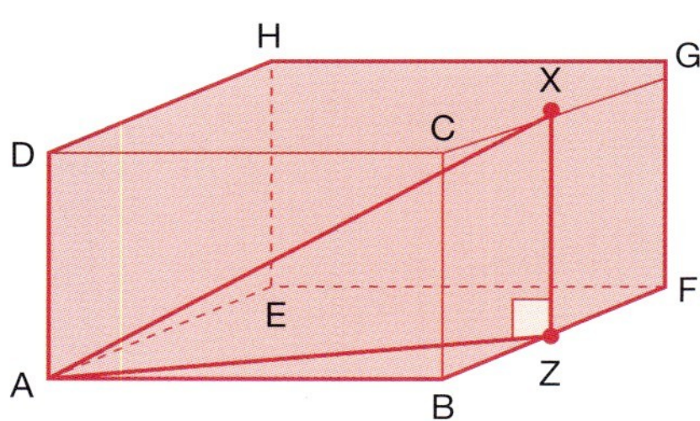

$XZ = 9\,\text{cm}$

Calculate AZ using triangle ABZ.

$AZ^2 = AB^2 + BZ^2$
$AZ^2 = 12^2 + 2.5^2$
$AZ^2 = 144 + 6.25$
$AZ = \sqrt{150.25} = 12.3\,\text{cm}$

Then find XA, using triangle AXZ.

$AX^2 = AZ^2 + XZ^2$
$AX^2 = 150.25 + 81$
$AX^2 = 231.25$
$AX = 15.2\,\text{cm}$

Exercise 5.5.1

Give all your answers to one decimal place.

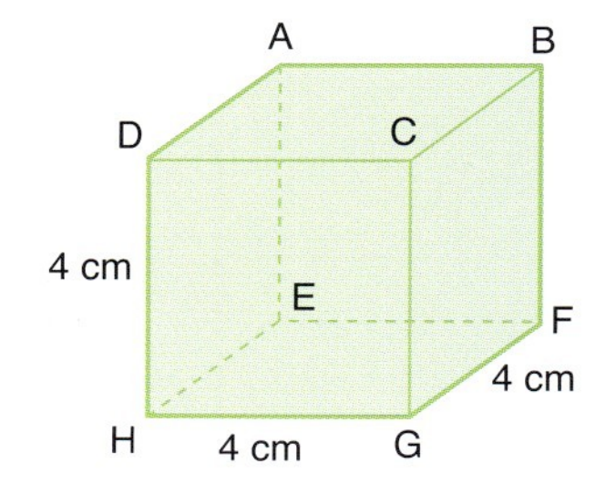

1 a Calculate the length HF.
b Calculate the length of HB.
c Calculate the angle BHG.
d Calculate XY, where X and Y are the midpoints of HG and FG respectively.

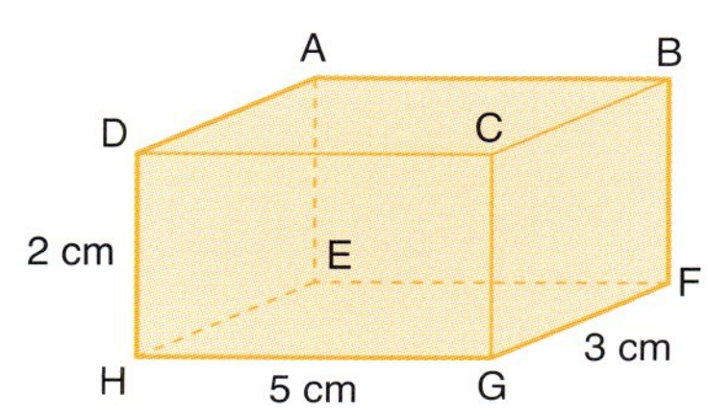

2 a Calculate the length CA.
b Calculate the length CE.
c Calculate the angle ACE.

3 **a** Calculate the length EG.
 b Calculate the length AG.
 c Calculate the angle AGE.

A B D C 12 cm E F 5 cm H 4 cm G

4 **a** Calculate the angle BCE.
 b Calculate the angle GFH.

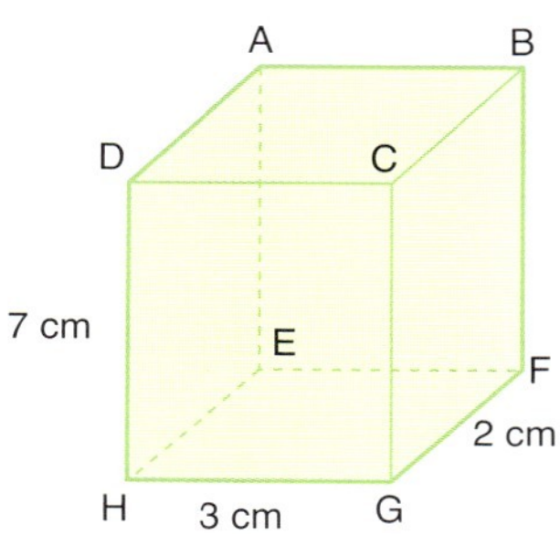

5 The diagram shows a right pyramid where A is vertically above X.
 a **i)** Calculate the length DB.
  ii) Calculate the angle DAX.
 b **i)** Calculate the angle CED.
  ii) Calculate the angle DBA.

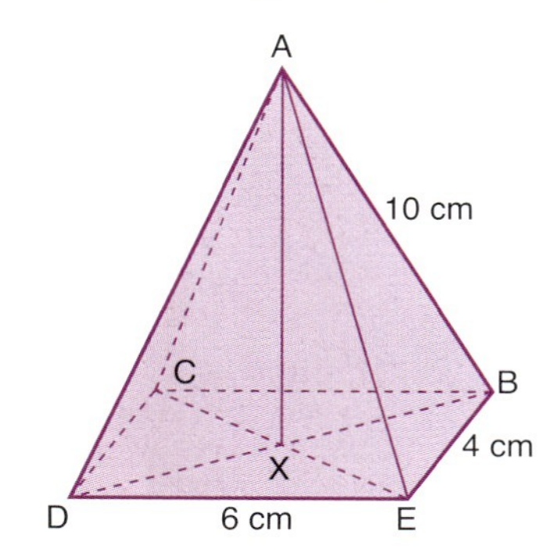

6 The diagram shows a right pyramid where A is vertically above X.
 a **i)** Calculate the length CE.
  ii) Calculate the angle CAX.
 b **i)** Calculate the angle BDE.
  ii) Calculate the angle ADB.

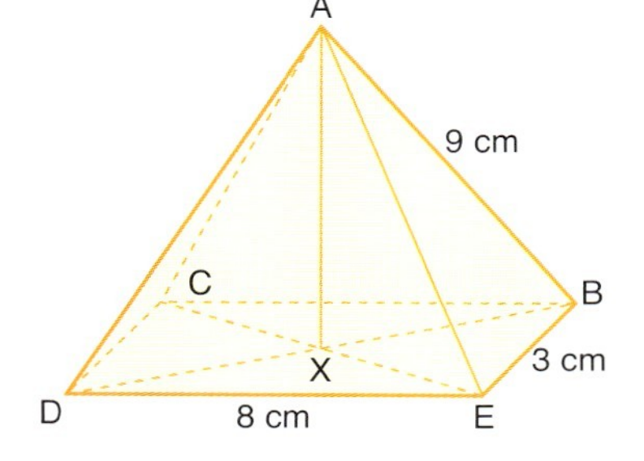

7 In this cone the angle YXZ = 60°.
  a Calculate the length XY.
  b Calculate the length YZ.
  c Calculate the circumference of the base.

8 In this cone, the angle XZY = 40°.
  a Calculate the length XZ
  b Calculate the length XY.

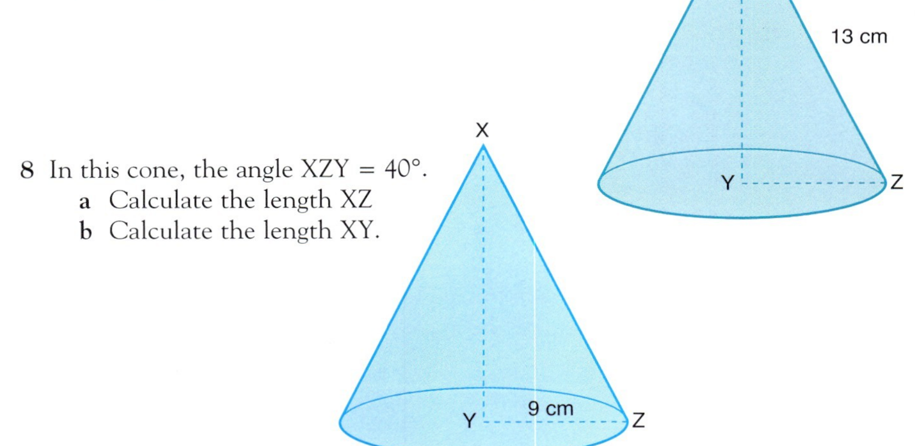

9 In the diagram, RS = 19.2 cm, SP = 16 cm and TQ = 7.2 cm. X is the midpoint of VW. Y is the midpoint of TW.

Calculate the following, drawing diagrams of right-angled triangles to help you.

  a PR b RV
  c PW d XY
  e SY

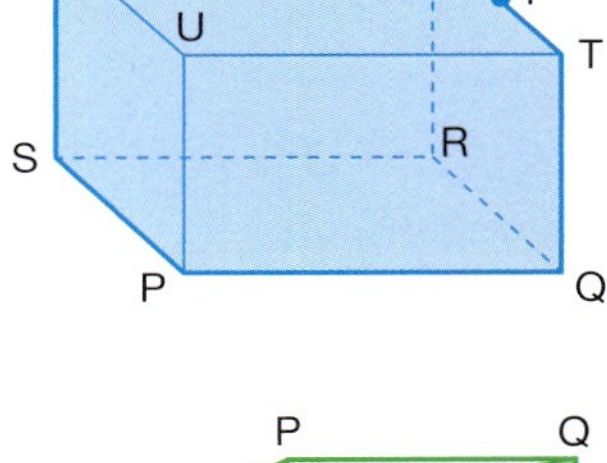

10 One corner of this cuboid has been sliced off along the plane QTU.

WU = 4 cm.

  a Calculate the length of the three sides of the triangle QTU.
  b Calculate the three angles Q, T and U in triangle QTU.
  c Calculate the area of triangle QTU.

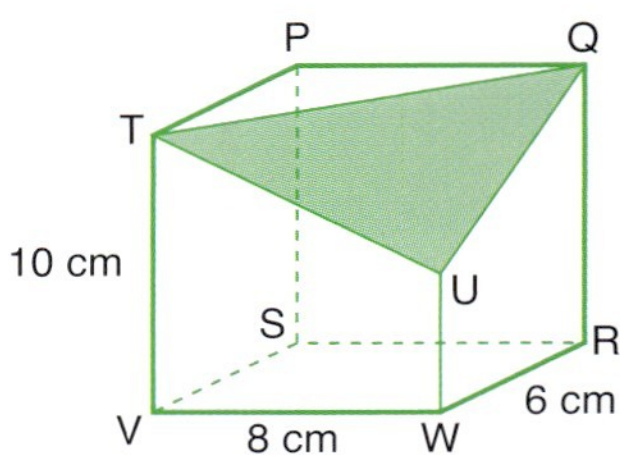

The angle between a line and a plane

To calculate the size of the angle between the line AB and the shaded plane, drop a perpendicular from B. It meets the shaded plane at C. Then join AC.

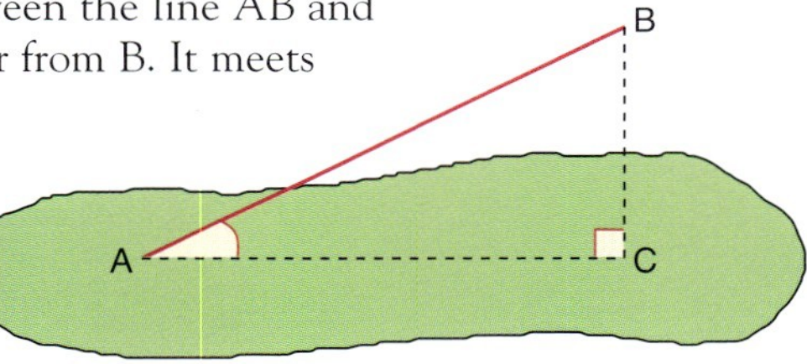

The angle between the lines AB and AC represents the angle between the line AB and the shaded plane.

The line AC is the **projection** of the line AB on the shaded plane.

Worked example

ABCDEFGH is a cuboid.

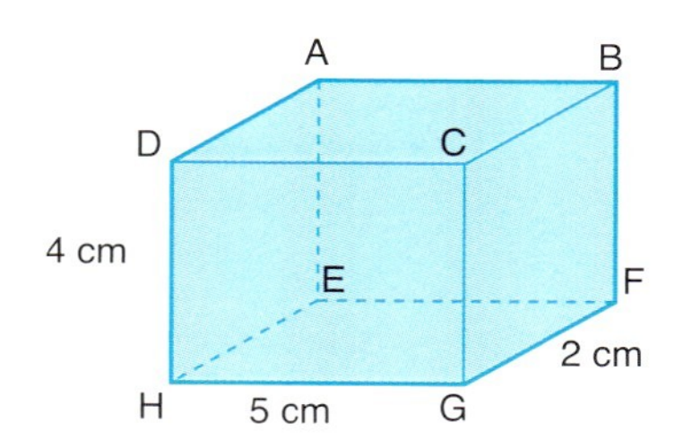

a Calculate the length CE.
b Calculate the angle between the line CE and the plane ADHE.

a First use Pythagoras' theorem to calculate the length EG:

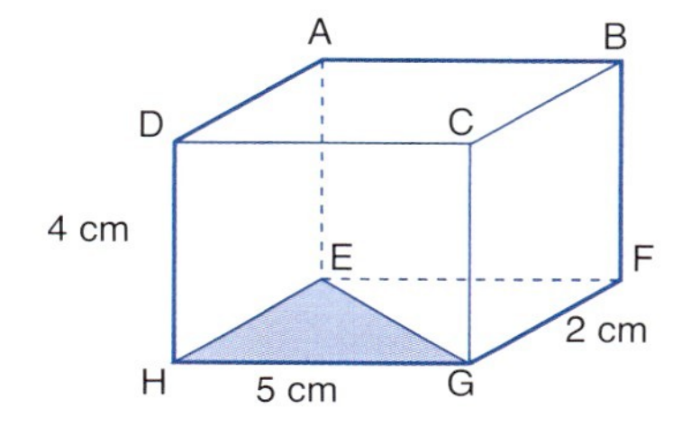

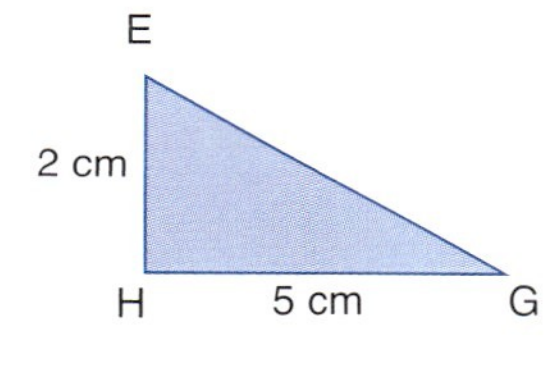

$EG^2 = EH^2 + HG^2$
$EG^2 = 2^2 + 5^2$
$EG^2 = 29$
$EG = \sqrt{29}$ m

Now use Pythagoras' theorem to calculate CE:

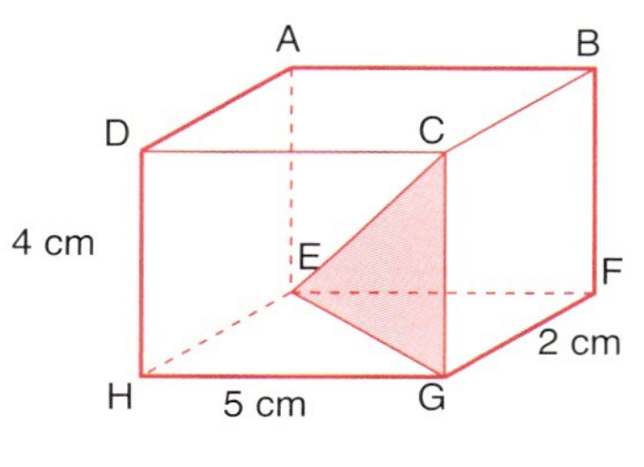

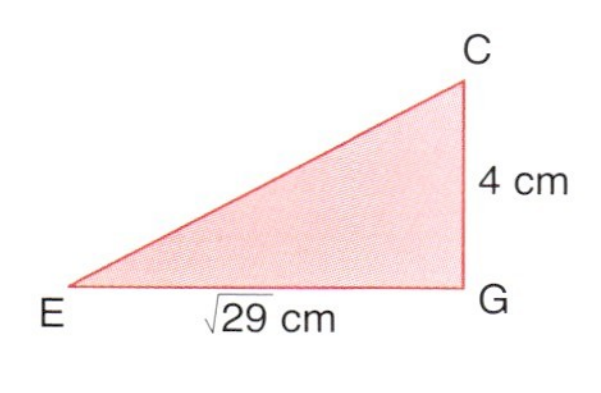

$CE^2 = EG^2 + CG^2$
$CE^2 = (\sqrt{29})^2 + 4^2$
$CE^2 = 29 + 16$
$CE = \sqrt{45}$ cm
$CE = 6.71$cm (3 s.f.)

b

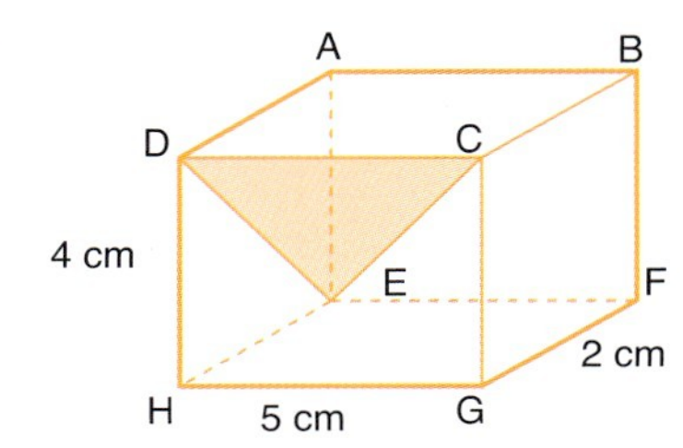

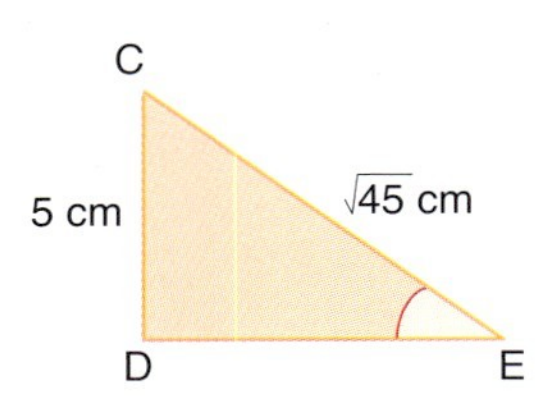

To calculate the angle between the line CE and the plane ADHE use the right-angled triangle CED and calculate the angle CED.

$$\sin E = \frac{CD}{CE}$$

$$\sin E = \frac{5}{\sqrt{45}}$$

$$E = \sin^{-1}\left(\frac{5}{\sqrt{45}}\right)$$

$$E = 48.2° \text{ (3 s.f.)}$$

Exercise 5.5.2

1 Name the projection of each line onto the given plane.

  a TR onto RSWV
  b TR onto PQUT
  c SU onto PQRS
  d SU onto TUVW
  e PV onto QRVU
  f PV onto RSWV

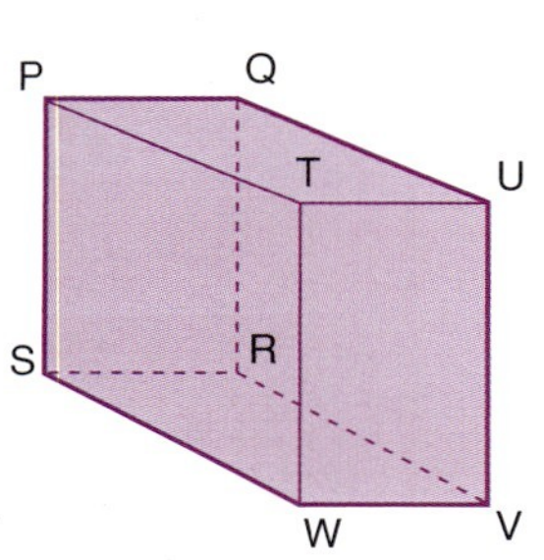

2 Name the projection of each line onto the given plane.

  a KM onto IJNM
  b KM onto JKON
  c KM onto HIML
  d IO onto HLOK
  e IO onto JKON
  f IO onto LMNO

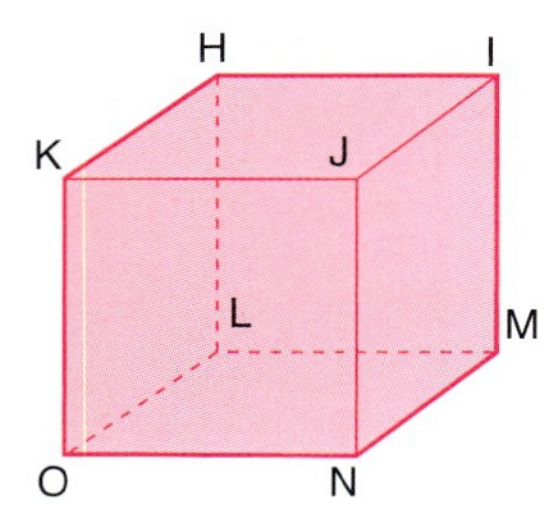

3 Name the angle between the given line and plane.

  a PT and PQRS
  b PU and PQRS
  c SV and PSWT
  d RT and TUVW
  e SU and QRVU
  f PV and PSWT

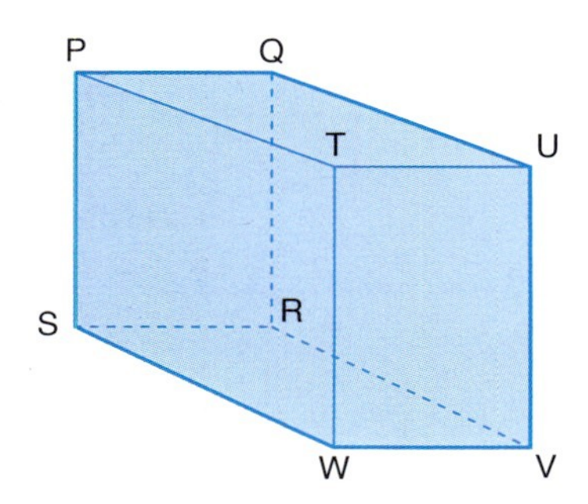

4 a Calculate the length BH.
b Calculate the angle between the line BH and the plane EFGH.

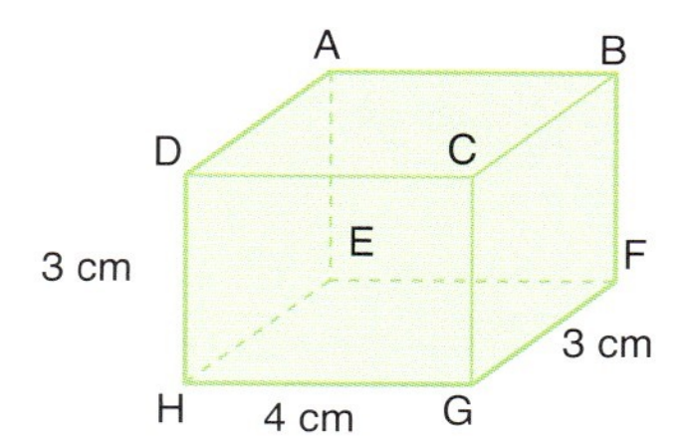

5 a Calculate the length AG.
b Calculate the angle between the line AG and the plane EFGH.
c Calculate the angle between the line AG and the plane ADHE.

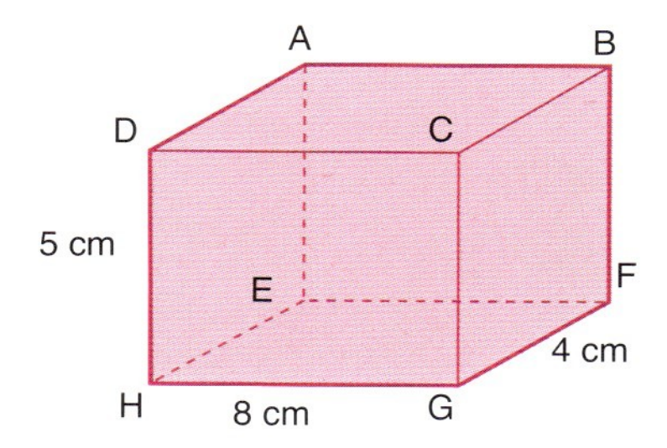

6 The diagram shows a right pyramid where A is vertically above X.
a Calculate the length BD.
b Calculate the angle between AB and CBED.

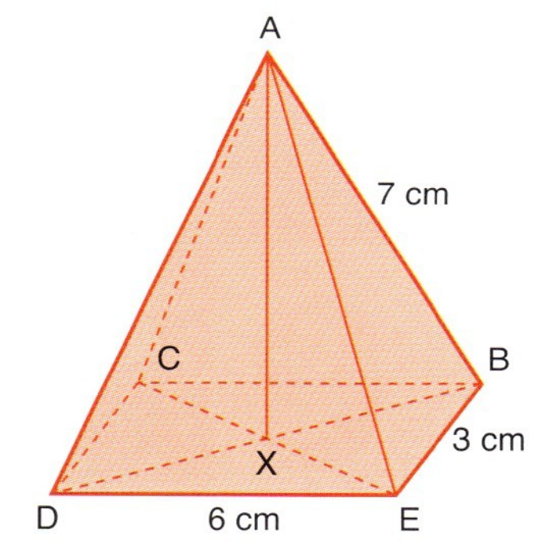

7 The diagram shows a right pyramid where U is vertically above X.
a Calculate the length WY.
b Calculate the length UX.
c Calculate the angle between UX and UZY.

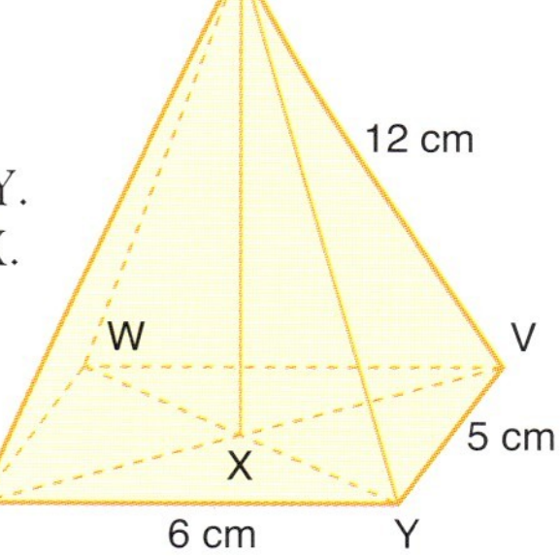

8 ABCD and EFGH are square faces lying parallel to each other.

Calculate:
a the length DB
b the length HF
c the vertical height of the object
d the angle DH makes with the plane ABCD.

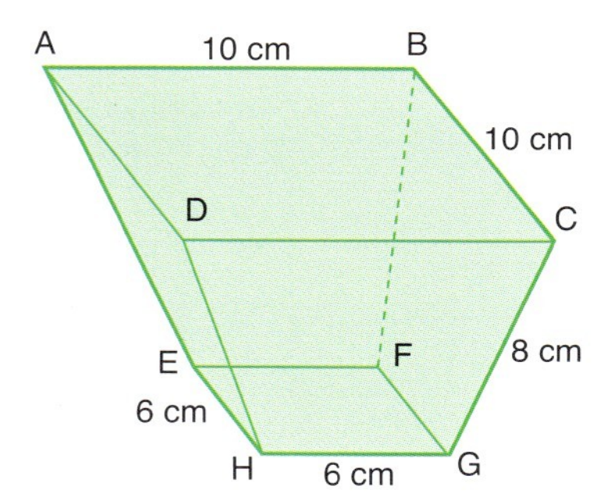

9 ABCD and EFGH are square faces lying parallel to each other.

Calculate:

a the length AC
b the length EG
c the vertical height of the object
d the angle CG makes with the plane EFGH.

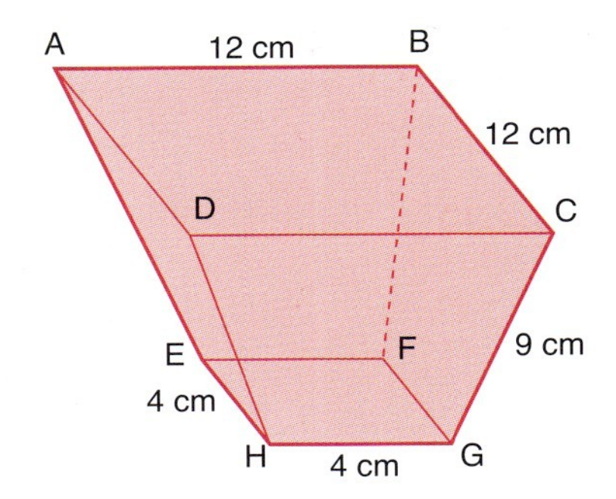

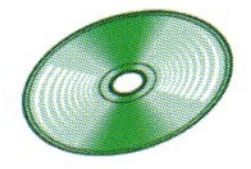

Volume and surface area of prisms

A **prism** is any three-dimensional object which has a constant cross-sectional area Below are a few examples of some of the more common types of prism:
When each of the shapes is cut parallel to the shaded face, the cross-section is constant and the shape is therefore classified as a prism.

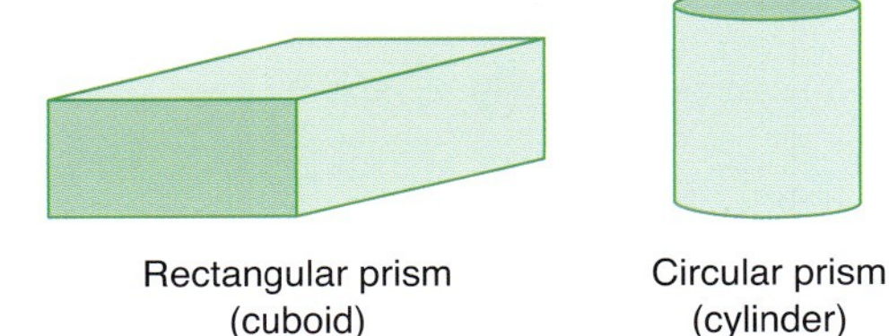

Rectangular prism (cuboid) Circular prism (cylinder) Triangular prism

Volume of a prism = area of cross-section × length

The surface area of a prism is the sum of the areas of its faces. The surface area of a cuboid can be calculated as:

Surface area of a cuboid $= 2(wl + lh + wh)$

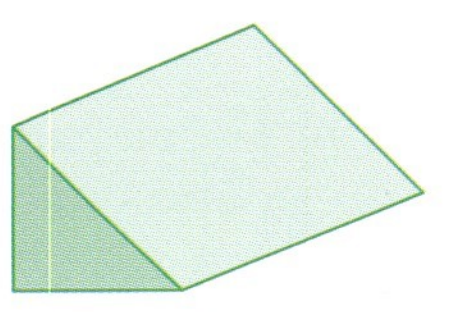

To calculate the surface area of other prisms it is best to visualize the net of the solid. A cylinder is made up of one rectangular piece and two circular pieces.

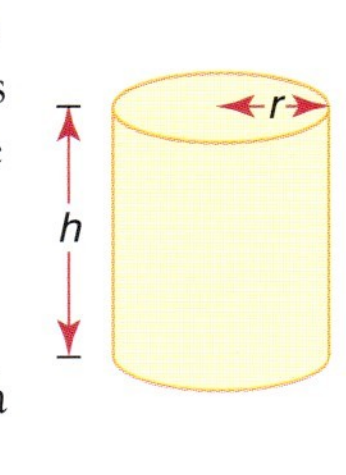

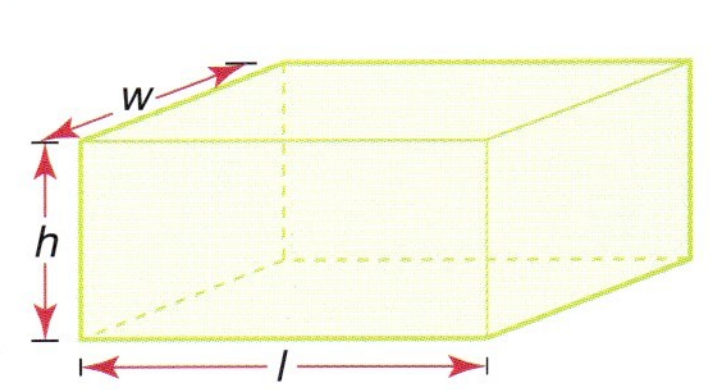

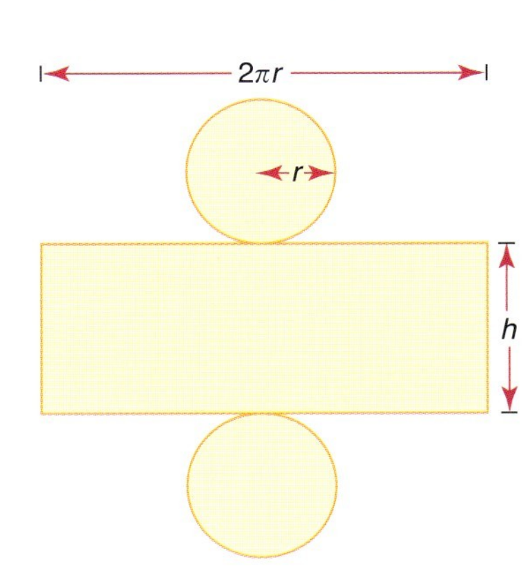

Area of circular pieces $= 2 \times \pi r^2$
Area of rectangular piece $= 2\pi r \times h$
Total surface area $= 2\pi r^2 + 2\pi rh$
$= 2\pi r(r + h)$

Worked examples

1 Calculate the volume and surface area of the L-shaped prism shown.

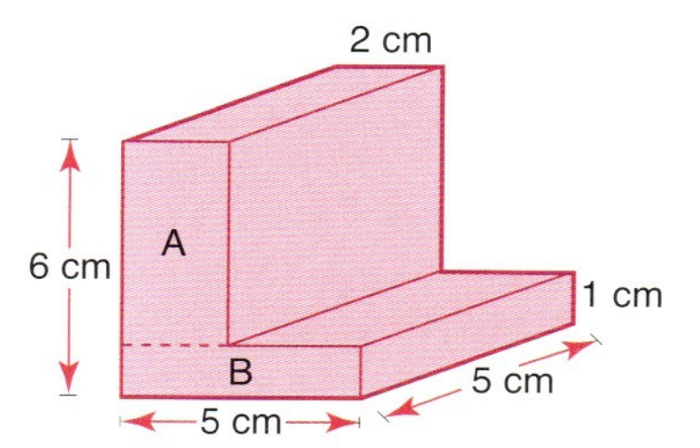

The cross-sectional area can be split into two rectangles:
Area of rectangle A = $5 \times 2 = 10\,\text{cm}^2$
Area of rectangle B = $5 \times 1 = 5\,\text{cm}^2$
Total cross-sectional area = $10\,\text{cm}^2 + 5\,\text{cm}^2 = 15\,\text{cm}^2$
Volume of prism = $15 \times 5 = 75$
The volume of the prism is $75\,\text{cm}^3$.
The surface area of the prism = $2 \times$ cross-sectional area + areas of other faces
$= 2 \times 15 + (6 + 2 + 5 + 3 + 1 + 5) \times 5$
$= 2 \times 15 + 22 \times 5 = 140\,\text{cm}^2$
The surface area of the prism is $140\,\text{cm}^2$.

2 Calculate the volume and surface area of the cylinder shown. Give your answers to one decimal place.

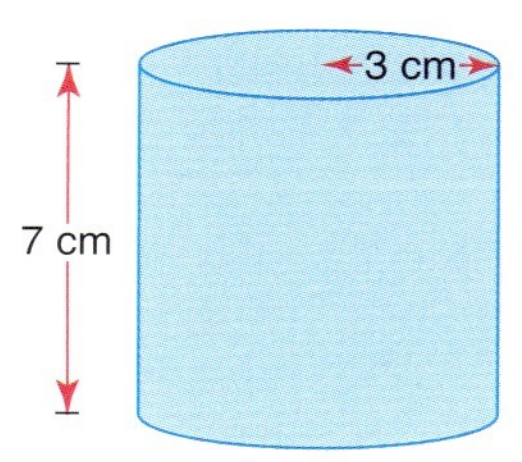

Volume of a prism = area of cross-section × length
$= \pi r^2 h$
$= \pi \times 3^2 \times 7$
$= 63\pi$
$= 197.9$ (1 d.p.)

The volume is $197.9\,\text{cm}^3$.

Total surface area = $2\pi r(r + h)$
$= 2\pi \times 3 \times (3 + 7)$
$= 6\pi \times 10$
$= 60\pi$
$= 188.5$ (1 d.p.)

The total surface area is $188.5\,\text{cm}^2$.

Exercise 5.5.3

1 Calculate the volume and surface area of each of the following prisms. Where necessary, give your answers to one decimal place.
  a A cuboid with these dimensions: $w = 6\,\text{cm}$, $l = 23\,\text{mm}$, $h = 2\,\text{cm}$.
  b A cylinder with a radius of 3.5 cm and a height of 7.2 cm.
  c A triangular prism where the base length of the triangular face is 5 cm, the perpendicular height is 24 mm and the length of the prism is 7 cm.

2 The diagram shows a plan view of a cylinder inside a box the shape of a cube. If the radius of the cylinder is 8 cm, calculate:
  a the height of the cube
  b the volume of the cube
  c the volume of the cylinder
  d the percentage volume of the cube not occupied by the cylinder.

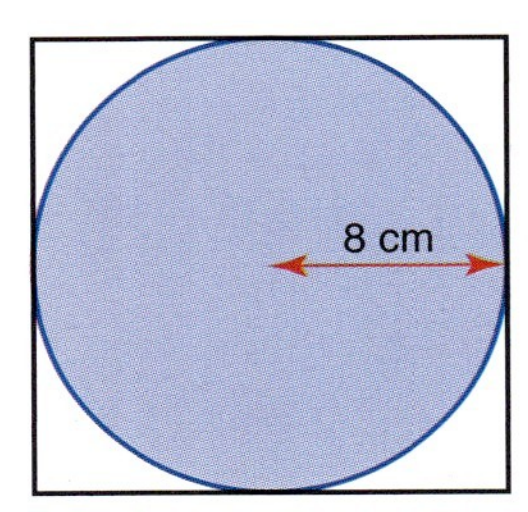

3 A chocolate bar is made in the shape of a triangular prism. The triangular face of the prism is equilateral and has an edge length of 4 cm and a perpendicular height of 3.5 cm. The manufacturer also sells these in special packs of six bars arranged as a hexagonal prism. If the prisms are 20 cm long, calculate:
  a the cross-sectional area of the pack
  b the volume of the pack
  c the surface area of the pack.

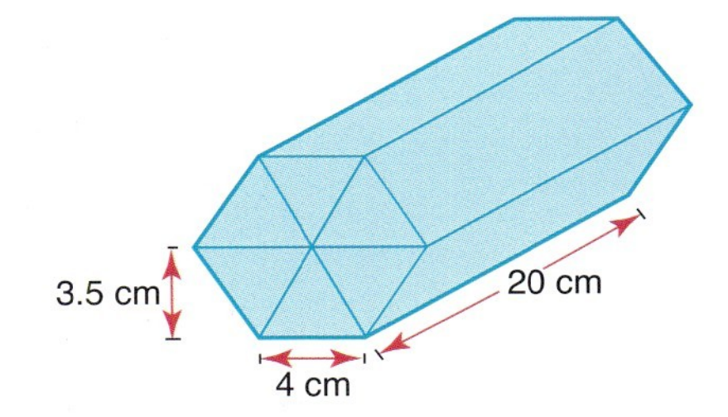

4 A cuboid and a cylinder have the same volume. The radius and height of the cylinder are 2.5 cm and 8 cm respectively. If the length and width of the cuboid are each 5 cm, calculate its height to one decimal place.

5 A section of steel pipe is shown in the diagram. The inner radius is 35 cm and the outer radius 36 cm. Calculate the volume of steel used in making the pipe if it has a length of 130 m.

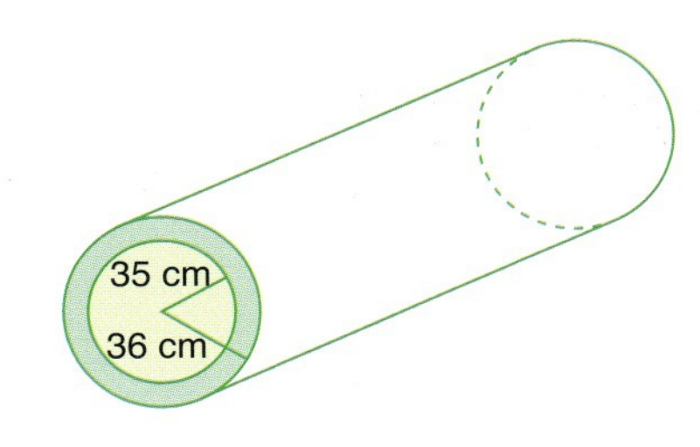

6 Two cubes are placed next to each other. The length of each of the edges of the larger cube is 4 cm. If the ratio of their surface areas is 1 : 4, calculate:

  a the surface area of the small cube
  b the length of an edge of the small cube.

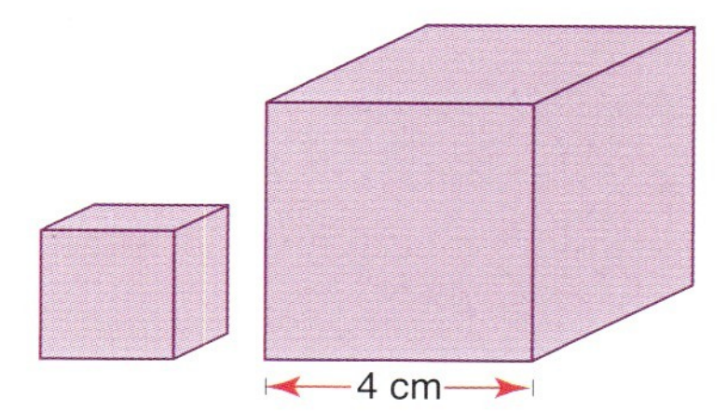

7 A cube and a cylinder have the same surface area. If the cube has an edge length of 6 cm and the cylinder a radius of 2 cm calculate:

  a the surface area of the cube
  b the height of the cylinder
  c the difference between the volumes of the cube and the cylinder.

8 Two cylinders have the same surface area. The shorter of the two has a radius of 3 cm and a height of 2 cm, and the taller cylinder has a radius of 1 cm. Calculate:

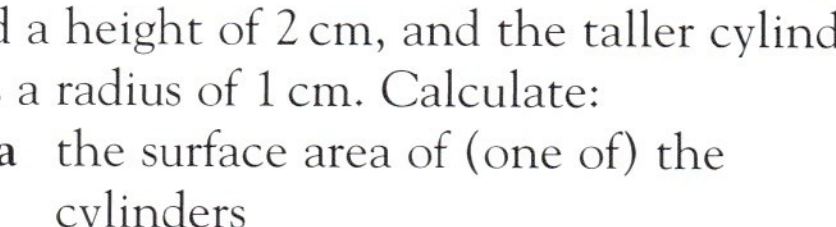

  a the surface area of (one of) the cylinders
  b the height of the taller cylinder
  c the difference between the volumes of the two cylinders.

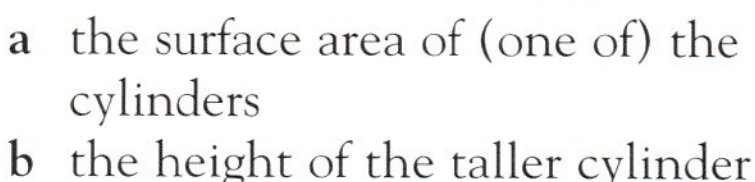

9 Two cuboids have the same surface area. The dimensions of one of the cuboids are

length = 3 cm, width = 4 cm and height = 2 cm.

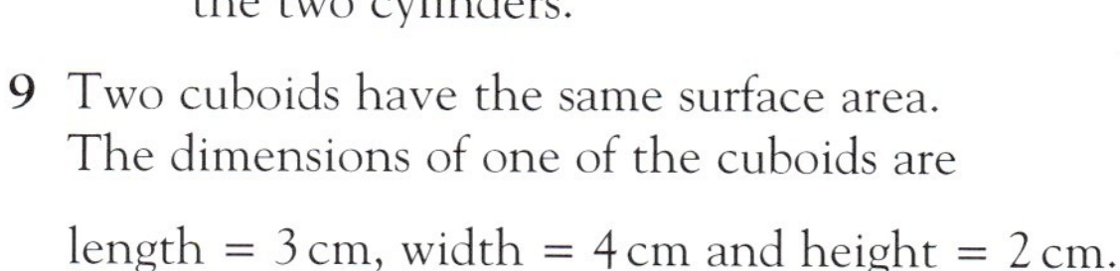

Calculate the height of the other cuboid if its length is 1 cm and its width is 4 cm.

Volume and surface area of a sphere

The volume of a sphere is given by the following formula:

Volume of sphere = $\frac{4}{3}\pi r^3$

The surface area of a sphere is given by the following formula:

Surface area of sphere = $4\pi r^2$

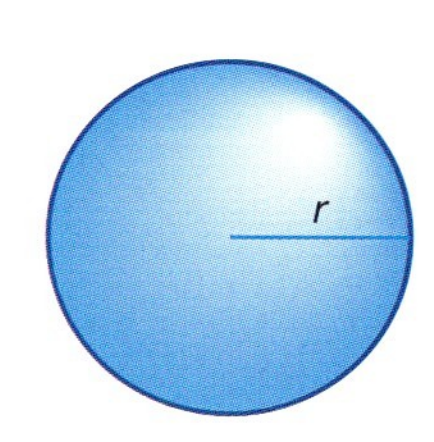

Worked example

1 Calculate the volume and surface area of the sphere shown, giving your answers to one decimal place.

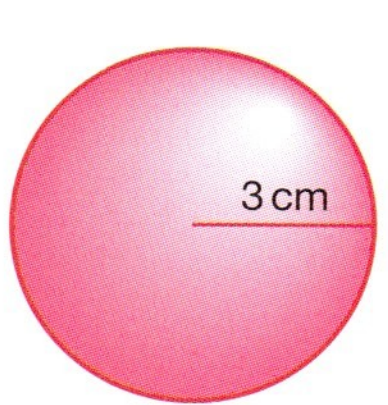

Volume of sphere $= \frac{4}{3}\pi r^3$

$= \frac{4}{3} \times \pi \times 3^3$

$= 113.1$

The volume is $113.1\,\text{cm}^3$.

Surface area of sphere $= 4\pi r^2$

$= 4 \times \pi \times 3^2$

$= 113.1$

The surface area is $113.1\,\text{cm}^2$.

2 Given that the volume of a sphere is $150\,\text{cm}^3$, calculate its radius to one decimal place.

$V = \frac{4}{3}\pi r^3$

$r^3 = \frac{3V}{4\pi}$

$r^3 = \frac{3 \times 150}{4 \times \pi} = 35.8$

$r = \sqrt[3]{35.8} = 3.3$

The radius is 3.3 cm.

Exercise 5.5.4

1 Calculate the volume and surface area of each of the following spheres. The radius r is given in each case.

a $r = 6\,\text{cm}$ b $r = 9.5\,\text{cm}$ c $r = 8.2\,\text{cm}$ d $r = 0.7\,\text{cm}$

2 Calculate the radius of each of the following spheres. Give your answers in centimetres and to one decimal place. The volume V is given in each case.

a $V = 720\,\text{cm}^3$ b $V = 0.2\,\text{m}^3$

3 Calculate the radius of each of the following spheres, given the surface area in each case.

a $A = 16.5\,\text{cm}^2$ b $A = 120\,\text{mm}^2$

4 Given that sphere B has twice the volume of sphere A, calculate the radius of sphere B.

5 cm A

r cm B

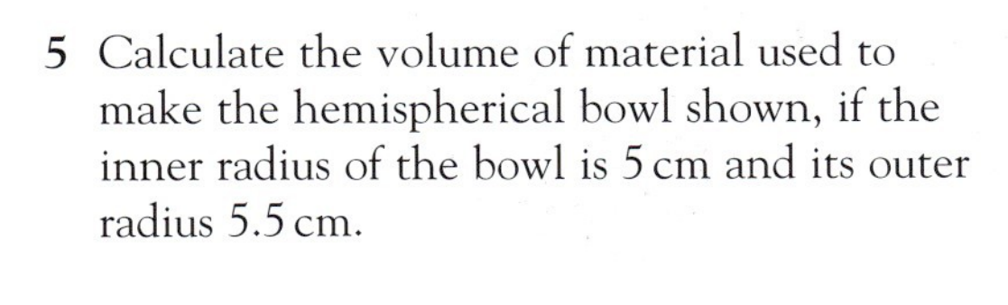

5 Calculate the volume of material used to make the hemispherical bowl shown, if the inner radius of the bowl is 5 cm and its outer radius 5.5 cm.

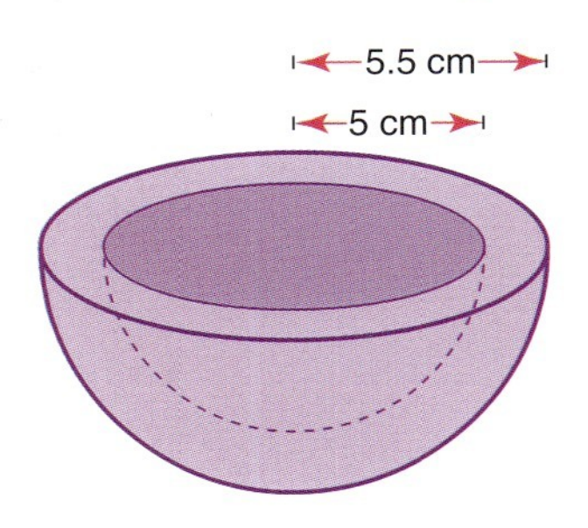

6 The volumes of the material used to make the sphere and hemispherical bowl shown are the same. Given that the radius of the sphere is 7 cm and the inner radius of the bowl is 10 cm, calculate the outer radius r cm of the bowl.

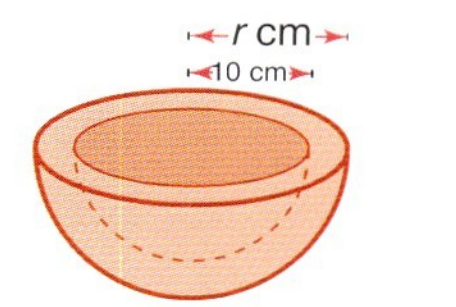

7 A ball is placed inside a box into which it will fit tightly. If the radius of the ball is 10 cm, calculate:

a the volume of the ball
b the volume of the box
c the percentage volume of the box not occupied by the ball.

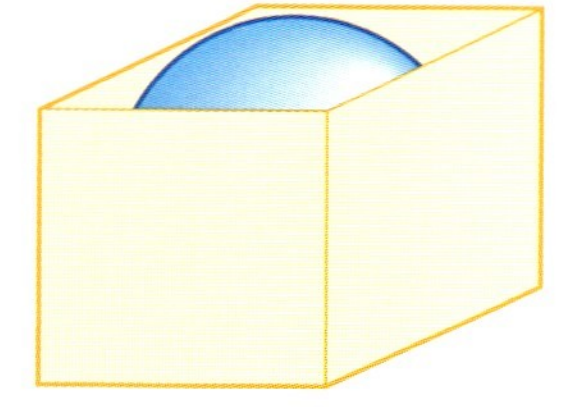

8 A steel ball is melted down to make eight smaller identical balls. If the radius of the original steel ball was 20 cm, calculate to the nearest millimetre the radius of each of the smaller balls.

9 A steel ball of volume 600 cm^3 is melted down and made into three smaller balls A, B and C. If the volumes of A, B and C are in the ratio 7 : 5 : 3, calculate to one decimal place the radius of each of A, B and C.

10 The cylinder and sphere shown have the same radius and the same height. Calculate the ratio of their volumes, giving your answer in the form:

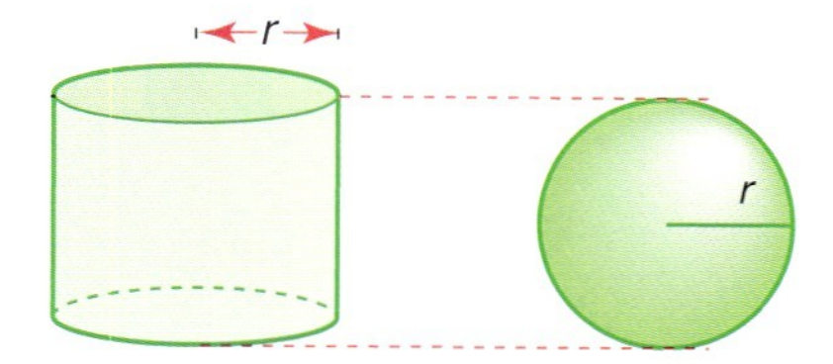

volume of cylinder : volume of sphere.

11 Sphere A has a radius of 8 cm and sphere B has a radius of 16 cm. Calculate the ratio of their surface areas in the form 1 : n.

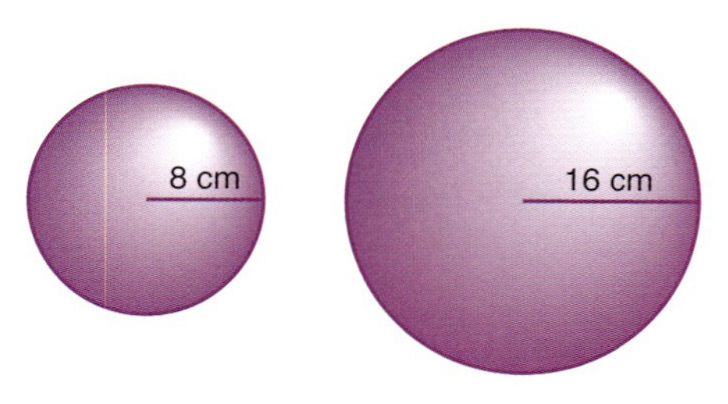

12 A hemisphere of diameter 10 cm is attached to a cylinder of equal diameter as shown.

If the total length of the shape is 20 cm, calculate:

a the surface area of the hemisphere
b the length of the cylinder
c the surface area of the whole shape.

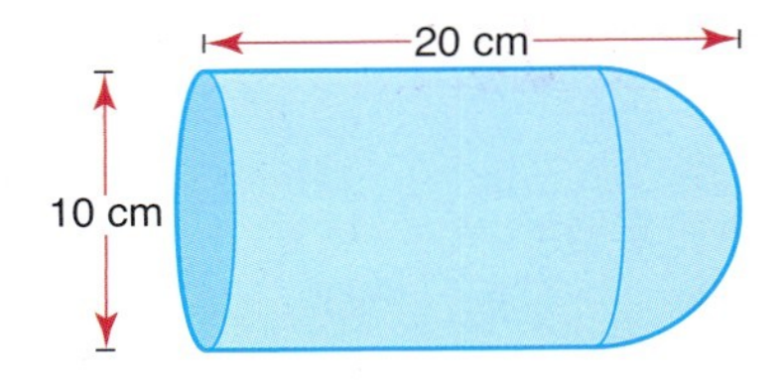

13 A sphere and a cylinder both have the same surface area and the same height of 16 cm.

Calculate, to one decimal place:

a the surface area of the sphere
b the radius of the cylinder.

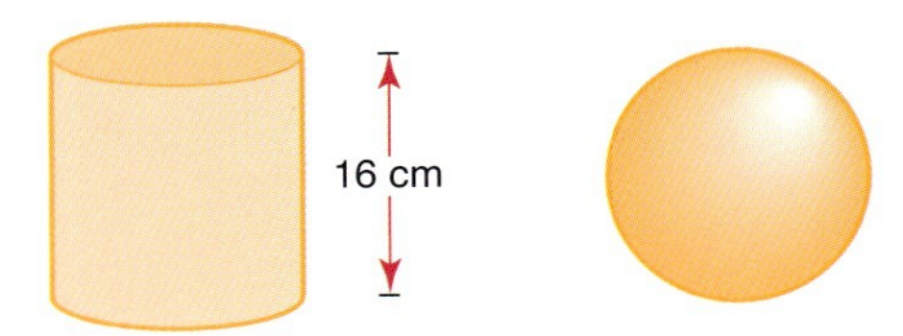

Volume and surface area of a pyramid

A **pyramid** is a three-dimensional shape. Each of the faces of a pyramid is planar (not curved). A pyramid has a polygon for its base and the other faces are triangles with a common vertex, known as the **apex**. A pyramid's individual name is taken from the shape of the base.

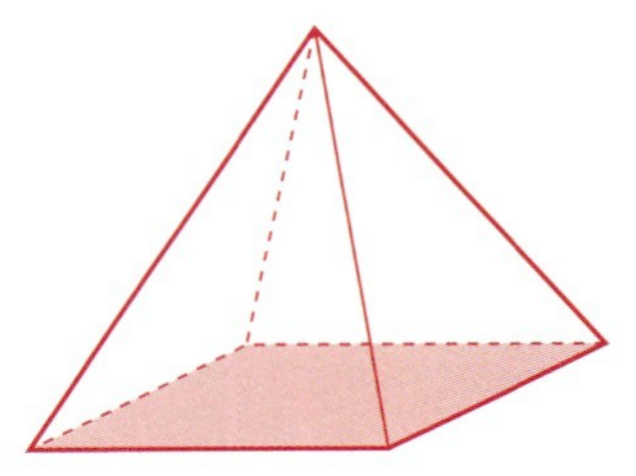

Square-based pyramid

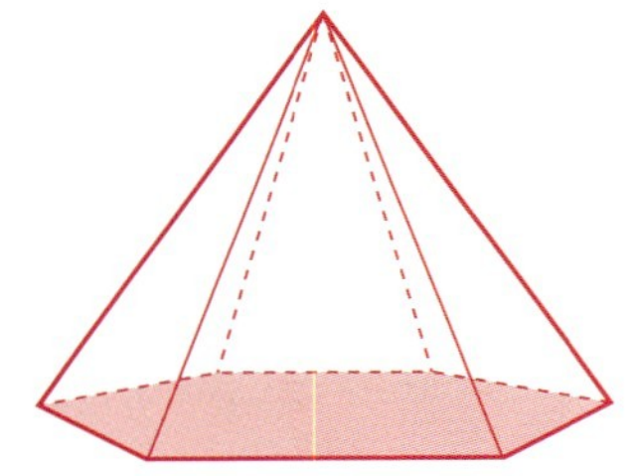

Hexagonal-based pyramid

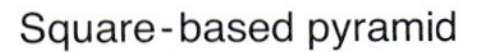

The volume of any pyramid is given by the following formula:

Volume of a pyramid $= \frac{1}{3} \times$ area of base $\times$ perpendicular height

The surface area of a pyramid is found simply by adding together the areas of all of its faces. You may need to use Pythagoras' theorem to work out the dimensions you need to calculate the volume and surface area.

Worked example

A rectangular-based pyramid has a perpendicular height of 5 cm and base dimensions as shown. Calculate the volume and surface area of the pyramid.

$$\text{Volume} = \tfrac{1}{3} \times \text{base area} \times \text{height}$$
$$= \tfrac{1}{3} \times 3 \times 7 \times 5$$
$$= 35$$

The volume is $35\,\text{cm}^3$.

To work out the surface area you need to know the perpendicular height of the triangular faces of the pyramid.

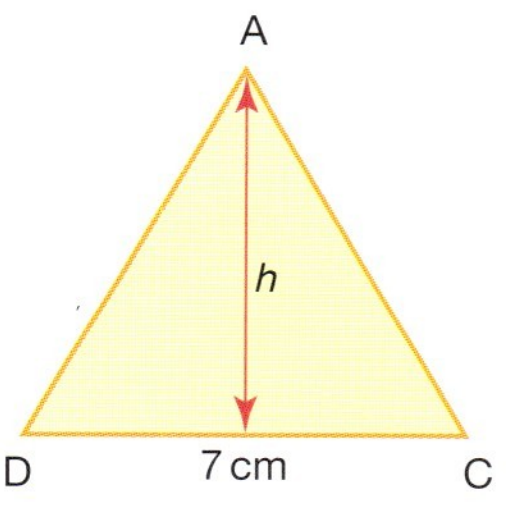

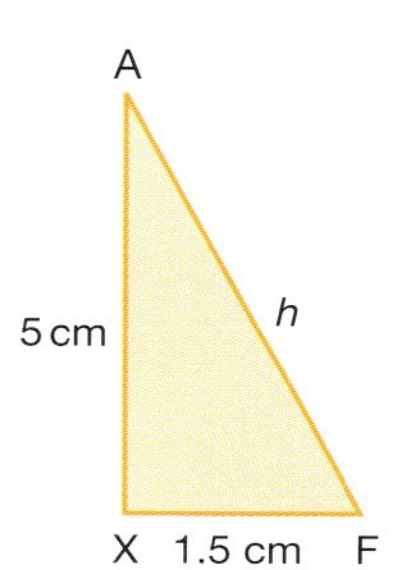

$$h^2 = 1.5^2 + 5^2 = 27.25$$
$$h = \sqrt{27.25}$$

$$\text{Area of two of the triangular faces} = 2 \times \tfrac{1}{2} \times 7 \times \sqrt{27.25}$$
$$= 36.541$$

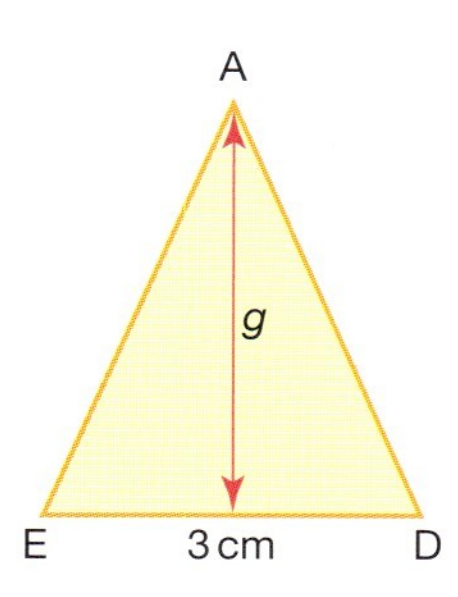

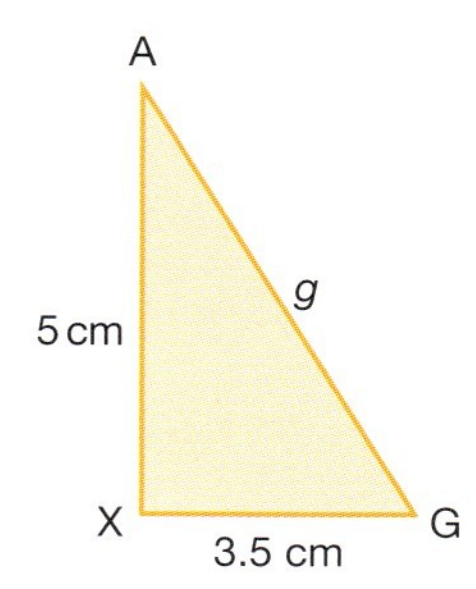

$$g^2 = 3.5^2 + 5^2 = 37.25$$
$$g = \sqrt{37.25}$$

$$\text{Area of other two of the triangular faces} = 2 \times \tfrac{1}{2} \times 3 \times \sqrt{37.25}$$
$$= 18.310$$

$$\text{Total surface area} = \text{area of base} + \text{area of triangular faces}$$
$$= 21 + 54.851$$
$$= 75.9$$

The surface area of the pyramid is $75.9\,\text{cm}^2$.

Exercise 5.5.5

1 Find the volume of each of the following pyramids.

a

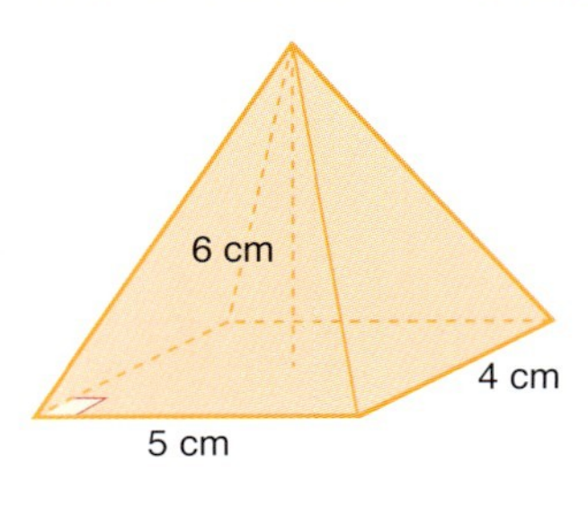

b

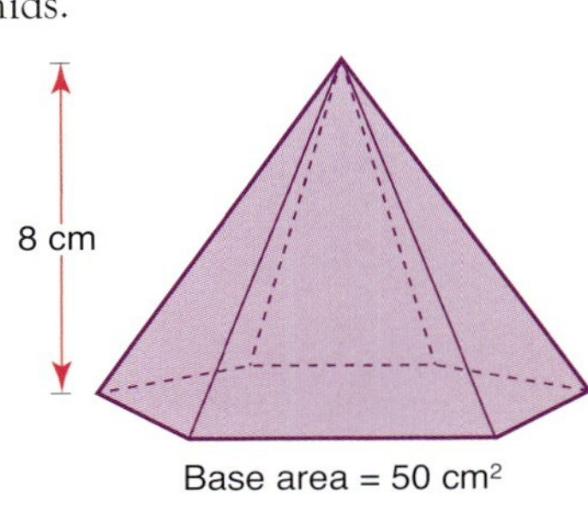

c

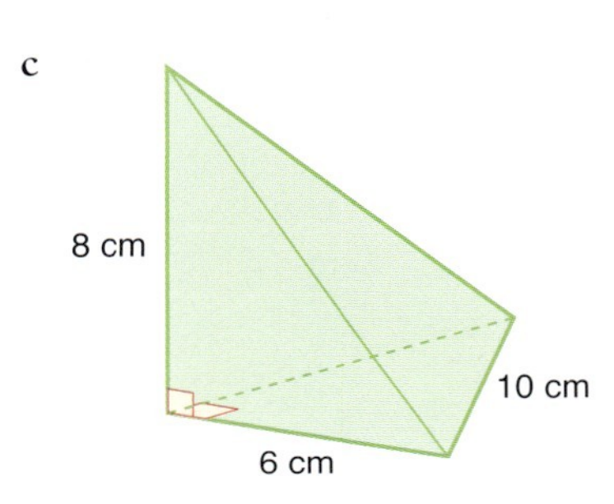

d

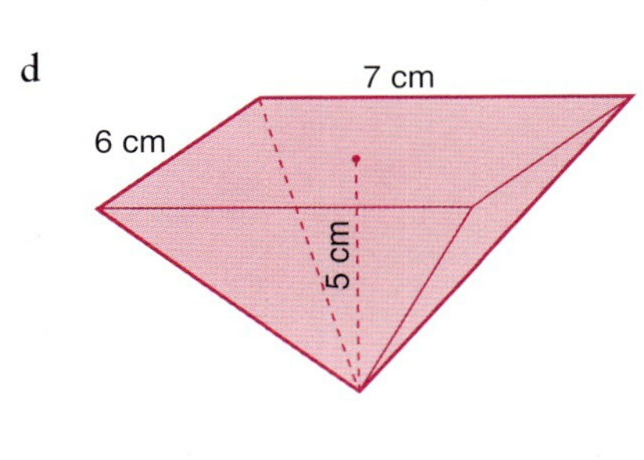

2 The rectangular-based pyramid shown has a sloping edge length of 12 cm. Calculate its volume and surface area, giving your answers to one decimal place.

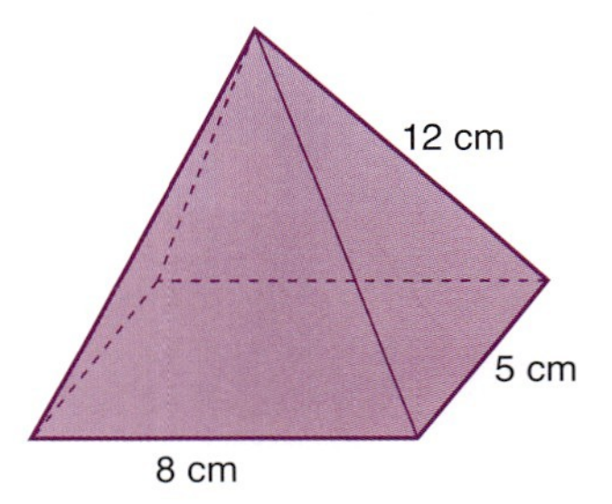

3 Two square-based pyramids are glued together as shown. Given that all the triangular faces are identical, calculate the volume and surface area of the whole shape.

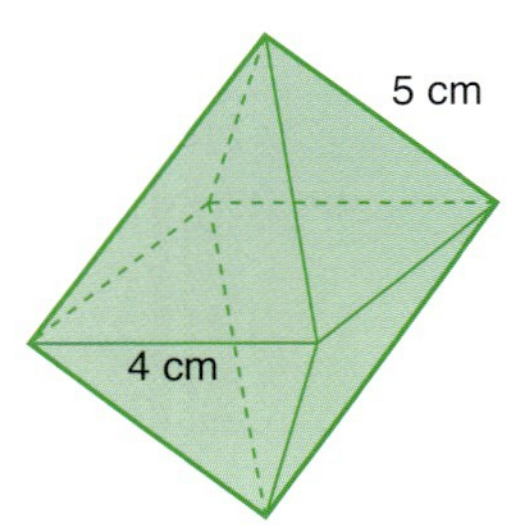

4 The top of a square-based pyramid is cut off. The cut is made parallel to the base. If the base of the smaller pyramid has a side length of 3 cm and the vertical height of the truncated pyramid is 6 cm, calculate:

  a the height of the original pyramid
  b the volume of the original pyramid
  c the volume of the truncated pyramid.

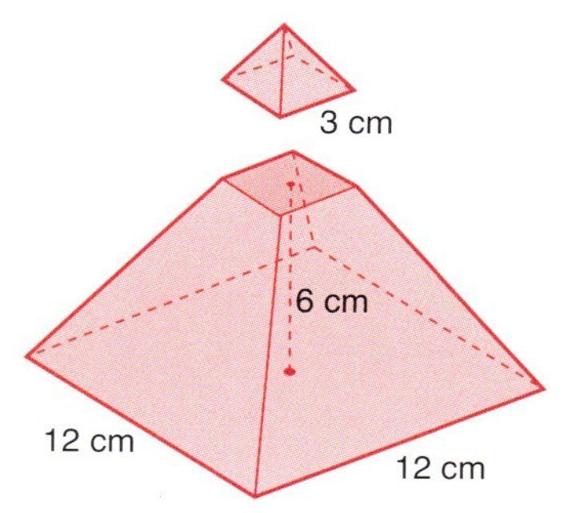

5 Calculate, to one decimal place, the surface area of the truncated square-based pyramid shown. Assume that all the sloping faces are identical.

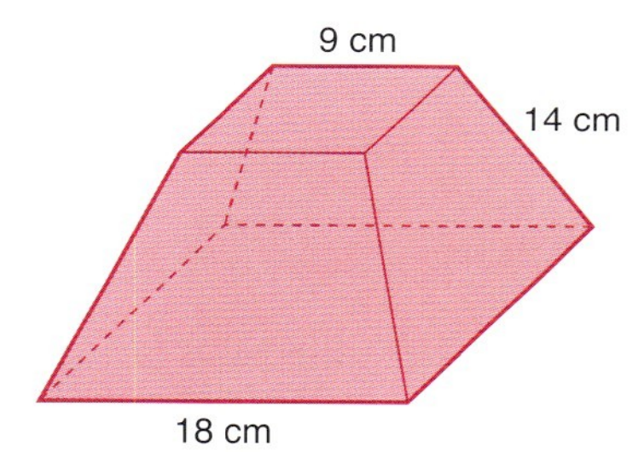

6 Calculate the perpendicular height h cm for the pyramid, given that it has a volume of 168 cm^3.

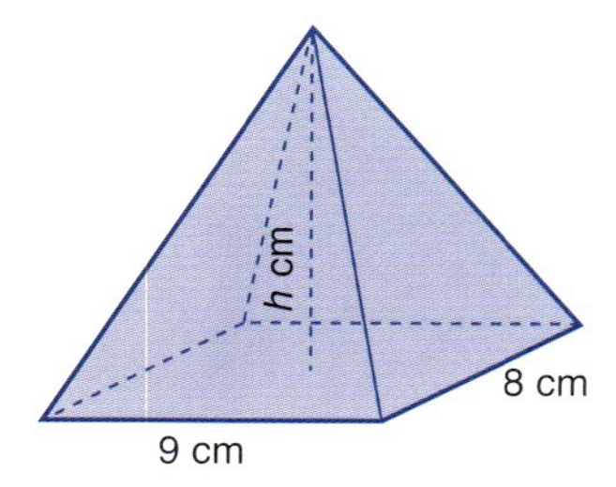

7 Calculate the length of the edge marked x cm, given that the volume of the pyramid is 14 cm^3.

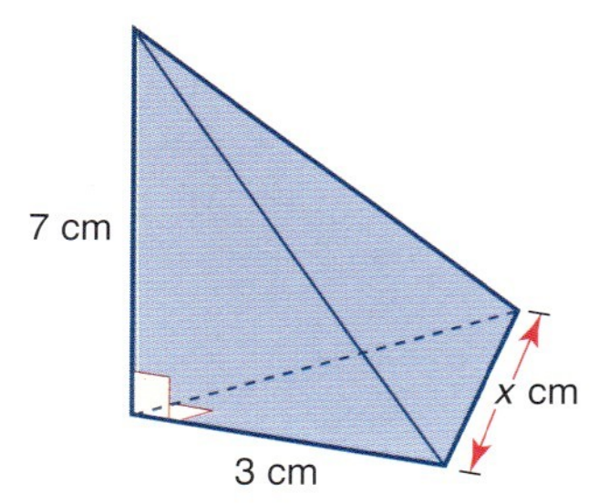

8 The top of a triangular-based pyramid (tetrahedron) is cut off. The cut is made parallel to the base. If the vertical height of the top is 6 cm, calculate to one decimal place:

a the height of the truncated piece
b the volume of the small pyramid
c the volume of the original pyramid.

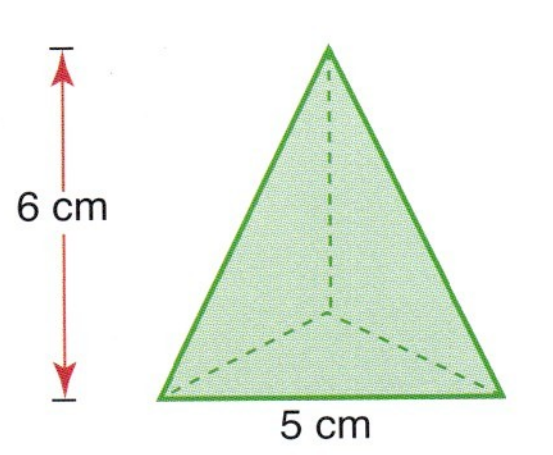

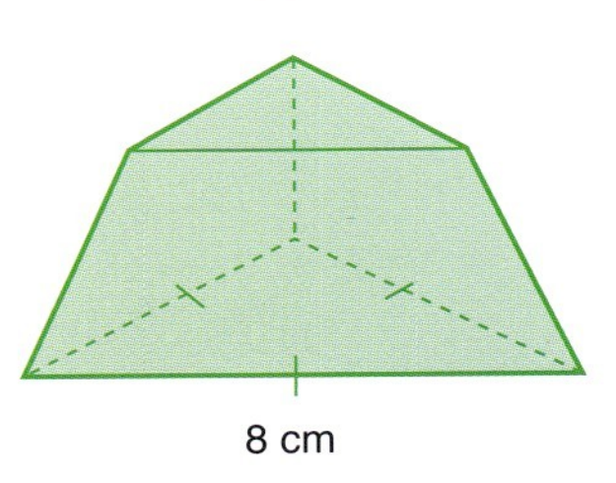

9 Calculate the surface area of a regular tetrahedron with edge length 2 cm.

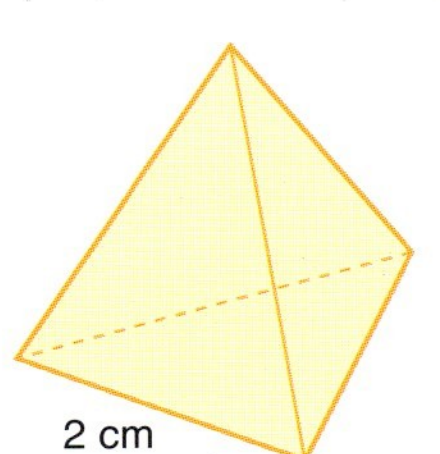

10 The two pyramids shown below have the same surface area.

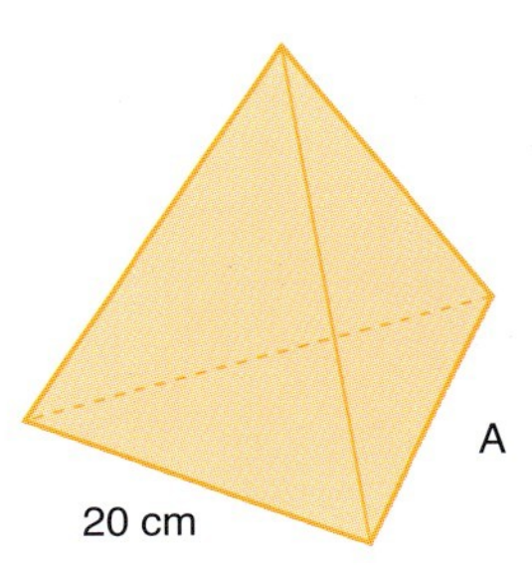

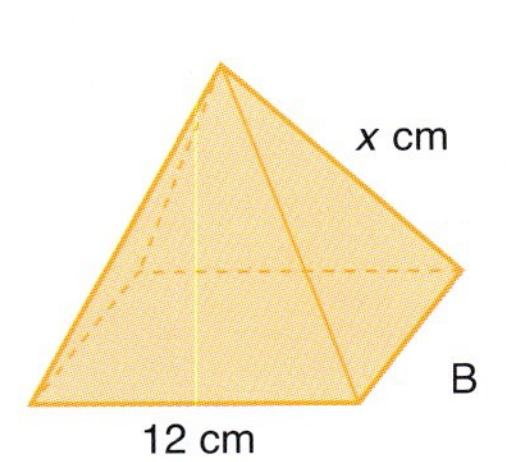

Calculate:

a the surface area of the tetrahedron
b the area of one of the triangular faces on the square-based pyramid
c the value of x.

Volume and surface area of a cone

A cone is a pyramid with a circular base. The formula for its volume is therefore the same as for any other pyramid.

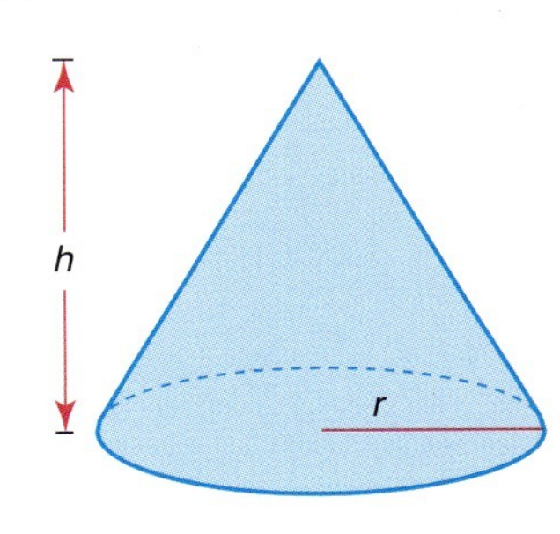

$$\text{Volume} = \frac{1}{3} \times \text{base area} \times \text{height}$$

$$= \pi r^2 h$$

Before we look at examples of finding the volume and surface area of a cone it is useful to look at how a cone is formed. A cone can be constructed from a sector of a circle.

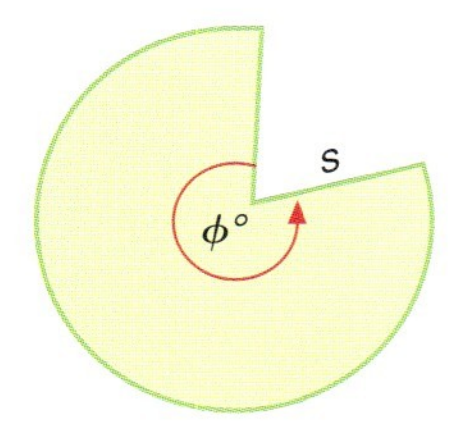

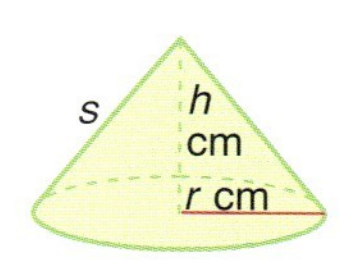

The length of the sloping side of the cone is equal to the radius of the sector. The base circumference is equal to the arc length of the sector. The curved surface area of the cone is equal to the area of the sector.

Arc length and sector area

An **arc** is part of the circumference of a circle between two radii. Its length is proportional to the size of the angle ϕ between the two radii. The length of the arc as a fraction of the circumference of the whole circle is therefore equal to the fraction that ϕ is of 360°.

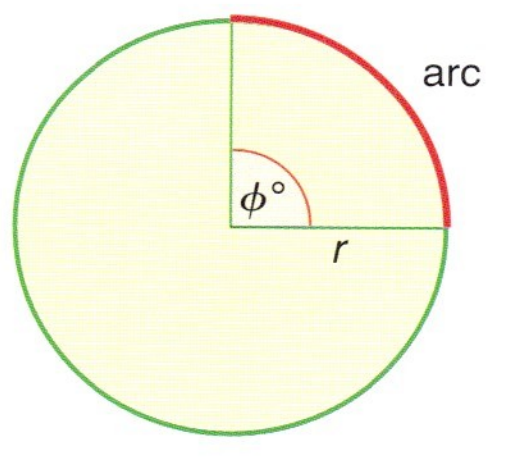

$$\text{Arc length} = \frac{\phi}{360} \times 2\pi r$$

A **sector** is the region of a circle enclosed by two radii and an arc. The area of a sector is proportional to the size of the angle ϕ between the two radii. As a fraction of the area of the whole circle, it is therefore equal to the fraction that ϕ is of 360°.

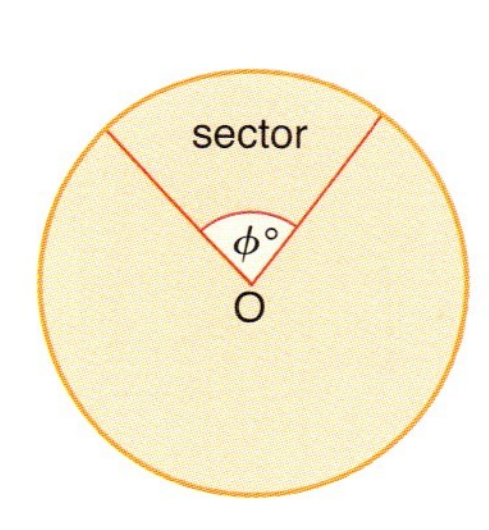

$$\text{Area of sector} = \frac{\phi}{360} \times \pi r^2$$

Calculating the volume of a cone

As we have seen, the formula for the volume of a cone is:

$$\begin{aligned}\text{Volume} &= \tfrac{1}{3} \times \text{base area} \times \text{height} \\ &= \tfrac{1}{3}\pi r^2 h\end{aligned}$$

Worked examples

1 Calculate the volume of the cone.

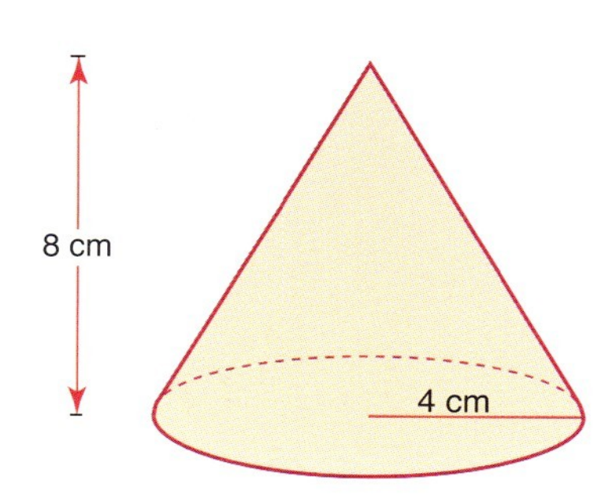

$$\begin{aligned}\text{Volume} &= \tfrac{1}{3}\pi r^2 h \\ &= \tfrac{1}{3} \times \pi \times 4^2 \times 8 \\ &= 134.0 \text{ (1 d.p.)}\end{aligned}$$

The volume is $134\,\text{cm}^3$

2 The sector below is assembled to form a cone as shown.

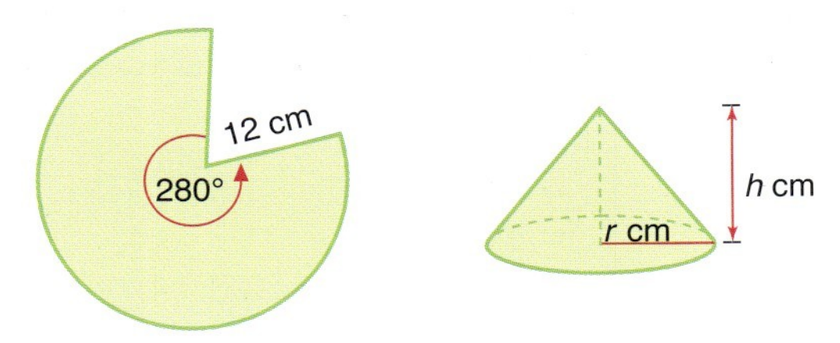

a Calculate the base circumference of the cone.
b Calculate the base radius of the cone.
c Calculate the vertical height of the cone.
d Calculate the volume of the cone.

a The base circumference of the cone is equal to the arc length of the sector.

$$\text{sector arc length} = \frac{\phi}{360} \times 2\pi r$$
$$= \frac{280}{360} \times 2\pi \times 12 = 58.6$$

So the base circumference is 58.6 cm.

b The base of a cone is circular, therefore:

$$C = 2\pi r$$
$$r = \frac{C}{2\pi} = \frac{58.6}{2\pi}$$
$$= 9.33$$

So the radius is 9.33 cm.

c The vertical height of the cone can be calculated using Pythagoras' theorem on the right-angled triangle enclosed by the base radius, vertical height and the sloping face, as shown below.

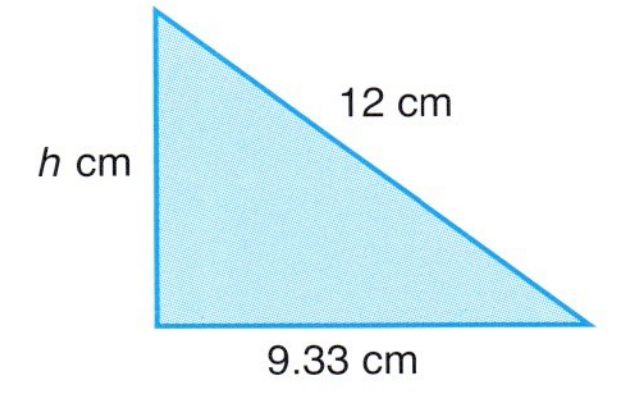

Note that the length of the sloping side is equal to the radius of the sector.

$$12^2 = h^2 + 9.33^2$$
$$h^2 = 12^2 - 9.33^2$$
$$h^2 = 56.9$$
$$h = 7.54$$

So the height is 7.54 cm.

d $$\text{Volume} = \tfrac{1}{3} \times \pi r^2 h$$
$$= \tfrac{1}{3} \times \pi \times 9.33^2 \times 7.54 = 688$$

So the volume is 688 cm^3.

Note: Although answers were given to three significant figures in each case, where the answer was needed in a subsequent calculation, the exact value was used and not the rounded one. By doing this we avoid introducing rounding errors into the calculations.

Calculating the surface area of a cone

The surface area of a cone comprises the area of the circular base and the area of the curved face. The area of the curved face is equal to the area of the sector from which it is formed.

Worked example

Calculate the total surface area of this cone.

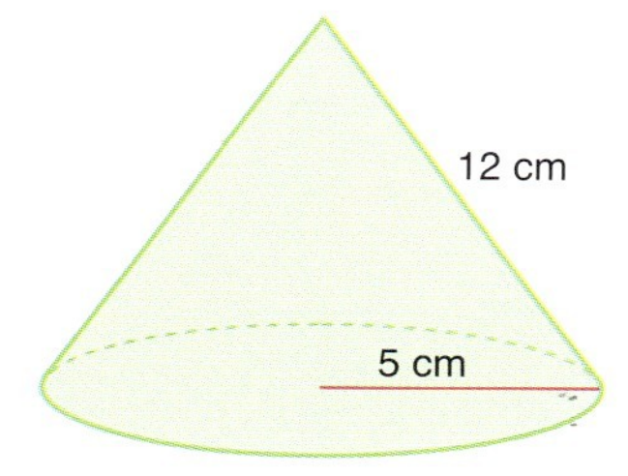

$$\text{Surface area of base} = \pi r^2$$
$$= 25\pi$$

The curved surface area can best be visualized if drawn as a sector as shown in the diagram:

12 cm

$\phi°$

10π cm

The radius of the sector is equivalent to the slant height of the cone. The arc length of the sector is equivalent to the base circumference of the cone.

$$\frac{\phi}{360} = \frac{10\pi}{24\pi}$$

$$\text{Therefore } \phi = 150°$$

$$\text{Area of sector} = \frac{150}{360} \times \pi \times 12^2 = 60\pi$$

$$\text{Total surface area} = 60\pi + 25\pi$$
$$= 85\pi$$
$$= 267$$

The total surface area is 267 cm^2.

Exercise 5.5.6

1 For each of the following, calculate the length of the arc and the area of the sector. O is the centre of the circle.

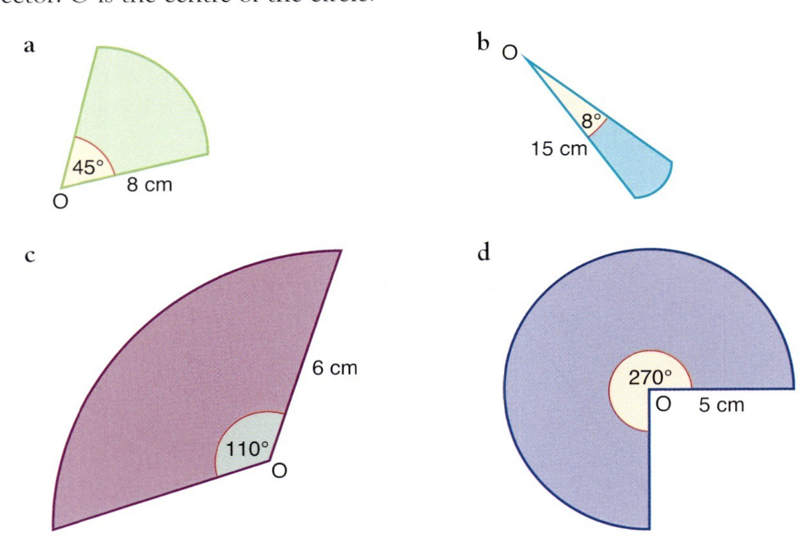

2 A circular cake is cut. One of the slices is shown.

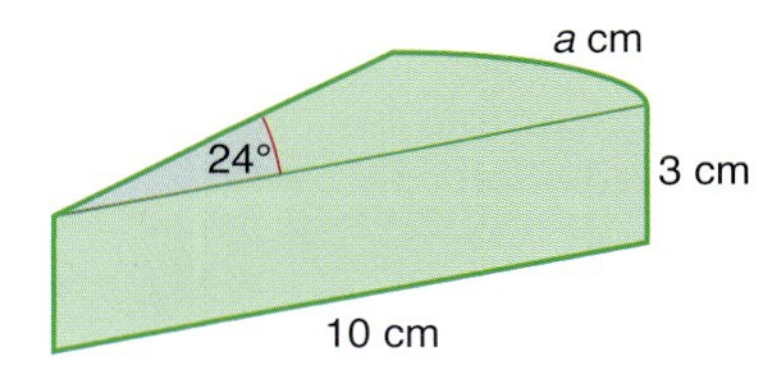

Calculate:

a the length a cm of the arc

b the total surface area of all the sides of the slice

c the volume of the slice.

3 Calculate the volume of each of the following cones. Use the values for the base radius r and the vertical height h given in each case.

a $r = 3$ cm, $h = 6$ cm

b $r = 6$ cm, $h = 7$ cm

c $r = 8$ mm, $h = 2$ cm

d $r = 6$ cm, $h = 44$ mm

4 Calculate the base radius of a cone with a volume of $600\,\text{cm}^3$ and a vertical height of 12 cm.

5 A cone has a base circumference of 100 cm and a slope height of 18 cm. Calculate:

a the base radius

b the vertical height

c the volume.

6 Calculate the volume and surface area of both of the following cones. Give your answers to one decimal place.

a

16 cm

6 cm

b

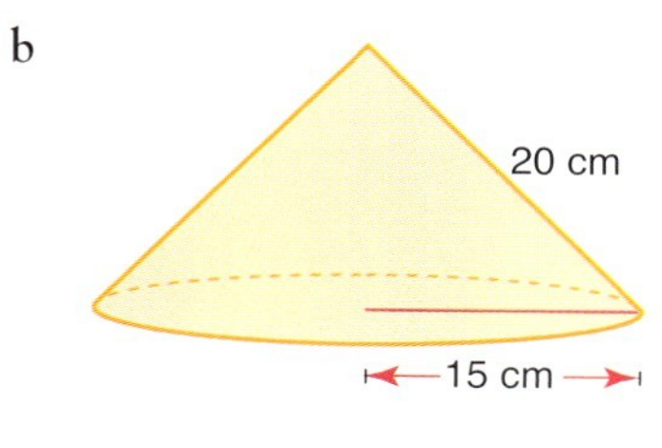

7 The two cones A and B shown below have the same volume. Using the dimensions shown and given that the base circumference of cone B is 60 cm, calculate the height h cm.

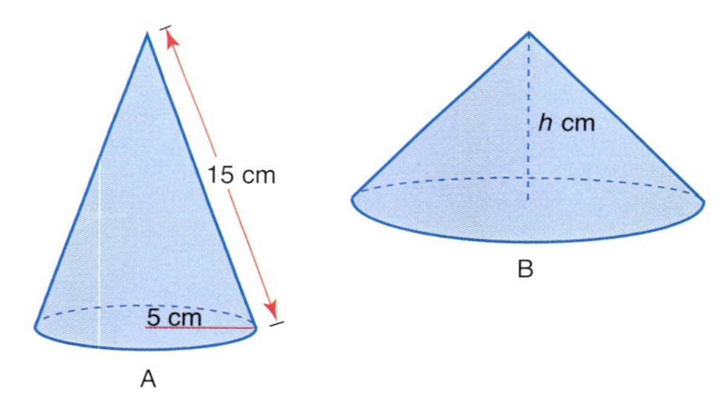

8 The sector shown is assembled to form a cone. Calculate:

a the base circumference of the cone
b the base radius of the cone
c the vertical height of the cone
d the volume of the cone
e the curved surface area of the cone.

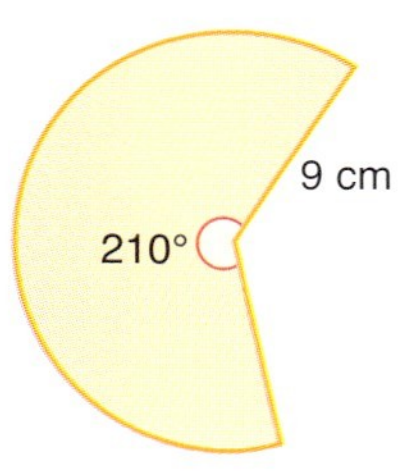

9 Two cones with the same base radius are stuck together as shown. Calculate the surface area of the shape.

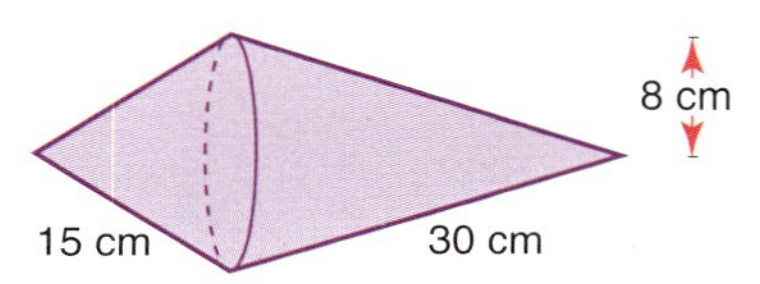

10 Two cones have the same total surface area.

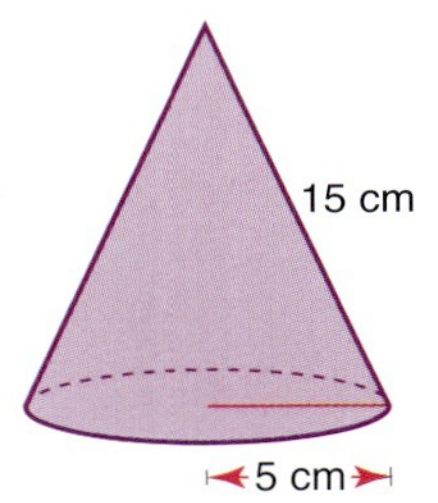

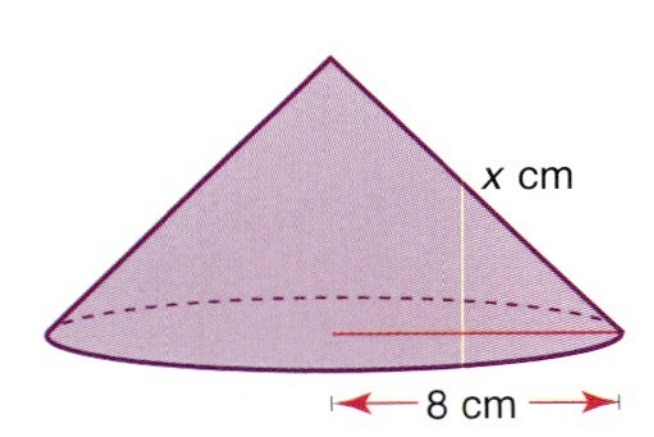

Calculate:

a the total surface area of each cone

b the value of x.

11 A cone is placed inside a cuboid as shown. If the base diameter of the cone is 12 cm and the height of the cuboid is 16 cm, calculate, to one decimal place where required:

a the volume of the cuboid

b the volume of the cone

c the volume of the cuboid not occupied by the cone.

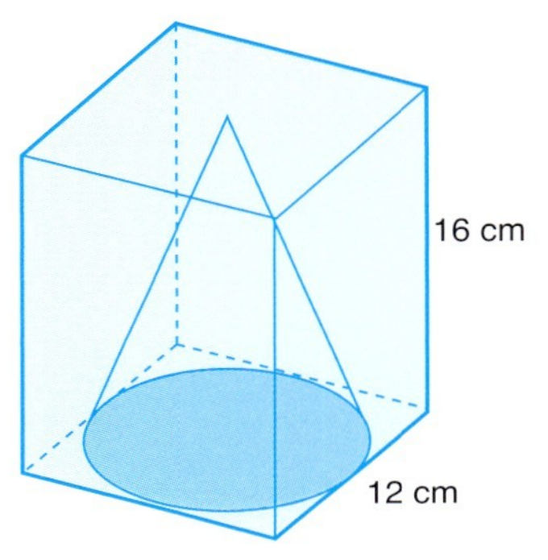

12 The diagram shows two similar sectors which are assembled into cones. Calculate:

a the volume of the smaller cone

b the volume of the larger cone

c the ratio of their volumes.

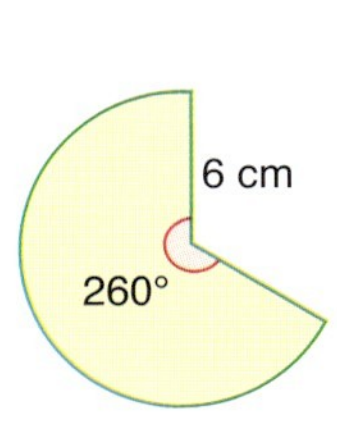

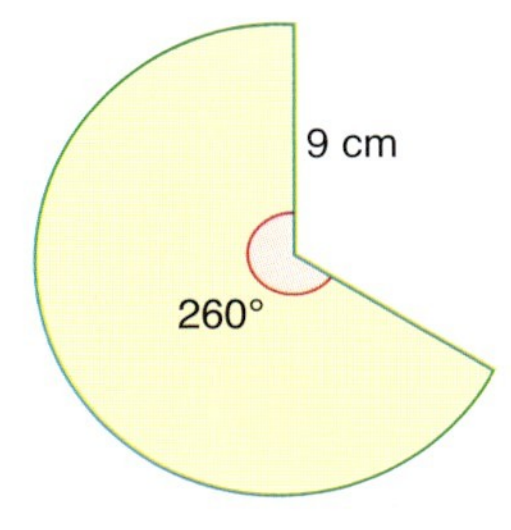

13 An ice cream consists of a hemisphere and a cone. Calculate:

a its total volume

b its total surface area.

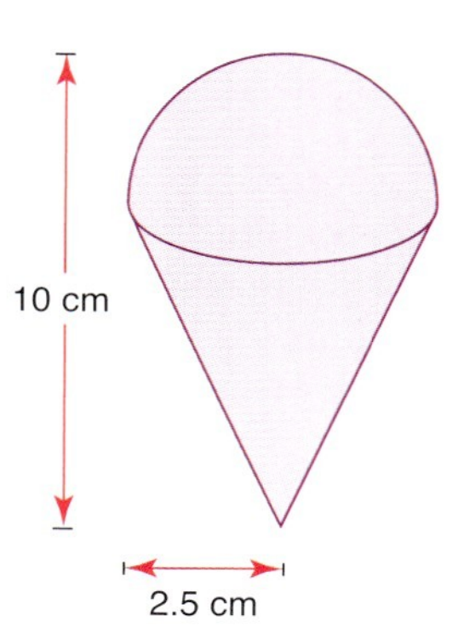

14 A cone is placed on top of a cylinder. Using the dimensions given, calculate

a the total volume of the shape

b its total surface area.

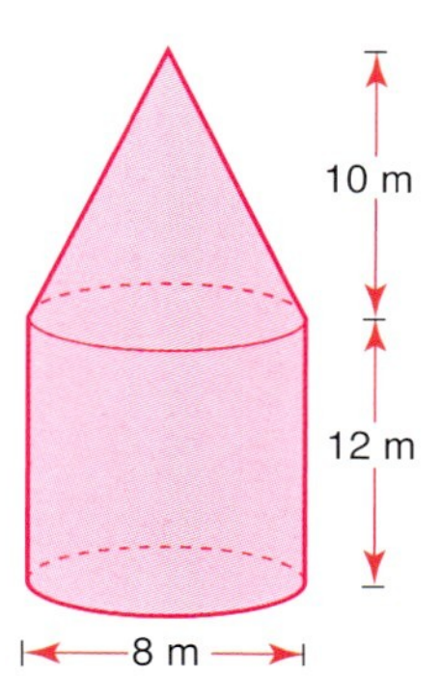

15 Two identical truncated cones are placed end to end as shown.

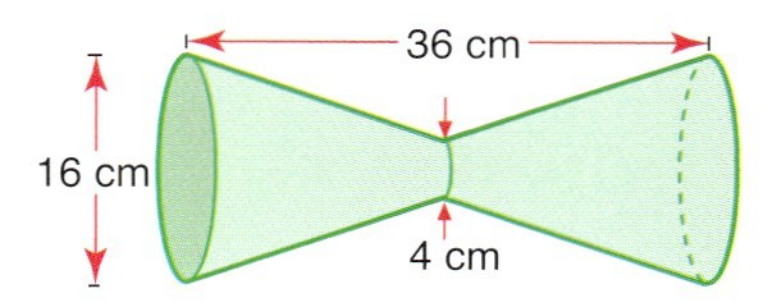

Calculate the total volume of the shape. Give your answer to one decimal place.

16 Two cones A and B are placed either end of a cylindrical tube as shown.

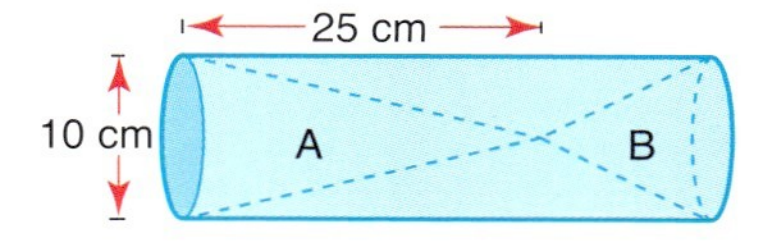

Given that the volumes of A and B are in the ratio 2 : 1, calculate, to one decimal place:

a the volume of cone A

b the height of cone B

c the volume of the cylinder.

Student assessment 1

1 The coordinates of the end points of two line segments are given below. In each case:
   i) calculate the distance between the two end points
   ii) give the coordinates of the midpoint.

   a $(-6, -1)$ $(6, 4)$
   b $(1, 2)$ $(7, 10)$
   c $(2, 6)$ $(-2, 3)$
   d $(-10, -10)$ $(0, 14)$

2 Sketch the following graphs on the same pair of axes, labelling each clearly.
   a $x = -2$
   b $y = 3$
   c $y = -3x$
   d $y = \frac{x}{4} + 4$

3 For each of the following linear equations:
   i) find the gradient and y-intercept
   ii) plot the graph.

   a $y = x + 1$ b $y = 3 - 3x$
   c $2x - y = -4$ d $2y - 5x = 8$

4 Find the equation of the straight line which passes through each of the following pairs of points. Express your answers in the form $y = mx + c$.
   a $(1, -1)$ $(4, 8)$ b $(0, 7)$ $(3, 1)$
   c $(-2, -9)$ $(5, 5)$ d $(1, -1)$ $(-1, 7)$

5 Solve the following pairs of simultaneous equations either algebraically or graphically.
   a $x + y = 4$
     $x - y = 0$
   b $3x + y = 2$
     $x - y = 2$
   c $y + 4x + 4 = 0$
     $x + y = 2$
   d $x - y = -2$
     $3x + 2y + 6 = 0$

Student assessment 2

Diagrams are not drawn to scale.

1 Calculate the length of the side marked x cm in each of the following. Give your answers correct to one decimal place.

a

x cm
8 cm
30°

b

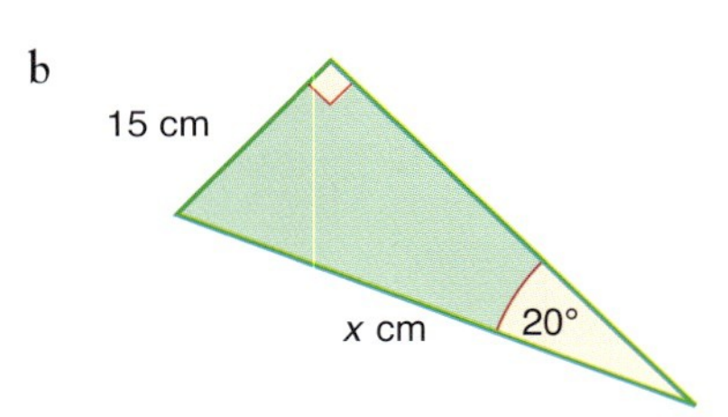

c

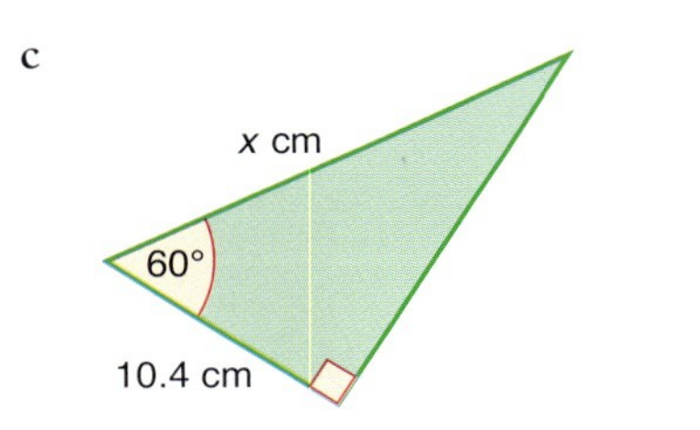

d

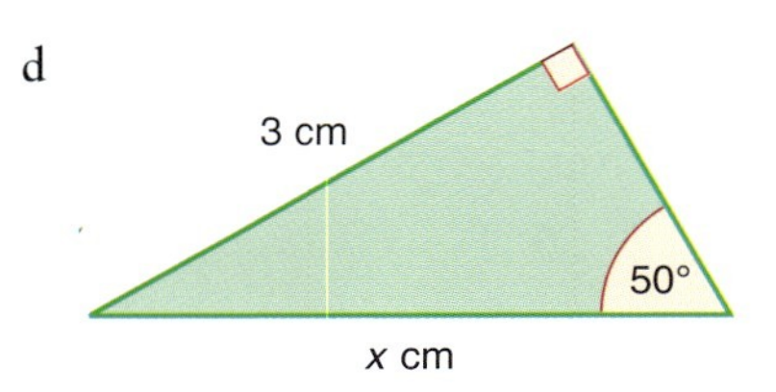

2 Calculate the angle marked $\theta°$ in each of the following. Give your answers correct to the nearest degree.

a

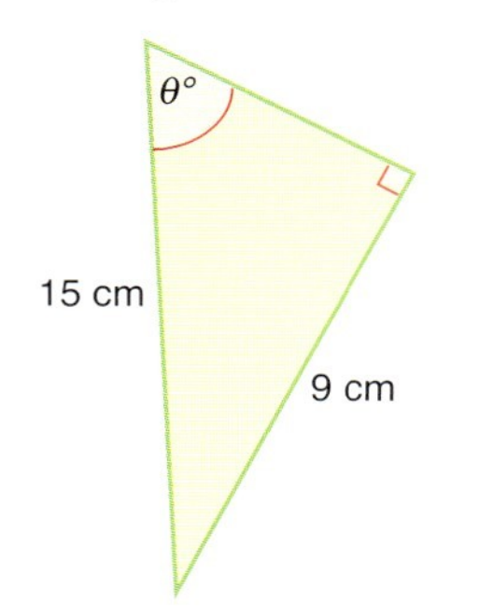

b

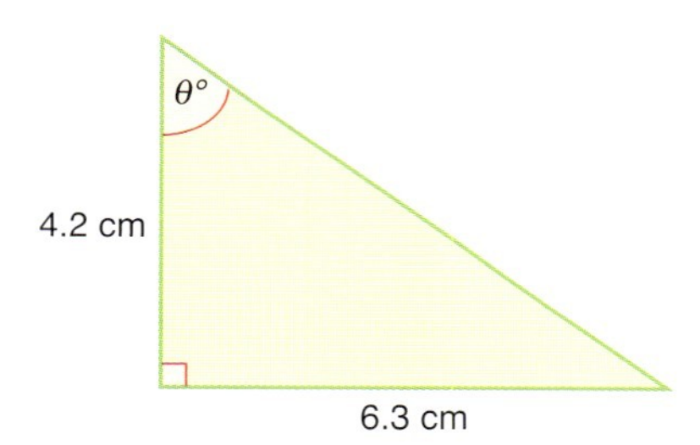

c

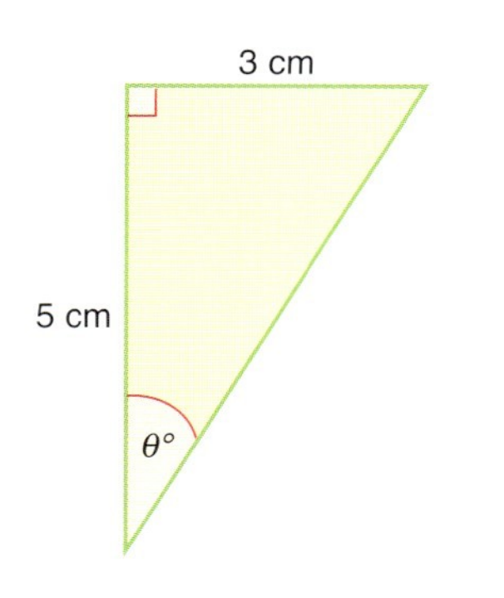

d

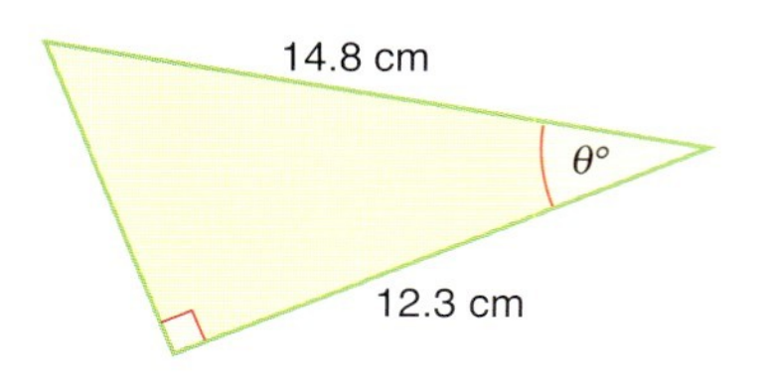

3 Calculate the length of the side marked q cm in each of the following. Give your answers correct to one decimal place.

a

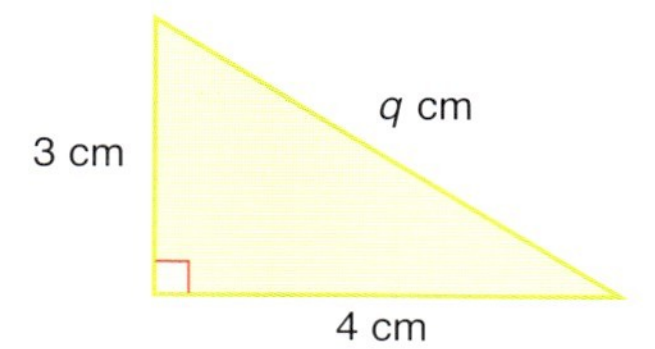

b

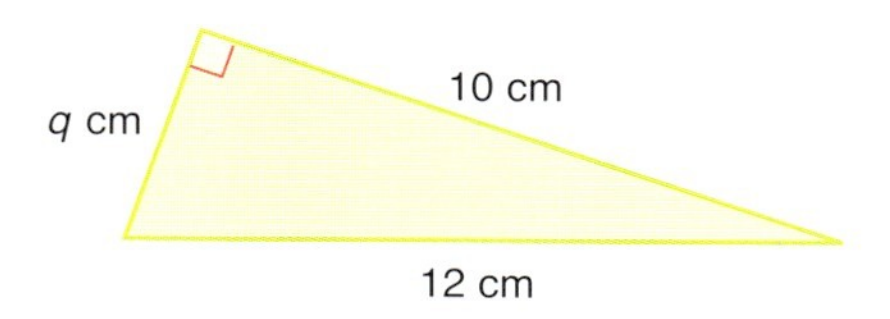

c

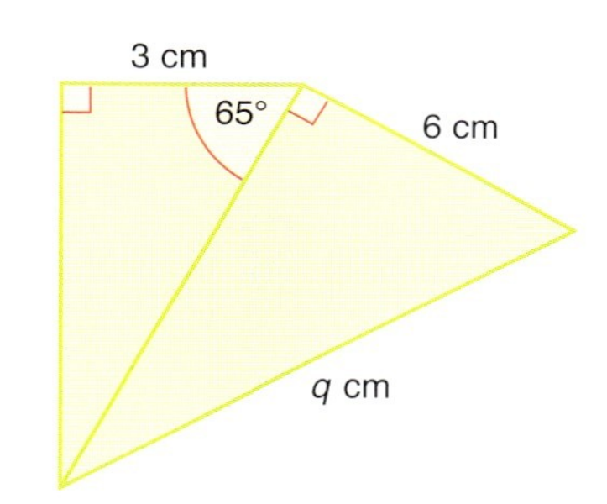

d

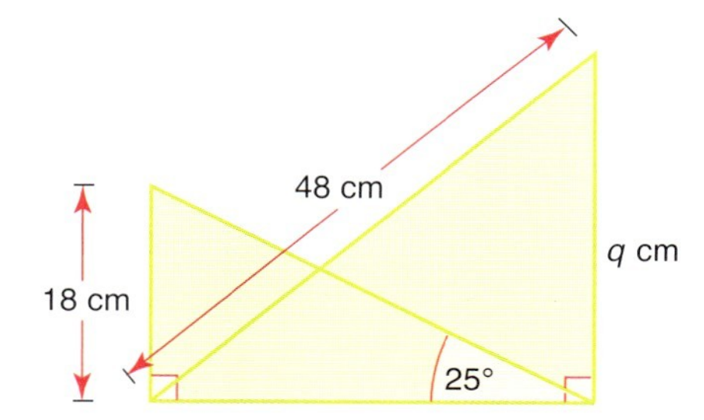

Student assessment 3

Diagrams are not drawn to scale.

1 A map shows three towns A, B and C. Town A is due north of C. Town B is due east of A. The distance AC is 75 km and the bearing of C from B is 245°.

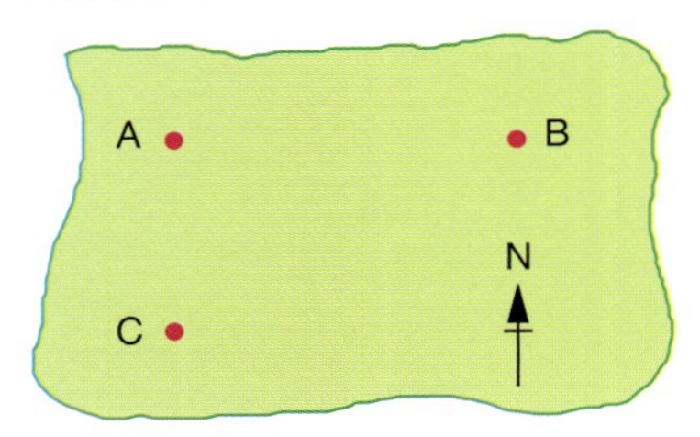

Calculate, giving your answers to the nearest 100 m:

a the distance AB

b the distance BC.

2 Two trees stand 16 m apart. Their tops make an angle of $\theta°$ at point A on the ground.

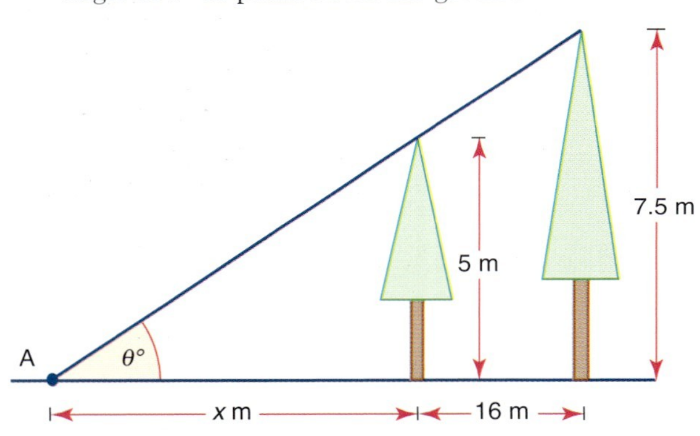

a Express $\theta°$ in terms of the height of the shorter tree and its distance x metres from point A.

b Express $\theta°$ in terms of the height of the taller tree and its distance from A.

c Form an equation in terms of x.

d Calculate the value of x.

e Calculate the value θ.

3 Two boats X and Y, sailing in a race, are shown in the diagram. Boat X is 145 m due north of a buoy B. Boat Y is due east of buoy B. Boats X and Y are 320 m apart.

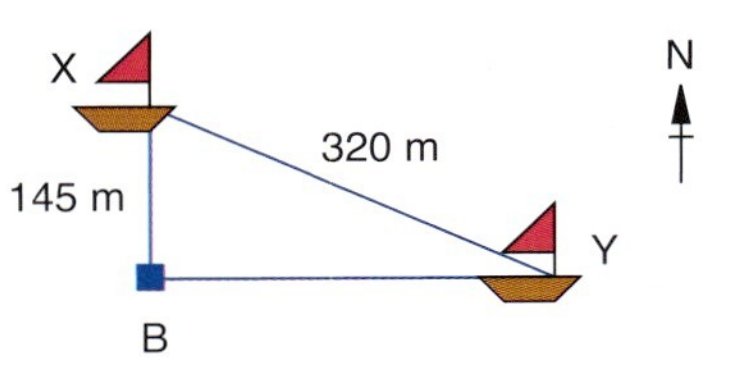

Calculate:

a the distance BY

b the bearing of Y from X

c the bearing of X from Y.

4 Two hawks P and Q are flying vertically above one another. Hawk Q is 250 m above hawk P. They both spot a snake at R.

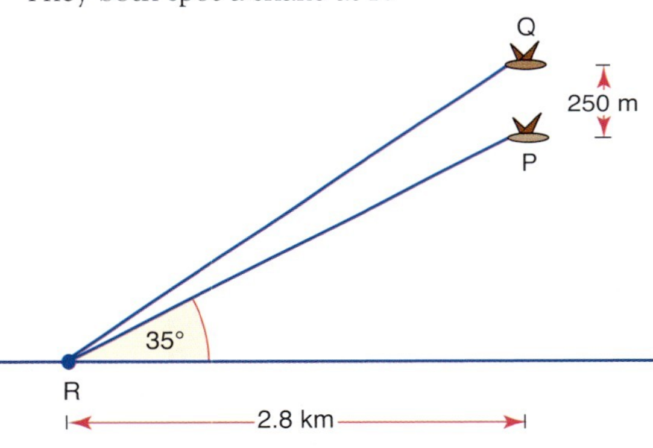

Using the information given, calculate:

a the height of P above the ground

b the distance between P and R

c the distance between Q and R.

5 A boy standing on a cliff top at A can see a boat sailing in the sea at B. The vertical height of the boy above sea level is 164 m, and the horizontal distance between the boat and the boy is 4 km.

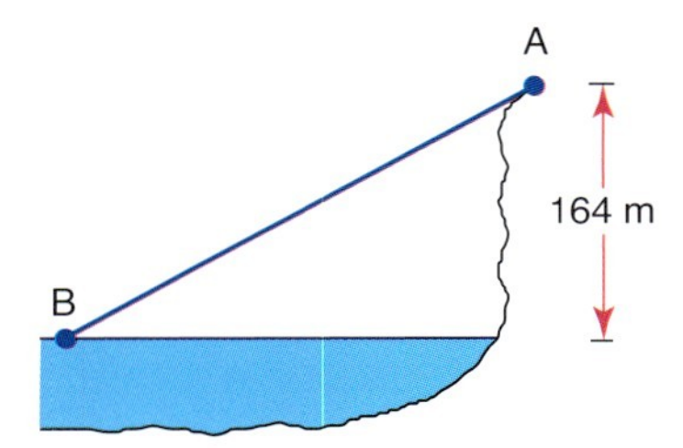

Calculate:

a the distance AB to the nearest metre

b the angle of depression of the boat from the boy.

6 Draw the graph of $y = \sin x°$ for $0° \le x° \le 180°$. Mark on the graph the angles 0°, 90°, 180°, and also the maximum and minimum values of y.

7 Express each of the following in terms of another angle between 0° and 180°.

a $\sin 50°$ b $\sin 150°$

c $\cos 45°$ d $\cos 120°$

8 Calculate the size of the obtuse angle marked $\theta°$ in this triangle.

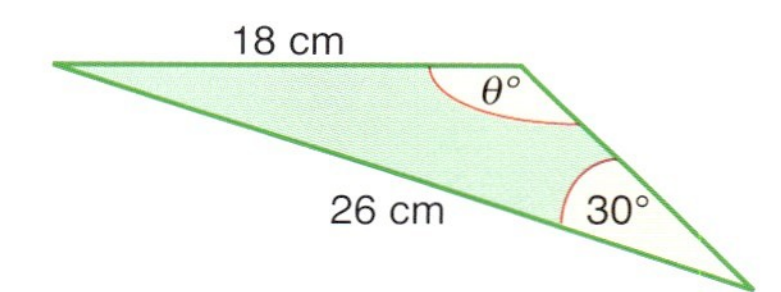

Student assessment 4

Give all lengths to one decimal place and all angles to the nearest degree.

1 For the cuboid shown, calculate:

a the length EG

b the length EC

c angle BEC.

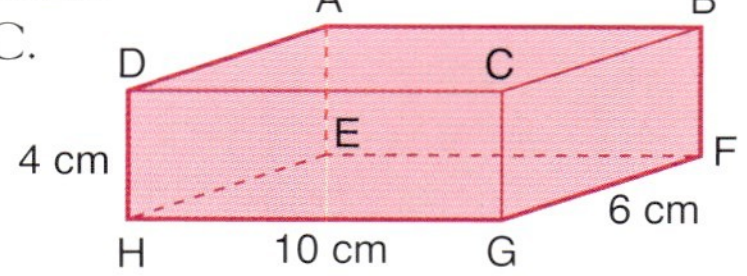

2 Using the triangular prism shown, calculate:

a the length AD

b the length AC

c the angle AC makes with the plane CDEF

d the angle AC makes with the plane ABFE.

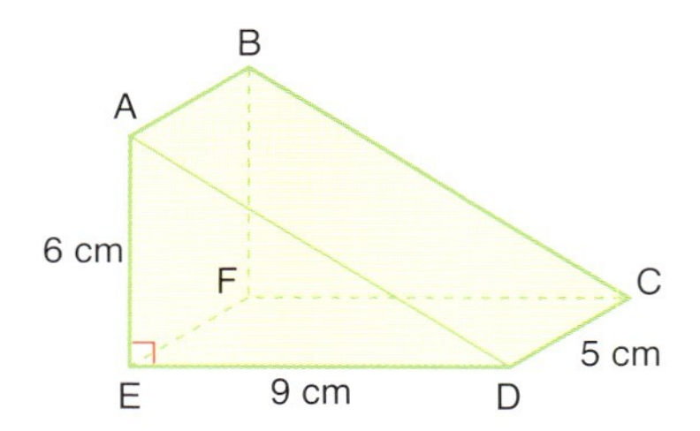

3 Draw a graph of $y = \cos\theta°$, for $0° \le \theta° \le 180°$. Mark on the angles 0°, 90°, 180°, and also the maximum and minimum values of y.

4 The cosine of which other angle between 0 and 180° has the same value as

a $\cos 128°$ b $-\cos 80°$?

5 For the triangle shown, calculate:

a the length PS

b angle QRS

c the length SR.

Round your answers to one decimal place, where necessary.

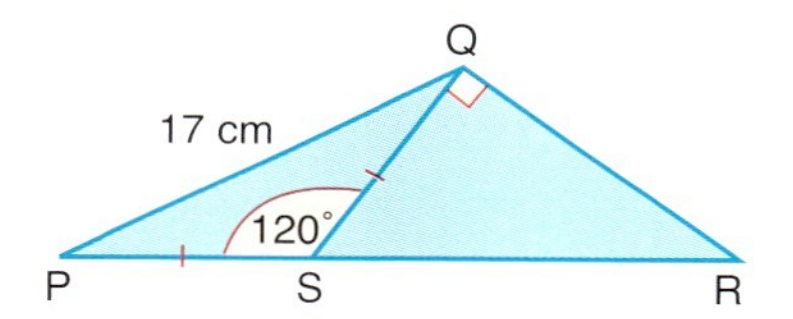

6 The Great Pyramid at Giza is 146 m high. Two people A and B are looking at the top of the pyramid. The angle of elevation of the top of the pyramid from B is 12°. The distance between A and B is 25 m. If both A and B are 1.8 m tall, calculate to one decimal place:

a the distance from B to the centre of the base of the pyramid

b the angle of elevation θ of the top of the pyramid from A

c the distance between A and the top of the pyramid.

Note: A, B and the top of the pyramid are in the same vertical plane.

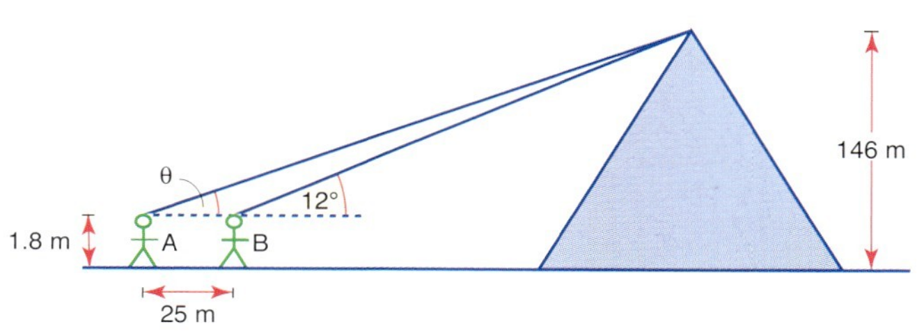

Student assessment 5

Give all lengths to one decimal place and all angles to the nearest degree.

Diagrams are not drawn to scale.

1 For this quadrilateral, calculate:
  a the length JL
  b angle KJL
  c the length JM
  d the area of JKLM.

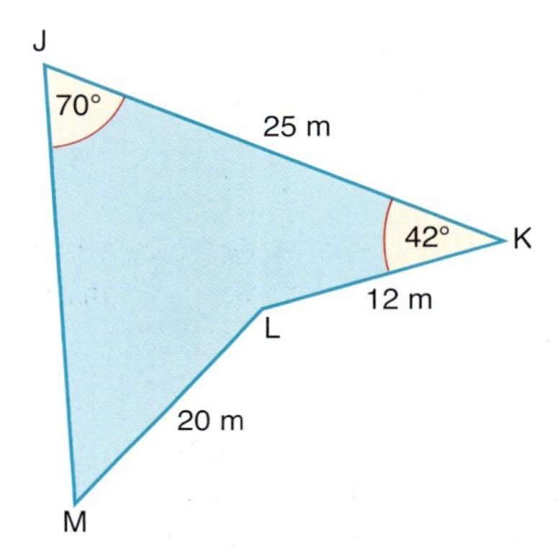

2 For the square-based right pyramid shown, calculate:
  a the length BD
  b angle ABD
  c the area of triangle ABD
  d the vertical height of the pyramid.

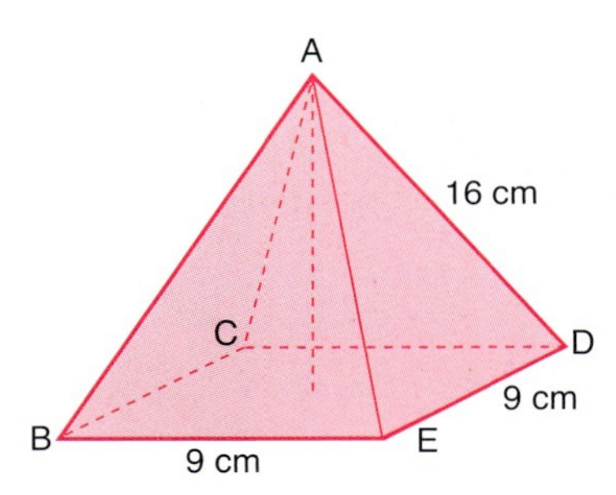

3 Find two angles between 0° and 360° which have the following cosine. Give each angle to the nearest degree.
  a 0.79 b −0.28

4 a On one diagram plot the graph of $y = \sin\theta°$ and the graph of $y = \cos\theta°$, for $0° \leq \theta° \leq 360°$.
  b Use your graph to find the angles for which $\sin\theta° = \cos°\theta°$.

5 The cuboid shown has one of its corners removed to leave a flat triangle BDC.

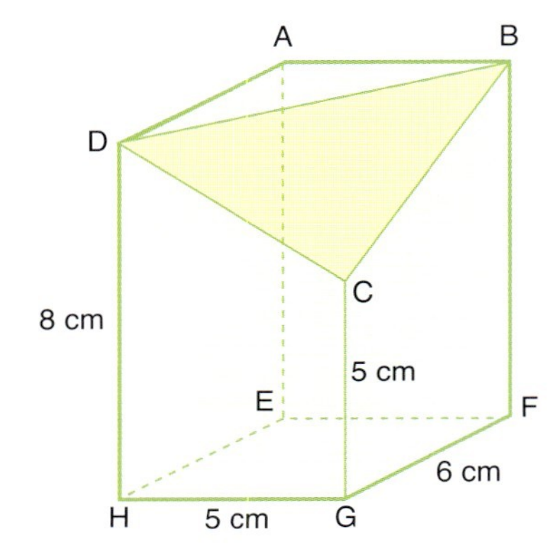

Calculate to two significant figures:
  a length DC
  b length BC
  c length DB
  d angle CBD
  e the area of triangle BDC
  f the angle AC makes with the plane AEHD.

6 In the diagram of the cuboid, X is the midpoint of VW and Y is the midpoint of TW.

RS = 24 cm
SP = 20 cm
TQ = 9 cm

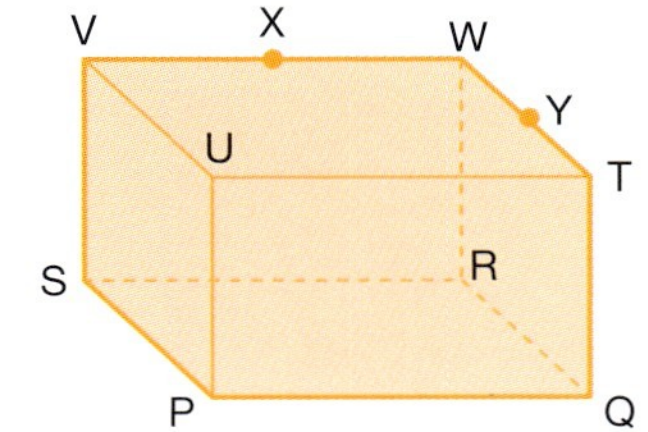

Calculate:
  a PR b RV
  c W d XY
  e SY.

Student assessment 6

Give all your answers to one decimal place.

1 A sphere has a radius of 6.5 cm. Calculate:
  a its total surface area
  b its volume.

2 A pyramid with a base the shape of a regular hexagon is shown. If the length of each of its sloping edges is 24 cm, calculate:
  a its total surface area
  b its volume.

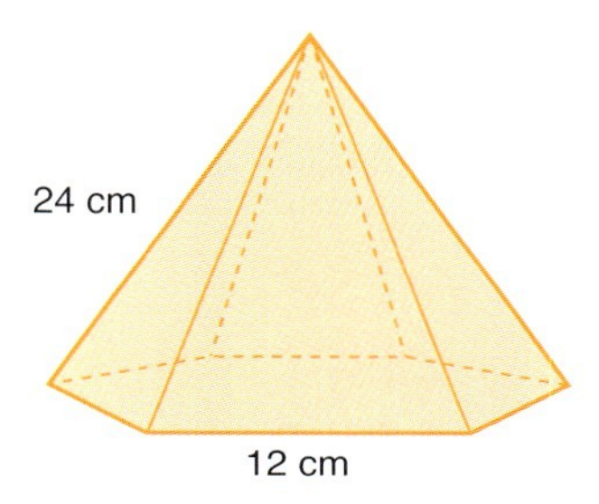

3 The prism here has a cross-sectional area in the shape of a sector.

  Calculate:
  a the radius r cm
  b the cross-sectional area of the prism
  c the total surface area of the prism
  d the volume of the prism.

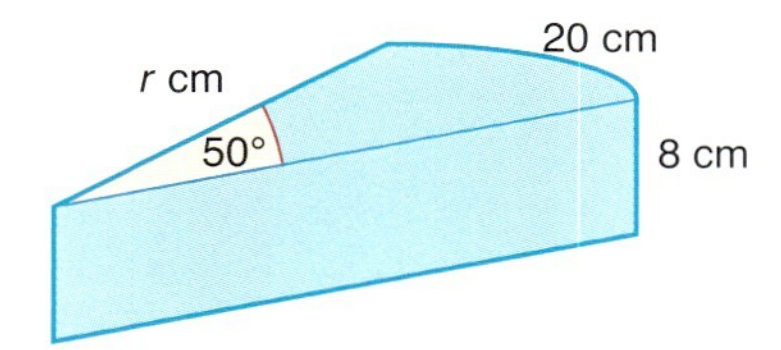

4 The cone and sphere shown here have the same volume.

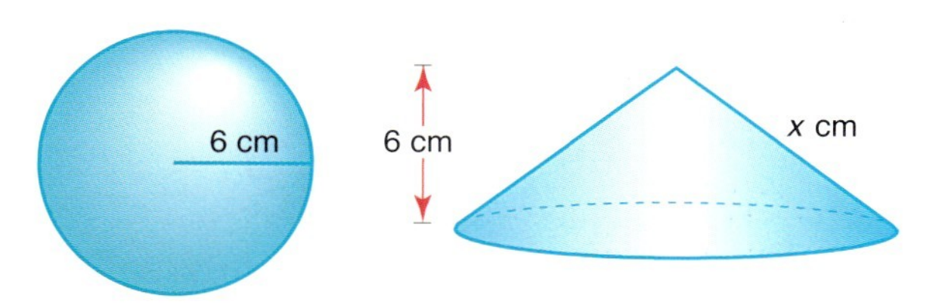

  If the radius of the sphere and the height of the cone are both 6 cm, calculate:
  a the volume of the sphere
  b the base radius of the cone
  c the slant height x cm
  d the surface area of the cone.

5 The top of a cone is cut off and a cylindrical hole is drilled out of the remaining truncated cone as shown.

  Calculate:
  a the height of the original cone
  b the volume of the original cone
  c the volume of the solid truncated cone
  d the volume of the cylindrical hole
  e the volume of the remaining truncated cone.

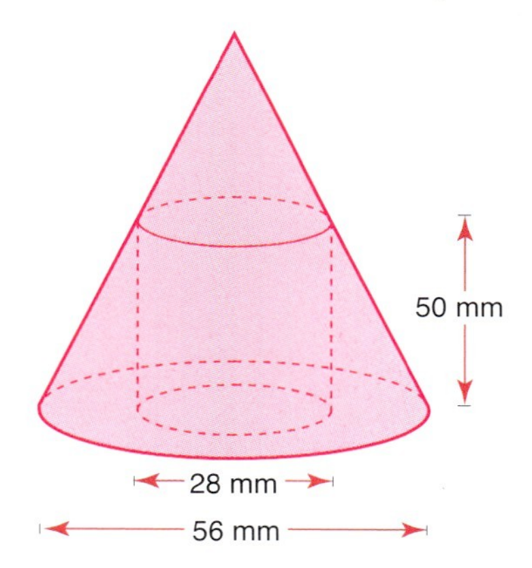

Discussion points, project ideas and theory of knowledge

1 The human brain has a longitudinal fissure (a long crack), which separates it into two hemispheres. The function of each hemisphere is different. Popular psychology talks of 'left brain, right brain thinking' where one side or another is responsible for logic, creativity, etc. Find out more about this idea and discuss its scientific basis.

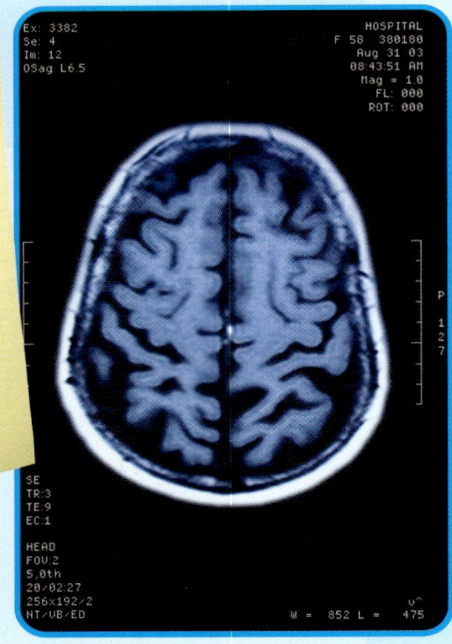

2 Pythagoras' theorem has a large number of proofs. These include Garfield's proof, proof using similar triangles, Euclid's proof and others. These types of proof could form the basis of a project. No proof that uses trigonometry is considered valid. Why do you think this is?

3 The terms arithmetic, algebra and geometry used to be studied in schools as separate subjects (and still are in parts of the USA). Discuss the statement that these terms are becoming redundant. What you think are now the most important areas of mathematics and should they, or can they, be studied in isolation?

4 Euclidean geometry is an axiomatic system. Discuss what this means and discover the main axioms of Euclidean geometry. Einstein's Theory of General Relativity maintains that space–time is non-Euclidean. Find out more about non-Euclidean geometry.

5 Investigate the mathematics of doodles like the one shown. This could be developed as a project.

6 'All the possible discoveries in Euclidean geometry have already been made.' Discuss this statement. Are other areas of mathematics becoming exhausted, and is there a natural limit to mathematical knowledge?

7 What are conic sections? The investigation of conic sections could form the basis of a project.

8 What is 'inner-tube' or 'doughnut' geometry? Its correct name is the geometry of the torus. Is the shortest distance between two points on a torus a straight line?

9 Look up the game of Hex. It was discovered by Piet Hein in 1942. It could form the basis of a project.

10 Euler's formula is concerned with regular solids. By using the interior angle of regular polygons it is possible to prove which regular solids can be constructed. Investigate this further.

Topic 6

Statistics

Syllabus content

6.1 Classification of data as discrete or continuous.

6.2 Simple discrete data: frequency tables; frequency polygons.

6.3 Grouped discrete or continuous data: frequency tables; mid-interval values; upper and lower boundaries.

Frequency histograms.

Stem and leaf diagrams (stem plots).

6.4 Cumulative frequency tables for grouped discrete data and for grouped continuous data; cumulative frequency curves.

Box and whisker plots (box plots).

Percentiles; quartiles.

6.5 Measures of central tendency.

For simple discrete data: mean; median; mode.

For grouped discrete and continuous data: approximate mean; modal group; 50th percentile.

6.6 Measures of dispersion range; interquartile range; standard deviation.

6.7 Scatter diagrams; line of best fit, by eye, passing through the mean point.

Bivariate data: the concept of correlation.

Pearson's product–moment correlation coefficient: use of the formula

$r = \dfrac{s_{xy}}{s_x s_y}$.

Interpretation of positive, zero and negative correlations.

6.8 The regression line for y on x: use of the formula $y - \bar{y} = \dfrac{s_{xy}}{(s_x)^2}(x - \bar{x})$.

Use of the regression line for prediction purposes.

6.9 The test for independence: formulation of null and alternative hypotheses; significance levels; contingency tables; expected frequencies; use of the formula $\chi^2 = \Sigma\dfrac{(f_o - f_e)^2}{f_e}$; degrees of freedom; use of tables for critical values; p-values.

Introduction

The word statistics comes from the Latin status meaning state. So statistics was related to information useful to the state.

Statistics is often not considered to be a branch of mathematics, and many universities have a separate statistics department.

Statistics developed out of studies of probability.

Societies such as the London Statistical Society, established in 1834, brought the study of statistics to new heights, but only the advent of computers has brought the ability to handle and analyze very large amounts of data.

6.1 Discrete and continuous data

Discrete data can only take specific values, for example the number of tickets sold for a concert can only be positive integer values.

Continuous data, on the other hand, can take any value within a certain range, for example the time taken to run 100 m will typically fall in the range 10–20 seconds. Within that range, however, the time stated will depend on the accuracy required. So a time stated as 13.8 s could have been 13.76 s, 13.764 s or 13.7644 s, etc.

Exercise 6.1.1

State whether the data below is discrete or continuous.

1 Your shoe size

2 Your height

3 Your house number

4 Your weight

5 The total score when two dice are rolled

6 A mathematics exam mark

7 The distance from the Earth to the moon

8 The number of students in your school

9 The speed of a train

10 The density of lead

6.2 Displaying simple discrete data

Data can be displayed in many different ways. It is therefore important to choose the method that displays the data most clearly and effectively.

The **frequency table** shows the shoe sizes of 20 students in a class.

Shoe size	6	$6\frac{1}{2}$	7	$7\frac{1}{2}$	8	$8\frac{1}{2}$	9
Frequency	2	3	3	6	4	1	1

This can be displayed as a **frequency histogram**.

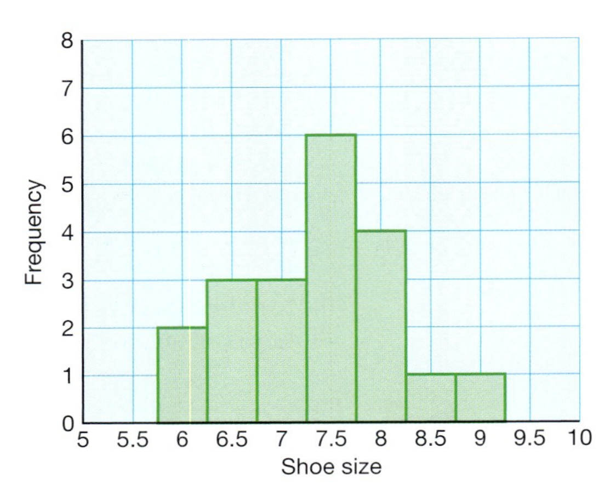

Shoe sizes are an example of discrete data as the data can only take certain values. As a result the frequency histogram has certain properties.

- Each bar is of equal width and its height represents the frequency.
- The bars touch (this is not the case with a bar chart).
- The value is written at the mid-width of each bar. This is because students with a foot size in the range 6.75–7.25, for example, would have a shoe size of 7.

If a line is drawn joining the midpoints of the top of each bar it forms a **frequency polygon**.

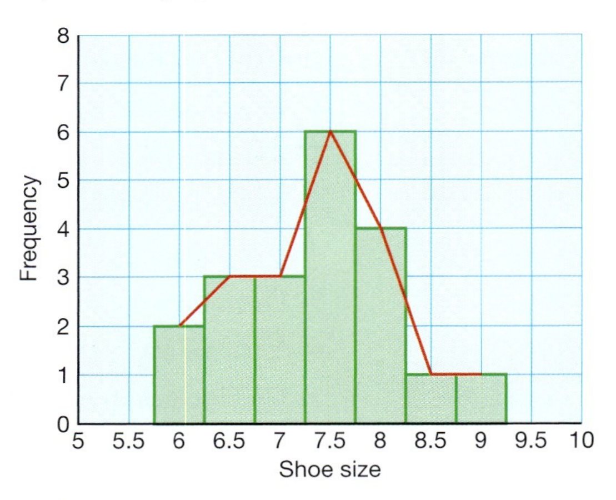

Note: A frequency polygon can be drawn without having to draw the frequency histogram.

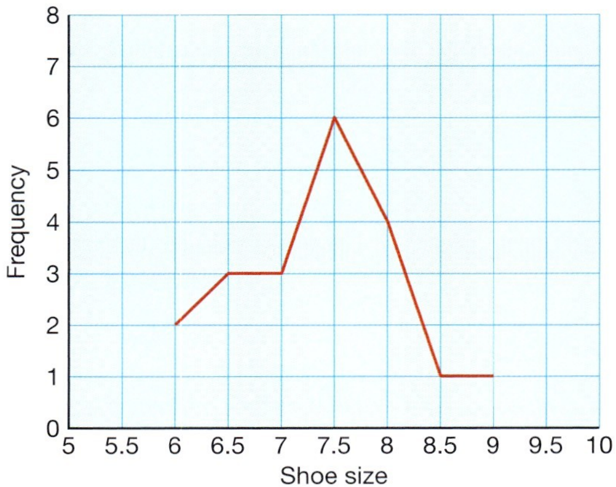

The GDC and graphing software can also be use to plot frequency histograms. The instructions below are for the shoe data given above.

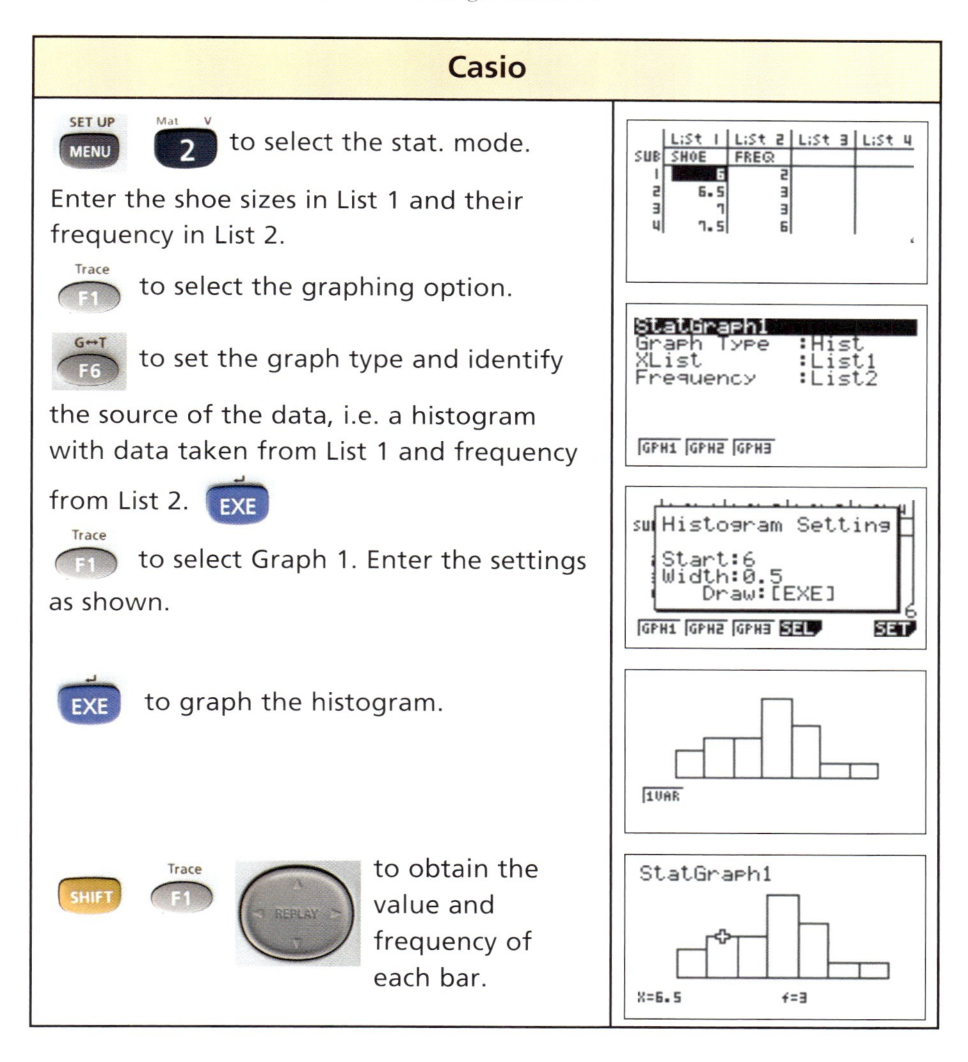

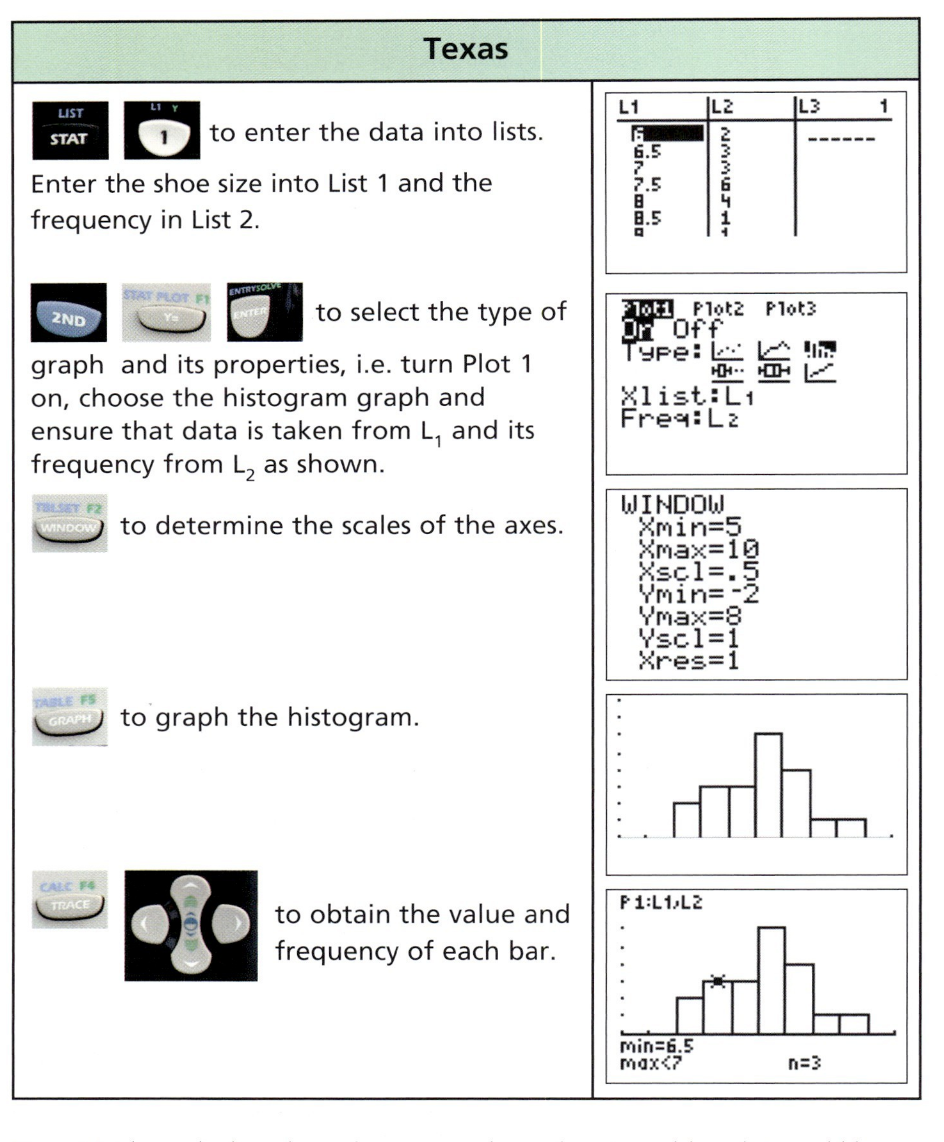

Note: Neither calculator has a frequency polygon function, although it could be drawn as a line graph.

Autograph

Select [Σ] to produce a statistics page.

 to enter the data as a frequency table.

Ensure the data type is marked as Discrete and the unit size as 0.5.

Select 'use (x, f) table' followed by [Edit]

Enter shoe size in column 'x' and frequency in column 'f'. Ensure an additional size is entered as zero as an upper bound. Click 'OK'.

Click on the histogram icon [histogram icon] and enter the histogram options. Click 'OK'.

To change the scale on the axes use [axes icon] .

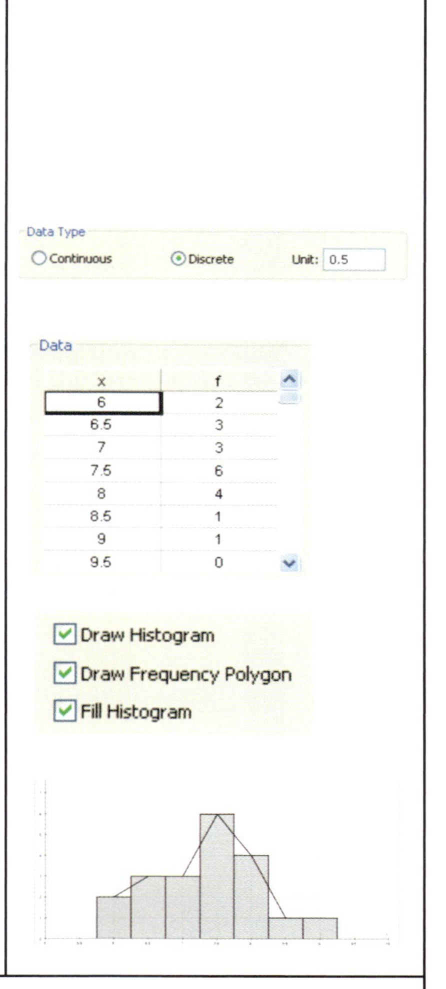

Note: Although the shoe sizes are discrete quantities, Autograph considers a shoe size of 6 to represent the group 5.75–6.25.

GeoGebra as yet does not have a spreadsheet facility for graphing statistical data. You should be able to use a spreadsheet program for graphing statistical data if Autograph is not available.

Exercise 6.2.1

1 The figures in the list below give the total number of chocolate sweets in each of 20 packets of sweets.

35, 36, 38, 37, 35, 36, 38, 36, 37, 35, 36, 36, 38, 36, 35, 38, 37, 38, 36, 38

a Present the data in a tally and frequency table.
b Present the data as a frequency histogram.
c Add a frequency polygon to the histogram drawn in part **b**.

2 Record the shoe sizes of everybody in your class.
a Present the results in a tally and frequency table.
b Present the data as a frequency polygon.
c What conclusions can you draw from your results?

6.3 Grouped discrete or continuous data

If there is a large range in the data, it is sometimes easier and more useful to group the data in a grouped frequency table.

The discrete data below shows the scores for the first round of a golf competition.

71 75 82 96 83 75 76 82 103 85 79 77 83 85 88
104 76 77 79 83 84 86 88 102 95 96 99 102 75 72

One possible way of grouping this data in a grouped frequency table is shown opposite.

Note: The groups are arranged so that no score can appear in two groups.

Score	Frequency
71–75	5
76–80	6
81–85	8
86–90	3
91–95	1
96–100	3
101–105	4

Exercise 6.3.1

1 The following data gives the percentage scores obtained by students from two classes, 12X and 12Y, in a mathematics exam.

12X

42 73 93 85 68 58 33 70 71 85 90 99 41 70 65
80 73 89 88 93 49 50 57 64 78 79 94 80 50 76 99

12Y

70 65 50 89 96 45 32 64 55 39 45 58 50 82 84
91 92 88 71 52 33 44 45 53 74 91 46 48 59 57 95

a Draw a grouped tally and frequency table for each of the classes.
b Comment on any similarities or differences between the results.

2 The number of apples collected from 50 trees is recorded below.

35	78	15	65	69	32	12	9	89	110	112	148	98
67	45	25	18	23	56	71	62	46	128	7	133	96
24	38	73	82	142	15	98	6	123	49	85	63	19
111	52	84	63	78	12	55	138	102	53	80		

Choose suitable groups for this data and represent it in a grouped frequency table.

With grouped continuous data, the groups are presented in a different way.

The results below are the times given (in h:min:s) for the first 50 people completing a marathon.

2:07:11	2:08:15	2:09:36	2:09:45	2:10:45
2:10:46	2:11:42	2:11:57	2:12:02	2:12:11
2:13:12	2:13:26	2:14:26	2:15:34	2:15:43
2:16:25	2:16:27	2:17:09	2:18:29	2:19:26
2:19:27	2:19:31	2:20:00	2:20:23	2:20:29
2:21:47	2:21:52	2:22:32	2:22:48	2:23:08
2:23:17	2:23:28	2:23:46	2:23:48	2:23:57
2:24:04	2:24:12	2:24:15	2:24:24	2:24:29
2:24:45	2:25:18	2:25:34	2:25:56	2:26:10
2:26:22	2:26:51	2:27:14	2:27:23	2:27:37

The data can be arranged into a grouped frequency table as follows.

Group	Frequency
2:05:00 ≤ t < 2:10:00	4
2:10:00 ≤ t < 2:15:00	9
2:15:00 ≤ t < 2:20:00	9
2:20:00 ≤ t < 2:25:00	19
2:25:00 ≤ t < 2:30:00	9

Note that, as with discrete data, the groups do not overlap. However, as the data is continuous, the groups are written using inequalities. The first group includes all times from 2 h 5 min *up to but not including* 2 h 10 min.

With continuous data, the upper and lower bound of each group are the numbers written as the limits of the group. In the example above, for the group 2:05:00 ≤ t < 2:10:00, the lower bound is 2:05:00; the upper bound is considered to be 2:10:00 despite it not actually being included in the inequality.

Frequency histograms for grouped data

A **frequency histogram** displays the frequency of either continuous or grouped discrete data in the form of bars. There are several important features of a frequency histogram for grouped data.

- The bars touch.
- The horizontal axis is labelled with a scale.
- The bars can be of varying width.
- The frequency of the data is represented by the area of the bar and not the height. (Note: In the case of bars of equal width, the area is directly proportional to the height of the bar and so the height is usually used as the measure of frequency.)

Worked example

The table shows the marks out of 100 in a mathematics exam for a class of 32 students.

Draw a histogram representing this data.

Test marks	Frequency
1–10	0
11–20	0
21–30	1
31–40	2
41–50	5
51–60	8
61–70	7
71–80	6
81–90	2
91–100	1

All the class intervals are the same. As a result the bars of the histogram will all be of equal width, and the frequency can be plotted on the vertical axis. The histogram is shown. Note that the upper and lower bounds are used to draw the bars.

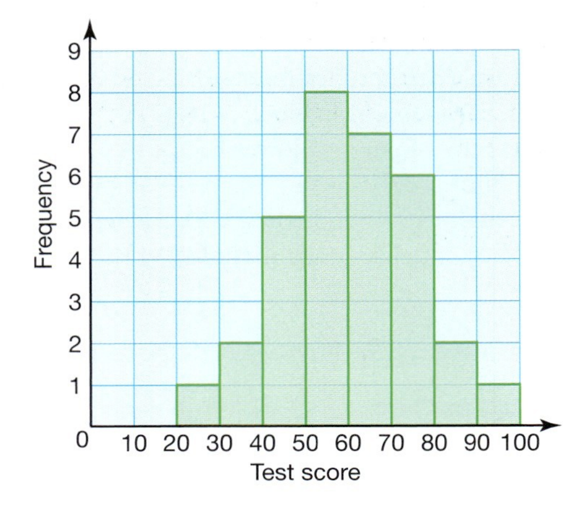

When the class intervals are of different widths, the frequency is proportional to the area of the bar and we plot **frequency density** on the vertical axis.

Worked example

The heights of 25 sunflowers were measured and the results recorded in the table.

Height (m)	Frequency
$0 \leq h < 1.0$	6
$1.0 \leq h < 1.5$	3
$1.5 \leq h < 2.0$	4
$2.0 \leq h < 2.25$	3
$2.25 \leq h < 2.50$	5
$2.50 \leq h < 2.75$	4

If a histogram were drawn with frequency plotted on the vertical axis, then it could look like the one shown.

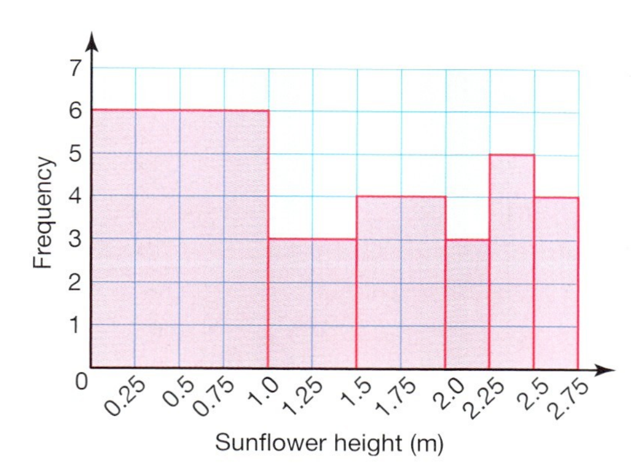

This graph is misleading because it leads people to the conclusion that most of the sunflowers were under 1 m, simply because the area of the bar is so great. In actual fact, only approximately one quarter of the sunflowers were under 1 m.

When class intervals are different, it is the area of the bar which represents the frequency not the height. Instead of frequency being plotted on the vertical axis, **frequency density** is plotted.

$$\text{Frequency density} = \frac{\text{frequency}}{\text{class width}}$$

The results of the sunflower measurements above can therefore be written as:

Height (m)	Frequency	Frequency density
$0 \le h < 1.0$	6	$6 \div 1 = 6$
$1.0 \le h < 1.5$	3	$3 \div 0.5 = 6$
$1.5 \le h < 2.0$	4	$4 \div 0.5 = 8$
$2.0 \le h < 2.25$	3	$3 \div 0.25 = 12$
$2.25 \le h < 2.50$	5	$5 \div 0.25 = 20$
$2.50 \le h < 2.75$	4	$4 \div 0.25 = 16$

The histogram can therefore be redrawn as shown giving a more accurate representation of the data.

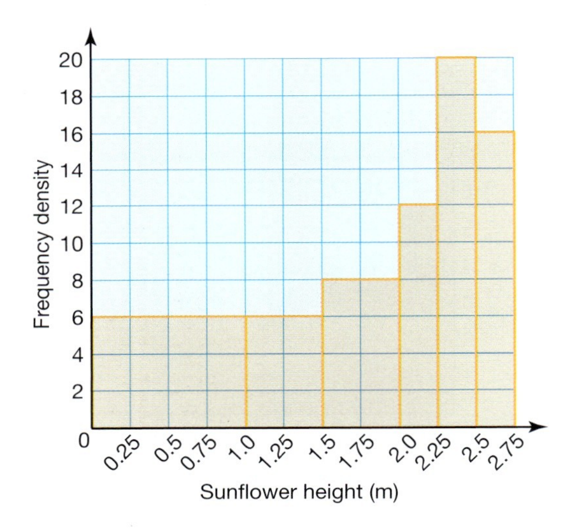

Notice that questions 1 and 2 in Exercise 6.3.2 both deal with **continuous data** but that the class intervals are represented in different ways. In question 2, 145– means the students whose heights fall in the range $145 \le h < 150$.

Exercise 6.3.2

1 The table shows the distances travelled to school by a class of 30 students. Represent this information on a histogram.

Distance (km)	Frequency
$0 \le d < 1$	8
$1 \le d < 2$	5
$2 \le d < 3$	6
$3 \le d < 4$	3
$4 \le d < 5$	4
$5 \le d < 6$	2
$6 \le d < 7$	1
$7 \le d < 8$	1

2 The heights of students in a class were measured. The results are shown in the table. Draw a histogram to represent this data.

Height (m)	Frequency
145–	1
150–	2
155–	4
160–	7
165–	6
170–	3
175–	2
180–185	1

3 The table shows the time taken, in minutes, by 40 students to travel to school.

Time (min)	Frequency
$0 \le t < 10$	6
$10 \le t < 15$	3
$15 \le t < 20$	13
$20 \le t < 25$	7
$25 \le t < 30$	3
$30 \le t < 40$	4
$40 \le t < 60$	4

a Copy the table and complete it by calculating the frequency density.
b Represent the information on a histogram.

4 Frances and Ali did a survey of the ages of the people living in their village. Part of their results are set out in the table.

Age (years)	Frequency
$0 \leq a < 1$	35
$1 \leq a < 5$	12
$5 \leq a < 10$	28
$10 \leq a < 20$	180
$20 \leq a < 40$	260
$40 \leq a < 60$	14
$60 \leq a < 90$	150

a Copy the table and complete it by calculating either the frequency or the frequency density.
b Represent the information on a histogram.

5 The table shows the ages of 150 people, chosen randomly, taking the 6:00 am train into a city.

Age (years)	Frequency
$0 \leq a < 15$	3
$15 \leq a < 20$	25
$20 \leq a < 25$	20
$25 \leq a < 30$	30
$30 \leq a < 35$	32
$40 \leq a < 50$	30
$50 \leq a < 80$	10

The histogram shows the results obtained when the same survey was carried out on the 11:00 am train.

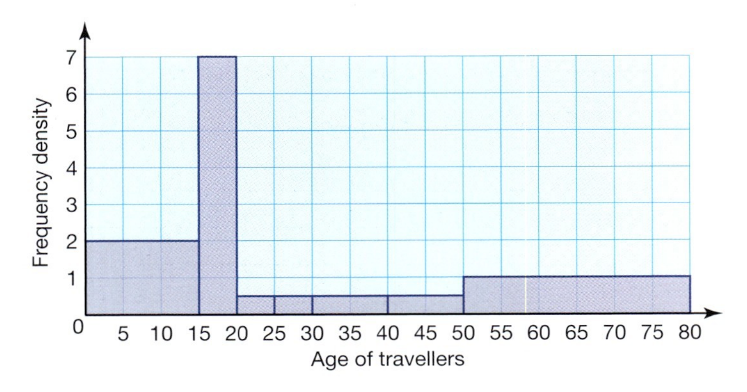

a Draw a histogram for the 6:00 am train.
b Compare the two sets of data and give two possible reasons for the differences.

Stem and leaf diagrams

Stem and leaf diagrams (also called **stem plots**) are a special type of bar chart, in which the bars are made from the data itself. This has the advantage that the original data can be recovered easily from the diagram.

The stem is the first digit of the numbers, so if the numbers are 63, 65, 67, 68, 69, the stem is 6. The leaves are the remaining digits written in order.

Worked example

The ages of people on a coach transferring them from an airport to a ski resort are shown below.

22	24	25	31	33	23	24	26	37	42
40	36	33	24	25	18	20	27	25	33
28	33	35	39	40	48	27	25	24	29

Display the data on a stem and leaf diagram.

Stem	Leaves
1	8
2	0 2 3 4 4 4 4 5 5 5 5 6 7 7 8 9
3	1 3 3 3 3 5 6 7 9
4	0 0 2 8

Key

2 | 5 means 25

Note that, on the right, the key states what the stem means. If the data were 1.8, 2.7, 3.2 etc., the key would state that 2 | 7 means 2.7.

Exercise 6.3.3

1 A test in mathematics is marked out of 40. The scores for a class of 32 students are shown below.

24	27	30	33	26	27	28	39
21	18	16	33	22	38	33	21
16	11	14	23	37	36	38	22
28	15	9	17	28	33	36	34

Display the data on a stem and leaf diagram.

2 A basketball team played 24 matches in the 2009 season. Their scores are shown below.

62	48	85	74	63	67	71	83
46	52	63	65	72	76	68	58
54	46	88	55	46	52	58	54

Display the scores on a stem and leaf diagram.

3 The 27 students in a class were each asked to draw a line 8 cm long with a straight edge rather than with a ruler. The lines were then measured and their lengths to the nearest millimetre were recorded.

8.8	6.2	8.3	7.9	8.0	5.9	6.2	10.0	9.7
7.9	5.4	6.8	7.3	7.7	8.9	10.4	5.9	8.3
6.1	7.2	8.3	9.4	6.5	5.8	8.8	8.0	7.3

Illustrate this data on a stem and leaf diagram.

Back-to-back diagrams

Stem and leaf diagrams are often used as an easy way to compare two sets of data. The leaves are usually put 'back-to-back' on either side of the stem.

Worked example

The stem and leaf diagram for the ages of people on a coach to a ski resort (as in the previous worked example) is shown below. The data is easily accessible.

1	8
2	0 2 3 4 4 4 4 5 5 5 5 6 7 7 8 9
3	1 3 3 3 3 5 6 7 9
4	0 0 2 8

Key
2 | 5 means 25

A second coach from the airport is taking people to a golfing holiday. The ages of the people are shown below.

43	46	52	61	65	38	36	28	37	45
69	72	63	55	46	34	35	37	43	48
54	53	47	36	58	63	70	55	63	64

Display the two sets of data on a back-to-back stem and leaf diagram.

Golf		Skiing
	1	8
8	**2**	0 2 3 4 4 4 4 5 5 5 5 6 7 7 8 9
8 7 7 6 6 5 4	**3**	1 3 3 3 3 5 6 7 9
8 7 6 6 5 3 3	**4**	0 0 2 8
8 5 5 4 3 2	**5**	
9 5 4 3 3 3 1	**6**	
2 0	**7**	

Key
3 | 5 means 35

Exercise 6.3.4

1 Write three sentences commenting on the back-to-back diagram in the worked example above.

2 The basketball team in question 2 of the previous exercise had replaced their team coach at the end of the 2008 season. Their scores for the 24 matches played in the 2008 season are shown below.

82	32	88	24	105	63	86	42	35	88	78	106
64	72	88	26	35	41	100	48	54	36	28	33

Display the scores from both seasons on a back-to-back stem and leaf diagram and comment on it.

3 The mathematics test results shown in question 1 of the previous exercise were for test B. Test A had already been set and marked and the teacher had gone over some of the questions with the class. The marks out of 40 for test A are shown below.

22	18	9	11	38	33	21	14	16	8	12
37	39	25	23	18	34	36	23	16	14	12
22	29	33	35	12	17	22	28	32	39	

Draw a back-to-back stem and leaf diagram for the scores from both tests and comment on it.

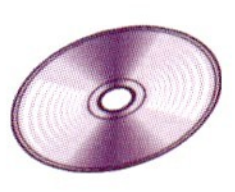

6.4 Cumulative frequency

Measures of central tendency and measures of dispersion are covered in Sections 6.5 and 6.6. In this section we look at cumulative frequency. A cumulative frequency graph is particularly useful when trying to calculate the median (the middle value) of a large set of grouped discrete data or continuous data, or when trying to establish how consistent a set of results are. Calculating the cumulative frequency is done by adding up the frequencies as we go along.

Worked example

The duration of two different brands of battery, A and B, is tested. Fifty batteries of each type are randomly selected and tested in the same way. The duration of each battery is then recorded. The results of the tests are shown in the table below.

Type A: duration (h)	Frequency	Cumulative frequency
$0 \le t < 5$	3	3
$5 \le t < 10$	5	8
$10 \le t < 15$	8	16
$15 \le t < 20$	10	26
$20 \le t < 25$	12	38
$25 \le t < 30$	7	45
$30 \le t < 35$	5	50

Type B: duration (h)	Frequency	Cumulative frequency
$0 \le t < 5$	1	1
$5 \le t < 10$	1	2
$10 \le t < 15$	10	12
$15 \le t < 20$	23	35
$20 \le t < 25$	9	44
$25 \le t < 30$	4	48
$30 \le t < 35$	2	50

a Plot a cumulative frequency curve for each brand of battery.
b Estimate the median duration for each brand.

a The points are plotted at the upper boundary of each class interval rather than at the middle of the interval. So, for type A, points are plotted at (5, 3), (10, 8), etc. The points are joined with a smooth curve which is extended to include (0, 0).

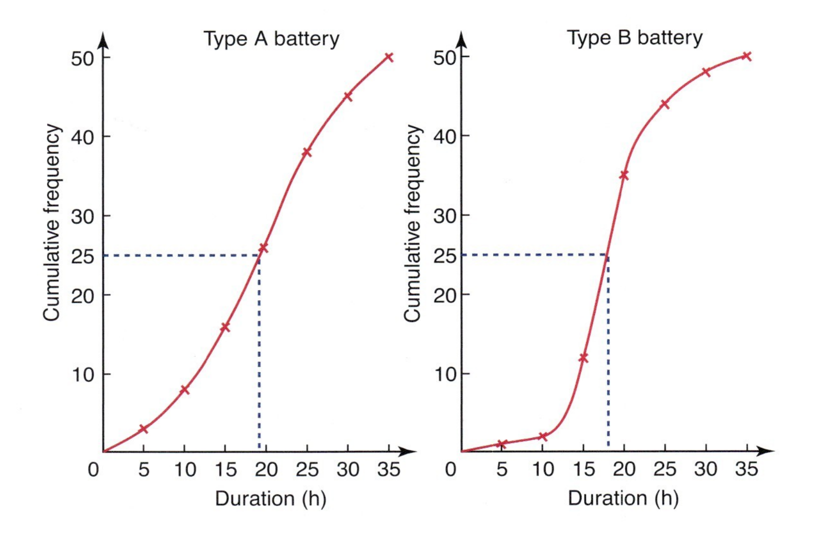

Both cumulative frequency curves are plotted above.

b The median value is the value which occurs half-way up the cumulative frequency axis. This is shown with broken lines on the graphs. Therefore:
median for type A batteries ≈ 19 h
median for type B batteries ≈ 18 h
This tells us that, on average, batteries of type A last longer (19 hours) than batteries of type B (18 hours).

Graphing software can produce a cumulative frequency curve and help with calculating the median value.

The example on the opposite page takes the data for battery A above.

Autograph

Select  to produce a statistics page.

to enter the grouped data properties as shown. Click 'OK'.

Click on the cumulative frequency graph icon .

Ensure 'cumulative frequency' and 'curve fit' are selected.

Click 'OK'.

To change the scale on the axes use .

To calculate the median, click on the 'cumulative

frequency diagram measurement' icon and select 'Median'. Click 'OK'.

A horizontal line is drawn at the middle value on the cumulative frequency axis. A vertical line can be dragged at its base until it intersects the horizontal line on the curve.

The median result is shown at the base of the screen.

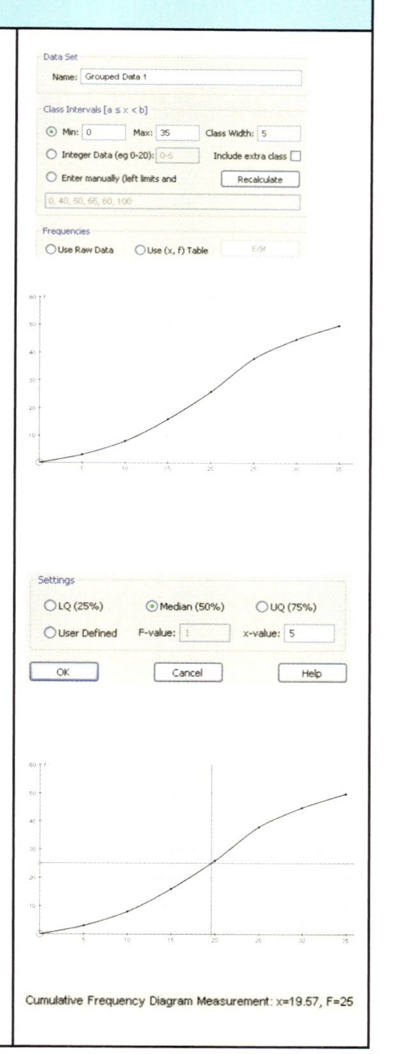

Exercise 6.4.1

1 Sixty athletes enter a cross-country race. Their finishing times are recorded and are shown in the table below.

Finishing time (h)	0–	0.5–	1.0–	1.5–	2.0–	2.5–	3.0–3.5
Frequency	0	0	6	34	16	3	1
Cumulative freq.							

a Copy the table and calculate the values for the cumulative frequency.
b Draw a cumulative frequency curve of the results.
c Show how your graph could be used to find the approximate median finishing time.
d What does the median value tell us?

2 Three mathematics classes take the same test in preparation for their final exam. Their raw scores are shown below.

Class A 12, 21, 24, 30, 33, 36, 42, 45, 53, 53, 57, 59, 61, 62, 74, 88, 92, 93
Class B 48, 53, 54, 59, 61, 62, 67, 78, 85, 96, 98, 99
Class C 10, 22, 36, 42, 44, 68, 72, 74, 75, 83, 86, 89, 93, 96, 97, 99, 99

a Using the class intervals $0 \leq x < 20$, $20 \leq x < 40$ etc., draw a grouped frequency table and cumulative frequency table for each class.
b Draw a cumulative frequency curve for each class.
c Show how your graph could be used to find the median score for each class.
d What does the median value tell us?

3 The table below shows the heights of students in a class over a three-year period.

Height (m)	**Frequency 2007**	**Frequency 2008**	**Frequency 2009**
$150 \leq h < 155$	6	2	2
$155 \leq h < 160$	8	9	6
$160 \leq h < 165$	11	10	9
$165 \leq h < 170$	4	4	8
$170 \leq h < 175$	1	3	2
$175 \leq h < 180$	0	2	2
$180 \leq h < 185$	0	0	1

a Construct a cumulative frequency table for each year.
b Draw the cumulative frequency curve for each year.
c Show how your graph could be used to find the median height for each year.
d What does the median value tell us?

Quartiles and the interquartile range

The cumulative frequency axis can also be represented in terms of **percentiles**. A percentile scale divides the cumulative frequency scale into hundredths. The maximum value of cumulative frequency is found at the 100th percentile. Similarly the median, being the middle value, is the 50th percentile. The 25th percentile is known as the **lower quartile**, and the 75th percentile is called the **upper quartile**.

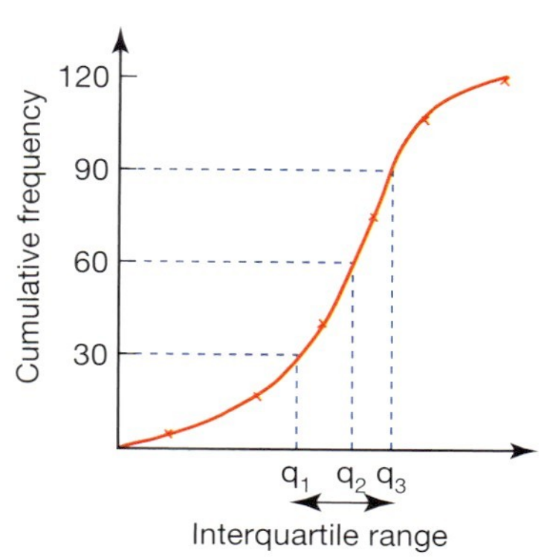

The range of a distribution is found by subtracting the lowest value from the highest value. Sometimes this will give a useful result, but often it will not. A better measure of dispersion is given by looking at the spread of the middle half of the results, i.e. the difference between the upper and lower quartiles. This result is known as the **interquartile range**.

Worked example

Consider again the two types of batteries A and B discussed earlier (page 247).

a Using the graphs, estimate the upper and lower quartiles for each battery.

b Calculate the interquartile range for each type of battery.

c Based on these results, how might the manufacturers advertise the two types of battery?

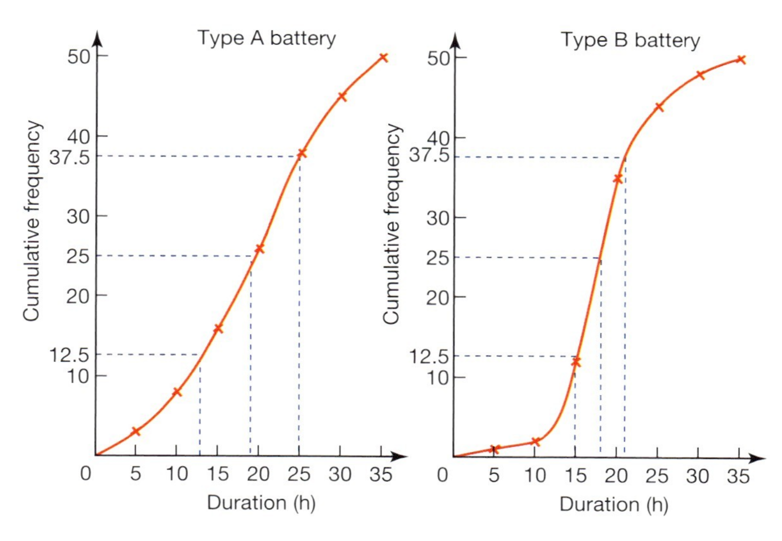

a Lower quartile of type A ≈ 13 h
Upper quartile of type A ≈ 25 h

Lower quartile of type B ≈ 15 h
Upper quartile of type B ≈ 21 h

b Interquartile range of type A ≈ 12 h

Interquartile range of type B ≈ 6 h

c Type A: on 'average' the longer-lasting battery
Type B: the more reliable battery

The interquartile range can be calculated using graphing software as shown below.

Autograph

Enter the data for battery A and produce the cumulative frequency graph as shown earlier.

To calculate each of the quartiles, click on the 'cumulative frequency diagram measurement' icon and select 'LQ(25%)'. Click 'OK'.

A horizontal line is drawn at the lower quartile value on the cumulative frequency axis. A vertical line can be dragged at its base until it intersects the horizontal line on the curve.

The lower quartile result is shown at the base of the screen.

The above procedure can be repeated for the upper quartile.

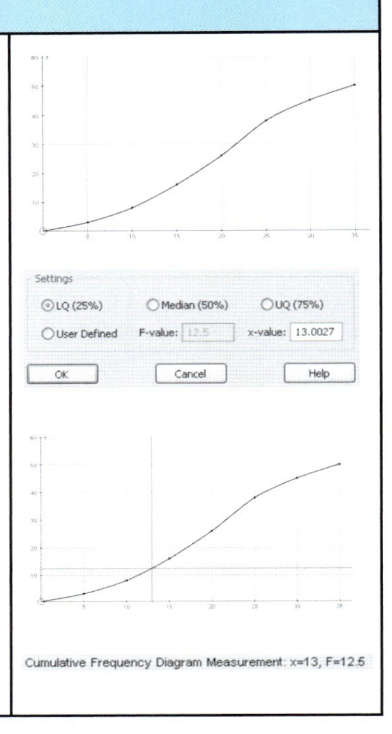

Exercise 6.4.2

1 Using the results obtained from question 2 of the previous exercise:
   a find the interquartile range of each of the classes taking the mathematics test
   b analyze your results and write a brief summary comparing the three classes.

2 Using the results obtained from question 3 of the previous exercise:
   a find the interquartile range of the students' heights each year
   b analyze your results and write a brief summary comparing the three years.

3 Forty boys enter for a school javelin competition. The distances thrown are recorded below.

Distance thrown (m)	0–	20–	40–	60–	80–100
Frequency	4	9	15	10	2

   a Construct a cumulative frequency table for the above results.
   b Draw a cumulative frequency curve.
   c If the top 20% of boys are considered for the final, estimate (using the graph) the qualifying distance.
   d Calculate the interquartile range of the throws.
   e Calculate the median distance thrown.

4 The masses of two different types of oranges are compared. Eighty oranges are randomly selected from each type and weighed. The results are shown below.

Type A	
Mass (g)	**Frequency**
75–	4
100–	7
125–	15
150–	32
175–	14
200–	6
225–250	2

Type B	
Mass (g)	**Frequency**
75–	0
100–	16
125–	43
150–	10
175–	7
200–	4
225–250	0

a Construct a cumulative frequency table for each type of orange.
b Draw a cumulative frequency graph for each type of orange.
c Calculate the median mass for each type of orange.
d Using your graphs estimate:
 i) the lower quartile
 ii) the upper quartile
 iii) the interquartile range
 for each type of orange.
e Write a brief report comparing the two types of orange.

5 Two competing brands of battery are compared. A hundred batteries of each brand are tested and the duration of each is recorded. The results of the tests are shown in the cumulative frequency graphs below.

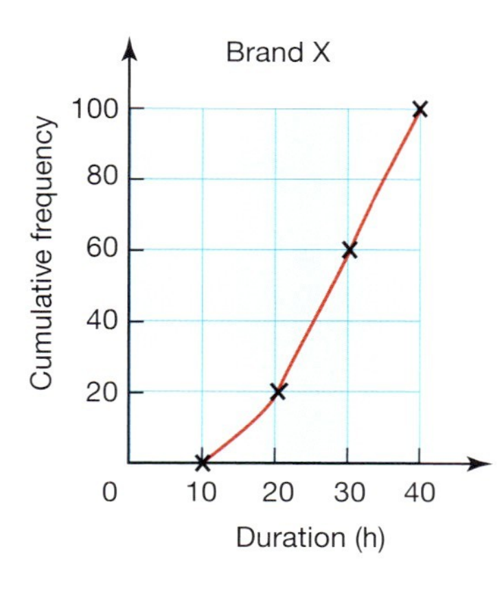

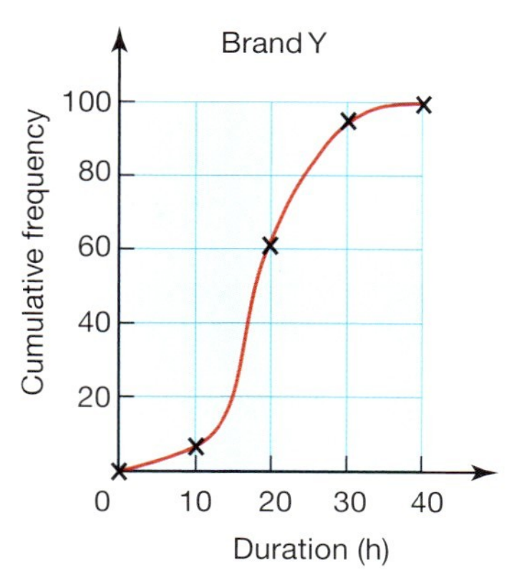

a The manufacturers of brand X claim that on average their batteries will last at least 40% longer than those of brand Y. Showing your method clearly, decide whether this claim is true.
b The manufacturers of brand X also claim that their batteries are more reliable than those of brand Y. Is this claim true? Show your working clearly.

Box and whisker plots

So far we have seen how cumulative frequency curves enable us to look at how data is **dispersed** (spread out) by working out the upper and lower quartiles and also the interquartile range.

Box and whisker plots (also known as **box plots**) provide another visual way of representing the spread of data. The diagram below demonstrates what a 'typical' box and whisker plot looks like and also highlights its main features.

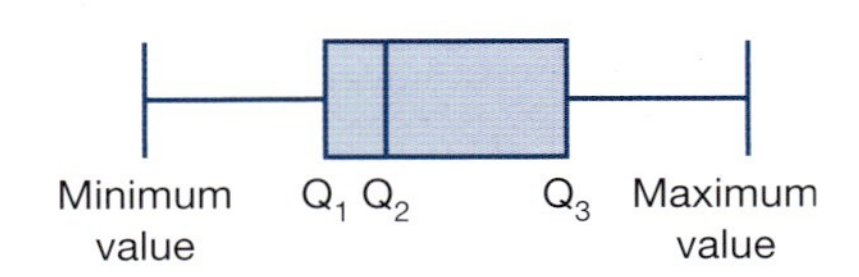

The box and whisker plot shows all the main features of the data, i.e. the minimum and maximum values, the upper and lower quartiles and the median. The box represents the middle 50% of the data (the interquartile range) and the whiskers represent the whole of the data (the range).

For discrete data, the median position is given by the formula $\frac{n+1}{2}$, where n represents the number of values. Similarly the position of the lower quartile can be calculated using the formula $\frac{n+1}{4}$ and the upper quartile by the formula $\frac{3(n+1)}{4}$.

Worked example

The shoe sizes of 15 boys and 15 girls from the same class are recorded in the frequency table below.

Shoe size	5	$5\frac{1}{2}$	6	$6\frac{1}{2}$	7	$7\frac{1}{2}$	8	$8\frac{1}{2}$	9	$9\frac{1}{2}$
Frequency (boys)	0	0	1	2	1	2	3	4	1	1
Frequency (girls)	1	3	4	4	1	1	1	0	0	0

a Calculate the lower quartile, median and upper quartile shoe sizes for the boys and girls in the class.
b Compare this data using two box and whisker plots (one for boys and one for girls).
c What conclusions can be drawn from the box and whisker plots?

a Lower quartile boy $= \frac{15+1}{4} = 4^{\text{th}}$ i.e. $q_1 = 7$

Median boy $= \frac{15+1}{2} = 8^{\text{th}}$ i.e. $q_2 = 8$

Upper quartile boy $= \frac{3(15+1)}{4} = 12^{\text{th}}$ i.e. $q_3 = 8\frac{1}{2}$

Lower Quartile girl is the 4th i.e. $q_1 = 5\frac{1}{2}$
Median girl is the 8th i.e. $q_2 = 6$
Upper quartile girl is the 12th i.e. $q_3 = 6\frac{1}{2}$

b For box and whisker plots it is necessary to know the maximum and minimum values.
Minimum boy shoe size is 6, maximum boy shoe size is $9\frac{1}{2}$.
Minimum girl shoe size is 5, maximum girl shoe size is 8.

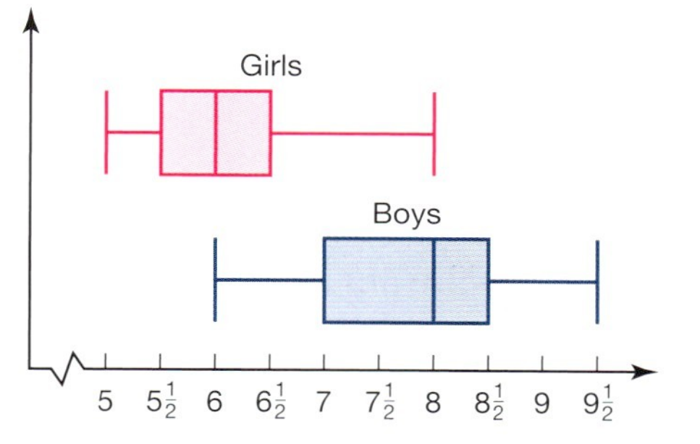

Note: There is no scale on the *y*-axis as it is not relevant in a box and whisker plot. Consequently, the height of a box and whisker plot does not have a particular value.

c
- The overall range of data is greater for boys than it is for girls.
- The interquartile range for girls is less than for boys, i.e. the middle 50% of girls have a narrower spread of shoe size than the middle 50% of boys.

A GDC can also produce a box and whisker plot. The instructions that follow are for the boys' data above.

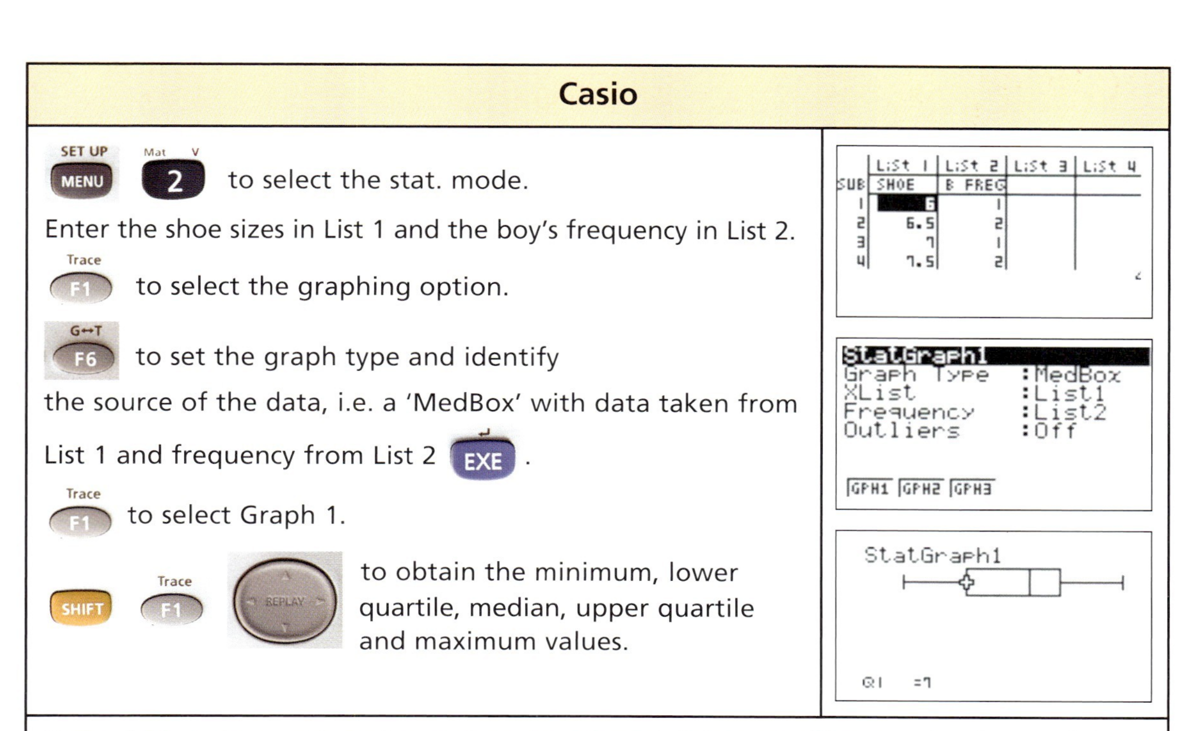

Note: Although a *y*-axis scale needs to be entered, it has no effect on the shape of the box and whisker plot.

Texas

to enter the data into lists.

Enter the boy's shoe size into List 1 and the frequency in List 2.

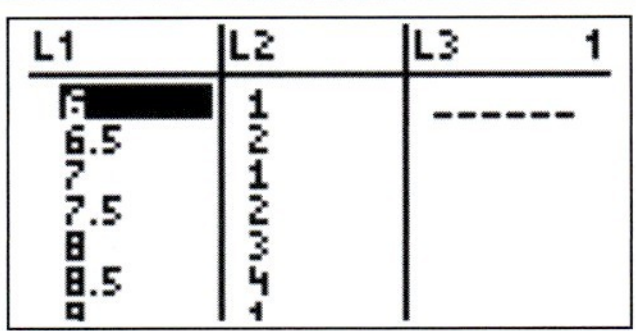

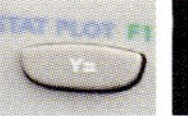

to select the type of graph and its properties, i.e. turn Plot 1 on, choose the box and whisker plot graph and ensure that data is taken from L1 and its frequency from L2 as shown.

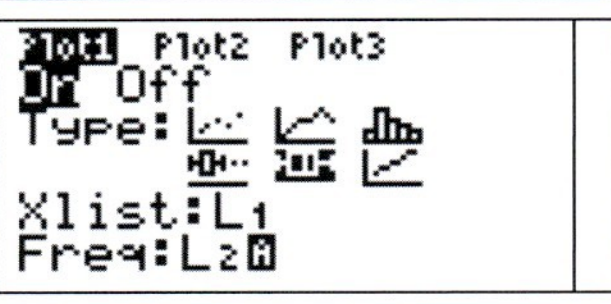

to determine the scale of the axes.

WINDOW
Xmin=5
Xmax=10
Xscl=1
Ymin=0
Ymax=1
Yscl=1
Xres=1

to graph the box and whisker plot.

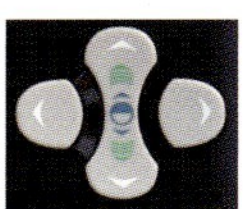
to obtain the minimum, lower quartile, median, upper quartile and maximum values.

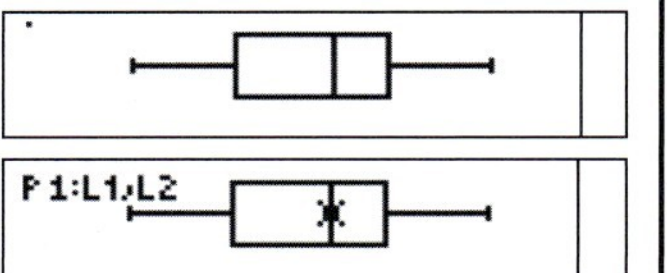

Note: Although a y-axis scale needs to be entered, it has no effect on the shape of the box and whisker plot.

Exercise 6.4.3

Using a GDC or otherwise, answer the following questions.

1 A football team records, over 20 matches, the number of goals it scored and the number of goals it let in in each match. The results are shown in the table below.

Number of goals	0	1	2	3	4	5
Frequency of goals scored	6	9	3	1	0	1
Frequency of goals let in	3	3	8	3	3	0

a For goals scored and goals let in, work out:
  i) the mean
  ii) the median
  iii) the lower quartile
  iv) the upper quartile
  v) the interquartile range.

b Using the same scale, represent the data using box and whisker plots.

c Write a brief report about what the box and whisker plots tell you about the team's results.

2 Two competing holiday resorts record the number of hours of sunshine they have each day during the month of August. The results are shown below.

Hours of sunshine	4	5	6	7	8	9	10	11	12
Resort A	1	2	3	5	5	4	4	4	3
Resort B	0	0	0	4	12	10	5	0	0

a For each resort work out the number of hours of sunshine represented by:
   i) the mean
   ii) the median
   iii) the lower quartile
   iv) the upper quartile
   v) the interquartile range.

b Using the same scale, represent the data using box and whisker plots.

c Based on the data and referring to your box and whisker plots, explain which resort you would choose to go to for a beach holiday in August.

3 A teacher decides to tackle the problem of students arriving late to his class. To do this, he records how late they are to the nearest minute. His results are shown in the table below.

Number of minutes late	0	1	2	3	4	5	6	7	8	9	10
Number of students	6	4	4	5	7	3	1	0	0	0	0

After two weeks of trying to improve the situation, he records a new set of results. These are shown below.

Number of minutes late	0	1	2	3	4	5	6	7	8	9	10
Number of students	14	7	4	1	1	1	0	0	1	0	1

The teacher decides to analyze these sets of data using box and whisker plots. By carrying out any necessary calculations and drawing the relevant box and whisker plots, decide whether his strategy has improved pupil punctuality. Give detailed reasons for your answer.

6.5 Measures of central tendency

'Average' is a word which, in general use, is taken to mean somewhere in the middle. For example, a woman may describe herself as being of average height. A student may think that he or she is of average ability in maths. Mathematics is more precise and uses three main methods to measure average.

- The **mode** is the value occurring most often.
- The **median** is the middle value when all the data is arranged in order of size.
- The **mean** is found by adding together all the values of the data and then dividing the total by the number of data values.

Worked example

The numbers below represent the number of goals scored by a team in the first 15 matches of the season. Find the mean, median and mode of the goals.

1 0 2 4 1 2 1 1 2 5 5 0 1 2 3

$$\text{Mean} = \frac{1+0+2+4+1+2+1+1+2+5+5+0+1+2+3}{15} = 2$$

Arranging all the data in order and then picking out the middle number gives the median: 0 0 1 1 1 1 1 (2) 2 2 2 3 4 5 5

The mode is the number that appears most often. Therefore the mode is 1.

Note: If there is an even number of data values, then there will not be one middle number, but a middle pair. The median is calculated by working out the mean of the middle pair.

Your GDC is also capable of calculating the mean and median of this set of data.

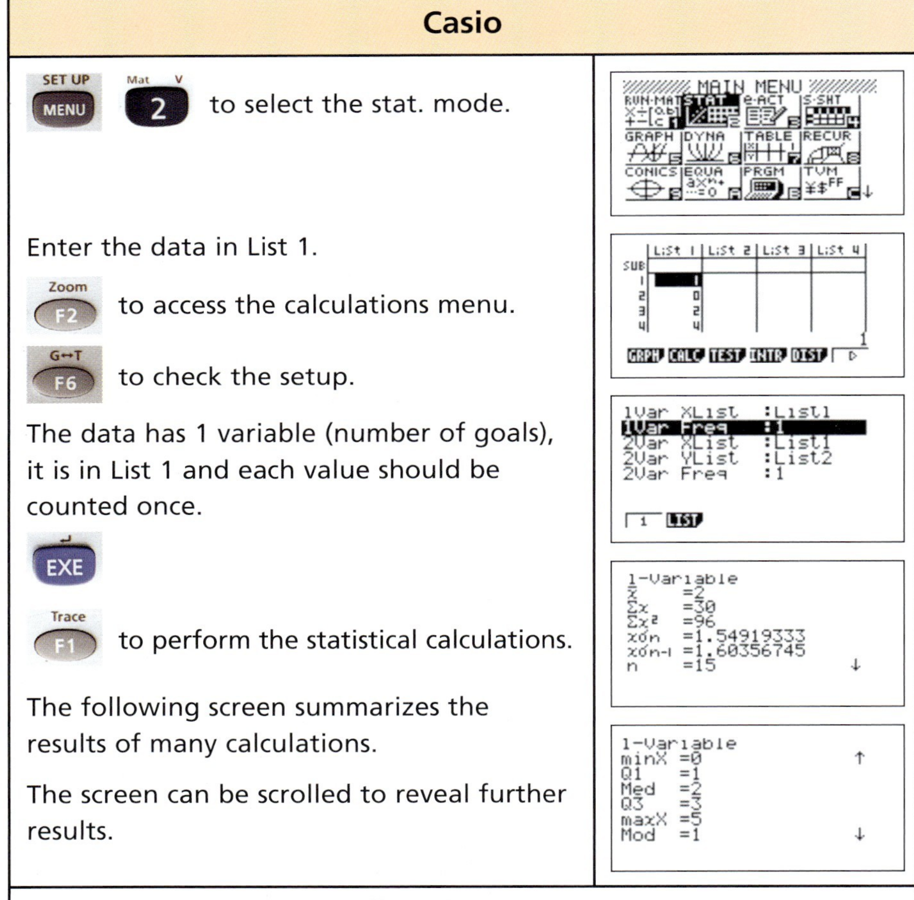

Note: The mean is given by $\bar{x}$ and the median by 'Med'.
The lower quartile, q_1, and upper quartile, q_3, are also displayed on this screen, as Q1 and Q3 respectively.

Texas

to enter the data into lists.

Enter the data in List 1.

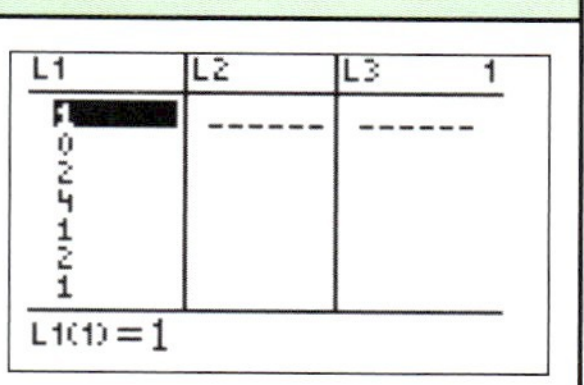

to select the 'Calc' menu.

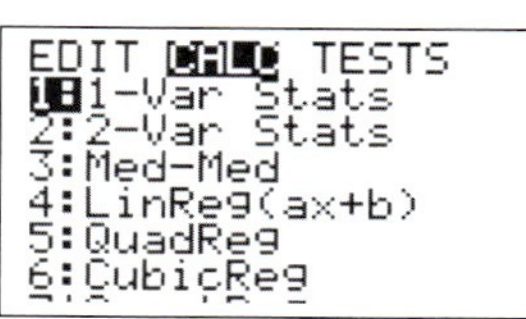

followed by

to perform statistical calculations on the 1–variable data in List 1.

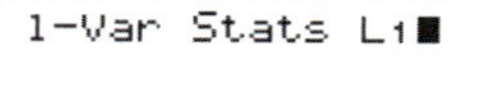

The following screen summarizes the results of many calculations.

1-Var Stats
$\bar{x}$=2
Σx=30
Σx²=96
Sx=1.603567451
σx=1.549193338
↓n=15

The screen can be scrolled to reveal further results.

1-Var Stats
↑n=15
minX=0
Q_1=1
Med=2
Q_3=3
maxX=5

Note: The mean is given by $\bar{x}$ and the median by 'Med'.
The lower quartile, q_1, and upper quartile, q_3, are also displayed on this screen, as Q_1 and Q_3 respectively.

Exercise 6.5.1

1 Find the mean, median and mode for each set of data.

a The number of goals scored by a hockey team in each of 15 matches
1 0 2 4 0 1 1 1 2 5 3 0 1 2 2

b The total scores when two dice are rolled
7 4 5 7 3 2 8 6 8 7 6 5 11 9 7 3 8 7 6 5

c The number of pupils present in a class over a three-week period
28 24 25 28 23 28 27 26 27 25 28 28 28 26 25

d An athlete's training times (seconds) for the 100 metres
14.0 14.3 14.1 14.3 14.2 14.0 13.9 13.8
13.9 13.8 13.8 13.7 13.8 13.8 13.8

2 The mean mass of the 11 players in a football team is 80.3 kg. The mean mass of the team plus a substitute is 81.2 kg. Calculate the mass of the substitute.

3 After eight matches a basketball player had scored a mean of 27 points. After three more matches his mean was 29. Calculate the total number of points he scored in the last three games.

Large amounts of data

When there are only three values in a set of data, the median value is given by the second value, i.e. 1 (2) 3.

When there are four values in a set of data, the median value is given by the mean of the second and third values, i.e 1 (2 3) 4.

When there are five values in a set of data, the median value is given by the third value.

If this pattern is continued, it can be deduced that for n sets of data, the median value is given by the value at $\frac{n+1}{2}$. This is useful when finding the median of large sets of data.

Worked example

The shoe sizes of 49 people are recorded in the table below. Calculate the median, mean and modal shoe size.

Shoe size	3	$3\frac{1}{2}$	4	$4\frac{1}{2}$	5	$5\frac{1}{2}$	6	$6\frac{1}{2}$	7
Frequency	2	4	5	9	8	6	6	5	4

As there are 49 data values, the median value is the 25th value. We can use the cumulative frequency to identify which class this falls within.

Shoe size	3	$3\frac{1}{2}$	4	$4\frac{1}{2}$	5	$5\frac{1}{2}$	6	$6\frac{1}{2}$	7
Cumulative frequency	2	6	11	20	28	34	40	45	49

The median occurs within shoe size 5. So the median shoe size is 5.
To calculate the mean of a large data set, we use the formula

$$\bar{x} = \frac{\sum fx}{n} \quad \text{where } n = \sum f.$$

Shoe size, *x*	3	$3\frac{1}{2}$	4	$4\frac{1}{2}$	5	$5\frac{1}{2}$	6	$6\frac{1}{2}$	7
Frequency, *f*	2	4	5	9	8	6	6	5	4
fx	6	14	20	40.5	40	33	36	32.5	28

Mean shoe size $= \frac{250}{49} = 5.10$

Note: The mean value is not necessarily a data value which appears in the set or a real shoe size.

The modal shoe size is $4\frac{1}{2}$.

These calculations can also be carried out on your GDC.

Casio

  to select the stat. mode.

Enter the shoe sizes in List 1 and their frequency in List 2.

 to access the calculations menu.

 to check the setup.

The data has one variable (shoe size), it is in List 1 and its frequency is in List 2.

 to perform the statistical calculations.

The screen summarizes the results of many calculations and can be scrolled to reveal further results.

Texas

 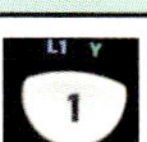 to enter the data into lists.

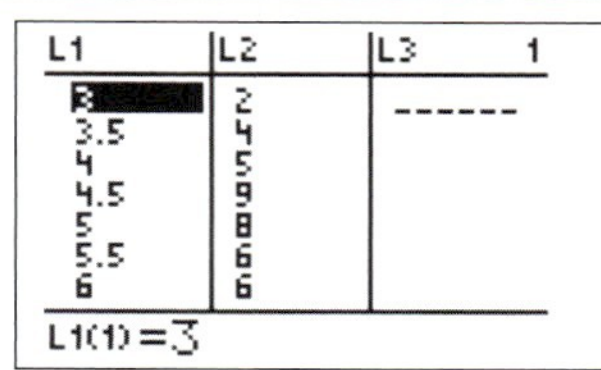

Enter the shoe size in List 1 and the frequency in List 2.

 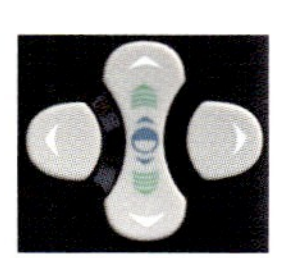 to select the 'Calc' menu.

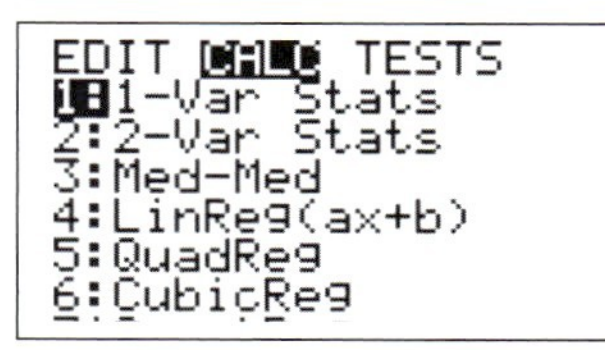

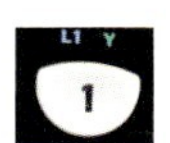 followed by

 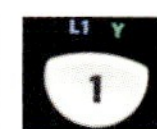   to perform statistical calculations on the 1–variable data in List 1 with frequency in List 2.

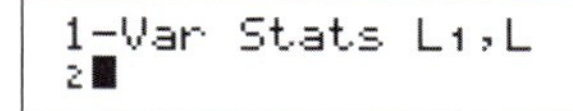

The screen summarizes the results of many calculations and can be scrolled to reveal further results.

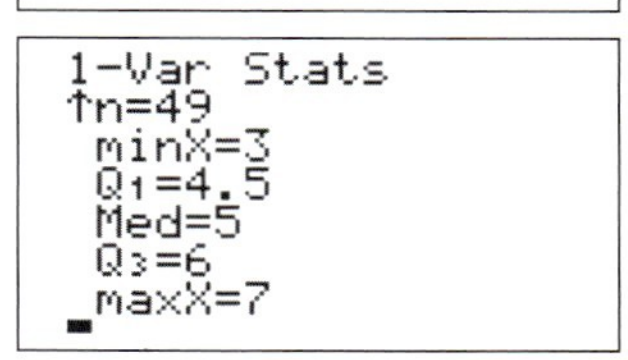

Exercise 6.5.2

1 An ordinary dice was rolled 60 times. The results are shown in the table below. Calculate the mean, median and mode of the scores.

Score	1	2	3	4	5	6
Frequency	12	11	8	12	7	10

2 Two dice were rolled 100 times. Each time their combined score was recorded. Below is a table of the results. Calculate the mean, median and mode of the scores.

Score	2	3	4	5	6	7	8	9	10	11	12
Frequency	5	6	7	9	14	16	13	11	9	7	3

3 Sixty flowering bushes are planted. At their flowering peak, the number of flowers per bush is counted and recorded. The results are shown in the table below.

Flowers per bush	0	1	2	3	4	5	6	7	8
Frequency	0	0	0	6	4	6	10	16	18

a Calculate the mean, median and mode of the number of flowers per bush.
b Which of the mean, median and mode would be most useful when advertising the bush to potential buyers?

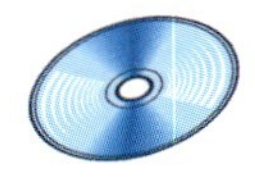

Mean and mode for grouped data

As has already been described, sometimes it is more useful to group data, particularly if the range of values is very large. However, by grouping data, some accuracy is lost.

The results below are the distances (to the nearest metre) run by twenty pupils in one minute.

256 271 271 274 275 276 276 277 279 280
281 282 284 286 287 288 296 300 303 308

Table 1: Class interval of 5

Group	250–254	255–259	260–264	265–269	270–274	275–279	280–284	285–289	290–294	295–299	300–304	305–309
Frequency	0	1	0	0	3	5	4	3	0	1	2	1

Table 2: Class interval of 10

Group	250–259	260–269	270–279	280–289	290–299	300–309
Frequency	1	0	8	7	1	3

Table 3: Class interval of 20

Group	250–269	270–289	290–309
Frequency	1	15	4

The three tables above highlight the effects of different group sizes. Table 1 is perhaps too detailed, whilst in Table 3 the group sizes are too big and consequently most of the results fall into one group. Table 2 is the most useful in that the spread of the results is still clear, although detail is still lost. In the 270–279 group we can see that there are eight pupils, but without the raw data we would not know where in the group they lie.

To find the mean of grouped data we assume that all the data within a group takes the mid-interval value.

$$\bar{x} = \frac{\Sigma fx}{n}$$

where $\bar{x}$ is the estimated mean, x is the mid-interval value and $n = \Sigma f$.

Group	250–259	260–269	270–279	280–289	290–299	300–309
Mid-interval value, *x*	254.5	264.5	274.5	284.5	294.5	304.5
Frequency, *f*	1	0	8	7	1	3
fx	254.5	0	2196	1991.5	294.5	913.5

$$\text{Estimated mean} = \frac{5650}{20} = 282.5$$

The **estimate** of mean distance run is 282.5 metres.
The modal group is 270–279.

The GDC can work out the mean, median and quartiles of grouped data. The mid-interval value should be entered in List 1 and the frequency in List 2. Then proceed as before.

Exercise 6.5.3

1 A pet shop has 100 tanks containing fish. The number of fish in each tank is recorded in the table below.

Number of fish	0–9	10–19	20–29	30–39	40–49
Frequency	7	12	24	42	15

a Calculate an estimate for the mean number of fish in each tank.
b Give the modal group size.

2 A school has 148 Year 12 pupils studying mathematics. Their percentage scores in their mathematics mock exam are recorded in the grouped frequency table below.

% Score	0–9	10–19	20–29	30–39	40–49	50–59	60–69	70–79	80–89	90–99
Frequency	3	2	4	6	8	36	47	28	10	4

a Calculate an estimate of the mean percentage score for the mock exam.
b What was the modal group score?

3 A stationmaster records how many minutes late each train is to the nearest minute. The table of results is given below.

No. of minutes late	0–4	5–9	10–14	15–19	20–24	25–29
Frequency	16	9	3	1	0	1

a Calculate an estimate for the mean number of minutes late a train is.
b What is the modal number of minutes late?
c The stationmaster's report concludes: 'Trains are, on average, less than five minutes late'. Comment on this conclusion.

6.6 Measures of dispersion

Range and interquartile range

The range is the difference between the greatest data value and the smallest data value. It is easy to calculate but often does not give a very good picture of how spread out the data is.

The interquartile range, which was covered in Section 6.4, often gives a better idea of the spread of the data. It is the spread of the middle 50% of the data and can be calculated as the difference between the upper and lower quartile.

Interquartile range $= q_3 - q_1$

It has an advantage over the simple calculation of range in that it isn't affected by extreme values, since it only deals with the middle 50% of the data. However, its drawback is that it doesn't take into account all the data.

Standard deviation

The **standard deviation** of data is a measure of dispersion that does take into account all of the data. It gives an average measure of difference (or deviation) from the mean of the data. The larger the value of the standard deviation, the more widely spread or dispersed the data is.

For example:
The ages of two groups of people are given below.

Group 1	2	6	10	16	16	22	26	30
Group 2	2	16	16	16	16	16	16	30

The mean, median, mode and range of the data are:

	Mean	Median	Mode	Range
Group 1	16	16	16	28
Group 2	16	16	16	28

These measures suggest that the data is the same, or at least very similar. However, if we look at the deviation from the mean, we will find that they are not.

The formula for the standard deviation of a set of data is given by $s_n = \sqrt{\frac{\sum(x-\bar{x})^2}{n}}$

where s_n is the standard deviation
x is each of the data values
$\bar{x}$ is the mean of the data
n is the number of data values.

The standard deviation of the ages of each group can be calculated as follows:

Group 1 $\bar{x} = 16$, $n = 8$		
x	$(x-\bar{x})$	$(x-\bar{x})^2$
2	−14	196
6	−10	100
10	−6	36
16	0	0
16	0	0
22	6	36
26	10	100
30	14	196
		$\sum(x-\bar{x})^2 = 664$
		$\frac{\sum(x-\bar{x})^2}{n} = 83$
		$\sqrt{\frac{\sum(x-\bar{x})^2}{n}} = 9.11$

Group 2 $\bar{x} = 16$, $n = 8$		
x	$(x-\bar{x})$	$(x-\bar{x})^2$
2	−14	196
16	0	0
16	0	0
16	0	0
16	0	0
16	0	0
16	0	0
30	14	196
		$\sum(x-\bar{x})^2 = 392$
		$\frac{\sum(x-\bar{x})^2}{n} = 49$
		$\sqrt{\frac{\sum(x-\bar{x})^2}{n}} = 7$

From the results it can be seen that the standard deviation for group 1 is greater than that for group 2. This implies that the data is more spread out.

The formula given above for the standard deviation can be cumbersome as the mean value has to be subtracted from each data value. A more efficient formula for the standard deviation is $s_n = \sqrt{\frac{\sum x^2}{n} - \bar{x}^2}$. With this formula the mean of the data is squared and subtracted only once.

Your GDC can also calculate the standard deviation of data. The instructions below relate to the data for group 1 above.

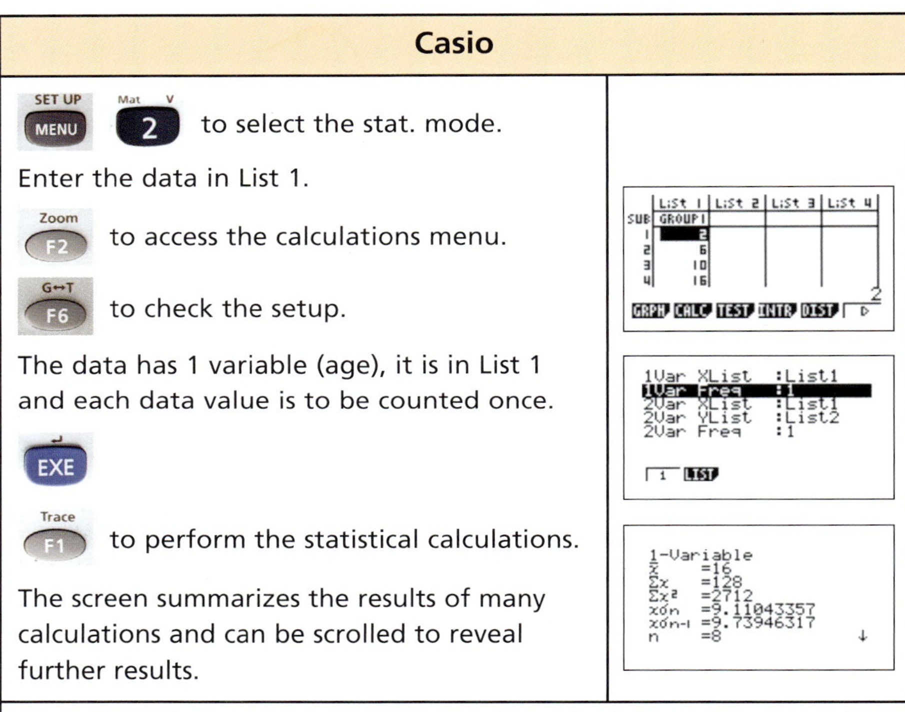

Casio

MENU 2 to select the stat. mode.

Enter the data in List 1.

F2 to access the calculations menu.

F6 to check the setup.

The data has 1 variable (age), it is in List 1 and each data value is to be counted once.

EXE

F1 to perform the statistical calculations.

The screen summarizes the results of many calculations and can be scrolled to reveal further results.

Note: The calculator gives two types of standard deviation, $x\sigma_n$ and $x\sigma_{n-1}$. The standard deviation you will need on this course is $x\sigma_n$.

Texas

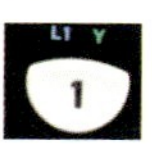

to enter the data into lists.

Enter the data in List 1.

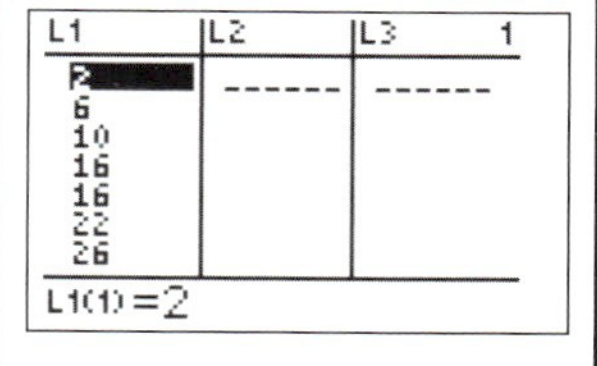

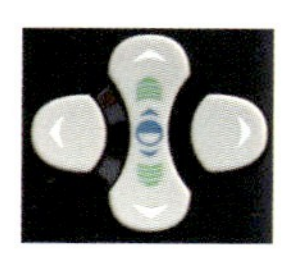
to select the 'Calc' menu.

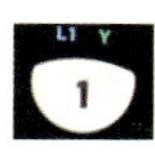

followed by

to perform statistical calculations on the 1–variable data in List 1.

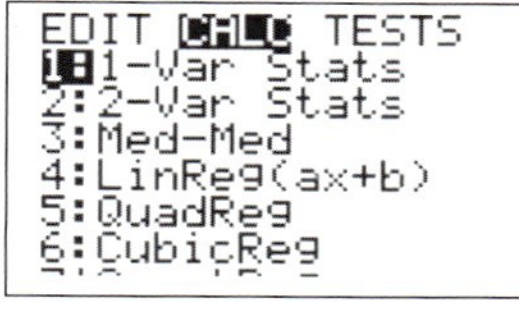

1-Var Stats L1

The screen summarizes the results of many calculations and can be scrolled to reveal further results.

1-Var Stats
$\bar{x}$=16
Σx=128
Σx²=2712
Sx=9.739463171
σx=9.110433579
↓n=8

Note: The calculator gives two types of standard deviation, s_x and σ_x. The standard deviation you will need on this course is σ_x.

You will notice that there appears to be different notation used for the standard deviation. It is important therefore for you to be familiar with the notation that your calculator uses and also what this course uses.

When the data is given in a frequency table, the formula for the standard deviation of a set of data is given by $s_n = \sqrt{\dfrac{\sum f\,(x - \bar{x})^2}{n}}$.

Worked example

A certain type of matchbox claims to contain 50 matches in each box. A sample of 60 boxes produced the following results.

Number of matches, x	Frequency, f	fx	$(x - \bar{x})$	$(x - \bar{x})^2$	$f(x - \bar{x})^2$
48	3	144	−2.05	4.2025	12.6075
49	11	539	−1.05	1.1025	12.1275
50	28	1400	−0.05	0.0025	0.07
51	16	816	0.95	0.9025	14.44
52	2	104	1.95	3.8025	7.605
Total	60	3003			48.85

Calculate: **a** the mean
b the standard deviation.

a $\bar{x} = \frac{\Sigma fx}{\Sigma f} = \frac{3003}{60} = 50.05$

b $s_n = \sqrt{\frac{\Sigma f\,(x - \bar{x})^2}{\Sigma f}} = \sqrt{\frac{46.85}{60}} = 0.884$

Exercise 6.6.1

Using a GDC or otherwise, calculate:
i) the mean
ii) the range
iii) the interquartile range
iv) the standard deviation
for the data given in questions 1–4.

1 **a** 6, 2, 3, 8, 7, 5, 9, 9, 2, 4
b 72, 84, 83, 81, 69, 77, 85, 79
c 1.6, 2.9, 3.7, 5.5, 4.2, 3.9, 2.8, 4.5, 4.2, 5.1, 3.9

2 The number of goals a hockey team scores during each match in a season

Number of goals	0	1	2	3	4	5	6
Frequency	3	8	11	7	4	2	1

3 The number of shots in a round for each player in a golf tournament

Number of shots	66	67	68	69	70	71	72	73	74
Frequency	1	2	1	4	11	17	43	18	3

4 The number of letters posted to 50 houses in a street

Number of letters	0	1	2	3	4	5	6
Frequency	5	8	12	8	7	8	2

5 A class of 25 students was asked to estimate the number of coins (x) in a jar. The results of their estimations are summarized below.

$\Sigma x = 2000$ $\Sigma x^2 = 163554$

Calculate:
a the mean of their estimates
b the standard deviation of their estimates.

6 The results for a series of experiments are given below.

Experiment	1	2	3	4	5	6	7	8	9	10
Result	6.2	6.1	6.3	6.3	6.7	6.1	6.2	6.3	6.1	5.9
Experiment	11	12	13	14	15	16	17	18	19	20
Result	6.0	6.2	6.1	6.3	6.0	6.1	6.2	6.3	6.1	

The result for experiment 20 is obscured. However it is known that the mean ($\bar{x}$) for all 20 experiments is 6.2.

Calculate: **a** the value of the 20th result
b the standard deviation of all the results.

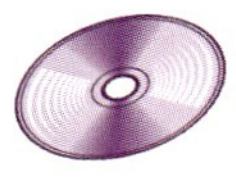

6.7 Scatter diagrams, bivariate data and linear correlation

When we record information about two different aspects (or **variables**) of a data item, such as height and mass of children or temperature and number of ice creams sold on particular days, we are collecting **bivariate data**. We can use the value of the two variables for a data item as the coordinates of a point to represent it on a graph called a **scatter diagram** (or scatter graph). Scatter diagrams are particularly useful if we wish to see if there is a relationship between the two variables. How the points lie when plotted indicates the type of relationship between the two variables.

Worked example

The heights and weights (masses) of 20 children under the age of five are recorded. The heights were recorded in centimetres and the weights in kilograms.

Height	32	34	45	46	52	59	63	64	71	73
Mass	5.834	3.792	9.037	4.225	10.149	6.188	9.891	16.010	15.806	9.929
Height	86	87	95	96	96	101	108	109	117	121
Mass	11.132	16.443	20.895	16.181	14.000	19.459	15.928	12.047	19.423	14.331

a Plot a scatter diagram for the data above.
b Comment on any relationship that you see.

a

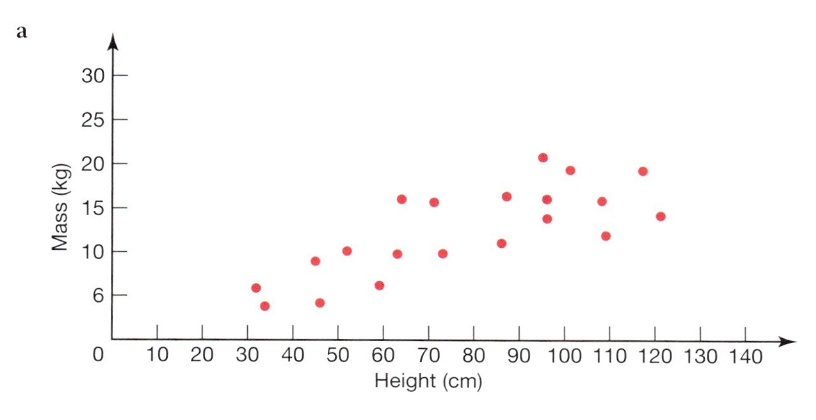

b The points tend to lie in a diagonal direction from bottom left to top right. This suggests that as height increases then, in general, weight increases too. Therefore there is a positive correlation between height and weight.

Lines of best fit

If the scatter diagram shows that there is a relationship between the two variables, we can use a line of best fit to estimate the value of one variable given a value of the other variable. To do this we draw a straight line passing through the data, i.e. a line of best fit. It will pass through the point $(\bar{x}, \bar{y})$ and leaves approximately half the points above the line and half the points below it. It does not need to pass through the origin.

Worked example

Using the data about the height and weight of children above, estimate the weight of a child with a height of 80 cm.

We have to assume that this child will follow the trend set by the other 20 children. To deduce an approximate value for the weight, we draw a line of best fit through the data, passing through the point $(\bar{x}, \bar{y})$.

$\bar{x} = 77.75$, $\bar{y} = 12.535$

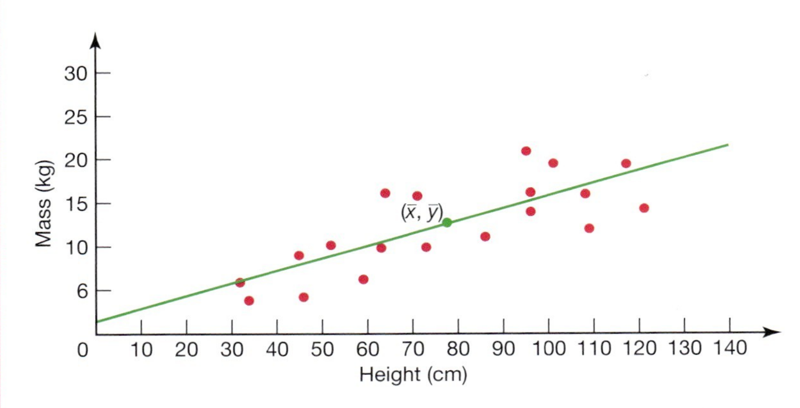

The line of best fit can now be used to give an approximate solution to the question. If a child has a height of 80 cm, you would expect his/her weight, by reading from the graph below, to be in the region of 13 kg.

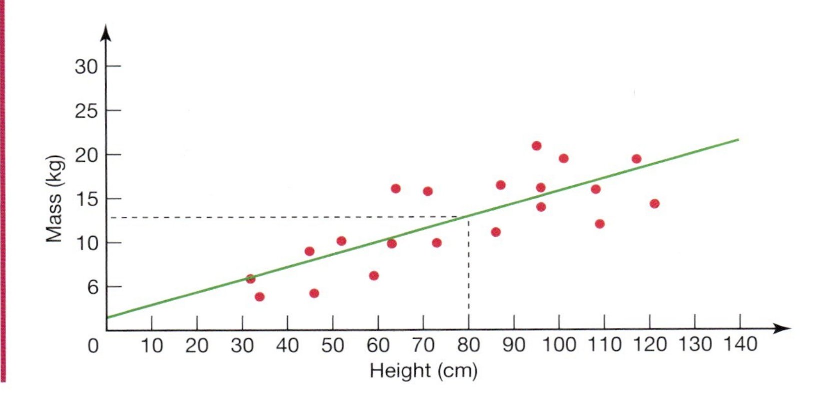

Your GDC and graphing software can plot scatter diagrams and analyze them.

For example, plot the following data for an ice cream vendor on a scatter diagram and, if appropriate, draw a line of best fit.

Temperature (°C)	15	24	18	24	19	26	22	24	27	28	30	25	22	17
Number of ice creams sold	8	34	20	38	28	37	32	29	33	35	44	28	30	25

Casio

to select the stat. mode.

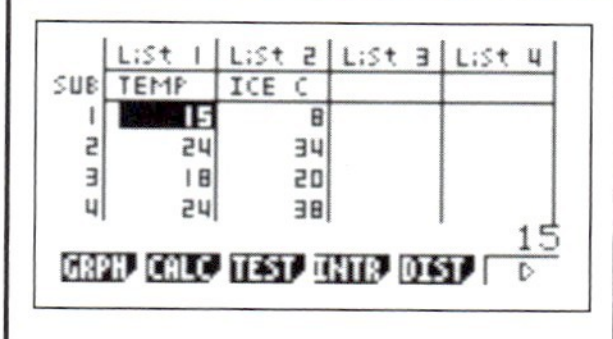

Enter the temperature data in List 1 and the number of ice creams sold in List 2.

Trace F1 to access the statistical graphing menu.

G↔T F6 to check the setup.

The graph type is 'scatter' with the x values from List 1 and the y values from List 2. Each data value is to be counted once.

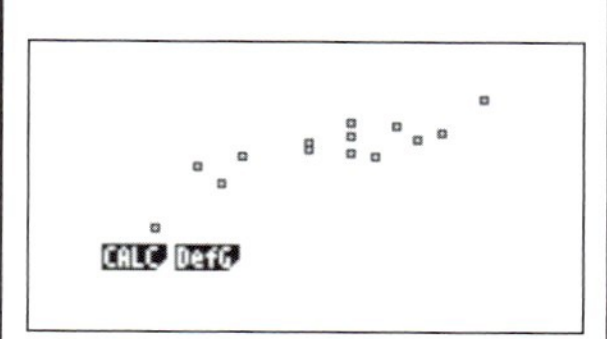

Trace F1 to plot the scatter diagram.

Trace F1 to select the graph calculation menu.

Zoom F2 as the line of best fit required is linear. The following screen summarizes the properties of the line of best fit in the form $y = ax + b$.

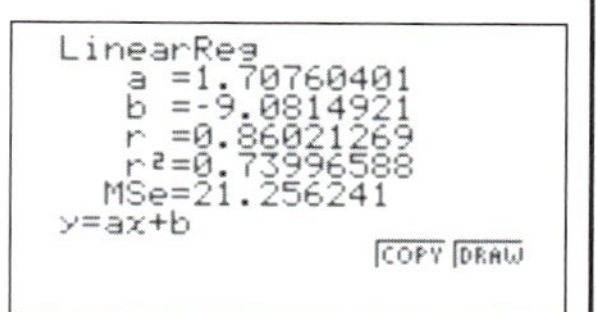

to plot the line of best fit.

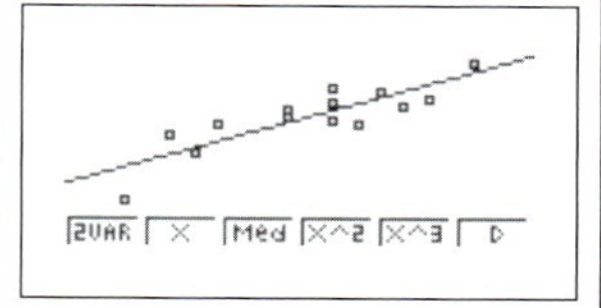

Note: The screen which gives the properties of the line of best fit also gives the value of r. This represents the product–moment correlation coefficient which is dealt with later in this topic.

Texas

to enter the data into lists.

Enter the temperature data in List 1 and the number of ice creams sold in List 2.

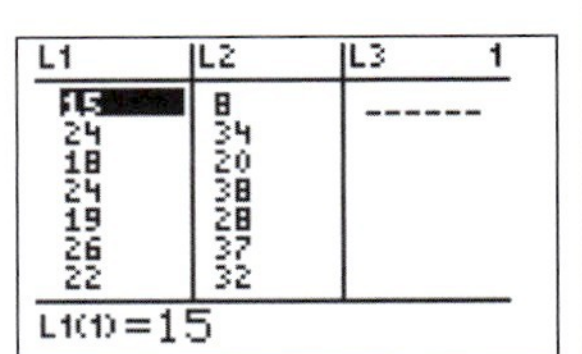

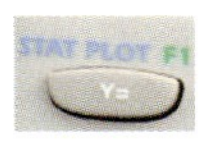

to enter the statistical plot setup.

Turn 'Plot 1' to 'On'. Choose the scatter diagram and ensure the *x* values are from List 1 and the *y* values from List 2.

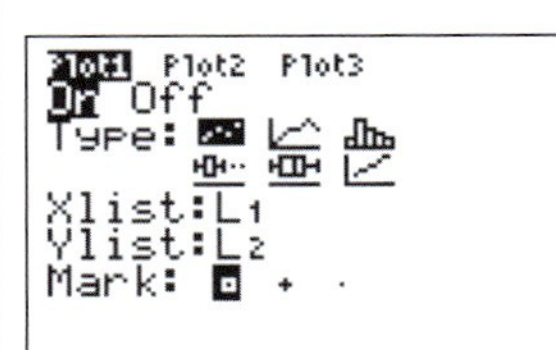

to set scale the scale for each axis.

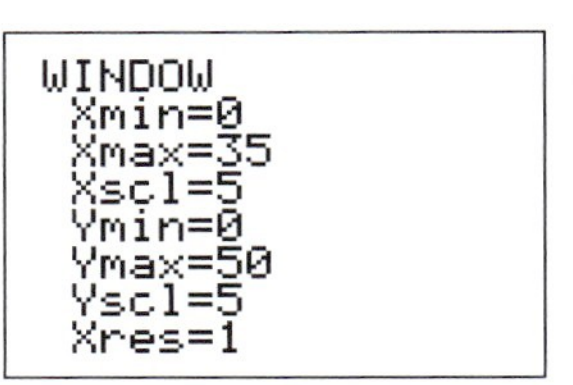

to plot the scatter diagram.

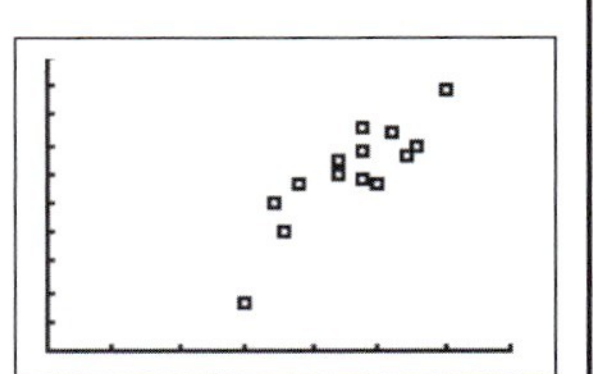

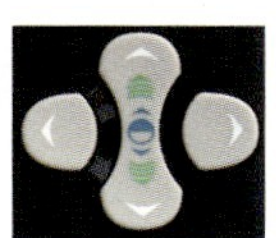
to select the 'Calc' menu.

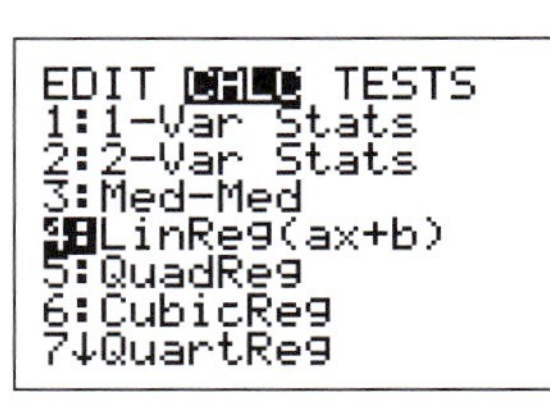

to find the linear equation of the line of best fit through the points.

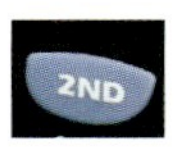

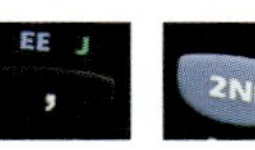

to calculate the equation of the line of best fit with *x* values from List 1 and *y* values from List 2.

LinReg(ax+b) L1,
L2

to display the equation in the form $y = ax + b$.

LinReg
y=ax+b
a=1.707604017
b=-9.081492109

Autograph

On the 2D-graph page option, click on to enter the data.

Enter the temperature in the x column and the number of ice creams sold in the y column.

Click 'OK' to graph the data.

To identify the coordinate $(\bar{x}, \bar{y})$, select 'Object' followed by 'centriod'.

To draw a linear line of best fit, select 'Object' followed by 'Best fit'. A straight line is a polynomial of order 1.

Click 'OK' to graph the line of best fit.

Select the line. Its equation will appear at the base of the screen.

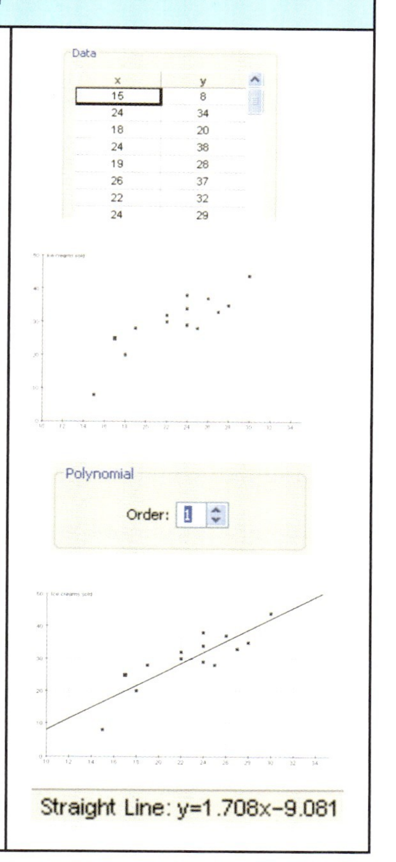

Types of correlation

There are several types of correlation depending on the arrangement of the points plotted on the scatter diagram. These are described below.

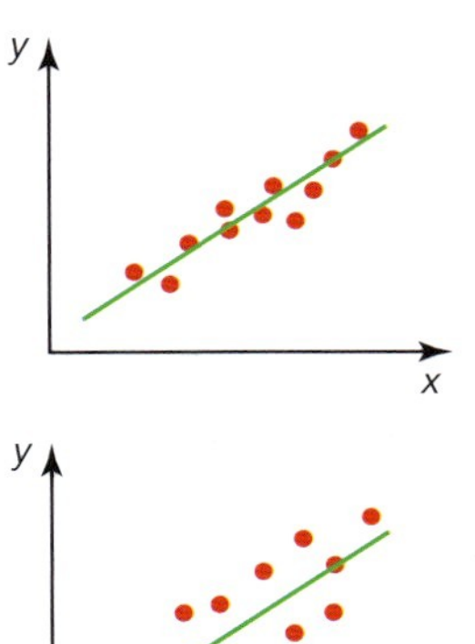

A **strong positive correlation**.
The points lie tightly around the line of best fit.
As x increases, so does y.

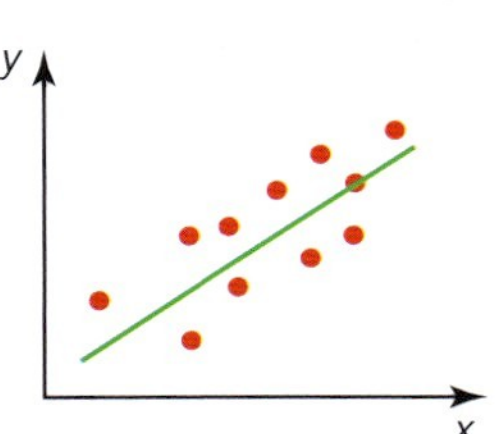

A **weak positive correlation**.
Although there is direction to the way the points are lying, they are not tightly packed around the line of best fit.
As x increases, y tends to increase too.

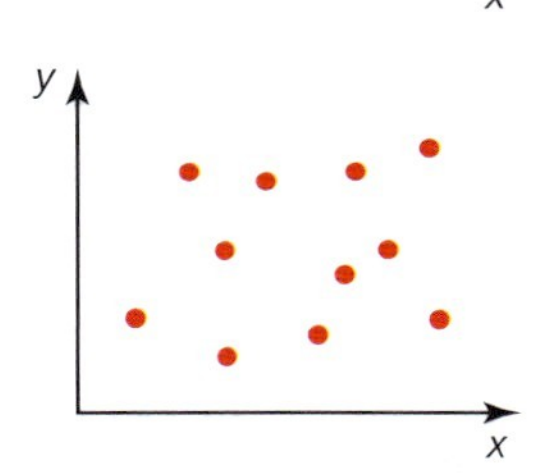

No correlation.
There is no pattern to the way in which the points are lying, i.e there is no correlation between the variables x and y. As a result, there can be no line of best fit.

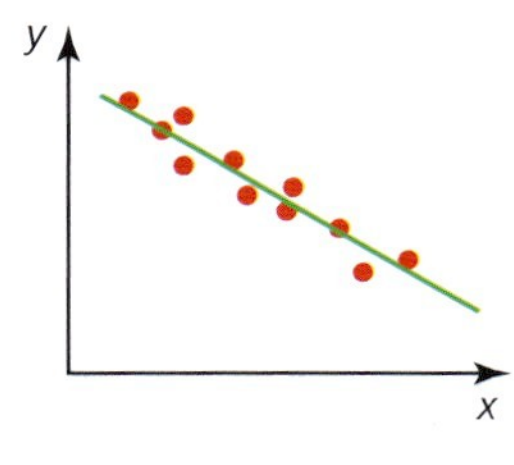

A **strong negative correlation**.
The points lie tightly around the line of best fit. As x increases, y decreases.

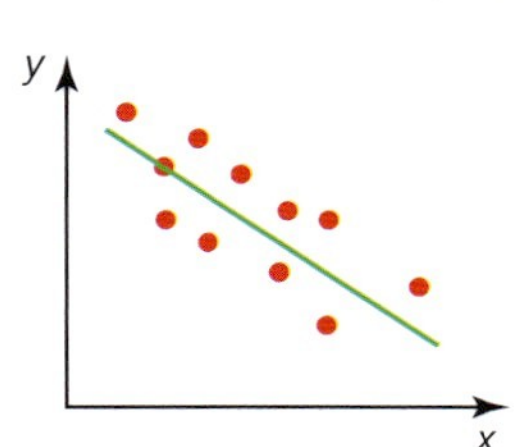

A **weak negative correlation**.
The points are not tightly packed around the line of best fit. As x increases, y tends to decrease.

Exercise 6.7.1

1 State what type of correlation you might expect, if any, if the following data was collected and plotted on a scatter diagram. Give reasons for your answer.
 a A student's score in a mathematics exam and their score in a science exam
 b A student's hair colour and the distance they have to travel to school
 c The outdoor temperature and the number of cold drinks sold by a shop
 d The number of goals your opponents score and the number of times you win
 e A person's height and the person's age
 f A car's engine size and its fuel consumption

2 The table shows the readings for the number of hours of sunshine and the amount of rainfall in millimetres for several cities and towns in the UK.

City	Hrs. of sunshine	Rainfall
Aberdeen	9.8	0.3
Aviemore	2.8	0
Belfast	4.1	6.6
Birmingham	6.4	0.8
Bournemouth	9.3	4.3
Bristol	10.2	1.8
Cardiff	10.8	2.8
Folkestone	10.5	0
Hastings	7.2	0.3
Isle of Man	9.3	2.5
Isle of Wight	10.7	0.3
London	10.1	1.5
Manchester	4.6	2.0
Margate	10.9	0
Newcastle	8.2	0.5
Newquay	4.1	2.3
Oxford	10.1	3.0
Scarborough	10.6	1.8
Skegness	12.0	3.3
Southport	8.2	7.4
Torquay	8.8	1.5

 a Plot a scatter diagram of hours of sunshine against amount of rainfall. Use a spreadsheet or graphing software if possible.
 b What type of correlation, if any, is there between the two variables? Comment on whether this is what you would expect.

3 The United Nations keeps an up-to-date database of statistical information on its member countries. The table below shows some of the information available.

Country	Life expectancy at birth (years, 1990–99)		Adult illiteracy rate (%, 1995)	Infant mortality rate (per 1,000 births, 1990–99)
	Female	Male	Total	
Australia	81	76	0	6
Barbados	79	74	2.6	12
Brazil	71	63	16.8	42
Chad	49	46	51.9	112
China	72	68	18.5	41
Colombia	74	67	9.6	30
Congo	51	46	25.6	90
Cuba	78	74	4.4	9
Egypt	68	65	48.9	51
France	82	74	0	6
Germany	80	74	0	5
India	63	62	48	72
Iraq	64	61	42	95
Israel	80	76	4.9	8
Japan	83	77	0	4
Kenya	53	51	22.7	66
Mexico	76	70	10.5	31
Nepal	57	58	64.1	83
Portugal	79	72	10	9
Russian Federation	73	61	0.9	18
Saudi Arabia	73	70	27.8	23
United Kingdom	80	75	0	7
United States of America	80	73	0	7

a By plotting a scatter diagram, decide if there is a correlation between the Adult illiteracy rate and the Infant mortality rate.

b Are your findings in part **a** above what you expected? Explain your answer.

c Without plotting a scatter diagram, decide if you think there is likely to be a correlation between male and female life expectancy at birth. Explain your reasons.

d Plot a scatter diagram to test if your predictions in part **c** were correct.

4 The table below gives the average time taken for thirty pupils in a class to get to school each morning and the distance they live from the school.

Distance (km)	2	10	18	15	3	4	6	2	25	23	3	5	7	8	2
Time (mins)	5	17	32	38	8	14	15	7	31	37	5	18	13	15	8
Distance (km)	19	15	11	9	2	3	4	3	14	14	4	12	12	7	1
Time (mins)	27	40	23	30	10	10	8	9	15	23	9	20	27	18	4

a Plot a scatter diagram of distance travelled against time taken.
b Describe the correlation between the two variables.
c Explain why some pupils who live further away may get to school quicker than some of those who live nearer.
d Draw a line of best fit on your scatter diagram.
e A new pupil joins the class. Use your line of best fit to estimate how far away she might live if she takes, on average, 19 minutes to get to school each morning.

Pearson's product–moment correlation coefficient

Karl Pearson was a statistician who was one of the few mathematicians who launched a totally new field. In his case he joined the zoologist Walter Weldon to study what he called 'biometry', that is, he applied statistical analysis to animal evolution, among other areas.

Pearson's product–moment correlation coefficient, r, measures the correlation between two variables x and y.

The range of values for r is $-1 \leq r \leq 1$

where -1 indicates a perfect negative correlation between x and y.
0 indicates no correlation between x and y
1 indicates a perfect positive correlation between x and y.

The value for r can be calculated by the following formula

$$r = \frac{s_{xy}}{s_x s_y}$$

where s_{xy} represents the covariance of x and y (a measure of how much two variables change together)
s_x represents the standard deviation of x
s_y represents the standard deviation of y.

Note: You will be given the value of the covariance s_{xy} if the formula is needed in an examination. However, it can be calculated using the formula

$$s_{xy} = \frac{\Sigma xy - \frac{(\Sigma x)(\Sigma y)}{n}}{n}.$$

Worked example

The table below shows the position of a football team in the English Premier League and the goal difference, that is the difference between the goals scored and goals conceded.

Position	1	2	3	4	5	6	7	8	9
Goal difference	+23	+20	+21	+18	+14	+16	+11	+9	+3
Position	10	11	12	13	14	15	16	17	18
Goal difference	0	−2	−4	−11	−9	−10	−14	−17	−33

a If $s_{xy} = -77.14$, calculate
  i) s_x
  ii) s_y
  iii) r.

b Interpret the value of r.

a i) $s_x = \sqrt{\frac{\Sigma x^2}{n} - \bar{x}^2}$, where $\bar{x} = \sqrt{\frac{\Sigma x}{n}}$

$s_x = \sqrt{\frac{2109}{18} - 9.50^2} = 5.19$

ii) $s_y = \sqrt{\frac{\Sigma y^2}{n} - \bar{y}^2}$, where $\bar{y} = \sqrt{\frac{\Sigma y}{n}}$

$s_y = \sqrt{\frac{4253}{18} - 1.94^2} = 15.25$

iii) $r = \frac{s_{xy}}{s_x s_y} = \frac{-77.14}{5.19 \times 15.25} = -0.975$

b r is very close to -1, implying a strong negative correlation. In this case this is misleading, as one would expect that the higher the position in the table the greater the goal difference. However, in this case a higher position in the table is represented by a lower number, i.e. the top position is 1 rather than 18.

The values for s_x, s_y and r can be calculated using the GDC.

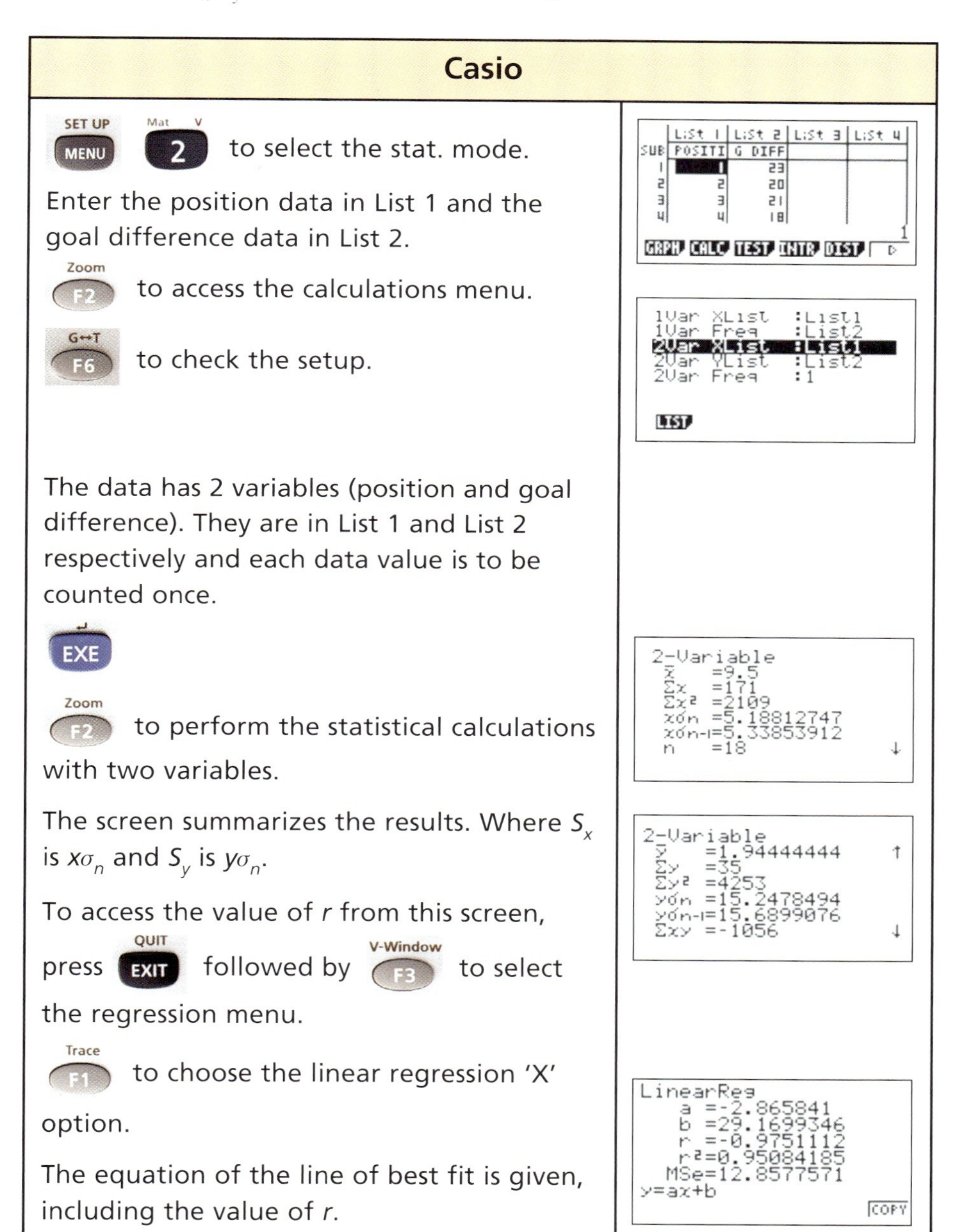

Casio

SET UP MENU, Mat 2 to select the stat. mode.

Enter the position data in List 1 and the goal difference data in List 2.

Zoom F2 to access the calculations menu.

G↔T F6 to check the setup.

The data has 2 variables (position and goal difference). They are in List 1 and List 2 respectively and each data value is to be counted once.

EXE

Zoom F2 to perform the statistical calculations with two variables.

The screen summarizes the results. Where S_x is $x\sigma_n$ and S_y is $y\sigma_n$.

To access the value of r from this screen, press QUIT EXIT followed by V-Window F3 to select the regression menu.

Trace F1 to choose the linear regression 'X' option.

The equation of the line of best fit is given, including the value of r.

Texas

 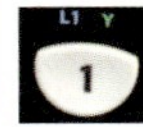 to enter the data into lists.

Enter the position data in List 1 and the goal difference in List 2.

```
L1    L2    L3    1
1     23    ------
2     20
3     21
4     18
5     14
6     16
7     11
L1(1)=1
```

  to select the 'Calc' menu.

  followed by   

 to perform statistical calculations on the 2−variable data with *x* values in List 1 and *y* values in List 2 .

```
2-Var Stats L1,L
2
```

The screen summarizes the results for both sets of data. s_x is σ_x and s_y is σ_y.

```
2-Var Stats
 x̄=9.5
 Σx=171
 Σx²=2109
 Sx=5.338539126
 σx=5.188127472
↓n=18
```

Access the equation for the line of best fit and the value of *r* from this screen by

  to select the 'Calc' menu.

```
2-Var Stats
↑ȳ=1.944444444
 Σy=35
 Σy²=4253
 Sy=15.68990767
 σy=15.24784947
↓Σxy=-1056
```

 to select 'LinReg (ax + b)' and type

 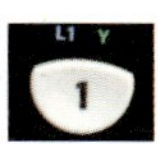    

to calculate the equation of the straight line and give the value of *r*.

```
LinReg
 y=ax+b
 a=-2.865841073
 b=29.16993464
 r²=.950841853
 r=-.9751112003
```

Note: The calculator gives two types of standard deviation, s_y and σ_x. The standard deviation you will need on this course is σ_x, even though the formula for standard deviation uses the notation s_x.

If *r* is not displayed on the screen, 'DiagnosticOn' needs to be set. This

is done via  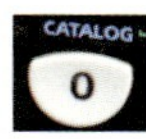 and scrolling down to select 'DiagnosticOn'.

Exercise 6.7.2

You are expected to use a GDC to calculate the value of the product–moment correlation coefficient.

1 The table below shows the height of women in an Olympic high jump event and the maximum height they jumped.

Height of competitor (m)	1.76	1.83	1.74	1.75	1.80	1.81	1.73	1.80	1.78	1.82
Height jumped (m)	1.83	2.06	1.75	1.88	2.04	2.02	1.78	1.90	1.78	1.90

a Plot a scatter diagram of the results.
b Calculate $\bar{x}$, the mean height of the competitors, and $\bar{y}$, the mean height jumped.
c Draw a line of best fit on your scatter diagram.
d Give the value of r, the product–moment correlation coefficient, and comment on its value.

2 The table below shows the percentage scored by a group of students in a mock English examination and what they scored in the final examination.

Mock %	72	68	83	81	54	59	77	82	69	81	32	54	37	88	28
Final %	75	66	90	81	48	52	81	85	70	90	27	52	34	96	17

a Plot a scatter diagram of the results.
b Calculate $\bar{x}$, the mean practice score, and $\bar{y}$, the mean final score.
c Draw a line of best fit on your scatter diagram.
d Give the value of r, the product–moment correlation coefficient, and comment on its value.

3 The same group of students also took a mock and final examination in mathematics. Their percentage scores are in the table below.

Practice %	72	68	84	78	53	59	77	82	55	62	30	51	40	88	43
Final %	76	71	90	80	58	81	77	90	54	67	50	58	60	87	64

a Calculate r and comment on its value.
b Comment on any differences between r calculated in this question and the previous question.

4 The product–moment correlation coefficient was calculated for data on each of the relationships listed below.

- The distance someone travels to work and the time it takes
- A group of students' results in an art examination and their results in a mathematics examination
- The value of a second-hand car and its age
- The size of a pumpkin and its weight

a Which relationship is **likely** to have the following values of r? Give reasons for your answers.

i) 0.98

ii) 0.2

iii) −0.8

b Estimate a value for r for the relationship not chosen in part **a** above. Give reasons for your answer.

5 For the students in your class, measure in centimetres their height and the length of their right foot.

a Plot a scatter diagram of the results.

b Calculate $\bar{x}$, the mean height, and $\bar{y}$, the mean foot size.

c Draw a line of best fit on your scatter diagram.

d Give the value of r, the product–moment correlation coefficient, and comment on its value.

e Would you expect the value of r be very different if the data had been collected from a class of 11-year-old pupils? Give reasons for your answer.

6.8 The regression line for *y* on *x*

Lines of best fit were drawn by eye in the earlier part of this work on correlation. Another, more accurate, line is called a regression line for y on x. This line is the optimum line of best fit.

The formula for the regression line for y on x is

$$y - \bar{y} = \frac{s_{xy}}{(s_x)^2}(x - \bar{x})$$

where $\bar{x}$ is the mean of x

$\bar{y}$ is the mean of y

s_x is the standard deviation of x

s_{xy} is the covariance.

Worked example

The table below shows the number of hours 20 students spent studying Spanish in the month before an examination and their percentage score.

Hours of study	21	32	13	40	15	26	27	18	19	10
Exam mark (%)	62	90	58	78	80	66	56	62	68	50
Hours of study	21	22	33	44	25	46	17	28	19	20
Exam mark (%)	74	70	76	80	63	95	57	69	58	64

a Find the correlation coefficient r, where $r = \dfrac{s_{xy}}{s_x s_y}$.

b Find the equation of the regression line for y on x.

c Estimate the score for the student who studies 30 hours in the month before the examination.

a $\bar{x} = \dfrac{\Sigma x}{n} = \dfrac{496}{200} = 24.8 \quad \bar{y} = \dfrac{\Sigma y}{n} = \dfrac{1378}{20} = 68.8$

$$s_x = \sqrt{\frac{\Sigma x^2}{n} - \bar{x}^2} = \sqrt{\frac{14\,174}{20} - 24.8^2} = 9.68$$

$$s_y = \sqrt{\frac{\Sigma y^2}{n} - \bar{y}^2} = \sqrt{\frac{97\,268}{20} - 68.8^2} = 11.40$$

$$s_{xy} = \frac{\Sigma xy - \dfrac{(\Sigma x)(\Sigma y)}{n}}{n} = \frac{35\,742 - \dfrac{496 \times 1376}{20}}{20} = 80.86$$

Therefore $r = \dfrac{s_{xy}}{s_x s_y} = \dfrac{80.86}{9.68 \times 11.40} = 0.733.$

Note: The values for $\bar{x}$, $\bar{y}$, Σx^2, Σy^2, Σxy, s_x, s_y, s_{xy} and r can all be taken from your calculator, once the data has been entered in two lists as shown before.

b $y - \bar{y} = \dfrac{s_{xy}}{(s_x)^2}(x - \bar{x})$

$$y - 68.8 = \frac{80.86}{9.68^2}(x - 24.8)$$

$$y = 0.863\,(x - 24.8) + 68.8$$

$$y = 0.863x + 47.39$$

Note: As there are many stages to these calculations, there is the possibility of rounding errors. The full answer at each stage should be saved in the memory of your calculator.

The equation of the regression line for y on x can be calculated on your GDC as shown before.

Autograph has the facility to produce the equation of the regression line for y on x efficiently as shown below.

Autograph

On the 2D-graph page option, click on

 to enter the data.

Enter the hours of study in the x column and the percentage score in the y column.

Click 'OK' to graph the data.

To plot the regression line for y on x, click 'Object' and select 'y-on-x Regression Line'.

Its equation will appear at the base of the screen.

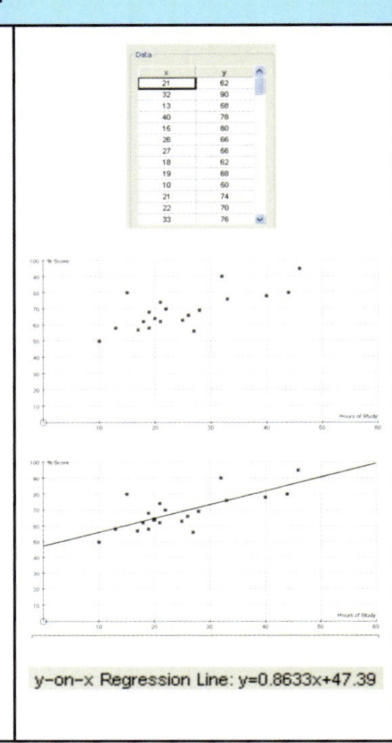

c As the equation for the regression line has been calculated, this can now be used to estimate further values.
Substitute $x = 30$ into the equation of regression line:

$y = 0.863x + 47.39$
$y = 0.863 \times 30 + 47.39 = 73.29$

Therefore the estimated score for a student who studied 30 hours is 73%.

When the predicted value falls within the range of the given data, then estimates are generally valid (interpolation). However, when the predicted value falls outside the range of the data (extrapolation), it is susceptible to large errors as there is the assumption that the data continues to behave linearly outside the given range.

For example, if using the data above we attempted to predict the examination score for a student who studied for 100 hours in the month prior to the exam, the extrapolation may produce invalid results as the data ranges from 10 to 46 hours of study, and 100 hours is clearly outside of this range.

Substituting $x = 100$ into the equation gives:

$y = 0.863 \times 100 + 47.39 = 133.7$

This indicates a score of 134%, which is clearly not possible.

Exercise 6.8.1

Use your GDC to answer the following questions.

1 A group of club cyclists take part in a 25 kilometre race. The table below shows the time taken and the mean number of hours' training per week done by each cyclist in the ten weeks prior to the race.

Hours training	12	20	3	6	28	22	16	9	1	11	24	17	13	9
Race time (mins)	80	55	90	86	52	66	70	80	88	74	56	69	74	78

a Calculate the correlation coefficient r.
b Calculate the equation of the regression line for y on x.
c Estimate how long, to the nearest minute, the race would have taken for someone who trained, on average, 18 hours per week.
d Comment on the validity of your answer to part **c** above.

2 The score (out of 161) for ten pupils in an IQ test at the age of 11 was compared with their percentage in an IB mathematics examination. The results are shown below.

IQ score	90	100	105	88	96	125	130	142	128	105
IB exam score	52	66	58	50	60	86	90	97	84	73

a Calculate the correlation coefficient r.
b Calculate the equation of the regression line for y on x.
c Two further students had IQ results of 95 and 155. Predict their respective IB scores.
d Comment on the validity of your answers to part **c** above.

3 The table shows the salary and the number of years experience of a group of fire fighters.

Salary $(000's)	32	27	40	40	36	32	25	30	27	40	37	33
Years of experience	7	4	20	18	11	7	1	5	3	17	15	12

a Calculate the correlation coefficient r and comment on its value.
b Calculate the equation of the regression line for y on x.
c Estimate the salary of a fire fighter with ten years' experience. Comment on the validity of your estimation.
d A firefighter has a salary of $100 000. Estimate his age using your equation of the regression line from part **b**. Comment on the validity of your answer.

4 The air temperature was taken at various heights by a meteorological balloon. The results are shown below.

Height (000's m)	4	8	12	16	20	24	28	32	36	40
Temperature (°C)	8	4	−20	−32	−40	−46	−48	−51	−57	−60

a Calculate the correlation coefficient r. Comment on its value.
b Calculate the equation of the regression line for y on x.
c Estimate the height of the balloon if the outside temperature is recorded as −70 °C. Comment on the validity of your answer.

6.9 The χ^2 test for independence

A chi-squared (χ^2) test for independence is used to assess whether or not paired observation, expressed in a contingency table (two-way table) are independent.

For example

Volunteers are testing a new drug in a clinical trial. It is claimed that the new drug will result in a more rapid improvement rate for sick patients than would happen if they did not receive the drug.

The **observed** results of the trials are presented in the contingency table below.

	Improved	**Did not improve**	**Total**
Given drug	55	40	95
Not given drug	42	43	85
Total	97	83	180

It is difficult to tell from the results whether or not the drug had a significant positive effect on improvement rate. Although the results show that more volunteers improved than did not improve when given the drug, it is not certain whether the difference in results is significant enough to justify the claim. In order to verify the claim, a chi-squared (χ^2) test for independence can be carried out.

The first step is to set up a **null hypothesis** (H_0). The null hypothesis is always that there is no link between the variables and it is contrasted against an **alternative hypothesis** (H_1) which states that there is a link between the variables. The null hypothesis is treated as valid unless the data contradicts it.

In the example above:

H_0: There is no link between patients being given the drug and improvement rates.
H_1: There is a link between patients being given the drug and improvement rates.

The observed results need to be compared with expected or theoretical population results.

Of the 180 people in the sample, 95 were given the drug. So, from this we estimate that, in the population, the probability of being given the drug is $\frac{95}{180}$.

A total of 97 patients in the trial improved. From the null hypothesis, we would expect $\frac{95}{180}$ of these to have been given the drug, i.e. the expected number of improved patients who had been given the drug is $\frac{95}{180} \times 97 = 51.19$.

The next step is to draw up a table of expected frequencies for a group of 180 patients under the null hypothesis that having the drug is independent of improvement.

	Improved	Did not improve	Total
Given drug	$\frac{95 \times 97}{180} = 51.19$	$\frac{95 \times 83}{180} = 43.81$	95
Not given drug	$\frac{85 \times 97}{180} = 45.81$	$\frac{85 \times 83}{180} = 39.19$	85
Total	97	83	180

Note: The full answer to each calculation should be stored in your calculator's memory.

The formula for calculating χ^2 is as follows:

$$\chi^2 = \Sigma\frac{(f_o - f_e)^2}{f_e}, \text{ where } f_o \text{ are the observed frequencies, } f_e \text{ are the expected frequencies}$$

$$\text{Therefore } \chi^2 = \frac{(55 - 51.19)^2}{51.19} + \frac{(40 - 43.81)^2}{43.81} + \frac{(42 - 45.81)^2}{45.81} + \frac{(43 - 39.19)^2}{39.19}$$

$$= 1.299$$

The importance of this number depends on two other factors:

- The percentage level of significance required (p)
- The number of degrees of freedom of the data (v).

The level of significance refers to the percentage of the data you would expect to be outside the normal bounds.

The number of degrees of freedom of the data relates to the amount of data that is needed in order for the contingency table to be completed once the totals for each row and column are known. In the example above, if any one of the four pieces of data are known, the rest can be deduced.

In general, if the data in a contingency table has c columns and r rows, then the number of degrees of freedom, $v = (c - 1)(r - 1)$.

The result for χ^2 above needs to be compared with the result in a χ^2 table. (This is provided in the information booklet in your examination.)

p	0.005	0.01	0.025	0.05	0.1	0.9	0.95	0.975	0.99	0.995
$v = 1$	0.00004	0.0002	0.001	0.004	0.016	2.706	3.841	5.024	6.635	7.875
2	0.010	0.020	0.051	0.103	0.211	4.605	5.991	7.378	9.210	10.597
3	0.072	0.115	0.216	0.352	0.584	6.251	7.815	9.348	11.345	12.838
4	0.207	0.297	0.484	0.711	1.064	7.779	9.488	11.143	13.277	14.860
5	0.412	0.554	0.831	1.145	1.610	9.236	11.070	12.883	15.086	16.750
6	0.676	0.872	1.237	1.635	2.204	10.645	12.592	14.449	16.812	18.548
7	0.989	1.239	1.690	2.167	2.833	12.017	14.67	16.013	18.475	20.278
8	1.344	1.646	2.180	2.733	3.490	13.362	15.507	17.535	20.090	21.955
9	1.735	2.088	2.700	3.325	4.168	14.864	16.919	19.203	21.666	23.589
10	2.156	2.558	3.247	3.940	4.865	15.987	18.307	20.483	23.209	25.188
11	2.603	3.053	3.816	4.575	5.578	17.275	19.675	21.920	24.725	26.757
12	3.074	3.571	4.404	5.266	6.304	18.549	21.026	23.337	26.217	28.300

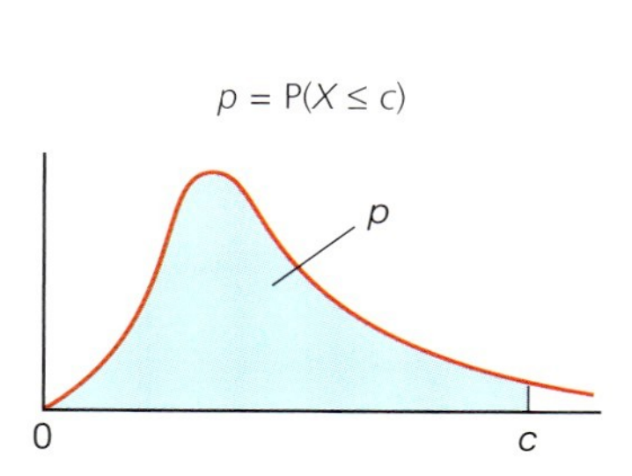

All the values in this table are known as critical values. The highlighted value of 3.841 is the critical value for the 5% level of significance for data with 1 degree of freedom.

Our calculation of χ^2 gave a value of 1.299. As $1.299 < 3.841$, the null hypothesis is supported, i.e. there is no evidence of a link between being given the drug and improvement rates in patients.

If the calculated value of χ^2 is greater than the critical value then the null hypothesis is rejected.

The table above has values for 1 degree of freedom ($v = 1$). Sometimes it is felt that these estimates are not sufficiently accurate and a different method known as Yates' correction can be used. However, this method is beyond the scope of this book.

The value of χ^2 can be calculated using your GDC for the example above.

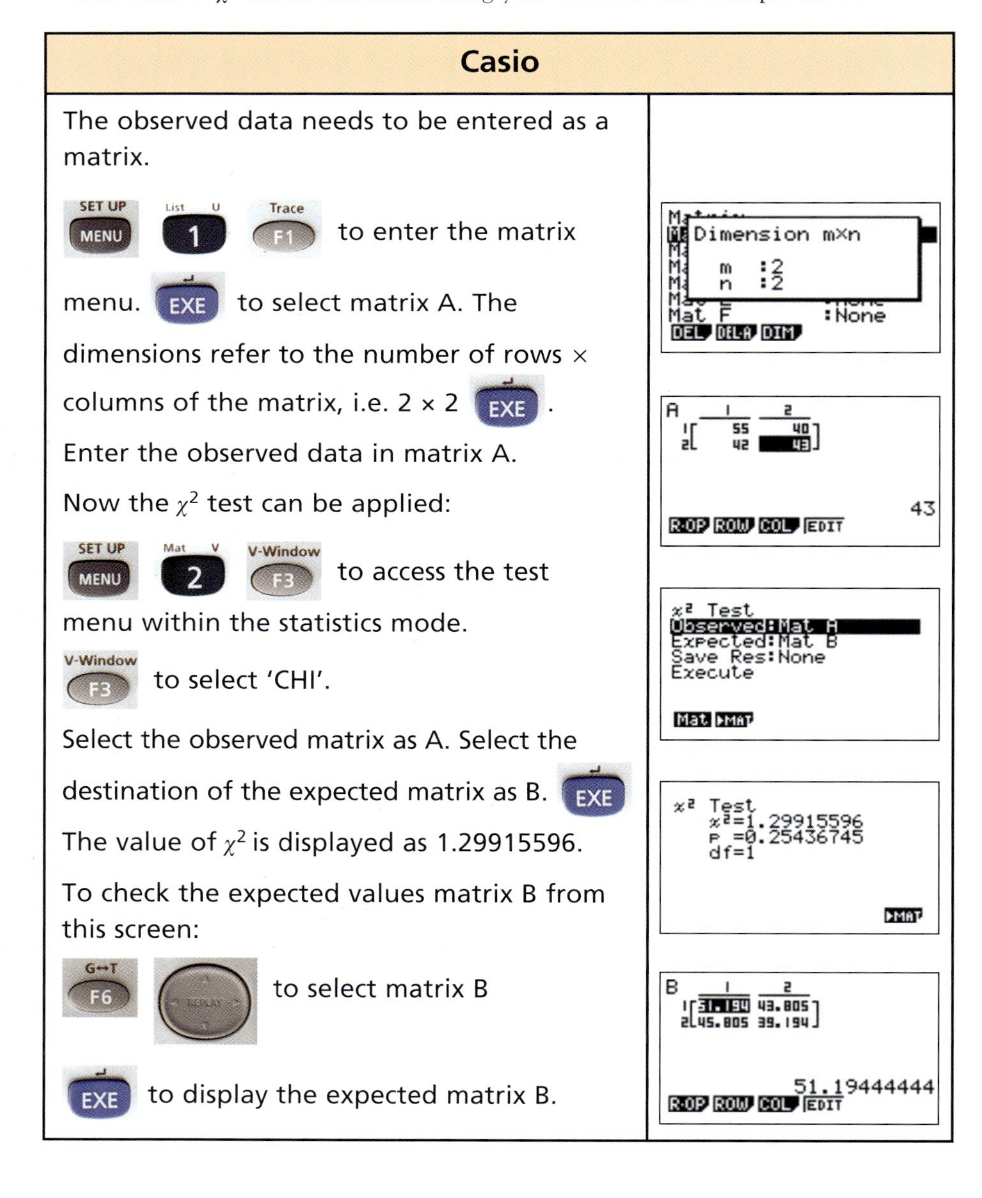

Texas

The observed data needs to be entered as a matrix.

  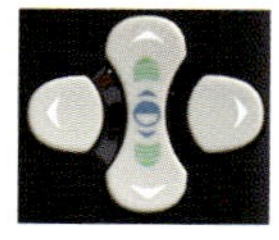 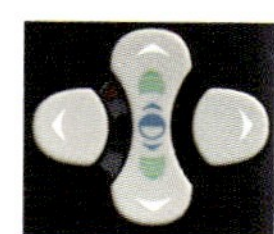

to enter the edit mode.

 to edit matrix A. Enter the dimensions of matrix A as rows × columns, i.e. 2 × 2.

```
MATRIX[A] 2 ×2
[0        0        ]
[0        0        ]
```

Enter the observed data into matrix A.

```
MATRIX[A] 2 ×2
[55       40       ]
[42       43       ]
```

Now the χ^2 test can be applied:

   to select the 'Tests' option within the statistics menu.

Scroll down to select 'C: χ^2–Test'. 

```
χ²-Test
 Observed:[A]
 Expected:[B]
 Calculate Draw
```

The default is for the observed data to be in matrix A and the expected matrix to be entered into matrix B.

Select 'Calculate'. 

```
χ²-Test
 χ²=1.299155961
 p=.2543674472
 df=1
```

The result for χ^2 is displayed on the screen.

To check the expected values of matrix B from this screen:

```
[B]
[[51.19444444 4…
 [45.80555556 3…
```

The matrix is displayed on the screen. Scroll across to view all the expected results.

Worked example

A television presenter on a motoring show claimed that 'men like cars with big engines and women do not'. In order to test the validity of this claim, data is collected from a random sample of men and women about the engine size of the cars that they own. This is presented in the contingency table below.

Engine size (litres)	1.6	1.8	2.0	2.3	2.8	3.0	5.0	Total
Male (f_o)	24	16	18	26	8	5	3	100
Female (f_o)	21	17	27	20	11	4	0	100
Total	45	33	45	46	19	9	3	200

a Set up a null and alternative hypothesis.
b Construct an expected frequency contingency table.
c Calculate the value of χ^2.
d State the number of degrees of freedom of the data.
e State, giving reasons, whether the presenter's statement was statistically valid at a 10% level of significance.

a H_0 Choice of engine size is independent of gender.
H_1 Choice of engine size is dependent on gender.

b

Engine size (litres)	1.6	1.8	2.0	2.3	2.8	3.0	5.0	Total
Male (f_o)	$\frac{100 \times 45}{200} = 22.5$	$\frac{100 \times 33}{200} = 16.5$	22.5	23	9.5	4.5	1.5	100
Female (f_o)	22.5	16.5	22.5	23	9.5	4.5	1.5	100
Total	45	33	45	46	19	9	3	200

c $\chi^2 = \Sigma\frac{(f_o - f_e)^2}{f_e} = \frac{(24 - 22.5)^2}{22.5} + \frac{(16 - 16.5)^2}{16.5} + \ldots\ldots + \frac{(0 - 1.5)^2}{1.5} = 6.398$

d $v = (2 - 1)(7 - 1) = 6$

e Look at the χ^2 table.

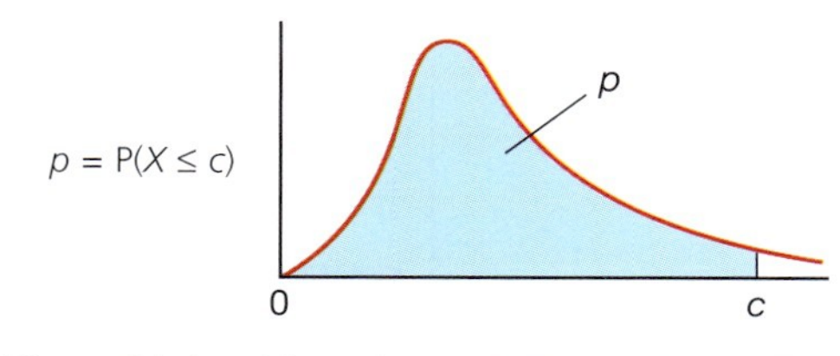

p	0.005	0.01	0.025	0.05	0.1	0.9	0.95	0.975	0.99	0.995
$v = 1$	0.00004	0.0002	0.001	0.004	0.016	2.706	3.841	5.024	6.635	7.875
2	0.010	0.020	0.051	0.103	0.211	4.605	5.991	7.378	9.210	10.597
3	0.072	0.115	0.216	0.352	0.584	6.251	7.815	9.348	11.345	12.838
4	0.207	0.297	0.484	0.711	1.064	7.779	9.488	11.143	13.277	14.860
5	0.412	0.554	0.831	1.145	1.610	9.236	11.070	12.883	15.086	16.750
6	0.676	0.872	1.237	1.635	2.204	10.645	12.592	14.449	16.812	18.548
7	0.989	1.239	1.690	2.167	2.833	12.017	14.67	16.013	18.475	20.278
8	1.344	1.646	2.180	2.733	3.490	13.362	15.507	17.535	20.090	21.955

As the χ^2 value of 6.398 is less than the critical value of 10.645, the null hypothesis is supported, i.e. the presenter's statement is **not** supported by statistical evidence.

Exercise 6.9.1

Check your answers using a GDC.

1 A survey of cyclists and non-cyclists was carried out to see whether their opinions differed as to whether helmets should be compulsory for cyclists.

The results of the survey are given below.

	Helmet compulsory	Helmet voluntary	Total
Cyclist	65	235	300
Non-cyclist	185	115	300
Total	250	350	600

a Set up a null and alternative hypothesis.
b Construct an expected frequency contingency table.
c Calculate the value of χ^2.
d State the number of degrees of freedom of the data.
e State, giving reasons, whether the null hypothesis is supported or rejected at a 5% level of significance.

2 Patients are given a new cancer drug. The patients selected for the trial were not expected to live for more than three months. For comparison purposes, a second group of similar patients were given a placebo (a tablet which contains no drug).

The results of the trial are given in the contingency table below.

	Alive after 3 months	Not alive after 3 months	Total
Given drug	87	53	140
Given placebo	65	65	130

a Set up a null and alternative hypothesis.
b Construct an expected frequency contingency table.
c Calculate the value of χ^2.
d State the number of degrees of freedom of the data.
e State, giving reasons, whether the null hypothesis is supported or rejected at a 1% level of significance.

3 A survey was carried out to establish whether or not smoking is related to high blood pressure.

The blood pressure of a random sample of smokers and non-smokers was taken and the results are summarized in the contingency table below.

	High blood pressure	Normal blood pressure	Total
Non-smoker	84	46	130
Smoker	362	108	470
Total	446	154	600

a Set up a null and alternative hypothesis.
b Construct an expected frequency contingency table.
c Calculate the value of χ^2.
d State the number of degrees of freedom of the data.
e State, giving reasons, whether the null hypothesis is supported or rejected at a 1% level of significance.

4 A survey is done to see if the type of holiday preferred by French people is gender dependent.

The results of the survey carried out on a random sample of French people are shown below.

	Beach	Walking	Cruise	Sail	Ski	Total
Male	62	51	22	31	44	210
Female	89	37	45	28	51	250
Total	151	88	67	59	95	460

a Set up a null and alternative hypothesis.
b Construct an expected frequency contingency table.
c Calculate the value of χ^2.
d State the number of degrees of freedom of the data.
e State, giving reasons, whether the null hypothesis is supported or rejected at a 10% level of significance.

5 A music magazine carried out a survey to see if a person's preference in music was dependent on their age. The results from a random sample of people are shown below.

Age \ Music	Rock	Pop	Blues	Jazz	Classical	Total
0–19	40	52	11	17	9	129
20–39	18	46	18	33	16	131
40+	12	14	28	24	11	89
Total	70	112	57	74	36	349

a Set up a null and alternative hypothesis.
b Calculate the value of χ^2.
c State the number of degrees of freedom of the data.
d State, giving reasons, whether the null hypothesis is supported or rejected at a 5% level of significance.

Student assessment 1

1 Identify which of the following types of data are discrete and which are continuous.
  a The number of goals scored in a hockey match
  b The price of a kilogram of carrots
  c The speed of a car
  d The number of cars passing the school gate each hour
  e The time taken to travel to school each morning
  f The wingspan of butterflies
  g The height of buildings

2 Maria helps her father feed their chickens every Sunday. Over a period of one year she kept a record of how long it took. Her results are shown in the table below.

Time (mins)	Frequency	Frequency density
$0 \leq t < 30$	8	
$30 \leq t < 45$	5	
$45 \leq t < 60$	8	
$60 \leq t < 75$	9	
$75 \leq t < 90$	10	
$90 \leq t < 120$	12	

  a Copy the table and complete it by calculating the frequency density correct to one decimal place.
  b Represent the information on a histogram.

3 Twenty students take three long jumps. The best result for each student (in metres) is recorded below.

4.3 5.4 4.3 4.0 3.8 5.1 3.6 5.5 6.2 4.7
5.2 3.8 2.4 4.7 3.9 5.6 5.8 4.7 3.3 2.9

The students were then coached in long jump technique and given three further jumps. Their individual best results are recorded below.

4.7 5.9 4.8 4.6 4.5 5.3 5.2 5.5 6.3 4.9
5.2 4.9 5.6 5.3 6.8 5.4 5.8 5.4 4.3 5.5

Draw a back-to-back stem and leaf diagram of their long jumps before and after coaching.

Comment on your diagram.

4 Four hundred students sit their mathematics IGCSE exam. Their marks (as percentages) are shown in the table below.

Mark (%)	Frequency	Cumulative frequency
31–40	21	
41–50	55	
51–60	125	
61–70	74	
71–80	52	
81–90	45	
91–100	28	

a Copy and complete the above table by calculating the cumulative frequency.
b Draw a cumulative frequency curve of the results.
c Using the graph, estimate a value for:
i) the median exam mark
ii) the upper and lower quartiles
iii) the interquartile range.

5 Eight hundred students sit an exam. Their marks (as percentages) are shown in the table below.

Mark (%)	Frequency	Cumulative frequency
1–10	10	
11–20	30	
21–30	40	
31–40	50	
41–50	70	
51–60	100	
61–70	240	
71–80	160	
81–90	70	
91–100	30	

a Copy and complete the above table by calculating the cumulative frequency.
b Draw a cumulative frequency curve of the results.
c An A grade is awarded to any pupil achieving at or above the upper quartile. Using your graph, identify the minimum mark required for an A grade.
d Any pupil below the lower quartile is considered to have failed the exam. Using your graph, identify the minimum mark needed so as not to fail the exam.
e How many pupils failed the exam?
f How many pupils achieved an A grade?

6 A businesswoman travels to work in her car each morning in one of two ways; either using the country lanes or using the motorway. She records the time taken to travel to work each day. The results are shown in the table below.

Time (mins)	Motorway frequency	Country lanes frequency
$10 \leq t < 15$	3	0
$15 \leq t < 20$	5	0
$20 \leq t < 25$	7	9
$25 \leq t < 30$	2	10
$30 \leq t < 35$	1	1
$35 \leq t < 40$	1	0
$40 \leq t < 45$	1	0

a Complete a cumulative frequency table for each of the sets of results shown above.
b Using your cumulative frequency tables, plot two cumulative frequency curves – one for the time taken to travel to work using the motorway, the other for the time taken to travel to work using country lanes.
c Use your graphs to work out the following for each method of travel:
i) the median travelling time
ii) the upper and lower quartile travelling times
iii) the interquartile range for the travelling times.
d With reference to your graphs or calculations, explain which is the most reliable way for the businesswoman to get to work.
e If she had to get to work one morning within 25 minutes of leaving home, which way would you recommend she goes? Explain your answer fully.

7 Two classes take a maths test. One class contains pupils of similar ability; the other is a mixed ability class. The results of the tests for each class are presented using the box and whisker plots below.

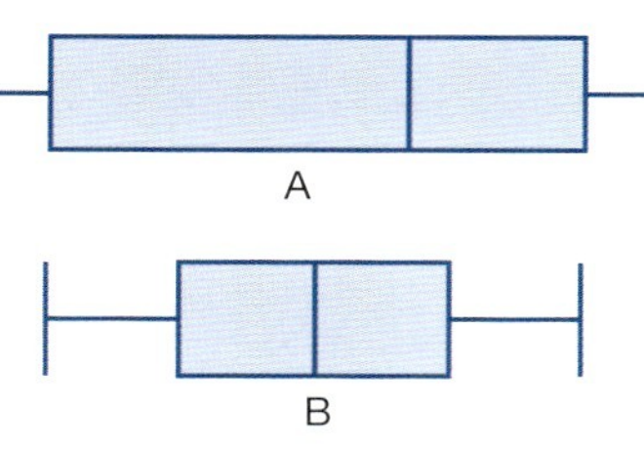

Explain clearly, giving your reasons, which of the two box and whisker plots is **likely** to belong to the mixed ability maths class and which is likely to belong to the other maths class.

8 Find the mean, median and mode of the following sets of data.
a 4 5 5 6 7
b 63 72 72 84 86
c 3 8 12 18 18 24
d 4 9 3 8 7 11 3 5 3 8

9 The mean mass of the 15 players in a rugby team is 85 kg. The mean mass of the team plus a substitute is 83.5 kg. Calculate the mass of the substitute.

10 Thirty children were asked about the number of pets they had. The results are shown in the table below.

Number of pets	0	1	2	3	4	5	6
Frequency	5	5	3	7	3	1	6

a Calculate the mean number of pets per child.
b Calculate the median number of pets per child.
c Calculate the modal number of pets.
d Draw a frequency polygon of the data.

11 Thirty families live in a street. The number of children in each family is given in the table below.

Number of children	0	1	2	3	4	5	6
Frequency	3	5	8	9	3	0	2

a Calculate the mean number of children per family.
b Calculate the median number of children per family.
c Calculate the modal number of children.
d Draw a frequency polygon of the data.

12 The number of people attending a disco at a club's over 30s evenings are shown below.

89 94 32 45 57 68 127 138
23 77 99 47 44 100 106 132
28 56 59 49 96 103 90 84
136 38 72 47 58 110

a Using groups 0–19, 20–39, 40–59, etc., present the above data in a grouped frequency table.
b Using your grouped data, calculate an estimate for the mean number of people going to the disco each night.

Student assessment 2

1 The number of people attending thirty screenings of a film at a local cinema is given below.

21 30 66 71 10 37 24 21
62 50 27 31 65 12 38 34
53 34 19 43 70 34 27 28
52 57 45 25 30 39

a Using groups 10–19, 20–29, 30–39, etc., present the above data in a grouped frequency table.
b Using your grouped data, calculate an estimate for the mean number of people attending each screening.

2 Find the standard deviation of the following set of numbers.

8, 8, 10, 10, 10, 12, 14, 15, 17, 20

3 A hockey team scores the following number of goals in their matches over a season.

Goals scored	0	1	2	3	4	5
Frequency	4	12	8	11	4	1

Calculate:
a the mean
b the range
c the standard deviation.

4 State what type of correlation you might expect, if any, if the following data was collected and plotted on a scatter diagram. Give reasons for your answer.
a The age of a motorcycle and its second-hand selling price
b The number of people living in a house and the number of rooms the house has

5 A department store decides to investigate whether there is a correlation between the number of pairs of gloves it sells and the outside temperature. Over a one-year period it records, every two weeks, how many pairs of gloves are sold and the mean daytime temperature during the same period. The results are given in the table below.

Mean temp. (°C)	3	6	8	10	10	11	12	14	16	16	17	18	18
Number of pairs of gloves	61	52	49	54	52	48	44	40	51	39	31	43	35
Mean temp. (°C)	19	19	20	21	22	22	24	25	25	26	26	27	28
Number of pairs of gloves	26	17	36	26	46	40	30	25	11	7	3	2	0

a Plot a scatter diagram of mean temperature against number of pairs of gloves.
b What type of correlation is there between the two variables?
c How might this information be useful for the department store in the future?

6 The popularity of a group of professional football players and their yearly salary is given in the table below.

Popularity	1	2	3	4	5	6	7	8	9	10
Salary (£ million)	4.8	3.6	4.5	3.1	7.7	6.3	2.9	3.1	4.1	1..8
Popularity	11	12	13	14	15	16	17	18	19	20
Salary (£ million)	4.5	3.1	2.7	3.9	6.2	5.8	4.1	5.3	7.2	6.5

a Calculate the equation of the regression line for y on x.
b Calculate the value of the correlation coefficient r.
c The statement is made in a newspaper 'Big money footballers are not popular with fans'. Comment on this statement in the light of your results above.

7 A group of dogs with kidney problems were given a new drug. A control group were not. The results were as follows.

	Improved	Did not improve	Total
Given drug	162	98	260
Not given drug	104	46	150
Total	266	144	410

a Set up a null and alternative hypothesis.
b Construct an expected frequency contingency table.
c Calculate the value of χ^2.
d State the number of degrees of freedom of the data.
e State, giving reasons, whether the null hypothesis is supported or rejected at a 5% level of significance.

8 A vote was taken for or against a ban on foxhunting. The results for town and country dwellers are recorded in the contingency table below.

	Town dwellers	Country dwellers	Total
Ban foxhunting	4240	1263	5503
Allow foxhunting	1360	2537	3897
Total	5600	3800	9400

a Set up a null and alternative hypothesis.
b Calculate the value of χ^2.
c State, giving reasons, whether the null hypothesis is supported or rejected at a 1% level of significance.

Topic 6 Discussion points, project ideas and theory of knowledge

1 Have statistics published about the African continent had a detrimental effect upon an ordinary person's view of that continent?

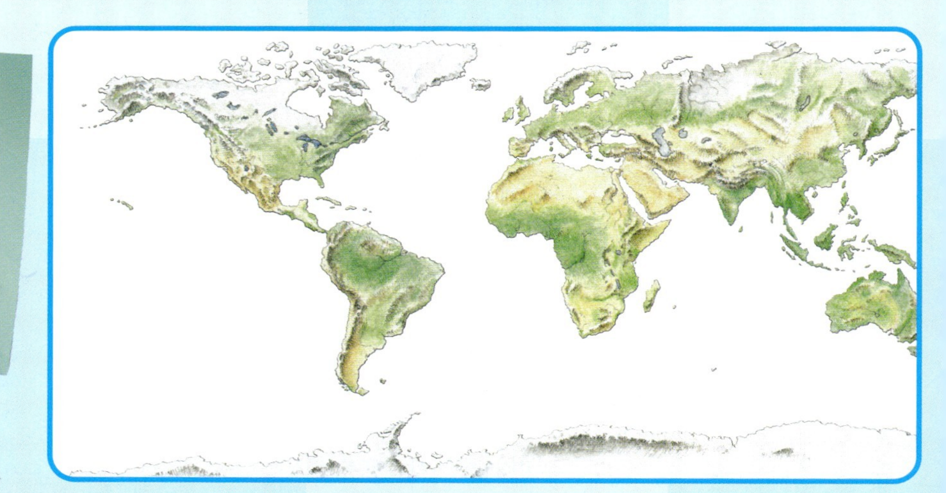

2 How far is it likely that the taking of opinion polls before an election influences the result of the election? If it does, should the taking of such polls be regulated?

3 Can any sample be an accurate reflection of a larger group? Can there be a perfect question in a questionnaire which is independent of both the interviewer and interviewee?

4 Early political opinion polls in the USA presidential election, which were conducted by telephone, turned out to be wildly inaccurate. Why were they so inaccurate?

5 Many statistical terms have a different meaning in normal conversation. Discuss some of these terms. Does the use of unusual symbols make it more or less easy to understand statistics?

6 Which form of statistical display is the easiest to understand? Does display depend upon context?

7 Charts and graphs can be used to inform or to confuse in advertising. The misuse of statistics could form the basis of a project.

8 'In a modern democracy whoever spends most gets elected.' Discuss this statement. Did money influence the choice of Roman Emperors?

9 A possible project would be to extend your knowledge of statistical analysis beyond the Mathematical Studies syllabus. Discuss some possible areas of study with your teacher.

10 Discuss the statement that a perfect correlation is the same as cause and effect.

11 Earlier it was mentioned that the GDC gave two different types of standard deviation. Investigate the difference between the two.

Topic 7 Introductory differential calculus

Syllabus content

7.1 Gradient of the line through two points, P and Q, that lie on the graph of a function.

Behaviour of the gradient of the line through two points, P and Q, on the graph of a function as Q approaches P.

Tangent to a curve.

7.2 The principle that

$$f(x) = ax^n \Rightarrow f'(x) = anx^{n-1}$$
$$\Rightarrow f''(x) = an(n-1)x^{n-2}.$$

The derivative of functions of the form

$f(x) = ax^n + bx^{n-1} + \ldots, n \in \mathbb{Z}$.

7.3 Gradients of curves for given values of x.

Values of x where $f'(x)$ is given.

Equation of the tangent at a given point.

7.4 Increasing and decreasing functions.

Graphical interpretation of

$f'(x) > 0, f'(x) = 0, f'(x) < 0$.

7.5 Values of x where the gradient of a curve is 0 (zero): solution of $f'(x) = 0$.

Local maximum and minimum points.

Introduction

Pierre de Fermat was a great French mathematician who proposed and solved many mathematical problems. He is most famous for 'Fermat's Last Theorem', which he proposed in 1637. Although seemingly simple, the theorem was not proved until 358 years later, in 1995.

Fermat's last theorem is an extension of Pythagoras' theorem. It states that: 'If an integer n is greater than 2, then the equation $a^n + b^n = c^n$ has no solutions.' Fermat suggested that he had found a simple proof of this, however it is not accepted by mathematicians today.

Andrew Wiles

Andrew Wiles, a Cambridge mathematician, worked secretly for many years to find a proof. This proof was extremely complex, building on work on elliptical curves by Eves Hellegouach, and required a proof of the Taniyama-Shimura conjecture. However, the final proof by Wiles is accepted as a work of genius because of the innovative way in which he brought together ideas.

Everyone accepts that Andrew Wiles proved Fermat's Last Theorem. However, this is not the case with the discovery of calculus. Sir Isaac Newton (1643–1727), another Cambridge mathematician, is accepted as one of the most influential men in human history. His work on gravitation and the Laws of Motion in his book, *Philosophae Naturalis Principia Mathematica*, influenced mathematics and science for hundreds of years. His work was only taken further by the great mind of Albert Einstein. Newton is credited by many as discovering calculus.

Gottfried Wilhelm Leibniz (1646–1716), the German mathematician and Philosopher, worked on what we now term calculus at the same time as Newton. The suggestion that Leibniz, who wrote poetry in addition to writing on maths, politics, law, theology, history and philology, had stolen Newton's ideas on calculus and merely improved them, caused a bitter argument that went on long after they had both died.

Claims as to who should get the credit for an invention are not unusual. Did Alexander Graham Bell or Antonio Meucci invent the telephone? Did the Scot John Logie Baird, the American Philo Taylor Farnsworth or the Russian Vladimir Kosma Zworkin, discover television? It often depends upon the country in which the book you read was published. It has been suggested that 'there is a time for a discovery' and that if one person had not made the breakthrough, someone else would have. This claim is supported by one of the great modern discoveries, the structure of the DNA molecule. Watson and Crick discovered the double helix structure but other scientists, particularly Rosalind Franklin, were very close to a solution.

Calculus is the cornerstone of much of the mathematics studied at a higher level. Differential calculus deals with finding the formula for the gradient of a function. In this topic, the functions will be of the form $f(x) = ax^n + bx^{n-1} + \dots$ where n is an integer.

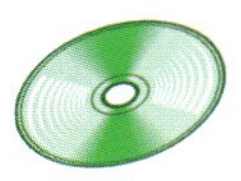

7.1 Gradient

You will already be familiar with finding the gradient of a straight line, shown below.

The gradient of the line passing through points (x_1, y_1) and (x_2, y_2) is calculated by $\frac{y_2 - y_1}{x_2 - x_1}$. Therefore the gradient of the line passing through points P and Q is $\frac{10 - 5}{11 - 1} = \frac{5}{10} = \frac{1}{2}$

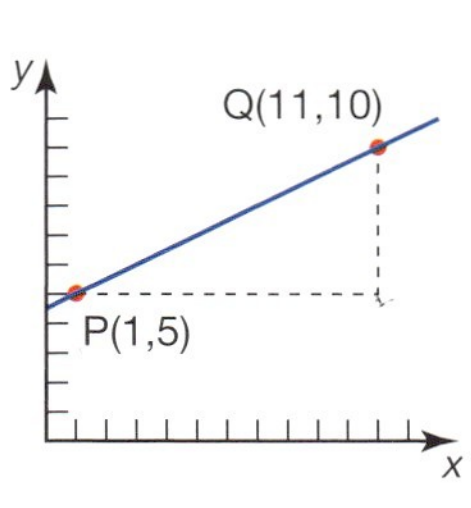

The gradient of a linear function (a straight line) is constant, i.e. the same at any point on the line. However non-linear functions, such as those that produce curves when graphed, are more difficult to work with as the gradient is not constant.

The graph opposite shows the function $f(x) = x^2$. Point P is on the curve at (3, 9). If P moves along the curve to the right, the gradient of the curve becomes steeper. If P moves along the curve towards the origin, the gradient of the curve becomes less steep.

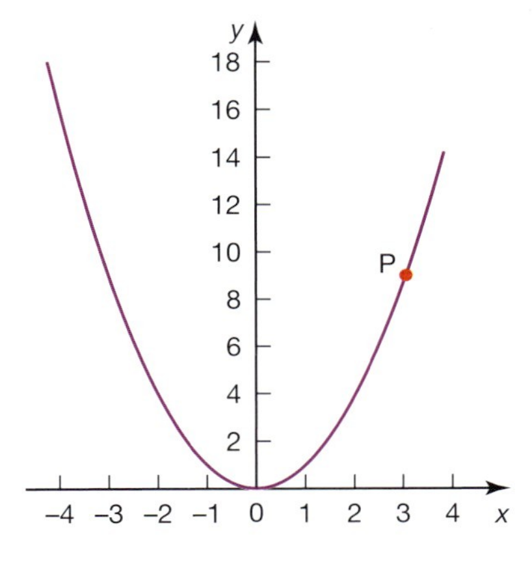

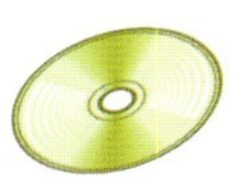

See GeoGebra file '**7.1 Gradient of $y = x^2$**'.

The gradient of the function $f(x) = x^2$ at the point P(1, 1) can be calculated as follows.

Mark a point $Q_1(3, 9)$ on the graph and draw the line segment PQ_1. The gradient of the line segment PQ_1 is only an approximation of the gradient of the curve at P.

$$\text{Gradient of } PQ_1 = \frac{9-1}{3-1} = 4$$

Mark a point Q_2 closer to P, for example (2, 4) and draw the line segment PQ_2. The gradient of the line segment PQ_2 is still only an approximation of the gradient of the curve at P, but it is a better approximation than the gradient of PQ_1.

$$\text{Gradient of } PQ_2 = \frac{4-1}{2-1} = 3$$

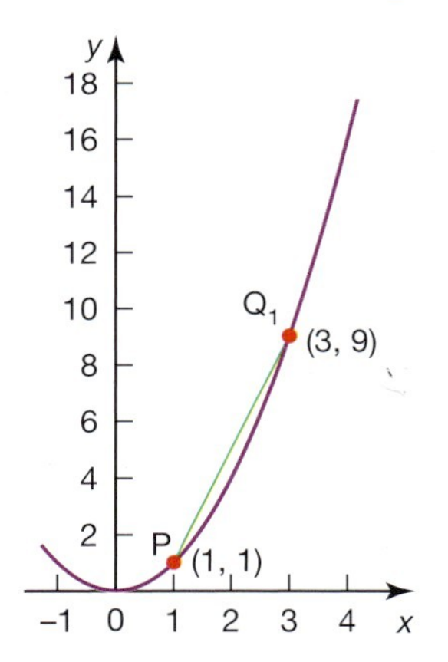

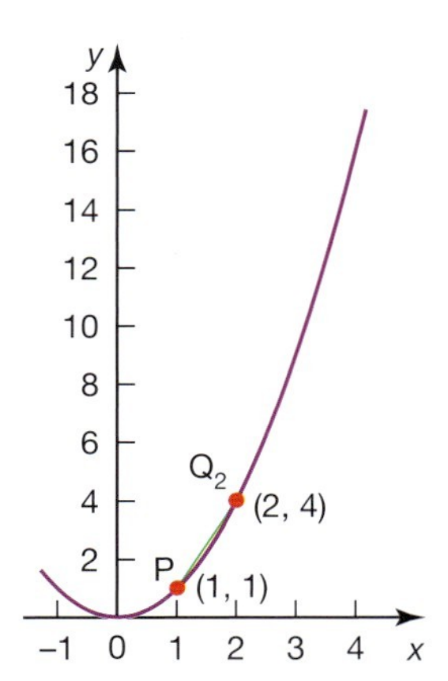

If a point $Q_3\ (1.5, 1.5^2)$ is chosen, the gradient PQ_3 will be an even better approximation.

$$\text{Gradient of } PQ_3 = \frac{1.5^2-1}{1.5-1} = 2.5$$

For the point $Q_4\ (1.25, 1.25^2)$, the gradient of $PQ_4 = \dfrac{1.25^2-1}{1.25-1} = 2.25$

For the point $Q_5\ (1.1, 1.1^2)$, the gradient of $PQ_5 = \dfrac{1.1^2-1}{1.1-1} = 2.1$

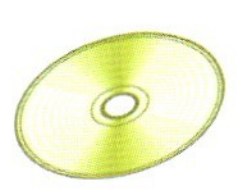

See GeoGebra file '**7.1 Gradient of line segment PQ**'.

The results above suggest that, as point Q gets closer to P, the gradient of the line segment PQ gets closer to 2.

It can be proved that the gradient of the function is $f(x) = x^2$ is 2 when $x = 1$.

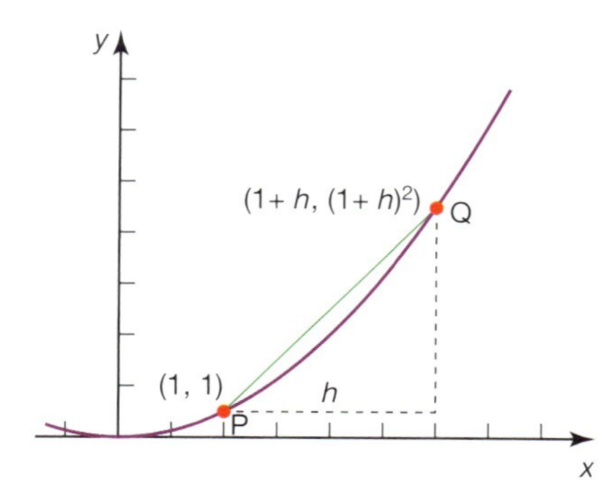

Consider points P and Q on the function $f(x) = x^2$. P is at (1, 1) and Q, h units from P in the x-direction has coordinates $(1+h, (1+h)^2)$.

$$\text{Gradient of line segment PQ} = \frac{(1+h)^2 - 1}{1+h-1} = \frac{1+2h+h^2-1}{h} = \frac{h(2+h)}{h} = 2 + h$$

As Q gets closer to P, h gets smaller and smaller (tends to 0), therefore the gradient $(2 + h)$ of the line segment PQ tends to 2.

$$\text{Therefore the gradient at P}(1, 1) = \lim_{h\to 0}(2 + h) = 2$$

i.e. the limit of $2 + h$ as h tends to 0 is 2.

In general:

The gradient of a curve at the point P is the same as the gradient of the tangent to the curve at P.

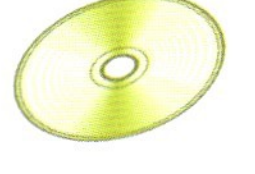

See GeoGebra file '**7.1 Line passing through P and Q**'.

Exercise 7.1.1

1 Using the proof above as a guide, find the gradient of the function $f(x) = x^2$ when:

  a $x = 2$ **b** $x = 3$ **c** $x = -1$

  d By looking at the pattern in your results, complete this sentence. For the function $f(x) = x^2$, the formula for the gradient is … .

2 Find the gradient of the function $f(x) = 2x^2$ when:

  a $x = 1$ **b** $x = 2$ **c** $x = -2$

  d By looking at the pattern in your results, complete this sentence. For the function $f(x) = 2x^2$, the formula for the gradient is … .

3 Find the gradient of the function $f(x) = \frac{1}{2}x^2$ when:

  a $x = 1$ **b** $x = 2$ **c** $x = 3$

  d By looking at the pattern in your results, complete this sentence. For the function $f(x) = \frac{1}{2}x^2$, the formula for the gradient is … .

In each case in Exercise 7.1.1, a rule was found for calculating the gradient at any point on the particular curve. This rule is known as the gradient function $f'(x)$ or $\frac{dy}{dx}$, i.e. the function $f(x) = x^2$ has a gradient function $f'(x) = 2x$

or $\frac{dy}{dx} = 2x$.

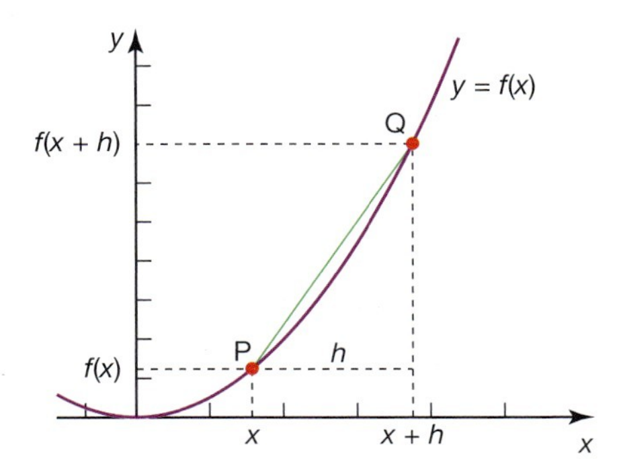

The above proof can be generalized for other functions $f(x)$.

Gradient of line segment PQ $= \dfrac{f(x+h) - f(x)}{(x+h) - x}$

$$\text{Gradient at P} = \lim_{h\to 0}(\text{Gradient of line segment PQ})$$

$$= \lim_{h\to 0} \frac{f(x+h) - f(x)}{h}$$

This is known as finding the gradient function from *first principles*.

Worked example

Find, from first principles, the gradient function of $f(x) = x^2 + x$.

$$\frac{dy}{dx} = \lim_{h\to 0} \frac{f(x+h) - f(x)}{h}$$

$$= \lim_{h\to 0} \frac{((x+h)^2+(x+h))-(x^2 + x)}{h}$$

$$= \lim_{h\to 0} \frac{x^2+2xh+h^2+x+h-x^2-x}{h}$$

$$= \lim_{h\to 0} \frac{2xh+h^2+h}{h}$$

$$= \lim_{h\to 0}(2x+h+1)$$

$$= 2x+1$$

i.e. the gradient at any point P(x, y) on the curve $y = x^2+x$ is given by $2x+1$.

You can check this by graphing a function and its gradient function simultaneously on a computer.

Autograph

- Type equation $y = x^2 + x$
- Select curve.
- Click on the gradient function icon .

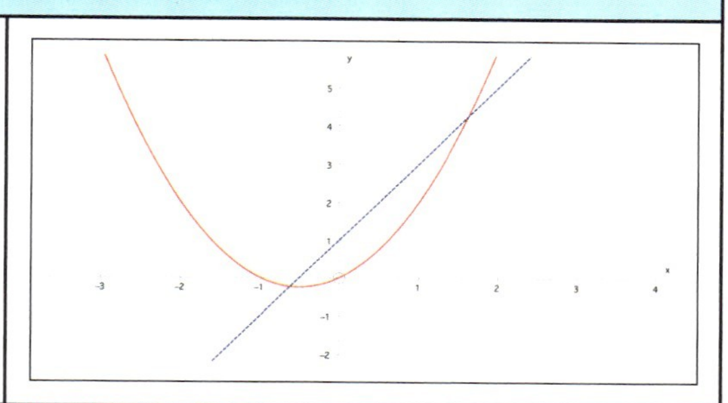

GeoGebra

- Type equation $f(x) = x^2 + x$
- Type Derivative [f]

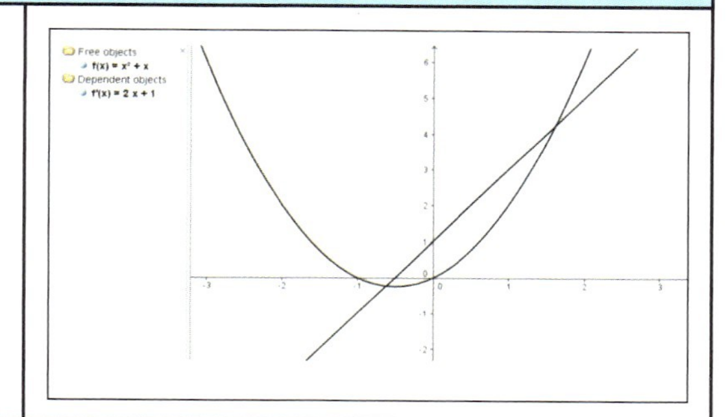

Note: The equation of the gradient function is given in the algebra window.

Exercise 7.1.2

1 Using the worked example above as a guide, find from first principles the gradient functions of each of the following functions.

a $f(x) = x^3$
b $f(x) = 3x^2$
c $f(x) = x^2 + 2x$
d $f(x) = x^2 - 2$
e $f(x) = 3x - 3$
f $f(x) = 2x^2 - x + 1$

2 Copy and complete the table by entering your gradient functions from question 1 above and from Exercise 7.1.1.

Function $f(x)$	Gradient function $f'(x)$
x^2	
$2x^2$	
$\frac{1}{2}x^2$	
$x^2 + x$	$2x + 1$
x^3	
$3x^2$	
$x^2 + 2x$	
$x^2 - 2$	
$3x - 3$	
$2x^2 - x + 1$	

3 For each of the functions given in question 2, use a computer to:
   i) graph, on the same axes, the function and gradient function
   ii) check that the equation of the gradient function is the same as the one you calculated.

4 Describe any patterns you notice in your table for question 2, between a function and its gradient function.

The functions used so far have all been polynomials. There is a relationship between a polynomial function and its gradient function. This is summarized below.

If $f(x) = ax^n$ then $\frac{dy}{dx} = anx^{n-1}$,

i.e. to work out the gradient function of a polynomial, multiply the coefficient of x by the power of x and then subtract 1 from the power.

Worked examples

1 Calculate the gradient function of the function $f(x) = 2x^3$.

$$\frac{dy}{dx} = 3 \times 2x^{(3-1)} = 6x^2$$

2 Calculate the gradient function of the function $f(x) = 5x^4$.

$$\frac{dy}{dx} = 4 \times 5x^{(4-1)} = 20x^3$$

Exercise 7.1.3

1 Calculate the gradient function of each of the following functions.

a $f(x) = x^4$ b $f(x) = x^5$ c $f(x) = 3x^2$
d $f(x) = 5x^3$ e $f(x) = 6x^3$ f $f(x) = 8x^7$

2 Calculate the gradient function of each of the following functions.

a $f(x) = \frac{1}{3}x^3$ b $f(x) = \frac{1}{4}x^4$ c $f(x) = \frac{1}{4}x^2$
d $f(x) = \frac{1}{2}x^4$ e $f(x) = \frac{2}{5}x^3$ f $f(x) = \frac{2}{9}x^3$

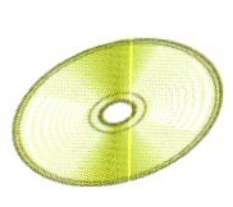

7.2 Differentiation

The process of finding the gradient function is known as **differentiation**. Differentiating a function produces the **derivative** or gradient function.

Worked examples

1 Differentiate the function $f(x) = 3$ with respect to x.

The graph of $f(x) = 3$ is a horizontal line as shown.

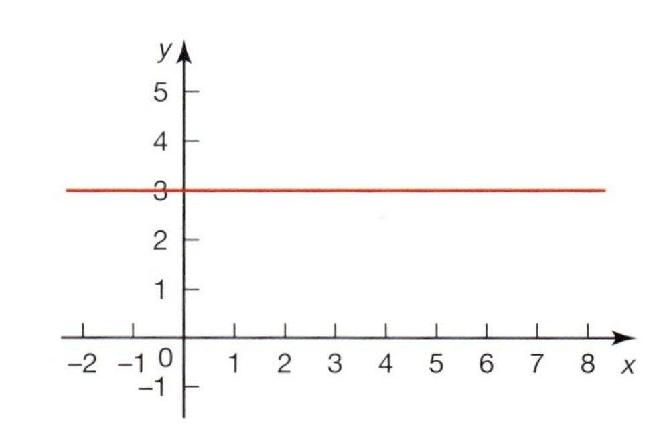

A horizontal line has a gradient of zero. Therefore,

$$f(x) = 3 \Rightarrow \frac{dy}{dx} = 0.$$

This can be calculated using the rule for differentiation:

$f(x) = 3$ can be written as $f(x) = 3x^0$

$$\frac{dy}{dx} = 0 \times 3x^{(0-1)} = 0$$

In general therefore, the derivative of a constant is zero.

If $f(x) = c \Rightarrow \frac{dy}{dx} = 0$.

2 Differentiate the function $f(x) = 2x$ with respect to x.

The graph of $f(x) = 2x$ is a straight line as shown.

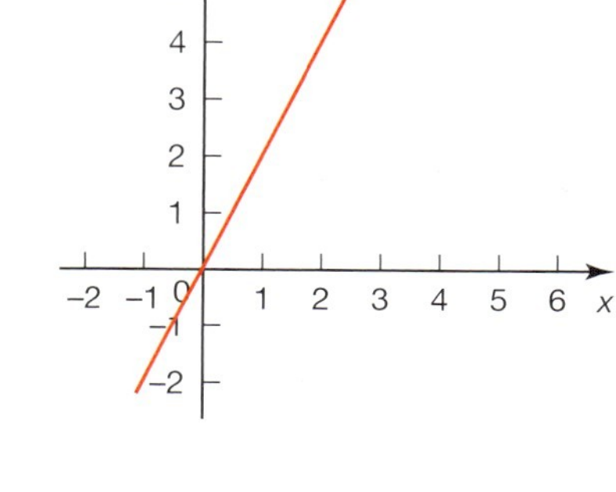

From work done on linear graphs, you know the gradient is 2. Therefore,

$f(x) = 2x \Rightarrow \frac{dy}{dx} = 2$.

This too can be calculated using the rule for differentiation:

$f(x) = 2x$ can be written as $f(x) = 2x^1$.

$$\frac{dy}{dx} = 1 \times 2x^{(1-1)} = 2x^0$$

But $x^0 = 1$, therefore $\frac{dy}{dx} = 2$.

In general, therefore, if $f(x) = ax \Rightarrow \frac{dy}{dx} = a$.

3 Differentiate the function $f(x) = \frac{1}{3}x^3 - 2x + 4$ with respect to x.

Graphically the function and its derivative are as shown.

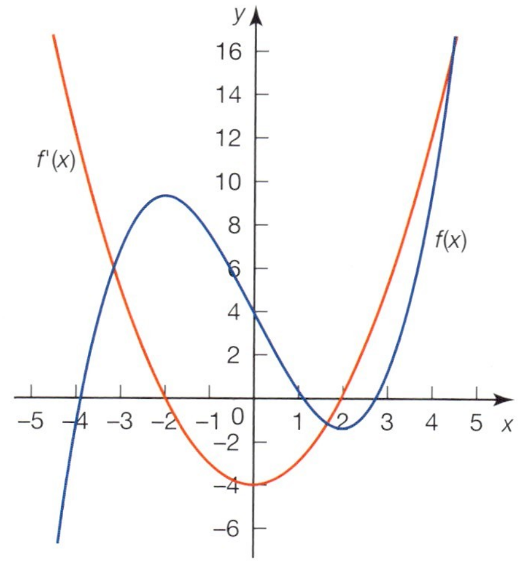

It can be seen that the derivative of the function $f(x)$ is a quadratic. The derivative can be calculated to be $f'(x) = x^2 - 2$.

This suggests that the derivative of a function with several terms can be found by differentiating each of the terms individually, which is indeed the case.

4 Differentiate the function $f(x) = \frac{2x^3 + x^2}{x}$ with respect to x.

Before differentiating, rewrite functions as sums of terms in powers of x.

$\frac{2x^3 + x^2}{x}$ can be written as $\frac{2x^3}{x} + \frac{x^2}{x}$

and simplified to $2x^2 + x$.

Therefore $f(x) = \frac{2x^3 + x^2}{x} = 2x^2 + x \Rightarrow \frac{dy}{dx} = 4x + 1.$

Note: A common error here is to differentiate each of the terms individually.

The derivative of $\frac{2x^3 + x^2}{x}$ is *not* $\frac{6x^2 + 2x}{1}$.

Exercise 7.2.1

1 Differentiate each of the following expressions with respect to x.

a $5x^3$ b $7x^2$ c $4x^6$

d $\frac{1}{4}x^2$ e $\frac{2}{3}x^6$ f $\frac{3}{4}x^5$

g 5 h $6x$ i $\frac{1}{8}$

2 Differentiate the following expressions with respect to x.

a $3x^2 + 4x$ b $5x^3 - 2x^2$ c $10x^3 - \frac{1}{2}x^2$

d $6x^3 - 3x^2 + x$ e $12x^4 - 2x^2 + 5$ f $\frac{1}{3}x^3 - \frac{1}{2}x^2 + x - 4$

g $-3x^4 + 4x^2 - 1$ h $-6x^5 + 3x^4 - x + 1$ i $-\frac{3}{4}x^6 + \frac{2}{3}x^3 - 8$

3 Differentiate the following expressions with respect to x.

a $\frac{x^3 + x^2}{x}$ b $\frac{4x^3 - x^2}{x^2}$ c $\frac{6x^3 + 2x^2}{2x}$

d $\frac{x^3 + 2x^2}{4x}$ e $3x(x + 1)$ f $2x^2(x - 2)$

g $(x + 5)^2$ h $(2x - 1)(x + 4)$ i $(x^2 + x)(x - 3)$

Negative powers of *x*

So far all the polynomials that have been differentiated have had positive powers of x. This section looks at the derivative of polynomials with negative powers of x, for example differentiating from first principles the function $f(x) = x^{-1}$ with respect to x.

You will know from your work on indices that x^{-1} can be written as $\frac{1}{x}$.

The function $f(x)$ is shown graphically with $P\left(x, \frac{1}{x}\right)$ and $Q\left(x + h, \frac{1}{x + h}\right)$.

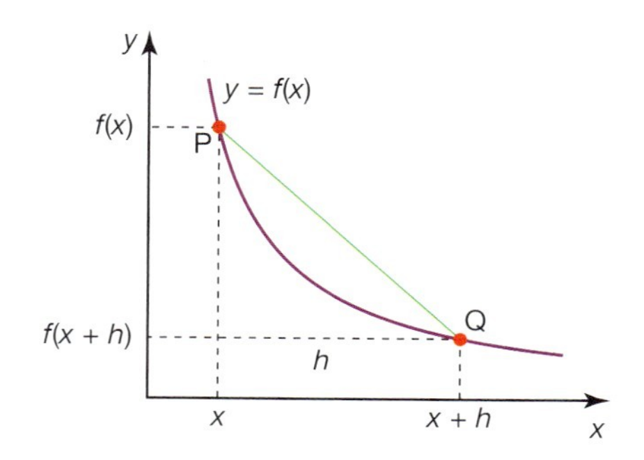

Gradient of line segment PQ

$$= \frac{\frac{1}{x} - \frac{1}{x+h}}{x - (x+h)} = \frac{\frac{x+h}{x(x+h)} - \frac{x}{x(x+h)}}{-h} = \frac{\frac{h}{x(x+h)}}{-h} = -\frac{1}{x(x+h)}$$

Therefore $\lim\limits_{h \to 0}\left(-\frac{1}{x(x+h)}\right) = -\frac{1}{x^2}$.

This tells us that for $f(x) = x^{-1}$, the derivative $f'(x) = -\frac{1}{x^2} = -x^{-2}$.

It can be seen that the original rule for the derivative of polynomials, namely if $f(x) = ax^n$ then $f'(x) = anx^{n-1}$, is still valid when n is negative.

$f(x) = x^{-1}$

$f'(x) = -1 \times x^{(-1-1)} = -x^{-2}$

Worked examples

1 Find the derivative of x^{-2}.

$$\frac{dy}{dx} = -2 \times x^{(-2-1)}$$
$$= -2x^{-3} = -\frac{2}{x^3}$$

2 Calculate $\frac{dy}{dx}$, when $y = 2x^{-1} + x^{-2} + 2$.

$$\frac{dy}{dx} = -1 \times 2x^{(-1-1)} + -2 \times x^{(-2-1)} + 0$$
$$= -2x^{-2} - 2x^{-3}$$
$$= -2\left(\frac{1}{x^2} + \frac{1}{x^3}\right)$$

3 Differentiate $\frac{2}{x^5}$ with respect to x.

First write the expression $\frac{2}{x^5}$ in the form ax^n, where a is a constant and n an integer:

$$\frac{2}{x^5} = 2 \times \frac{1}{x^5} = 2x^{-5}$$

$$\frac{dy}{dx} = -5 \times 2x^{(-5-1)}$$
$$= -10x^{-6} = -\frac{10}{x^6}$$

Exercise 7.2.2

1 Find the derivative of each of the following expressions.

a x^{-1} b x^{-3} c $2x^{-}$ d $-x^{-2}$ e $-\frac{1}{3}x^{-3}$ f $-\frac{2}{5}x^{-5}$

2 Write the following expressions in the form ax^n, where a is a constant and n an integer.

a $\frac{1}{x}$ b $\frac{2}{x}$ c $\frac{3}{x^2}$ d $\frac{2}{3x^3}$ e $\frac{3}{7x^2}$ f $\frac{2}{9x^3}$

3 Calculate $f'(x)$ for the following curves.

a $f(x) = 3x^{-1} + 2x$ b $f(x) = 2x^2 + x^{-1} + 1$

c $f(x) = 3x^{-1} - x^{-2} + 2x$ d $f(x) = \frac{1}{x^3} + x^3$

e $f(x) = \frac{2}{x^4} - \frac{1}{x^3} + 1$ f $f(x) = -\frac{1}{2x^2} + \frac{1}{3x^3}$

So far we have only used the variables x and y when finding the gradient function. This does not always need to be the case. Sometimes it is more convenient or appropriate to use other variables.

If a stone is thrown vertically upwards from the ground, with a speed of $10\,\text{m}\,\text{s}^{-1}$, its distance ($s$) from its point of release is given by the formula $s = 10t - 4.9t^2$, where t is the time in seconds after the stone's release.

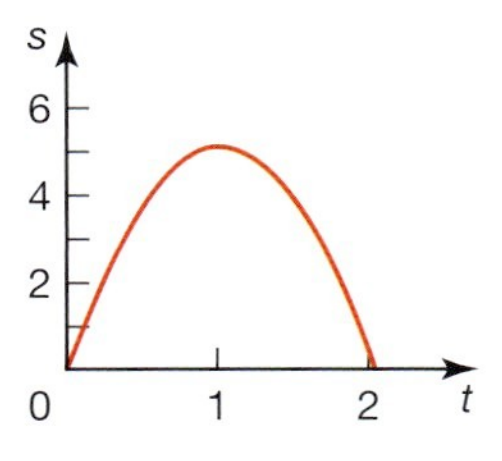

This is represented graphically as shown.

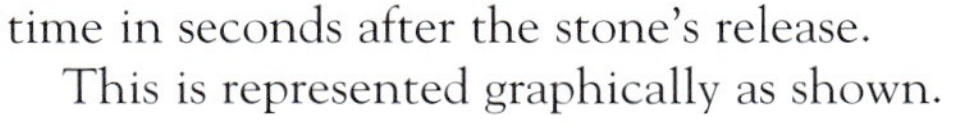

The velocity (v) of the stone at any point can be found by calculating the rate of change of distance with respect to time, i.e. $\frac{ds}{dt}$.

Therefore, if $s = 10t - 4.9t^2$ $\quad v = \frac{ds}{dt} = 10 - 9.8t$

Worked examples

1 Calculate $\frac{ds}{dt}$ for the function $s = 6t^2 - 4t + 1$

$$\frac{ds}{dt} = 12t - 4$$

2 Calculate $\frac{dr}{dt}$ for the function $r = \frac{6}{t^2} + 2t^2 - 1$.

Rewriting the function as $r = 6t^{-2} + 2t^2 - 1$,

$$\frac{dr}{dt} = -12t^{-3} + 4t = -\frac{12}{t^3} + 4t$$

3 Calculate $\frac{dv}{dt}$ for the function $v = (r^3 + 1)\left(\frac{2}{r^2} - 1\right)$

Expanding the brackets gives: $v = 2r - r^3 + \frac{2}{r^2} - 1$

Therefore $\frac{dv}{dr} = 2 - 3r^2 - \frac{4}{r^3}$

Exercise 7.2.3

1 Differentiate each of the following with respect to t.

a $y = 3t^2 + t$ b $v = 2t^3 + t^2$ c $m = 5t^3 - t^2$

d $y = 2t^{-1}$ e $r = \frac{1}{2}t^{-2}$ f $s = t^4 - t^{-2}$

2 Calculate the derivative of each of the following functions.

a $y = 3x^{-1} + 4$ b $s = 2t^{-1} - t$ c $v = r^{-2} - \frac{1}{r}$

d $P = \frac{l^{-4}}{2} + 2l$ e $m = \frac{n}{2} - \frac{n^{-3}}{3}$ f $a = \frac{2t^{-2}}{5} - t^3$

3 Calculate the derivative of each of the following functions.

a $y = x(x + 4)$ b $r = t(l - t)$ c $v = t\left(\frac{1}{t} + t^2\right)$

d $p = r^2\left(\frac{2}{r} - 3\right)$ e $a = x\left(x^{-2} + \frac{x}{2}\right)$ f $y = t^{-1}\left(t - \frac{1}{t^2}\right)$

4 Differentiate each of the following with respect to t.

a $y = (t + 1)(t - 1)$ b $r = (t - 1)(2t + 2)$

c $p = \left(\frac{1}{t} + 1\right)\left(\frac{1}{t} - 1\right)$ d $a = (t^{-2} + t)(t^2 - 2)$

e $v = \left(\frac{2t^2}{3} + 1\right)(t - 1)$ f $y = \left(\frac{3}{2t^4} - t\right)\left(2t - \frac{3}{t}\right)$

Calculating the second derivative

In the previous section we looked at the position of a stone thrown vertically upwards. Its velocity (v) at any point was found by differentiating the equation for the distance (s) with respect to t, i.e. $v = \frac{ds}{dt}$.

However, acceleration (a) is the rate of change of velocity with time, i.e. $a = \frac{dv}{dt}$. Therefore, as $s = 10t - 4.9t^2$

$$v = \frac{ds}{dt} = 10 - 9.8t$$

$$a = \frac{dv}{dt} = -9.8 \text{ (acceleration due to gravity)}$$

You will have noticed that the equation for the distance was differentiated twice to get the acceleration, i.e. the second derivative was obtained. Calculating the second derivative is a useful operation as will be seen later.

The notation used for the second derivative follows on from that used for the first derivative.

$$f(x) = ax^n \qquad\qquad y = ax^n$$

$$\Rightarrow f'(x) = anx^{n-1} \qquad \text{or} \qquad \frac{dy}{dx} = anx^{n-1}$$

$$\Rightarrow f''(x) = an(n-1)x^{n-2} \qquad\qquad \frac{d^2y}{dx^2} = an(n-1)x^{n-2}$$

So either $f''(x)$ or $\frac{d^2y}{dx^2}$ is used to represent the second derivative when differentiating with respect to x, depending on whether the original function is given in the form $f(x) = ax^n$ or $y = ax^n$.

Worked examples

1 Find $\frac{d^2y}{dx^2}$ when $y = x^3 - 2x^2$.

$$\frac{dy}{dx} = 3x^2 - 4x$$

$$\frac{d^2y}{dx^2} = 6x - 4$$

2 Find $f''(x)$ when $f(x) = \frac{2}{x} + \frac{1}{2x^2}$

$$f(x) = 2x^{-1} + \tfrac{1}{2}x^{-2}$$

$$f'(x) = -2x^{-2} - x^{-3}$$

$$f''(x) = 4x^{-3} + 3x^{-4} = \frac{4}{x^3} + \frac{3}{x^4}$$

3 Find $\frac{d^2s}{dt^2}$ if $s = 3t + \frac{1}{2}t^2$.

$$\frac{ds}{dt} = 3 + t$$

$$\frac{d^2s}{dt^2} = 1$$

Exercise 7.2.4

1 Find the second derivative of each of the following.

a $y = 2x^3$
b $y = x^4 - \frac{1}{2}x^2$
c $y = \frac{1}{3}x^6$
d $y = 3x^2 - 2$
e $y = \frac{x^2}{4}$
f $y = 3x$

2 Find $\frac{d^2y}{dx^2}$ for each of the following.

a $y = x^{-1}$
b $y = x^{-2} + x^3$
c $y = \frac{x^{-3}}{6}$
d $y = \frac{2x^{-2}}{3}$
e $y = \frac{3}{x} - \frac{5}{x^2}$
f $y = \frac{2 - x}{x^2}$

3 Differentiate the following twice with respect to x.

a $v = x^2(x - 3)$
b $P = \frac{1}{2}x^2(x^2 + x)$
c $t = x^{-1}(1 + x^3)$
d $u = \frac{2 - x^4}{x^2}$
e $y = (x^2 + 1)(x^3 - x)$
f $r = \frac{x(2x - 1)}{x^4}$

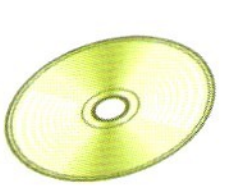

7.3 The gradient of a curve at a given point

You have seen that differentiating the equation of a curve gives the general equation for the gradient of any point on the curve. Using the general equation of the gradient, gradients at specific points on the curve can be calculated.

For the function $f(x) = \frac{1}{2}x^2 - 2x + 4$, the gradient function $f'(x) = x - 2$. The gradient at any point on the curve can be calculated using this.

For example, when $x = 4$, $f'(x) = 4 - 2$
$= 2$

i.e. the gradient of the curve $f(x) = \frac{1}{2}x^2 - 2x + 4$ is 2 when $x = 4$, as shown below.

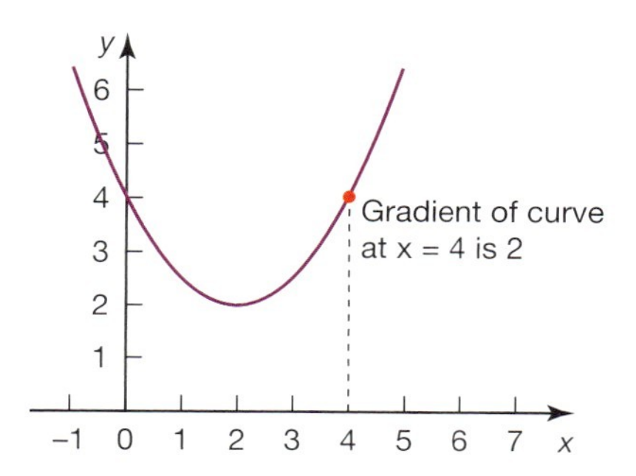

GDCs and graphing software can also help to visualize the question and check the solution.

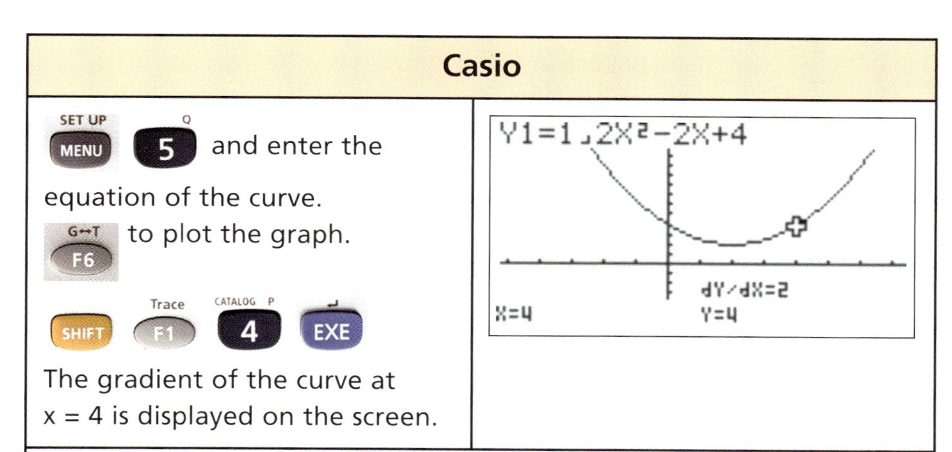

Note: If the $\frac{dy}{dx}$ derivative feature is not displayed on the screen, it can be turned on via the set up menu.

Texas

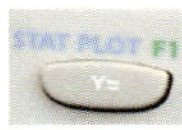 and enter the equation of the curve.

 to plot the graph.

  and select $\frac{dy}{dx}$.

  to calculate the gradient when $x = 4$.

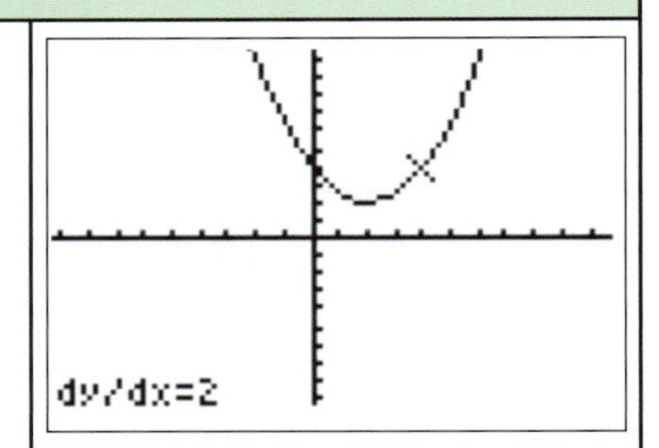

Autograph

- Type equation $y = \frac{1}{2}x^2 - 2x + 4$
- Select the curve.
  Click on coordinate icon (,) and enter the value $x = 4$.
  OK
  A point is plotted on the curve at $x = 4$.
- Click on the point. Select 'object' followed by 'tangent'.

A tangent is drawn to the curve at $x = 4$. Its equation and the coordinate of the point are displayed at the base of the screen.

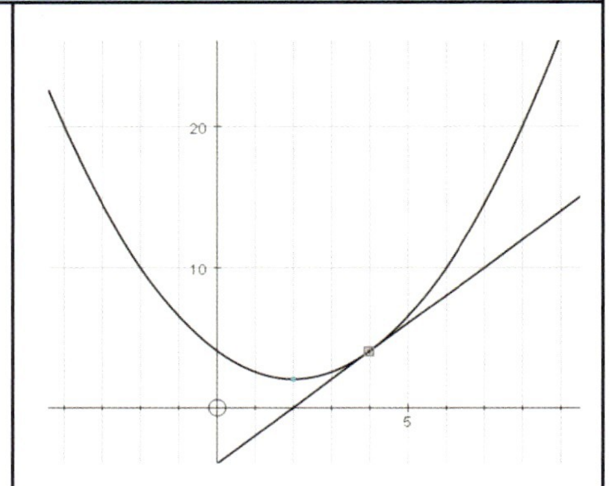

Note: The equation of the tangent is displayed at the base of the screen, i.e. $y = 2x - 4$. The gradient can therefore be deduced from this.

GeoGebra

- Type equation
  $f(x) = \frac{1}{2}x\text{^}2 - 2x + 4$
- Type 'Tangent[4, f]'. This draws a tangent to the curve at $x = 4$.

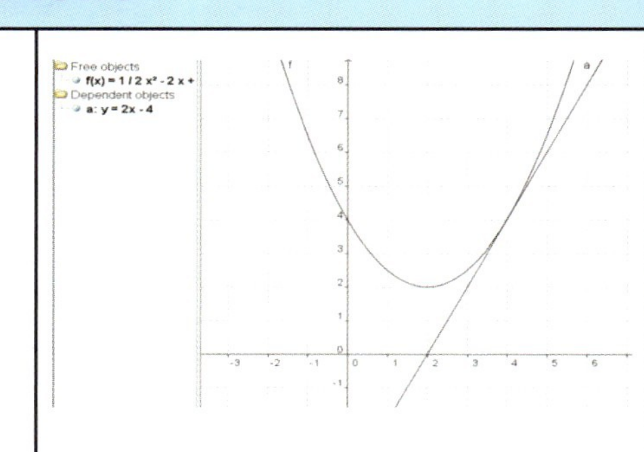

Note: The equation of the tangent is displayed in the algebra window, i.e. $y = 2x - 4$. The gradient can therefore be deduced from this.

Worked example

Calculate the gradient of the curve $f(x) = x^3 + x - 6$ when $x = -1$.

The gradient function $f'(x) = 3x^2 + 1$.

When $x = -1$, $f'(-1) = 3(-1)^2 + 1 = 4$,

i.e. the gradient is 4.

Exercise 7.3.1

1 Find the gradient of each of the following functions at the given values of x.

a $f(x) = x^2$; $x = 3$

b $f(x) = \frac{1}{2}x^2 - 2$; $x = -3$

c $f(x) = 3x^3 - 4x^2 - 2$; $x = 0$

d $f(x) = -x^2 + 2x - 1$; $x = 1$

e $f(x) = -\frac{1}{2}x^3 + x - 3$; $x = -1$, $x = 2$

f $f(x) = 6x$; $x = 5$

2 Find the gradient of each of the following functions at the given values of x.

a $f(x) = \frac{1}{x}$; $x = 2$

b $f(x) = \frac{1}{x^2}$; $x = 1$

c $f(x) = \frac{1}{x^3} - 3x$; $x = 2$

d $f(x) = x^2 - \frac{1}{2x^2}$; $x = -1$

e $f(x) = \frac{1}{6x^3} + x^2 - 1$; $x = 2$

f $f(x) = \frac{1}{x} - \frac{1}{x^2} + \frac{1}{x^3}$; $x = \frac{1}{2}$, $x = -\frac{1}{2}$

3 The number of people, N, newly infected on day t of a stomach bug outbreak is given by $N = 5t^2 - \frac{1}{2}t^3$.

a Calculate the number of new infections, N, when:

i) $t = 1$ **ii)** $t = 3$ **iii)** $t = 6$ **iv)** $t = 10$.

b Calculate the rate of new infections with respect to t, i.e. calculate $\frac{dN}{dt}$.

c Calculate the rate of new infections when:

i) $t = 1$ **ii)** $t = 3$ **iii)** $t = 6$ **iv)** $t = 10$.

d Using a GDC, sketch the graph of N against t.

e Using your graph as a reference, explain your answers to part **a**.

f Using your graph as a reference, explain your answers to part **c**.

4 A weather balloon is released from the ground. Its height h (m) after time t (hours) is given by the formula $h = 30t^2 - t^3$, $t \leq 20$.

a Calculate the balloon's height when:

i) $t = 3$ **ii)** $t = 10$.

b Calculate the rate at which the balloon is climbing with respect to time t.

c Calculate the rate of ascent when:
   i) $t = 2$ ii) $t = 5$ iii) $t = 20$.

d Using a GDC, sketch the graph of h against t.

e Using your graph as a reference, explain your answers to part **c**.

f Use your graph to estimate the time when the balloon was climbing at its fastest rate. Explain your answer.

Calculating *x*, when the gradient is given

So far we have calculated the gradient of a curve for a given value of x. It is possible to work backwards and calculate the value of x, when the gradient at a point is given.

Consider the function $f(x) = x^2 - 2x + 1$. It is known that the gradient at a particular point on the curve is 4. It is possible to calculate the coordinate of the point.

The gradient function of the curve is $f'(x) = 2x - 2$.
As the gradient at this particular point is 4 (i.e. $f'(x) = 4$), an equation can be formed:

$$2x - 2 = 4$$
$$2x = 6$$
$$x = 3$$

Therefore, when $x = 3$, the gradient of the curve is 4.

Once again a GDC and graphing software can help solve this type of problem.

Worked example

The function $f(x) = x^3 - x^2 - 5$ has a gradient of 8 at a point P on the curve.

Calculate the possible coordinates of point P.

The gradient function $f'(x) = 3x^2 - 2x$

At P, $3x^2 - 2x = 8$.

This can be rearranged into the quadratic $3x^2 - 2x - 8 = 0$ and solved either algebraically or graphically as shown in Topic 2.

Graphically

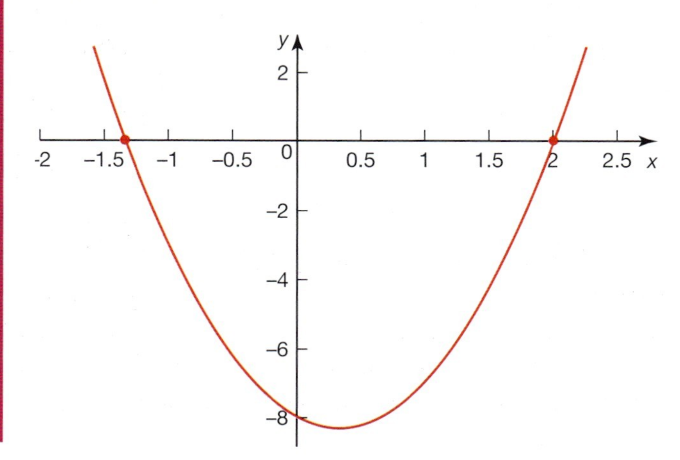

From the graph it can be seen that $x = -1\frac{1}{3}$ and $x = 2$, i.e. there are two possible positions for point P.

By substituting these values of x into the equation of the curve, the y-coordinates of P can be calculated.

$$f\left(-1\tfrac{1}{3}\right) = \left(-1\tfrac{1}{3}\right)^3 - \left(-1\tfrac{1}{3}\right)^2 - 5 = -\frac{247}{27} = -9\frac{4}{27}$$

$$f(2) = 2^3 - 2^2 - 5 = -1$$

Therefore, the possible coordinates of P are $\left(-1\frac{1}{3}, -9\frac{4}{27}\right)$ and $(2, -1)$.

Algebraically

The quadratic equation $3x^2 - 2x - 8 = 0$ can be solved algebraically by factorizing.

$(3x + 4)(x - 2) = 0$

Therefore, $(3x + 4) = 0 \Rightarrow x - \frac{4}{3}$ or $(x - 2) = 0 \Rightarrow x = 2$.

Your GDC can also be used to solve (quadratic) equations.

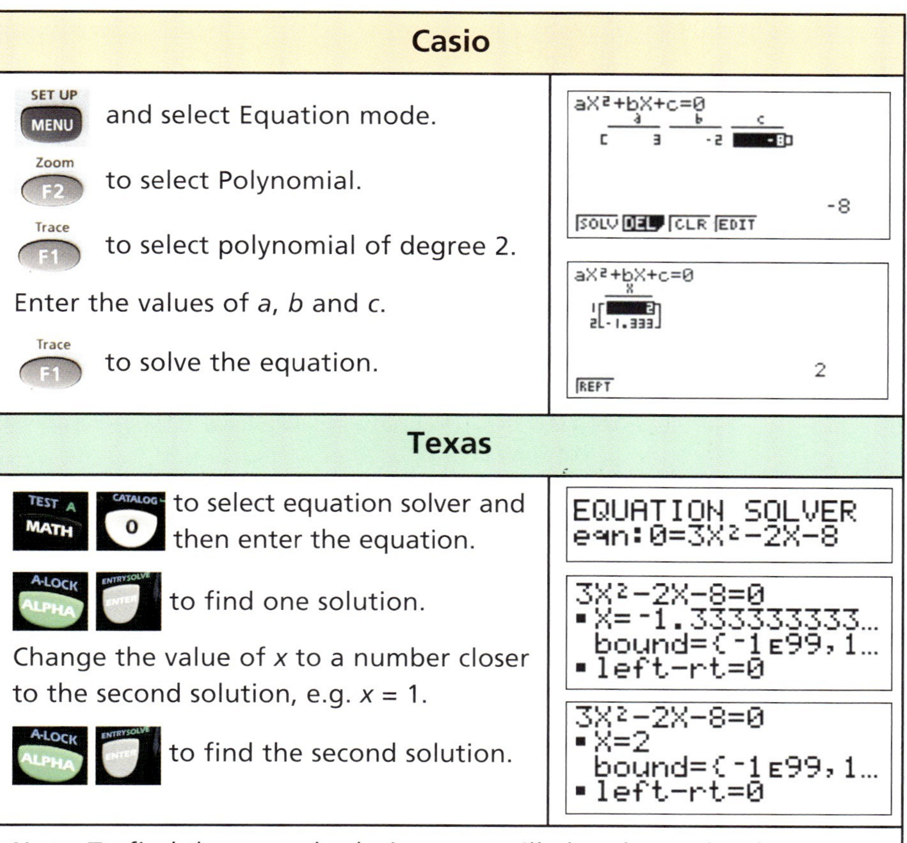

Note: To find the second solution, you will already need to know approximately where it is. This can be done by graphing the equation first.

Once the values of x have been calculated, the values of y can be calculated as before.

Exercise 7.3.2

1 Find the coordinate of the point P on each of the following curves, at the given gradient.

a $f(x) = x^2 - 3$, gradient $= 6$

b $f(x) = 3x^2 + 1$, gradient $= 15$

c $f(x) = 2x^2 - x + 4$, gradient $= 7$

d $f(x) = \frac{1}{2}x^2 - 3x - 1$, gradient $= -3$

e $f(x) = \frac{1}{3}x^2 + 4x$, gradient $= 6$

f $f(x) = -\frac{1}{5}x^2 + 2x + 1$, gradient $= 4$

2 Find the coordinate(s) of the point(s) on each of the following curves, at the given gradient.

a $f(x) = \frac{1}{3}x^3 + \frac{1}{2}x^2 + 4x$, gradient $= 6$

b $f(x) = \frac{1}{3}x^3 + 2x^2 + 6x$, gradient $= 3$

c $f(x) = \frac{1}{3}x^3 - 2x^2$, gradient $= -4$

d $f(x) = x^3 - x^2 + 4x$, gradient $= 5$

3 A stone is thrown vertically downwards off a tall cliff. The distance (s) it travels in metres is given by the formula $s = 4t + 5t^2$, where t is the time in seconds after the stone's release.

a What is the rate of change of distance with time $\frac{ds}{dt}$? (This represents the velocity.)

b How many seconds after its release is the stone travelling at a velocity of $9\ \text{m s}^{-1}$?

c The stone hits the ground travelling at $34\ \text{m s}^{-1}$. How many seconds did the stone take to hit the ground?

d Using your answer to part **c**, calculate the distance the stone falls and hence the height of the cliff.

4 The temperature (T °C) inside a pressure cooker is given by the formula $T = 20 + 12t^2 - t^3$; $t \leq 8$, where t is the time in minutes after the cooking started.

a Calculate the temperature at the start.

b What is the rate of temperature increase with time?

c What is the rate of temperature increase when:

i) $t = 1$ ii) $t = 4$ iii) $t = 8$?

d The pressure cooker was switched off when $\frac{dT}{dt} = 36$.

How long after the start could the pressure cooker have been switched off?

e What was the temperature of the pressure cooker if it was switched off at the greater of the two times calculated in part **d**?

Equation of the tangent at a given point

As was seen in Section 7.1, the gradient of a tangent drawn at a point on a curve is equal to the gradient of the curve at that point.

Worked examples

What is the equation of the tangent to $f(x) = \frac{1}{2}x^2 + 3x + 1$ in the graph below?

The function $f(x) = \frac{1}{2}x^2 + 3x + 1$ has a gradient function of $f'(x) = x + 3$

At point P, where $x = 1$, the gradient of the curve is 4.

The tangent drawn to the curve at P also has a gradient of 4.

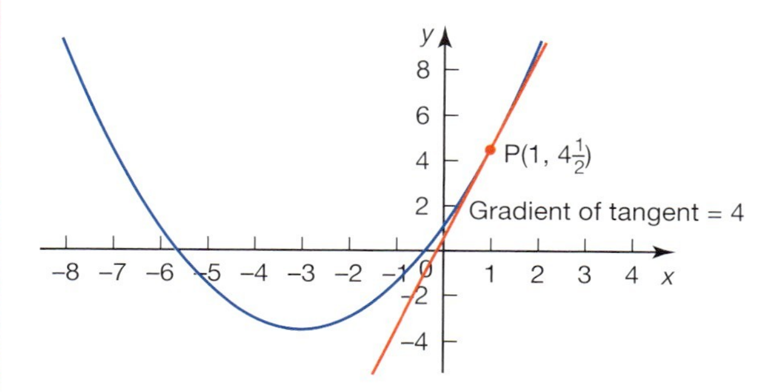

As it is a straight line, it must take the form $y = mx + c$. The gradient m is 4 as shown above.

Therefore, $y = 4x + c$.

As the tangent passes through the point $P(1, 4\frac{1}{2})$, these values can be substituted for x and y so that c can be calculated.

$4\frac{1}{2} = 4 + c$

$\Rightarrow c = \frac{1}{2}$

The equation of the tangent is therefore $y = 4x + \frac{1}{2}$.

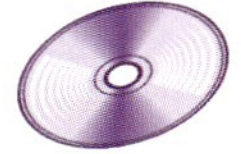

Exercise 7.3.3

1 For the function $f(x) = x^2 - 3x + 1$:
   a calculate the gradient function
   b calculate the gradient of the curve at the point A (2, 1).

   A tangent is drawn to the curve at A.
   c What is the gradient of the tangent?
   d Calculate the equation of the tangent in the form $y = mx + c$.

2 For the function $f(x) = 2x^2 - 4x - 2$:
   a calculate the gradient of the curve where $x = 2$.

   A tangent is drawn to the curve at the point (2, −2).
   b Calculate the equation of the tangent in the form $y = mx + c$.

3 A tangent is drawn to the curve $f(x) = \frac{1}{2}x^2 - 4x - 2$ at the point P $(0, -2)$.
  a Calculate the gradient of the tangent at P.
  b Calculate the equation of the tangent in the form $y = mx + c$.

4 A tangent T_1 is drawn to the curve $f(x) = -x^2 + 4x + 1$ at the point A $(4, 1)$.
  a Calculate the gradient of the tangent at A.
  b Calculate the equation of the tangent in the form $y = mx + c$.

  Another tangent T_2 is drawn at the point B $(2, 5)$.
  c Calculate the equation of T_2.

5 A tangent T_1 is drawn to the curve $f(x) = -\frac{1}{4}x^2 - 3x + 1$ at the point P $(-2, 6)$.
  a Calculate the equation of T_1.

  Another tangent, T_2, with equation $y = 10$, is also drawn to the curve at a point Q.
  b Calculate the coordinates of point Q.
  c T_1 and T_2 are extended so that they intersect. Calculate the coordinates of the point of intersection.

6 The equation of a tangent T, drawn on the curve $f(x) = -\frac{1}{2}x^2 - x - 4$ at P, has an equation $y = -3x - 6$.
  a Calculate the gradient function of the curve.
  b What is the gradient of the tangent T?
  c What are the coordinates of the point P?

7.4 Increasing and decreasing functions

The graph shows the heart rate of an adult male over a period of time.

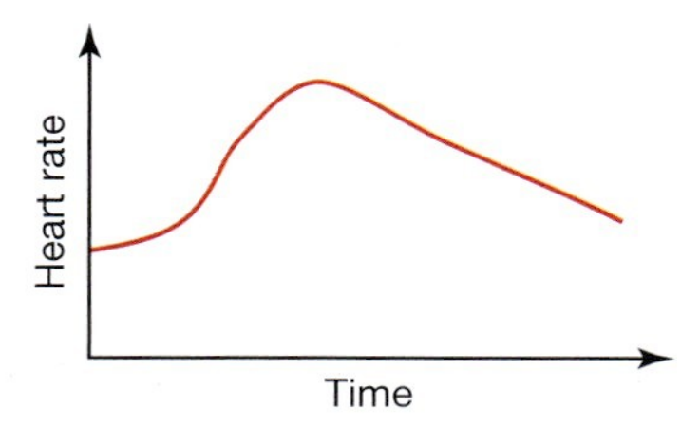

By simply looking at the graph, it is easy to see when his heart rate is increasing, decreasing and when it is at a maximum.

By comparing the shape of the graph with its gradient, we can see that when the gradient is positive, the heart rate is increasing. When the gradient is negative, the heart rate is decreasing. When the gradient is zero, the heart rate is at its maximum.

This section will look at the properties of increasing and decreasing functions.

For the function $f(x) = x^2$ the following observations can be made.

When $x > 0$ the gradient is positive, therefore $f(x)$ is an increasing function for this range of values of x.

When $x < 0$ the gradient is negative, therefore $f(x)$ is a decreasing function for this range of values of x.

When $x = 0$ the gradient is zero, therefore $f(x)$ is stationary at this value of x.

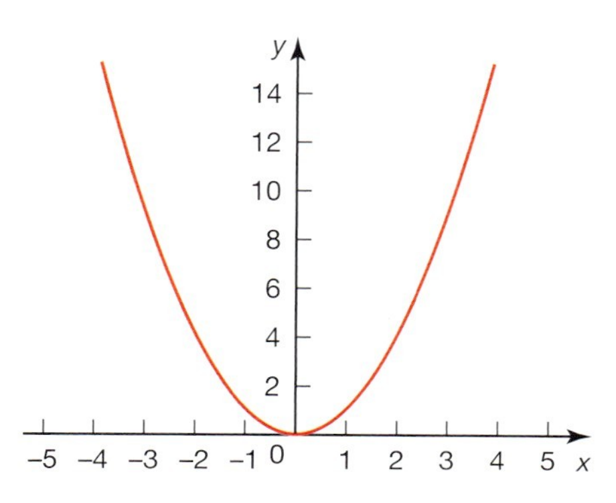

Worked examples

1 a Using your GDC, sketch the function $f(x) = x^3 - 5x^2 + 8$.

b Calculate the range of values of x for which $f(x)$ is a decreasing function.

a

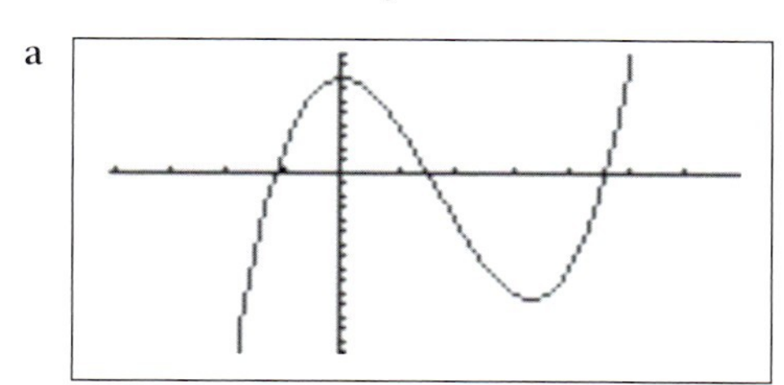

b $f(x) = x^3 - 5x^2 + 8$, therefore $f'(x) = 3x^2 - 10x$.

For a decreasing function, $f'(x) < 0$.

Therefore, either solve the inequality $3x^2 - 10x < 0$ to find the range of values of x.

Or, from the graph it can be deduced that the function is decreasing between the stationary points A and B.

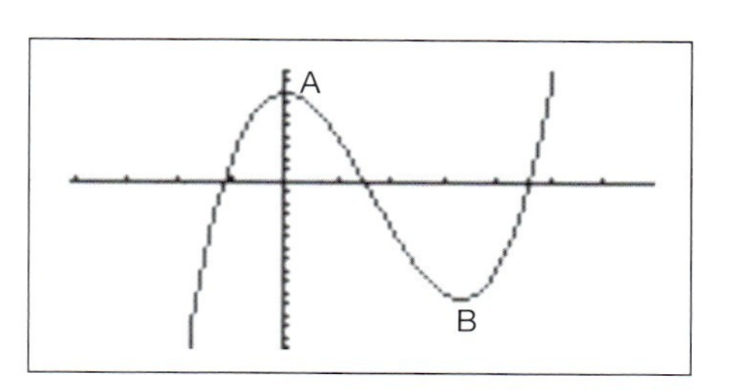

At the stationary points, $f'(x) = 0$ so the quadratic $3x^2 - 10x = 0$ is solved to find the stationary points.

$3x^2 - 10x = 0$
$x(3x - 10) = 0$
$x = 0$ or $\frac{10}{3}$

Therefore $f(x) = x^3 - 5x^2 + 8$ is a decreasing function in the range $0 < x < \frac{10}{3}$.

2 Show that $f(x) = x^3 + x - 8$ is an increasing function for all values of x.

$f'(x) = 3x^2 + 1$

For $f(x)$ to be an increasing function, $f'(x) > 0$.

$3x^2 + 1 > 0$ for all values of x as x^2 is never negative.

Therefore $f(x)$ is an increasing function for all values of x.

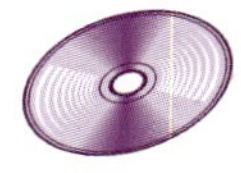

Exercise 7.4.1

1 For each of the following, calculate:

i) $f'(x)$

ii) the range of values of x for which $f(x)$ is increasing.

a $f(x) = x^2 - 4$

b $f(x) = x^2 - 3x + 10$

c $f(x) = -x^2 + 10x - 21$

d $f(x) = x^3 - 12x^2 + 48x - 62$

e $f(x) = -x^3 + 25x$

f $f(x) = \frac{1}{4}x^4 - \frac{1}{2}x^2$

2 Using the functions and your calculations in question 1 above find, in each case, the range of values of x for which $f(x)$ is a decreasing function.

3 a Prove that $f(x) = \frac{1}{3}x^3 + \frac{1}{3}x$ is an increasing function for all values of x.

b Prove that $f(x) = \frac{1}{3}x^3 - x - \frac{1}{5}x^5$ is a decreasing function for all values of x.

4 Calculate the range of values of k in the function $f(x) = x^3 + x^2 - kx$, given that $f(x)$ is an increasing function for all values of x.

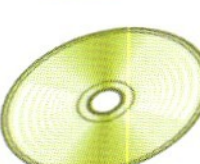

7.5 Stationary points

In the previous section we looked at the concept of increasing and decreasing functions. If the gradient of a curve is positive (i.e. $f'(x) > 0$), then it is an increasing function for that value of x. If the gradient of a curve is negative (i.e. $f'(x) < 0$), then the function is a decreasing one for that value of x.

However, there are cases where the gradient function is zero (i.e. $f'(x) = 0$). This section looks at those cases in more detail.

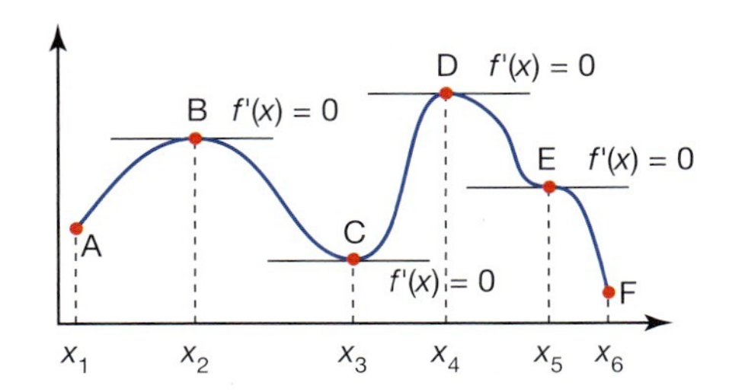

The curve on the previous page is an increasing function between A and B and between C and D (where x takes values in the range $x_1 < x < x_2$ and $x_3 < x < x_4$ respectively). It is a decreasing function between B and C, between D and E and between E and F (where x takes values in the range $x_2 < x < x_3$, $x_4 < x < x_5$ and $x_5 < x < x_6$ respectively). However, at points B, C, D and E the gradient function is zero. These are known as **stationary points**.

There are different types of stationary point. Points B and D are called local maxima. Point C is a local minima, whilst point E is a point of inflexion.

It is not necessary to sketch a graph in order to find the position of any stationary points or to find out what type of stationary point they are.

Worked example

a Find the coordinates of the stationary points on the curve with equation $y = \frac{1}{3}x^3 - 4x + 5$.

b Determine the nature of each of the stationary points.

a If $y = \frac{1}{3}x^3 - 4x + 5$, $\frac{dy}{dx} = x^2 - 4$.

At a stationary point, $\frac{dy}{dx} = 0$.

Therefore, solve $x^2 - 4 = 0$ to find the x coordinate of any stationary point.

$x^2 - 4 = 0$
$x^2 = 4$
$x = 2$ or $x = -2$

When $x = 2$ and $x = -2$ are substituted into the equation of the curve, the y-coordinates are found.

When $x = 2$: $\quad y = \frac{1}{3}(2)^3 - 4(2) + 5 = -\frac{1}{3}$

When $x = -2$: $\quad y = \frac{1}{3}(-2)^3 - 4(-2) + 5 = 10\frac{1}{3}$

The coordinates of the stationary points are $\left(2, -\frac{1}{3}\right)$ and $\left(-2, 10\frac{1}{3}\right)$.

b There are several methods that can be used to establish the type of stationary point.

i) Graphical deduction

As the curve is a cubic of the form $y = ax^3 + bx^2 + cx + d$ and a, the coefficient of x^3, is positive, the shape of the curve is of the form

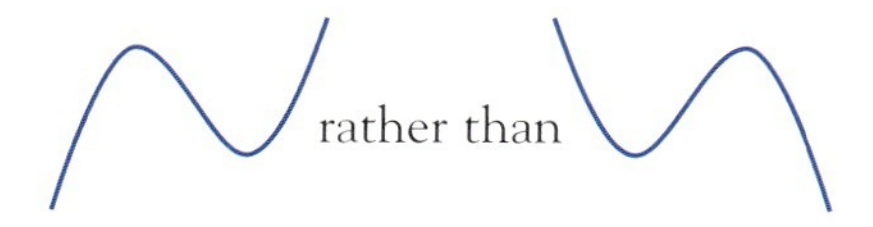

therefore it can be deduced that the coordinates of the stationary points are

$\left(-2, 10\frac{1}{3}\right)$

$\left(2, -\frac{1}{3}\right)$

Hence $\left(-2, 10\frac{1}{3}\right)$ is a maximum point and $\left(2, -\frac{1}{3}\right)$ a minimum point.

ii) Gradient inspection

The gradient of the curve either side of a stationary point can be calculated.

At the stationary point where $x = 2$, consider the gradient at $x = 1$ and $x = 3$.

$\frac{dy}{dx} = x^2 - 4$ when $x = 1$, $\frac{dy}{dx} = -3$

when $x = 3$, $\frac{dy}{dx} = 5$

The gradient has changed from negative to positive as x increased, therefore the stationary point must be a minimum.

At the stationary point where $x = -2$, consider the gradient at $x = -3$ and $x = -1$.

$\frac{dy}{dx} = x^2 - 4$ when $x = -3$, $\frac{dy}{dx} = 5$

when $x = -1$, $\frac{dy}{dx} = -3$

The gradient has changed from positive to negative as x increased, therefore the stationary point must be a maximum.

In general:

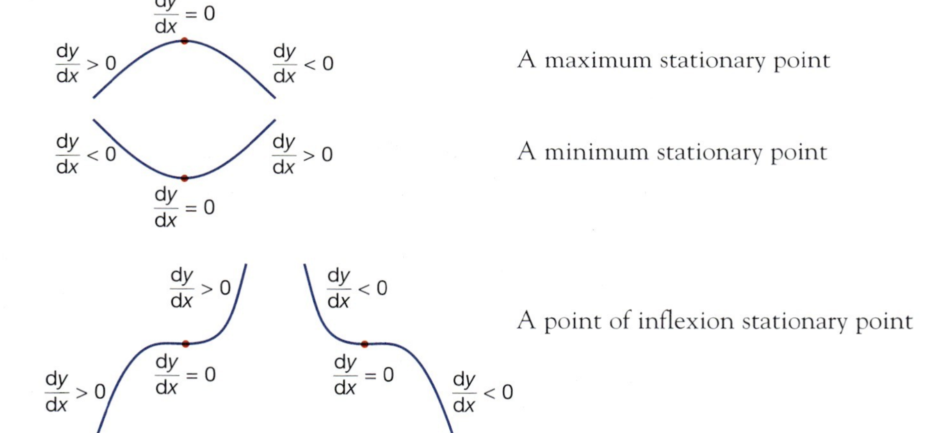

A GDC and graphing software can also be used to find the position of any stationary points. For example, find the coordinates of the stationary points for the graph of $y = \frac{1}{3}x^3 - 4x + 5$.

Casio

to select the graph mode.

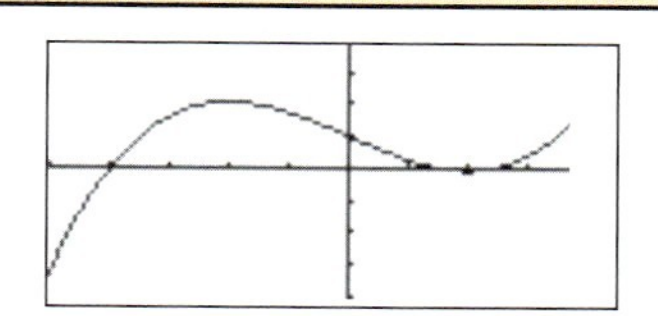

Enter the equation $y = \frac{1}{3}x^3 - 4x + 5$

to graph the function.

to access the graph solve menu.

to find any maximum points.

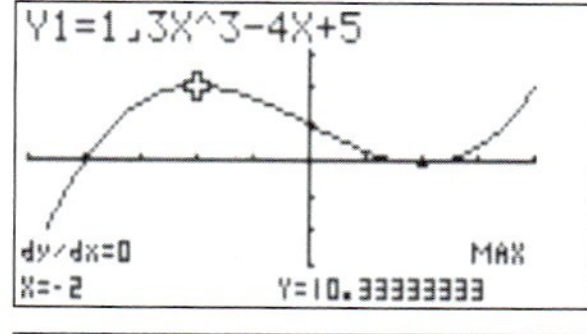

The results are displayed on the graph.

to find any minimum points.

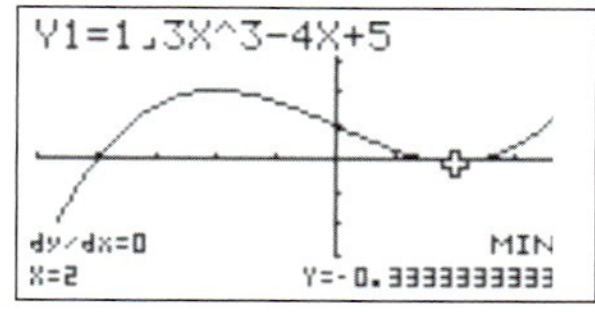

Texas

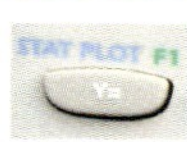

to enter the function.

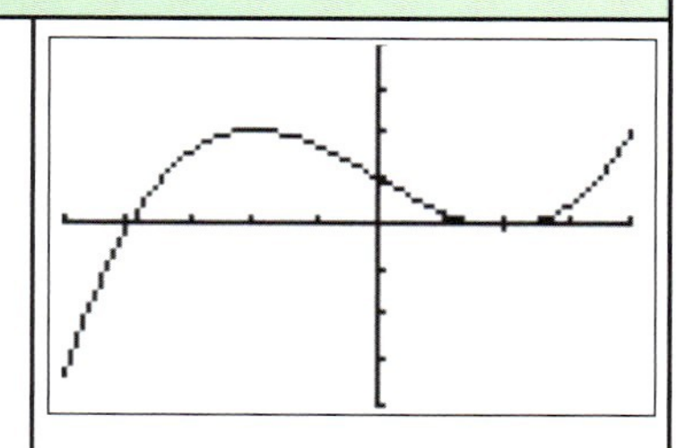

Enter the equation $y = \frac{1}{3}x^3 - 4x + 5$

to graph the function.

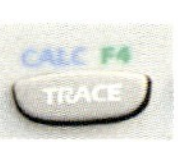

to search for the maximum point.

to move to the left of the maximum point.

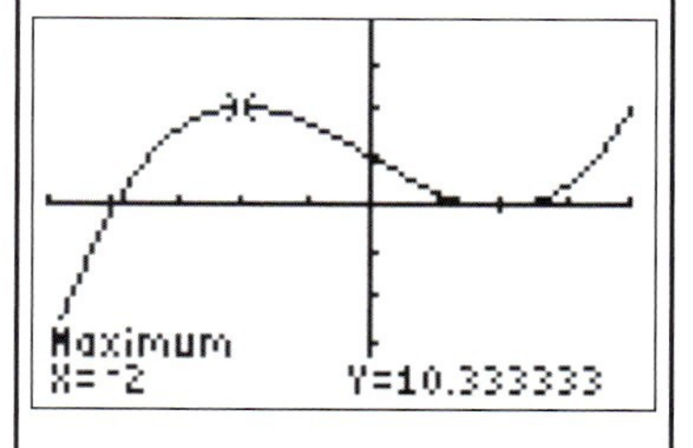

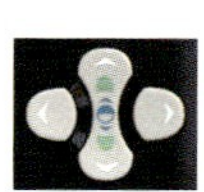
to move to the right of the maximum point.

to find the maximum point.

to search for the minimum point, followed by the same procedure described above for the maximum point.

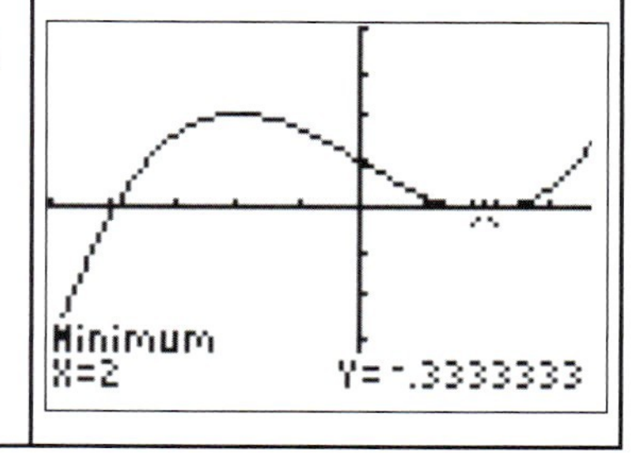

Autograph

Select and enter the equation

$y = \frac{1}{3}x^3 - 4x + 5$

To change the scale on the axes use .

Select the curve then click 'Object' followed by '$f'(x) = 0$'.

The solutions appear at the base of the screen or more clearly in the results box by selecting .

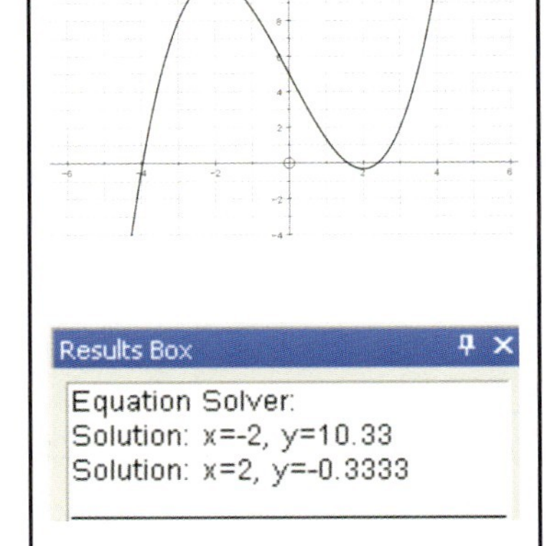

GeoGebra

Enter the equation
$y = \frac{1}{3}x\text{^}3 - 4x + 5$ into the input field.

To find the position of any stationary points type 'Extremum (f)' into the input field. The points are plotted on the graph and their coordinates displayed in the algebra window.

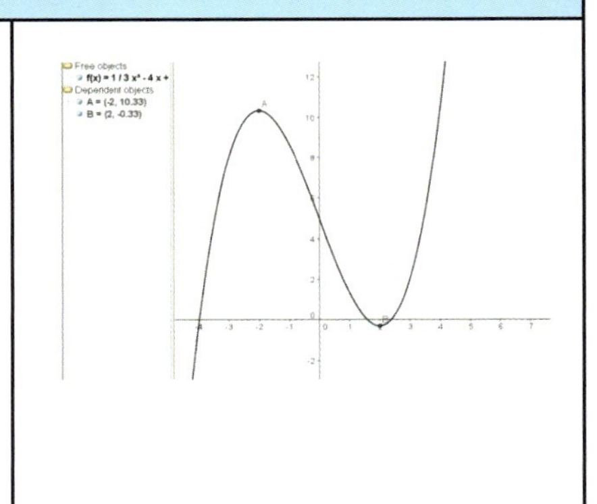

There is also a way of finding stationary points using the second derivative $\frac{d^2y}{dx^2}$. Although it is beyond the scope of this textbook, you may wish to investigate it as it is, in most cases, an efficient method.

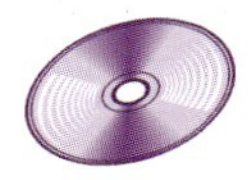

Exercise 7.5.1

For questions 1 and 2, calculate:
i) the gradient function
ii) the coordinates of any stationary points.

1 **a** $f(x) = x^2 - 6x + 13$ **b** $f(x) = x^2 + 12x + 35$
c $f(x) = -x^2 + 8x - 13$ **d** $f(x) = -6x + 7$

2 **a** $f(x) = x^3 - 12x^2 + 48x - 58$ **b** $f(x) = x^3 - 12x$
c $f(x) = x^3 - 3x^2 - 45x + 8$ **d** $f(x) = \frac{1}{3}x^3 + \frac{3}{2}x^2 - 4x - 5$

For questions 3 and 4:

i) calculate the gradient function
ii) calculate the coordinates of any stationary points
iii) determine the type of stationary point
iv) calculate the value of the y-intercept
v) sketch the graph of the function.

3 a $f(x) = 1 - 4x - x^2$ b $f(x) = \frac{1}{3}x^3 - 4x^2 + 12x - 3$
c $f(x) = -\frac{2}{3}x^3 + 3x^2 - 4x$ d $f(x) = x^3 - \frac{9}{2}x^2 - 30x + 4$

4 a $f(x) = x^3 - 9x^2 + 27x - 30$ b $f(x) = x^4 - 4x^3 + 16x$

Student assessment 1

1 Find the gradient function of each the following.
a $y = x^3$ b $y = 2x^2 - x$
c $y = -\frac{1}{2}x^2 + 2x$ d $y = \frac{2}{3}x^3 + 4x^2 - x$

2 Differentiate the following functions with respect to x.
a $f(x) = x(x + 2)$
b $f(x) = (x + 2)(x - 3)$
c $f(x) = \dfrac{x^3 - x}{x}$ d $f(x) = \dfrac{x^3 + 2x^2}{2x}$
e $f(x) = \dfrac{3}{x}$ f $f(x) = \dfrac{x^2 + 2}{x}$

3 Find the second derivative of both of these functions.
a $y = x^4 - 3x^2$ b $f(x) = 2t^5 - t^3$

4 Find the gradient of the following curves at the given values of x.
a $f(x) = \frac{1}{2}x^2 + x$; $x = 1$
b $f(x) = -x^3 + 2x^2 + x$; $x = 0$
c $f(x) = \dfrac{1}{2x^2} + x$; $x = -\frac{1}{2}$
d $f(x) = (x - 3)(x + 8)$; $x = \frac{1}{4}$

5 A stone is dropped from the top of a cliff. The distance it falls (s) is given by the equation $s = 5t^2$, where s is the distance in metres and t the time in seconds.
a Calculate the velocity v, by differentiating the distance s with respect to time t.
b Calculate the stone's velocity after 3 seconds.
c The stone hits the ground travelling at 42 m s^{-1}. Calculate:
i) how long it took for the stone to hit the ground
ii) the height of the cliff.

Student assessment 2

1 The function $f(x) = x^3 + x^2 - 1$ has a gradient of zero at points P and Q, where the x-coordinate of P is less than that of Q.
a Calculate the gradient function $f'(x)$.
b Calculate the coordinates of P.
c Calculate the coordinates of Q.
d Determine which of the points P or Q is a maximum. Explain your method clearly.

2 a Explain why the point A (1, 1) lies on the curve $y = x^3 - x^2 + x$.
b Calculate the gradient of the curve at A.
c Calculate the equation of the tangent to the curve at A.

3 $f(x) = (x - 2)^2 + 3$
a Calculate $f'(x)$.
b Determine the range of values of x for which $f(x)$ is a decreasing function.

4 $f(x) = x^4 - 2x^2$
a Calculate $f'(x)$.
b Determine the coordinates of any stationary points.
c Determine the nature of any stationary point.
d Find where the graph intersects or touches:
i) the y-axis
ii) the x-axis.
e Sketch the graph of $f(x)$.

Topic 7 Discussion points, project ideas and theory of knowledge

1 If you look up the history of calculus, you will find that Leibniz and Newton both discovered calculus at the same time. Their work followed on from the earlier work of the Iraqi mathematician Ibn al Haytham. Is there a 'readiness' for major discoveries to be made? Discuss with reference to the telephone, television, radioactivity and the structure of DNA.

2 Terms used in calculus, such as 'differential', 'derivative' and 'integration', have other meanings elsewhere. How did these words become mathematical terms?

3 Discuss the statement that calculus is easier to do than to understand.

4 How far was China's 'one child' policy influenced by calculations of the probable future population of China? Did these predictions use calculus?

5 'Calculus is the point at which mathematics and real life part company'. Discuss with reference to calculus and other advanced mathematics.

6 Was calculus invented or discovered? Discuss with reference to point 1 and to the history of mathematics.

7 A possible project would be to extend your knowledge of calculus beyond the syllabus requirements of Mathematical Studies. Consult your teacher for some possible areas of study.

8 'A good GDC means everyone is capable of doing higher maths.' Discuss with particular reference to calculus.

9 Areas of mathematics which were once the frontiers of knowledge, (think Euclidean geometry and trigonometry) are in time studied by young children. Will calculus be studied in primary schools by the end of the century?

10 The spread of a pandemic can be calculated using calculus. This could form the basis of a project.

11 The introduction to this topic suggests that a genius is one who 'brings ideas together in a new way'. Do you agree with this definition or do you have another?

12 Look up Fermat's Last Theorem. The BBC Horizon series made a programme on Andrew Wiles' work to prove the theorem. It may be available and is a brilliant example of making a very difficult proof understandable. Watch it if you can.

Topic 8

Financial mathematics

Syllabus content

8.1 Currency conversions.

8.2 Simple interest: use of the formula $I = \frac{Crn}{100}$, where C = capital, r = % rate, n = number of time periods and I = interest.

8.3 Compound interest: use of the formula $I = C \times \left(1 + \frac{r}{100}\right)^n - C$.

Depreciation.

The value of r can be positive or negative.

8.4 Construction and use of tables: loan and repayment schemes; investment and saving schemes; inflation.

Introduction

Juno Moneta (Juno the Alone) was the Roman God of Finance. The word money is derived from her name. In Rome, financial contracts and loans, etc. were sworn before this god and contracts were lodged in her temple.

The kingdom of the Lydian King Croesus (Lydia is now part of Turkey) is credited as being the first to mint coins in about 560 BC. The expression 'As rich as Croesus' came to describe people of vast wealth.

Numismatics is the study of money and its history. Numismatists are coin and note collectors.

One hundred thousand years ago in what is now Swaziland in Africa, trading was done using red ochre as a form of money. In other parts of the world conch shells and precious metals, particularly gold and silver, were used as currency by traders to supersede the straight exchange of goods.

For a substance to be used as money it must be a 'scarce good'. It may be red ochre, diamonds or, in some circumstances, cigarettes. Scarce goods like gold are called 'commodity money'. Bank notes, which came later, are called 'representative money' as the paper itself has no value, but it can be exchanged for other goods. One pound sterling is so called because it could, on demand, be exchanged for one pound weight of sterling silver.

What are the advantages and disadvantages of paper money compared to commodity money?

8.1 Currency conversion

The Eurozone is a term that describes those European countries that replaced their previous currencies with one common currency, i.e. the euro.

In 2002, just after its launch, its exchange rate against the US dollar (USD or $) was below parity, i.e one euro was worth less than one dollar.

In June 2008 before the banking crisis, one euro was worth approximately one dollar and sixty cents. The value of a bank note relative to other currencies can change, sometimes very rapidly.

When changing currencies, banks take a commission (a fee). This meant that when Europe had a large number of different currencies, traders lost money paying these commissions and currency fluctuations meant long-term planning was difficult for exporters.

Commission can either be a fixed sum or a percentage of the money exchanged. In addition, when you exchange money, there are two rates; one for when selling and another for buying. For example, a bank might buy £1 sterling (GBP) for $1.30 and sell for $1.35. So if you changed £1000 into dollars you would receive $1300 but if you changed your dollars back into pounds you would receive 1300 ÷ 1.35 = £963, a cost of £37. Had they charged a percentage commission of 3%, it would have cost £30 to make the original exchange. So you would get £970 × 1.30 = $1261.

The table below shows the rate at which countries entering the euro in 2002 changed their old currency for one euro.

Country	Currency	Exchange for 1 euro
France	Franc	6.56 francs
Germany	Deutsche mark	1.96 Deutsche marks
Italy	Lira	1940 lire
Spain	Peseta	166 pesetas
Holland	Guilder	2.20 guilders
Austria	Schilling	13.76 schillings
Belgium	Franc	40 francs
Finland	Markka	5.95 markkas
Greece	Drachma	341 drachmas
Ireland	Punt	0.79 punts
Portugal	Escudo	200 escudos
Luxembourg	Franc	40 francs

The euro became the single currency for the twelve countries above on 1st January 2002. It was the biggest change in currencies Europe had ever seen. By 2008 Andorra, Cyprus, Malta, Monaco, Montenegro, San Marino, Slovenia and Vatican City had joined the euro. European countries still to join include Great Britain, Sweden and Denmark. Why do you think these countries have not joined the euro?

Note: There is no universal agreement about whether the € sign should go before or after the number. This depends to a great extent on the conventions that were in place in each country with its previous currency.

Worked example

Using the exchange rate in the table above, change 15 000 pesetas into guilders.

From the table: 166 pesetas = 1 euro = 2.20 guilders

$$166 \text{ pesetas} = 2.20 \text{ guilders}$$

$$1 \text{ peseta} = \frac{2.20}{166} \text{ guilders}$$

$$15\,000 \text{ pesetas} = \frac{2.20}{166} \times 15\,000 \text{ guilders}$$

Therefore, 15 000 pesetas = 198.80 guilders.

Exercise 8.1.1

Using the exchange rates in the table above, convert each of the following.

1 100 Deutsche marks into French francs

2 500 guilder into drachmas

3 20 000 lira into schillings

4 7500 pesetas into escudos

5 1000 Belgian francs into French francs

6 1 million punt into markkas

7 200 000 lira into pesetas

8 3000 Deutsche marks and 1000 schillings into punts

9 5000 Deutsche marks into lira

10 2500 guilder into shillings

This gives you some idea of the many changes occurring every day in Europe before 2002.

The pound sterling was a 'reserve currency'. Many economists think that the euro could become a 'reserve currency'. What does the term mean?

Currencies change their value with respect to each other because of the change in economic circumstances in each country.

For example, in 2002 1€ could be exchanged for $1.04. In 2008 1 euro could be exchanged for $1.54.

The exchange rate today can be found from what is called the currency exchange market.

In 2008, visitors to the USA from Germany would have found the country cheap to visit. Americans visiting Europe would have had the opposite experience, since their dollars did not buy as many euros as in previous years. In theory the external exchange rate does not affect the internal worth of the currency. In practice, because countries need to import and export goods and services, the strength or weakness of a currency will directly affect people. For example if your currency weakens compared to the dollar, then the cost of oil and gas, which are priced in dollars, will increase. Sometimes a currency becomes so weak that it is impossible to import goods. In 2008, the currency of Zimbabwe was not accepted for overseas trade payments. The economy suffered badly and many Zimbabweans could not find work.

Worked example

In 2001 Kurt and his family travelled to the USA. The flights cost $1650, hotels cost $2200 and other expenses were $4000. At that time, the exchange rate was 1€ = $1.05.

In 2008 they took the same vacation but the costs in dollars had increased by 10%. The exchange rate was now 1 euro = 1.55 dollars.

a What was the total cost in euros in 2001?

b What was the total cost in euros in 2008?

a Total cost in dollars = $7850

$\$1.05 = 1€ \Rightarrow \$1 = 0.95€$

Therefore total cost in euro was $7850 \times 0.95 = 7476€$.

b A 10% increase is equivalent to a multiplier of 1.10.

Total cost in dollars = $7850 \times 1.10 = \$8635$

$\$1.55 = 1€ \Rightarrow \$1 = 0.65€$

Therefore total cost in euros was $8635 \times 0.65 = 5613€$.

Exercise 8.1.2

1 The rate of exchange of the South African rand is £1 = 12.8 rand.
£1 = 1.66 US dollars.
  a How many pounds could be exchanged for 100 rand?
  b How many rand could be exchanged for $1000?

2 An apartment in Barcelona is offered for sale for 350 000€. The exchange rate is £1 =1.24€.
  a What was the cost of the apartment in pounds?
  b The value of the pound against the euro increases by 10%. What is the new cost of the apartment in pounds?

3 The Japanese yen trades at 190¥ to £1. The dollar trades at $1.90 to the pound.
  a What is the yen–dollar exchange rate?
  b What is the dollar–yen exchange rate?

4 The Russian rouble trades at 24 roubles to $1. The Israeli shekel trades at 3.5 shekels to $1.
  a What is the rouble–shekel exchange rate?
  b What is the shekel–rouble exchange rate?

5 Gold is priced at $930 per ounce. 16 ounces = 1 pound weight.
  a What is the cost of 1 ton (2240 pounds) of gold?
  1 dollar = 0.62 euros and 1 dollar = 41 Indian rupees.
  b What does 1 ton of gold cost in rupees?
  c What does 1 ton of gold cost in euros?
  d What weight (in ounces) of gold could be bought for 1 million euros?

8.2 Simple interest

Interest is money added by a bank or building society to sums deposited by customers, or money charged to customers for borrowing. The money deposited or borrowed is called the capital. The **percentage interest** is the given rate and the money is usually left or borrowed for a fixed period of time.

The following formula can be used to calculate **simple interest**:

$$I = \frac{Crn}{100}$$

where I = the simple interest paid
C = the capital (the amount borrowed or lent)
n = number of time periods (often years)
r = percentage rate.

Worked examples

1 Find the simple interest earned on €250 deposited for 6 years at 8% p.a.

$$I = \frac{Crn}{100}$$

$$I = \frac{250 \times 8 \times 6}{100}$$

$$I = 120$$

The interest paid is €120.

2 How long will it take for a sum of €250 invested at 8% to earn interest of €80?

$$I = \frac{Crn}{100}$$

$$80 = \frac{250 \times 8 \times n}{100}$$

$$8000 = 2000n$$

$$n = 4$$

It takes 4 years for €250 to earn €80 of interest.

3 What rate per year must be paid for a capital of £750 to earn interest of £180 in 4 years?

$$I = \frac{Crn}{100}$$

$$180 = \frac{750 \times r \times 4}{100}$$

$$180 = 30r$$

$$r = 6\%$$

A rate of 6% must be paid for £750 to earn interest of £180 in 4 years.

The total amount A after simple interest is added is given by the formula

$A = C + \frac{Crn}{100}$.

This can also be written as $A = C + n \times \frac{Cr}{100}$.

Compare this with the formula for the nth term of an arithmetic sequence used in Section 2.5:

$u_n = u_1 + (n - 1)d$

In simple interest calculations, the final amounts after each year (A) form an arithmetic sequence with first term $u_1 = C$ and common difference $d = \frac{Cr}{100}$.

Exercise 8.2.1

All rates of interest are annual rates.

1 Find the simple interest paid in each of the following cases.

	Capital	Rate	Time period
a	NZ$300	6%	4 years
b	£750	8%	7 years
c	425¥	6%	4 years
d	2800 baht	4.5%	2 years
e	HK$880	6%	7 years

2 How long will it take for the following amounts of interest to be earned?

	C	R	I
a	500 baht	6%	150 baht
b	5800¥	4%	96¥
c	AU$4000	7.5%	AU$1500
d	£2800	8.5%	£1904
e	900€	4.5%	243€
f	C = 400 Ft	9%	252 Ft

3 Calculate the rate of interest per year which will earn the given amount of interest in the given time period.

	Capitals	Time periods	Interest
a	400€	4 years	1120€
b	US$800	7 years	US$224
c	2000 baht	3 years	210 baht
d	£1500	6 years	£675
e	850€	5 years	340€
f	AU$1250	2 years	AU$275

4 Calculate the capital required to earn the interest stated in the number of years and with the rates given.

	Interest	Time period	Rate
a	80 Ft	4 years	5%
b	NZ$36	3 years	6%
c	340€	5 years	8%
d	540 baht	6 years	7.5%
e	540€	3 years	4.5%
f	US$348	4 years	7.25%

5 What rate of interest is paid on a deposit of £2000 that earns £400 interest in 5 years?

6 How long will it take a capital of 350€ to earn 56€ interest at 8% per year?

7 A capital of 480 Ft earns 108 Ft interest in 5 years. What rate of interest was being paid?

8 A capital of 750€ becomes a total of 1320€ in 8 years. What rate of interest was being paid?

9 AU$1500 is invested for 6 years at 3.5% per year. What is the interest earned?

10 500 baht is invested for 11 years and becomes 830 baht in total. What rate of interest was being paid?

8.3 Compound interest

In this section we move on from simple interest to look at compound interest. For the first time period, there is no difference between the two types: a percentage of the capital is paid as interest. However, simple interest is calculated using the original amount whereas compound interest is paid on the total amount, which includes the interest paid in the first time period. Simple interest has few applications in real life: when we talk about interest in real life, it will usually be compound interest. If you have a savings account the money in it will earn compound interest and when people take out a loan they will pay compound interest on the money they have borrowed.

For example, a builder is going to build six houses on a plot of land in Spain. He borrows 500 000 euros at 10% p.a. and will pay off the loan in full after 3 years, when he expects to have finished building the houses and to have sold them. At the end of the first year he will owe:

500 000€ + 10% of 500 000€ i.e. 500 000€ × 1.10 = 550 000€

At the end of the second year he will owe:

550 000€ + 10% of 550 000€ i.e. 550 000€ × 1.10 = 605 000€

At the end of the third year he will owe:

605 000€ + 10% of 605 000€ i.e. 605 000€ × 1.10 = 665 500€

The amount of interest he has to pay is:

665 500€ − 500 000€ = 165 500€

The simple interest is 50 000€ per year, i.e. a total of 150 000€.
The difference of 15 500€ is the compound interest.
The time taken for a debt to grow at compound interest can be calculated as shown in the example below.

Worked example

How long will it take for a debt to double with a compound interest rate of 27% p.a.?
An interest rate of 27% implies a multiplier of 1.27.

Time (years)	0	1	2	3
Debt	C	$1.27C$	$1.27^2C = 1.61C$	$1.27^3C = 2.05C$

× 1.27 × 1.27 × 1.27

The debt will have more than doubled after 3 years.

Using the example of the builder's loan above, if C represents the capital he borrows, then after 1 year his debt will be given by the formula:

$D = C\left(1 + \frac{r}{100}\right)$, where r is the rate of interest.

After 2 years: $D = C\left(1 + \frac{r}{100}\right)\left(1 + \frac{r}{100}\right)$

After 3 years: $D = C\left(1 + \frac{r}{100}\right)\left(1 + \frac{r}{100}\right)\left(1 + \frac{r}{100}\right)$

After n years: $D = C\left(1 + \frac{r}{100}\right)^n$

This formula for the debt includes the original capital loan. By subtracting C, the compound interest is calculated.

$$I = C\left(1 + \frac{r}{100}\right)^n - C$$

Compound interest is an example of a geometric sequence. You studied geometric sequences in more detail in Topic 2.6.

The nth term of a geometric sequence is given by:

$$u_n = u_1 r^{n-1}$$

Compare this with the formula for the amount of money earning compound interest at r%:

$$A = C\left(1 + \frac{r}{100}\right)^n$$

Note: The differences between this and the formula for calculating compound interest. u_n is analogous to the amount in an account, A, but n is used differently in the two formulae. The initial amount in an account, C, is when $n = 0$ whereas the first term of a geometric sequence, u_1, is when $n = 1$. r represents the common difference in the formula for the nth term of a geometric sequence. The common difference in a sequence of the amount of money in an account earning compound interest is $\left(1 + \frac{r}{100}\right)$, where r is the rate of interest.

The interest is usually calculated annually, but there can be other time periods. Compound interest can be charged or credited yearly, half-yearly, quarterly, monthly or daily. (In theory, any time period can be chosen.)

Worked examples

1 Alex deposits 1500€ in his savings account. The interest rate offered by the savings account is 6% compound interest each year for a 10-year period. Assuming Alex leaves the money in the account, calculate how much interest he has gained after 10 years.

$$I = 1500\left(1 + \frac{6}{100}\right)^{10} - 1500$$

$$I = 2686.27 - 1500 = 1186.27$$

The amount of interest gained is 1186.27€.

2 Adrienne deposits £2000 in her savings account. The interest rate offered by the bank for this account is 8% compound interest per year. Calculate the number of years Adrienne needs to leave the money in her account for it to double in value.

An interest rate of 8% implies a common ratio of 1.08.
This can be found by generating the geometric sequence using the recurrence rule $u_{n+1} = u_n \times 1.08$ on your calculator as shown on page 37. Think of the initial amount as u_0.

$u_1 = 2000 \times 1.08 = 2160$
$u_2 = 2160 \times 1.08 = 2332.80$
$u_3 = 2332.80 \times 1.08 = 2519.42$
...
$u_9 = 3998.01$
$u_{10} = 4317.85$

Adrienne needs to leave the money in the account for 10 years in order for it to double in value.

3 Use your GDC to find the compound interest paid on a loan of $600 for 3 years at an annual percentage rate (A.P.R.) of 5%.

The total payment is $694.58 so the interest due is $694.58 – $600 = $94.58.

4 Use a GDC to calculate the compound interest when $3000 is invested for 18 months at an APR of 8.5%. The interest is calculated every 6 months.

Note: The interest for each time period of 6 months is $\frac{8.5}{2}$%. There will therefore be three time periods of 6 months each.

$3000 \times 1.0425^3 = £3398.99$

The final sum is $3399, so the interest is $3399 – $3000 = $399.

Exercise 8.3.1

1 A shipping company borrows £70 million at 5% p.a. compound interest to build a new cruise ship. If it repays the debt after 3 years, how much interest will the company pay?

2 A woman borrows 100 000€ for home improvements. The interest rate is 15% p.a. and she repays it in full after 3 years. Calculate the amount of interest she pays.

3 A man owes $5000 on his credit cards. The APR is 20%. If he doesn't repay any of the debt, calculate how much he will owe after 4 years.

4 A school increases its intake by 10% each year. If it starts with 1000 students, how many will it have at the beginning of the fourth year of expansion?

5 8 million tonnes of fish were caught in the North Sea in 2005. If the catch is reduced by 20% each year for 4 years, what weight is caught at the end of this time?

6 How many years will it take for a debt to double at 42% p.a. compound interest?

7 How many years will it take for a debt to double at 15% p.a. compound interest?

8 A car loses value at a rate of 15% each year. How long will it take for its value to halve? Give your answer in years and months.

9 $3600 is invested for 18 months at 9.5% APR compound interest. What is the total interest paid when interest is calculated:
   a annually
   b half-yearly
   c monthly?

10 960€ is invested for two years at 7.5% APR. What is the total interest paid when it is compounded:
   a annually
   b every 6 months
   c monthly?

8.4 Your money

Loans and repayments

Banks and other lenders often supply customers with the kind of repayment tables shown below.

The table shows the repayments due each calendar month on a loan of $100 000 borrowed over different time periods.

		Interest rate				
		6%	**6.5%**	**7%**	**7.5%**	**8%**
Time in years	**5**	1933	1956	1980	2003	2027
	10	1110	1135	1161	1187	1213
	15	844	871	899	927	956
	20	644	746	775	805	836

Worked example

A customer borrows \$200 000 at 6% interest at a repayment term of 20 years.

a What is the monthly repayment?

b What is the total repayment over the term of the loan?

a \$200 000 at 6% has a repayment of 2 × \$100 000 at 6%, therefore using the table above, the monthly repayment is 2 × \$644 = \$1288.

b Repayment will be \$1288 × 12 × 20 = \$309 120.

Note: The table is the same for all currencies.

Savings schemes

If you put money into a savings scheme or bank, the interest is usually paid annually. Some schemes, for example those intended to be used for a pension, may calculate monthly interest (although the annual amount is published).

The table below shows interest paid on an investment of 100 000€ (not including capital).

This is compound interest with no withdrawals.

		Percentage rate					
		5%		**6%**		**7%**	
		Yearly	**Monthly**	**Yearly**	**Monthly**	**Yearly**	**Monthly**
Time in years	**1**	500	511	600	617	700	723
	2	1025	1050	1236	1272	1450	1500
	5	2763	2834	3382	3849	4025	4176
	10	6290	6470	7 909	8194	9672	10100
	20	16533	17126	22071	22310	28700	30387

Worked example

What is the interest on 500 000€ invested at 6% for 5 years calculated:

a annually

b monthly?

a 3382 × 5 = 16 910€

b 3849 × 5 = 19 245€

Inflation

The price of goods and services tends to increase over time, as do salaries. Which comes first is a 'chicken and egg' situation.

Most governments keep a check on prices to produce a measure of price increases. In the UK this is called The Retail Price Index. This indicates the rate of inflation in the economy.

The rate of inflation is compounded year on year.

Worked example

In London, from 2002 to 2007, house price inflation was 18% compounded.
In the same period, average salaries increased by 4% annually.
In 2002 a house cost £240 000 and a couple with a combined income of £60 000 could just afford to buy it.

a What was the value of the house in 2007?
b What was the couple's combined income in 2007, if their salaries increased in line with the average?
c What is the price to earnings ratio in:
 i) 2002   ii) 2007?

a $240\,000 \times 1.18^5 = £549\,062$
b $60\,000 \times 1.04^5 = £72\,999$
c i) $240\,000 : 60\,000 = 4 : 1$
 ii) $549\,062 : 72\,999 \approx 7.5 : 1$

Exercise 8.4.1

For questions 1–4, use the loan and repayments table on page 340.

1 Calculate the total repayment on a loan of $250 000 at 6.5% interest for 20 years?

2 At what rate of interest will a loan of 100 000€ cost 161 820€ to repay in 15 years?

3 The repayments on a loan taken for 20 years at 6% were £386 400. How much was borrowed?

4 $9962.50 was paid monthly on an investment for 5 years at 6%. How much was invested?

5 a Inflation in the Eurozone is 3.5%. How long before prices double?
 b If inflation doubles to 7%, how long will it take before prices double?

6 At what rate of inflation will prices double in 5 years?

7 An economy has a rate of inflation of 3%. Savings are paid at 4.5% compound interest. What is the 'real' return on $50 000 of savings over 10 years?

Student assessment 1

1 Vincent Van Gogh painted his *Sunflower series* of still life paintings. One was called *Vase with 15 Sunflowers*. His brother bought it in 1887 for the equivalent of \$10. In 1987, the Japanese Insurance millionaire Yasuo Goto bought it for the equivalent of \$40 million. What was the compound rate at which the painting increased in value?

2 €1 is equivalent to \$1.35 and £1 is equivalent to 1.32€. What is 1 million dollars worth in pounds?

3 What is the difference in percentage terms between 10% simple interest for 10 years and 10% compound interest for 10 years?

4 The population of a town increases by 5% each year. If the population was 86 000 in 1997, in which year would you expect the population to exceed 100 000 for the first time?

5 €3 million is borrowed for two years at an interest rate of 8%. Find:

a the simple interest

b the compound interest.

6 A house increases in value by 20% each year. How long will it take to double in value?

7 The population of a type of insect increases by approximately 10% each day. How many days will it take for the population to double?

8 A man borrows €5 million for 3 years at an interest rate of 6%. Find:

a the simple interest

b the compound interest

c the compound interest calculated quarterly.

9 A boat loses 15% of its value each year. How long does it take to halve in value?

10 A couple borrows \$250 000 over 20 years at 7.5% interest to buy a house. Use the table on page 340 to calculate:

a the monthly repayment

b the total repayment

c the value of the house if prices increase by 11% annually.

Student assessment 2

1 Jackson Pollock sold his abstract painting *1948 number 5* in 1949 for the sum of \$100 000. It was most recently bought by David Martinez at an auction in 2006 for \$140 million. What was the compound rate at which the painting increased in value?

2 €1 is worth \$1.35 and £1 is worth 1.32€. What is one million pounds worth in dollars?

3 What is the difference in percentage terms between 12.5% simple interest for 20 years and 12.5% compound interest for 20 years?

4 The population of a city increases by 15% each year. If the population was 800 000 in 1997, in which year would you expect the population to exceed 3 000 000 for the first time?

5 €5 million is borrowed for 12 years at an interest rate of 5%. Find:

a the simple interest

b the compound interest

6 A house increases in value by 12.5% each year. How long will it take to double in value?

7 The population of a type of insect increases by approximately 7% each day. How many days will it take for the population to double?

8 A man borrows four million dollars for three years at an interest rate of 8.5%. Find:

a the simple interest

b the compound interest

c the compound interest calculated quarterly.

9 A car loses 12% of its value each year. How long will it take before it is only worth 25% of its original value?

10 A couple borrows \$350 000 over 15 years at 7% interest to buy a house. Use the table on page 340 to calculate:

a the monthly repayment

b the total repayment

c the value of the house if prices increase by 11% annually.

Topic 8 Discussion points, project ideas and theory of knowledge

1 Find the current exchange rate of the US dollar to the euro. Find websites offering currency exchange with and without commission and compare the exchange rates. How many euros would be given by each for $1000? How many dollars do you have to exchange for the number of euros to be the same?

2 In 2008 the world financial markets froze in what was termed 'The Credit Crunch'. What was this and how was it caused?

3 Mark Twain said in 1900: 'Put your money in land. They have stopped making it.' What did he mean and is it true? John Maynard Keynes said: 'The markets can remain irrational longer than you can remain solvent.' Who were these two men? Discuss the quotes.

4 Why do investors 'turn to gold' in times of financial uncertainty?

5 Compound interest has been described as the eighth wonder of the world. With reference to question 1 of the student assessments about the sale of the Van Gogh and Jackson Pollock paintings, what do you think that means?

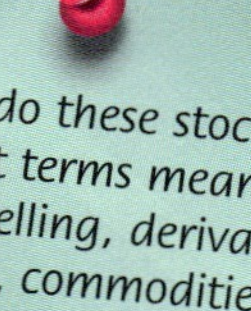

6 What do these stock market terms mean: short selling, derivatives, futures, commodities, gilts, rights issue? The mathematics of the stock exchange could form the basis of a project.

7 Discuss the advantages and disadvantages of a single monetary system controlled by a World Central Bank.

8 What is the difference between the 'stated return' on an investment and the 'real return'?

9 What are 'hedge funds' and how do they function?

10 What was the purpose of a meeting at Bretton Woods after the Second World War? What did the meeting lead to, and why was it significant?

Revision exercises

2.2 Approximation

Approximation: decimal places; significant figures. Percentage errors. Estimation.

1 Round each of the following numbers to the nearest:

i 1000   ii 100   iii 10

a 2842   b 12 938   c 9581   d 496

2 Round each of the following numbers to:

i 1 decimal place   ii 2 decimal places   iii 3 decimal places.

a 2.1827   b 0.9181   c 9.9631   d 0.0386

3 Round each of the following numbers to:

i 1 significant figure   ii 2 significant figures   iii 3 significant figures.

a 3.9467   b 20.36   c 0.015 48   d 0.9752

4 Without using a calculator estimate the answer to the following calculations. Show your method clearly.

a 305×9 b 19.2^2 c $\frac{26.1 \times 3.8}{11}$

d $408 \div 18.8$ e $\frac{8.7^2 \times 1.9^2}{21.3}$ f $(32.2 \times 3.1)^2$

5 Estimate the area of each of the following shapes. Show your method clearly.

a

12.3 cm

4.9 cm

b

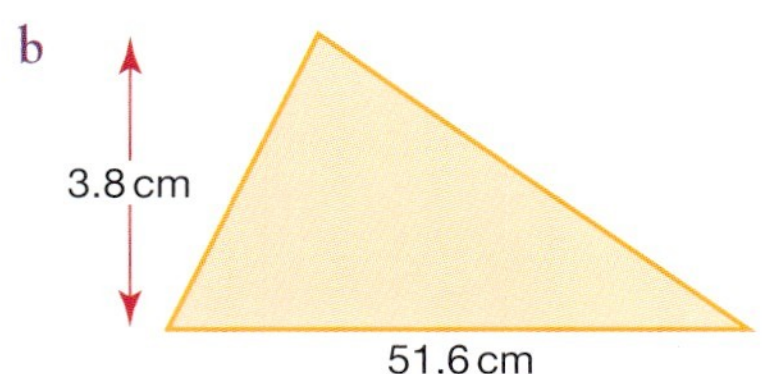

c

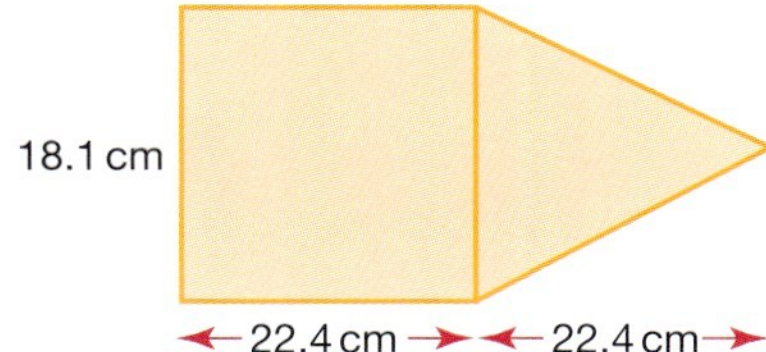

d

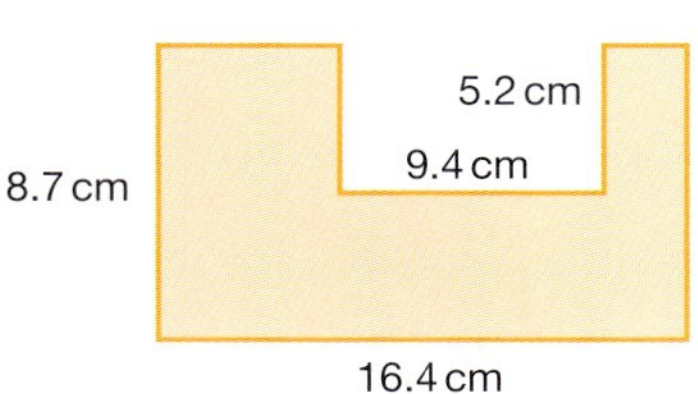

6 a Calculate the actual area of each of the shapes in question 5.

b Calculate the percentage error in your estimates for each of the areas.

7 A certain brand of weighing scales claim that they are accurate to within ±3% of the actual mass being weighed.
A suitcase is weighed and the scales indicate a mass of 18.5 kg. Calculate
a the maximum possible mass of the case
b the minimum possible mass of the case.

8 The value π is sometimes approximated to either 3 or $\frac{22}{7}$.
A circle has a radius of 8 cm.
a Using the π button on your calculator, calculate the area of the circle, giving your answer correct to 5 decimal places.
b Calculate the percentage error in the area if π is approximated to 3.
c Calculate the percentage error in the area if π is approximated to $\frac{22}{7}$.

9 The formula for converting temperatures given in degrees Celsius (C) to temperatures in degrees Fahrenheit (F) is $F = \frac{9}{5}C + 32$.
a The temperature in a classroom is recorded as 18 °C. Convert this to degrees Fahrenheit using the formula above.
An approximate conversion is to use the formula $F = 2C + 30$.
b Calculate the temperature of the classroom in degrees Fahrenheit using the approximation above.
c Calculate the percentage error in using this approximation for a temperature of 18 °C.
d What would the percentage error be for a temperature of 30 °C?
e At what temperature would the percentage error be zero?

10 The formula for calculating the velocity of a stone dropped from rest off a cliff is given by $v = gt$, where v is the velocity in m s^{-1}, g the acceleration in m s^{-2} and t the time in seconds.
a Taking g as $9.81\ \text{m s}^{-2}$, calculate the velocity of the stone after 6 seconds.
b Calculate the velocity of the stone if g is approximated to $10\ \text{m s}^{-2}$.
c Calculate the percentage error in the approximation.

2.3 Standard form

Expressing numbers in the form $a \times 10^k$ where $1 \le a < 10$ and $k \in \mathbb{Z}$.
Operations with numbers expressed in the form $a \times 10^k$ where $1 \le a < 10$ and $k \in \mathbb{Z}$.

1 Which of the following numbers are not in the form $a \times 10^k$ where $1 \le a < 10$ and $k \in \mathbb{Z}$?
a 7.3×10^3 b 60.4×10^2 c 1.0×10^{-2}
d 0.5×10^3 e 3.874×10^5 f 8×10^{-6}

2 Write the following numbers in the form $a \times 10^k$ where $1 \le a < 10$ and $k \in \mathbb{Z}$.
a 32 000 b 620 c 777 000 000
d 90 000 e 8 million f 48.5 million

3 The distance (in kilometres) from London to five other cities in the world is given below.

London to Tokyo	9567 km
London to Paris	343 km
London to Wellington	18 831 km
London to Cambridge	78 km
London to Cairo	3514 km

Write each of the distances in the form $a \times 10^k$ where $1 \le a < 10$ and $k \in \mathbb{Z}$ correct to two significant figures.

4 Calculate each of the following, giving your answers in the form $a \times 10^k$ where $1 \le a < 10$ and $k \in \mathbb{Z}$.
a 500×6000 b $20 \times 450\,000$
c 3 million $\times$ 26 d 5 million $\times$ 8 million

5 Write the following in the form $a \times 10^k$ where $1 \le a < 10$ and $k \in \mathbb{Z}$.
a 0.04 b 0.0076 c 0.000 005 d 0.030 40

6 Write the following numbers in ascending order of magnitude.
3.6×10^{-3} 2.5×10^{-2} 7.4×10^{-2}
9.8×10^{-1} 8.7×10^{-4} 1.4×10^{-2}

7 Calculate each of the following, giving your answer in the form $a \times 10^k$ where $1 \le a < 10$ and $k \in \mathbb{Z}$.
a $6.3 \times 10^2 \div 8.4 \times 10^5$ b $400 \div 800\,000$
c $7 \times 10^4 \div 4.2 \times 10^8$ d $\dfrac{1.5 \times 10^2}{9 \times 10^{10}}$

8 Deduce the value of n in each of the following.
a $0.0003 = 3 \times 10^n$ b $0.000\,046 = 4.6 \times 10^n$
c $0.005^2 = 2.5 \times 10^n$ d $0.0006^n = 2.16 \times 10^{-10}$

9 A boy walks 40 km at a constant rate of $2\,\text{m}\,\text{s}^{-1}$.
Calculate how long, in seconds, the boy takes to walk the 40 km.
Give your answer in the form $a \times 10^k$ where $1 \le a < 10$ and $k \in \mathbb{Z}$.

10 The Earth's radius is approximately 6370 km.
Calculate the Earth's circumference in metres, giving your answer in the form $a \times 10^k$ where $1 \le a < 10$ and $k \in \mathbb{Z}$ correct to three significant figures.

2.4 SI units of measurement

SI (*Système International*) and other basic units of measurement: for example, gram (g), metre (m), second (s), litre (l), metre per second (m s^{-1}), Celsius and Fahrenheit scales.

1 Write an estimate for the following using the appropriate unit.
 a The mass of a large suitcase
 b The length of a basketball court
 c The height of a two-storey building
 d The capacity of a car's fuel tank
 e The distance from the North Pole to the South Pole
 f The mass of a table-tennis ball

2 Convert the following distances.
 a 20 cm into millimetres
 b 35 km into metres
 c 46 mm into centimetres
 d 60 m into kilometres
 e 320 m into millimetres
 f 95 mm into kilometres

3 Convert the following masses.
 a 100 kg into tonnes
 b 60 g into kilograms
 c 3.6 tonnes into kilograms
 d 14 g into milligrams
 e 8.67 kg into milligrams
 f 2560 g into tonnes

4 Convert the following capacities.
 a 2600 ml into litres
 b 80 ml into litres
 c 1.65 litres into millilitres
 d 0.085 litres into millilitres

5 The masses of four containers are as follows:

 25 kg 0.35 t 650 g 0.27 kg

 Calculate the total mass of the four containers in kilograms.

6 The lengths of five objects are as follows:

 56 mm 24 cm 0.672 m 1030 mm 1.5 cm

 Calculate the length, in metres, of the five objects if they are laid end to end.

7 The liquid contents of four containers are emptied into a tank with a capacity of 30 litres. The capacities of the four containers are as follows:

 3250 ml 1.05 litres 26000 ml 762 ml

 Calculate the overspill, in litres, after the liquids have been poured in.

8 The following formula converts temperatures in degrees Celsius (C) to temperatures in degrees Fahrenheit (*F*).

 $F = \frac{9}{5}C + 32$

 a Rearrange the formula to make C the subject.
 b Convert the following temperatures in degrees Fahrenheit to degrees Celsius.
  i 120 °F ii 65 °F iii 255 °F

continued on the next page…

9 Using the formula $F = \frac{9}{5}C + 32$ for converting temperatures in degrees Celsius (C) to temperatures in degrees Fahrenheit (F), calculate the temperature that gives the same number in degrees Fahrenheit as it does in degrees Celsius.

10 The formulae for converting temperatures between degrees Celsius (C), Fahrenheit (F) and Kelvin (K) are $F = \frac{9}{5}C + 32$ and $K = C + 273$.

Convert the following temperatures.

a 25 °C to degrees Fahrenheit

b 300 °K to degrees Celsius

c 650 °K to degrees Fahrenheit

d 125 °F to Kelvin

2.5 Arithmetic sequences and series

Arithmetic sequences and series, and their applications.
Use of the formulae for the nth term and the sum of the first n terms.

1 In each of the following sequences, the recurrence relation and u_1 are given.
 i Calculate and u_2, u_3 and u_4.
 ii State whether the sequence is arithmetic or not.
 a $u_{n+1} = u_n - 6,\ u_1 = 15$
 b $u_{n+1} = 12 - u_n,\ u_1 = 15$
 c $u_{n+1} = 3u_n + 2,\ u_1 = \frac{1}{2}$
 d $u_{n+1} = \dfrac{2u_n - 5}{2},\ u_1 = 5$

2 In each of the following arithmetic sequences
 i deduce the formula for the nth term
 ii calculate the 20th term.
 a 4, 9, 14, 19, 24
 b 3, −5, −13, −21, −29
 c −4.5, −2, 0.5, 3, 5.5
 d 3.5, 3.25, 3, 2.75, 2.5

3 In the following arithmetic sequences:
 i deduce the common difference d
 ii the formula for the nth term.
 a −25, ..., ..., ..., −1
 b 7, ..., ..., ..., ..., ..., −14
 c $u_4 = -12,\ u_{20} = 100$
 d $u_7 = 19,\ u_{42} = -128$

4 Write in full the terms of the following series.
 a $\sum_{1}^{5} 2n - 1$
 b $\sum_{3}^{7} -n + 6$
 c $\sum_{2}^{8} 6 - \frac{1}{2}n$
 d $\sum_{1}^{4} 3(-n + 2)$

5 Write the following arithmetic series using the $\sum$ notation. Each series starts at $n = 1$.
 a $2 + 6 + 10 + 14 + 18$
 b $5 + 3 + 1 + -1 + -3 + -5$
 c $-\frac{1}{2} + 1 + 2\frac{1}{2} + 4 + 5\frac{1}{2} + 7$
 d $-4.1 + -4.2 + -4.3 + -4.4$

6 Evaluate the following.
 a $\sum_{1}^{50} 3n$
 b $\sum_{1}^{25} 20 - n$
 c $\sum_{20}^{30} n + 1$
 d $\sum_{4}^{32} -2n + 50$

7 The 5th and 15th terms of an arithmetic series are −10 and 10 respectively.
 Calculate:
 a the common difference d
 b the first term
 c the 20th term
 d S_{20}.

8 The 11th term of an arithmetic series is 65. If $S_{11} = 495$, calculate:
 a the first term
 b the common difference
 c S_{20}.

9 The 7th term of an arithmetic series is 2.5 times the 2nd term, x. If the 10th term is 34, calculate:
 a the common difference in terms of x
 b the first term
 c the sum of the first 10 terms.

10 The first term of an arithmetic series is 24. The last term is −12. If the sum of the series is 150, calculate the number of terms in the series.

2.6 Geometric sequences and series

Geometric sequences and series, and their applications.
Use of the formula for the *n*th term and the sum of *n* terms.

1 Describe in words the difference between an arithmetic and a geometric sequence.

2 In each of the following geometric sequences, the recurrence relation and u_1 are given.
   i Calculate the values of u_2, u_3 and u_4.
   ii State whether the sequence is geometric or not.
   a $u_{n+1} = 4u_n + 2, u_1 = 0$
   b $u_{n+1} = -3u_n, u_1 = 1$
   c $u_{n+1} = \frac{5}{2}u_n, u_1 = 6$
   d $u_{n+1} = \dfrac{6 - 2u_n}{2}, u_1 = 4$

3 For the geometric sequences below calculate:
   i the common ratio r
   ii the next two terms
   iii the formula for the *n*th term.
   a 5, 15, 45, 135
   b 1296, 216, 36, 6
   c 36, 24, 16, $10\frac{2}{3}$
   d 4, −10, 25, $-62\frac{1}{2}$

4 The *n*th term of a geometric sequence is given by the formula $u_n = -3 \times 4^{n-1}$.
   Calculate:
   a u_1, u_2 and u_3
   b the value of n if $u_n = -12\,288$.

5 Part of a geometric sequence is given as …, 27, …, …, 1, … where u_2 and u_5 are 27 and 1 respectively. Calculate:
   a the common ratio r
   b u_1
   c u_{10}.

6 A homebuyer takes out a loan with a mortgage company for £300 000. The interest rate is fixed at 5.5% per year. If she is unable to repay the loan during the first four years, calculate the amount extra she will have to pay by the end of the fourth year, due to interest.

7 Evaluate the following sums.
   a $\sum_{1}^{5} 3^n$
   b $\sum_{1}^{6} -2(3)^{n-1}$
   c $\sum_{1}^{10} \frac{1}{2}(4)^n$
   d $\sum_{2}^{7} 9\left(-\frac{1}{3}\right)^n$

8 In a geometric series, $u_3 = 10$ and $u_6 = \frac{16}{25}$. Calculate:
   a the common ratio r
   b the first term
   c S_7.

9 Four consecutive terms of a geometric series are $(p - 5)$, (p), $(2p)$ and $(3p + 10)$.
   a Calculate the value of p.
   b Calculate the two terms before $(p - 5)$.
   c If $u_3 = (p - 5)$, calculate S_{10}.

10 In a geometric series $u_1 + u_2 = 5$. If $r = \frac{2}{3}$, find the sum of the infinite series.

2.7 Graphical solution of equations

Solutions of pairs of linear equations in two variables by use of a GDC.
Solutions of quadratic equations: by factorizing; by use of a GDC.

For these questions, use of a GDC or graphing software is expected.

1 Sketch the following straight-line graphs on the same axes, labelling each clearly. Write the coordinate of the point at which they intercept the y-axis.

a $y = x - 5$ b $y = 2x - 5$ c $y = -x - 5$

2 The diagram shows four straight-line graphs. The line $y = x + 2$ is marked. Write down possible equations for the other three graphs.

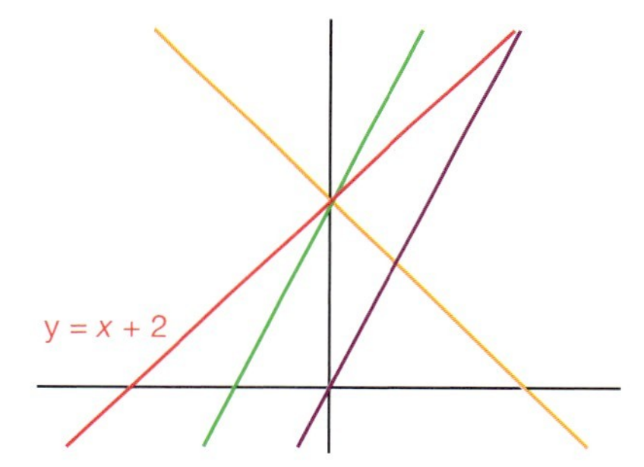

3 Find the coordinates of the points of intersection of the following pairs of linear graphs.

a $y = 8 - x$
$y = 2x - 1$

b $3x + 2y = 2$
$3y = x + 14$

c $y = 3 - 4x$
$3y + 10x = 16$

d $y = \frac{1}{2}x - 3$
$x - 4y = 6$

4 a Sketch the following linear graphs on the same axes.
$y = \frac{1}{2}x + 3$ $x = 2y + 8$
b Explain why there are no points of intersection.

5 Sketch each of the following quadratic equations on separate axes.
a $y = x^2 + 6x + 8$ b $y = x^2 - 16$
c $y = 9 - x^2$ d $y = -(x - 3)(x - 5)$

6 For each of the following quadratic equations:
i sketch the graph
ii find the coordinates of any roots.
a $y = x^2 - 10x + 21$ b $y = 12 + 4x - x^2$
c $y = -x^2 + 10x - 25$ d $y = 2x^2 - 12x + 20$
e $y = 8x^2 - 2x - 1$

3.1 Set theory

Basic concepts of set theory: subsets; intersection; union; complement

1 In the following questions:
   i describe the set in words
   ii write another two elements in the set.
   a {Egypt, Morocco, Zimbabwe, Nigeria, ...}
   b {3, 6, 9, 12, ...}
   c {Amazon, Nile, Mississippi, ...}
   d $\left\{\frac{1}{2}, \frac{2}{3}, \frac{3}{4}, \frac{4}{5}, \ldots\right\}$

2 In the following, set A = {integers between 20 and 50}.
   a List the subset B {even numbers}.
   b List the subset C {prime numbers}.
   c List the subset D {square numbers}.

3 In the following, set P = {a, b, c}.
   If $Q \subset P$, list all the possible sets Q.

4 State whether each of the following statements is true or false.
   a {odd numbers} $\subseteq$ {real numbers}
   b {1, 3, 5, 7, 9} $\subset$ {prime numbers}
   c {New York, Paris, Tokyo} $\not\subset$ {cities}
   d {euro, dollar, yen, rupee} $\subseteq$ {currencies}

5 If the universal set U = {5, 10, 15, 20, 25, 30, 35, 40, 45, 50}.
   A = {35, 40, 45, 50} where $A \subset U$. Deduce the set defined by A'.

6 U = {days in the week} and P = {Monday, Sunday}. Deduce the set defined by P'.

7 The set M = {Alex, Johanna, Sarah, Vicky, Asif, Gabriella, Pedro} and the set N = {Alex, Gabriella, Frances, Raul, Luisa}.
   If the universal set $U = M \cup N$, write the following sets.
   a $M \cup N$ b $M \cap N$ c $M \cap N'$

8 The set E = {even numbers}, F = {odd numbers} and the universal set U = {positive integers}. Describe the contents of the following sets.
   a $E \cup F$ b $E \cap F$ c $E' \cap F$

3.2 Venn diagrams

Venn diagrams and simple applications.

1 a Copy the Venn diagram below.

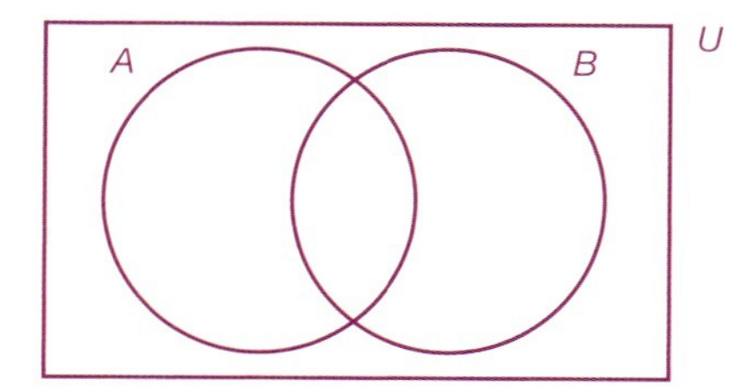

U = {Integers from 1 to 20},
A = {1, 3, 5, 7, 9, 11, 13, 15, 17, 19}
B = {2, 3, 5, 7, 11, 13, 17, 19}

b Enter the information above in the Venn diagram.

2 The Venn diagram below shows three sets of numbers.

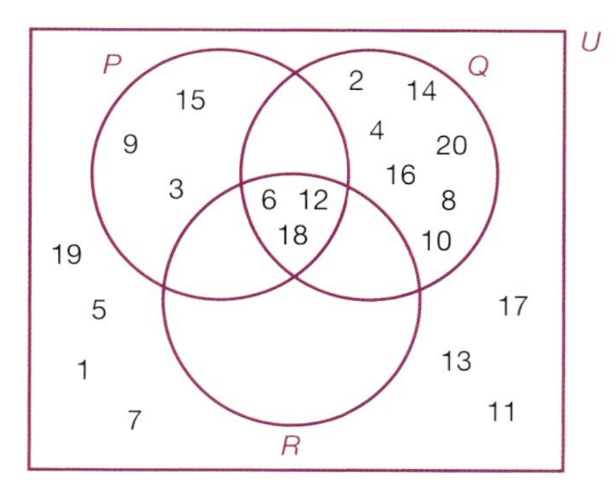

Complete the following.

a $P = \{...\}$ **b** $R = \{...\}$ **c** $P \cap Q = \{...\}$
d $P \cup Q = \{...\}$ **e** $P' \cap Q = \{...\}$

3 The Venn diagram below shows three sets of numbers. Complete the following.

a $L \cap N = \{...\}$
b $N \cup M = \{...\}$
c $L \cap M \cap N = \{...\}$
d $N' \cap L = \{...\}$
e $N' \cup L' = \{...\}$
f $M' \cup L \cap N = \{...\}$

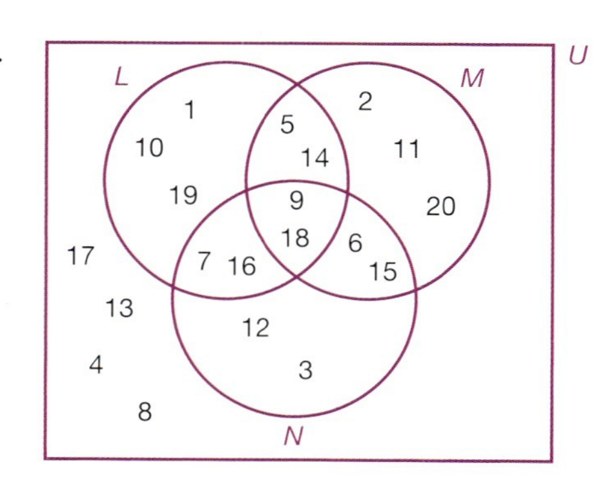

continued on the next page...

4 In the Venn diagram below, the numbers shown represent the number of members in each set. For example, $n(E) = 3$.

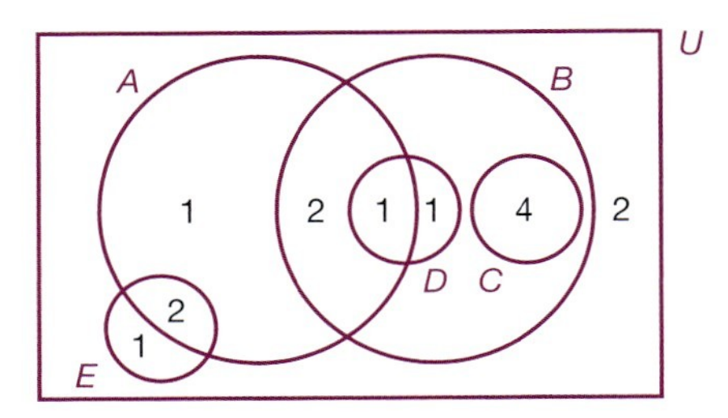

a State whether the following statements are true or false.
i $C \subset B$ **ii** $E \subset A$ **iii** $D \cap C = \varnothing$

b Give the number of members in each of the following statements.
i $n(B)$ **ii** $n(A)$ **iii** $n(A \cap B)$
iv $n(A \cup E)$ **v** $n(A \cap B \cap D)$ **vi** $n(A' \cap D)$

5 Represent the following sets in a Venn diagram.
$A = \{3, 4, 7, 8\}$ $B = \{1, 2, 4, 5, 6, 7, 9\}$ $C = \{1, 2, 6\}$

6 In a college 60% of students study mathematics and 40% study science.
75% of students study either maths or science or both.
Represent this information in a Venn diagram.

7 A language school offers three languages for its students to study: English, Spanish and Chinese. Each student is required to study at least two languages.
85% study English, 50% study Spanish and 20% study all three.
Copy and complete the following Venn diagram for the information above.

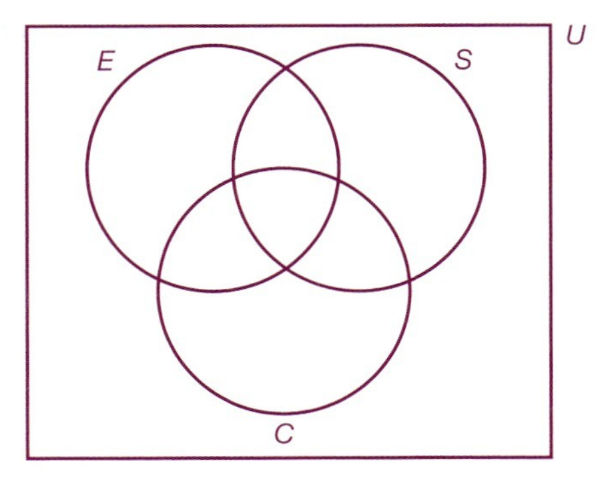

3.3 Sample space

Sample space: event, A; complementary event, A'.

1 Two ordinary dice are rolled and their scores added together.
Write down the sample space of the combined scores.

2 A packet of sweets contains sweets of three different colours: red, yellow and green.
Write down the sample space if two sweets are picked at random.

3 Two students sit a maths exam.
a What are possible complementary results?
b What is the sample space?

4 In a football match the total number of goals scored is 5.
a What are the possible complementary results?
b What is the sample space for the number of goals scored by each team?

5 A mother gives birth to triplets.
a What is the number of events for the sex of the babies?
b What is the sample space for the sex of the babies?

6 A pack of cards is shuffled thoroughly. Assuming that, for a particular game, three cards are picked at random and only their colour is important:
a what are the complementary event
b what is the sample space for the colour of the three cards?

3.4 and 3.5 Logic and Sets and logical reasoning

Basic concepts of symbolic logic: definition of a proposition; symbolic notation of propositions.
Compound statements: implication, $\Rightarrow$; equivalence, $\Leftrightarrow$; negation, $\neg$; conjunction, $\wedge$; disjunction, $\vee$; exclusive disjunction, $\veebar$.
Translation between verbal statements, symbolic form and Venn diagrams.
Knowledge and use of 'exclusive disjunction' and the distinction between it and 'disjunction'.

1 State whether the following are propositions.
For each proposition state whether it is true, false or indeterminate.
  a Five squared is twenty-five.
  b Linear equations can include values of x^2.
  c y equals plus nine.
  d No dogs can talk.
  e It is snowing today.
  f How many students are there in your class?

2 Write the following compound propositions using the symbols for conjunction (and), disjunction (or, or both) and exclusive conjunction (or but not both).
p: Anna has a brother.
q: Petra has a sister.

  a Anna has a brother and Petra has a sister.
  b Anna has a brother or Petra has a sister or both are true.
  c Anna has a brother or Petra has a sister but not both are true.

3.6 and 3.7 Truth tables and Implication; converse; inverse; contrapositive and logical equivalence

Truth tables: the use of truth tables to provide proofs for the properties of connectives; concepts of logical contradiction and tautology.
Definition of implication: converse; inverse; contrapositive.
Logical equivalence.

1 Draw a truth table for the three propositions p, q and r.
Compare it with the sample space for the result of tossing three coins.

2 Why is $p \wedge q$ and $p \veebar q$ a contradiction?

3 A truth table for the propositions p and q is given below.
Copy and complete the table.

p	q	$\neg p$	$p \vee q$	$p \wedge q$	$p \veebar q$
T	T				

4 Construct a truth table to show that $\neg(p \vee q)$ is logically equivalent to $(\neg p) \wedge (\neg q)$.

5 For the following propositions,
i rewrite the propositions using 'if … then'
ii state the converse, inverse and contrapositive of the propositions
iii state whether the propositions are true or false.
a An odd number is divisible by two.
b An octagon has eight sides.
c An icosahedron has twelve faces.
d Congruent triangles are also similar.

6 a Draw a truth table for $(p \Rightarrow q) \veebar (q \Rightarrow p)$.
b Comment on the meaning of $(p \Rightarrow q) \veebar (q \Rightarrow p)$ by referring to your table.

7 a Illustrate the proposition $(p \Rightarrow q) \veebar (q \Rightarrow p)$ on a Venn diagram by shading the correct region(s).
b Describe the shaded region(s) using set notation.

3.8 and 3.9 Probability and Combined events

Equally likely events.
Probability of an event A given by P(A) given by $P(A) = \frac{n(A)}{n(U)}$.
Probability of a complementary event, P(A′) = 1 − P(A).
Venn diagrams; tree diagrams; tables of outcomes. Solution of problems using 'with replacement' and 'without replacement'.

1 There are 10 red, 6 blue and 8 green sweets in a packet.
 a If a sweet is picked at random, calculate the probability that it is:
  i red ii red or blue.
 b If the first sweet taken from the packet is blue and not put back, calculate the probability that the second sweet is:
  i red ii blue or green.

2 A four-sided dice (numbered 1 to 4) and a six-sided dice (numbered 1 to 6) are rolled and their scores added together.
 a Copy and complete the two-way table below showing all the possible outcomes.

		Six-sided dice					
		1	2	3	4	5	6
Four-sided dice	1					6	
	2						
	3		5				
	4						

 b Calculate the probability of getting a total greater than 8.
 c Calculate the probability of getting a total score of 6.

3 A hexagonal spinner is split into sixths as shown.

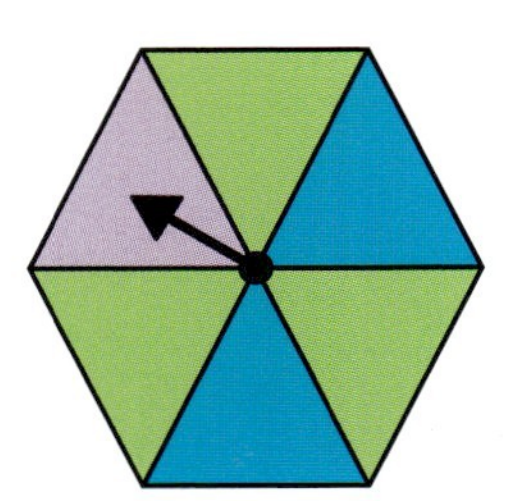

The spinner is spun twice.
 a Draw a tree diagram to show all the possible outcomes.
 b Write the probability of each outcome on each branch.
 c Calculate the probability that the spinner lands on blue on both occasions.
 d Calculate the probability that the spinner lands on blue at least once out of the two spins.

4 A football team plays three matches. The team can either win, draw or lose. The results of each match are independent of each other. The probability of winning is $\frac{2}{3}$ and the probability of drawing is $\frac{1}{4}$.
  a Calculate the probability of losing.
  b Calculate the probability that the team wins all three matches.
  c Calculate the probability that the team does not lose all three matches.

5 A student buys 15 tickets for a raffle. 300 tickets are sold in total. Two tickets are drawn at random. Calculate the probability that:
  a she wins both prizes
  b she wins at least one prize.

6 A college offers three sports clubs for its students to attend after school. They are volleyball (V), basketball (B) and Football (F). The number of students attending each is shown in the Venn diagram below.

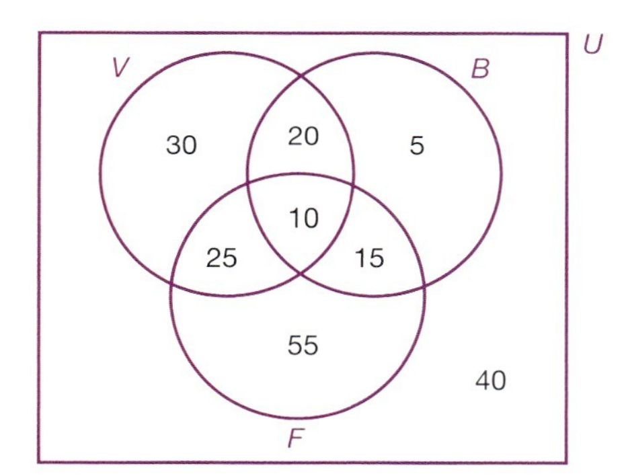

  a How many students attend none of the sports clubs?
  A student is picked at random. Calculate the following probabilities.
  b The probability that the student plays volleyball
  c $P(V \cap B)$ d $P(V \cap B \cap F)$ e $P(V \cup F)$ f $P(F')$

7 The following are three sets of numbers.
  A = {2, 4, 6, 8, 10, 12, 14, 16, 18, 20, 22, 24, 26, 28, 30}
  B = {3, 6, 9, 12, 15, 18, 21, 24, 27, 30}
  C = {5, 10, 15, 20, 25, 30}

  a Draw a Venn diagram showing the three sets of numbers.
  b A number is picked at random. Calculate the following probabilities.
    i $P(A)$ ii $P(B \cup C)$ iii $P(A' \cap B)$

8 In a class of 30 students, 24 study Biology, 14 study Chemistry and 1 studies neither.
  a Draw a Venn diagram to show this information.
  b A student is picked at random. Calculate the following probabilities.
    i $P(B')$ ii $P(B \cup C)$ iii $P(B \cap C')$

3.10 Laws of probability

Laws of probability.
Combined events: $P(A \cup B) = P(A) + P(B) - P(A \cap B)$.
Mutually exclusive events: $P(A \cup B) = P(A) + P(B)$.

1 In a 100 m sprint, the record holder has a probability of 0.85 of winning. He has a 0.08 probability of coming second.
  a What is his probability of finishing in the first two?
  b Given that he hasn't come first, what is the probability that he has come second?

2 I spin a coin and throw a dice. What is the probability of getting:
  a a tail and a multiple of 2
  b a tail or a multiple of 2
  c a tail or a multiple of 2, but not both?

3 Three friends have a birthday in the same week. Assuming that they are independent events, calculate the probability that they are all on different days.

4 Raul takes a bus followed by a train to work. On a particular day the probability of him catching the bus is 0.65 and the probability of catching the train is 0.6.
The probability of catching neither is 0.2. A represents catching the bus and B the train.
  a State $P(A \cup B)'$
  b Find $P(A \cup B)$
  c Given that $P(A \cup B) = P(A) + P(B) - P(A \cap B)$ calculate $P(A \cap B)$.
  d Calculate $P(B \mid A)$, the probability of catching the train given that he caught the bus.

5 Julie revised for a multiple choice science exam. Unfortunately she only managed to revise 60% of the facts necessary. During the exam, if there is a question on any of the topics she revised she gets the answer correct.
If there is a question on any of the topics she hasn't revised, she has a $\frac{1}{5}$ chance of getting it right.
  a A question is chosen at random. What is the probability that she got the answer correct?
  b If she got a question correct, what is the probability that it was on one of the topics she had revised?

6 Miguel has a driving test on one day and a drama exam the next. The probability of him passing the driving test is 0.82. The probability of him passing the drama exam is 0.95. The probability of failing both is 0.01. Given that he has passed the driving test, what is the probability that he passed his drama exam too?

4.3 Quadratic functions and their graphs

The graph of the quadratic function: $f(x) = ax^2 + bx + c$.
Properties of symmetry; vertex; intercepts.

1 For each of the quadratic functions below:
  i use a GDC to sketch the function
  ii write the coordinates of the points where the graph intercepts the x-axis
  iii write down the value of the y-intercept.
  a $f(x) = x^2 - 9x + 20$
  b $f(x) = x^2 - 3x - 18$
  c $f(x) = (x - 4)^2$
  d $f(x) = x^2 + 10x + 27$

2 Write the equation of the axis of symmetry of each of the following quadratic functions.
  a $y = x^2 - 2x$
  b $y = -x^2 - 4x$
  c $y = x(5 - x)$
  d $y = -x^2 + 3x - 10$

3 Give a possible equation of a quadratic function with each of the following axes of symmetry.
  a $x = 6$
  b $x = -5$

4 Factorize the following quadratic functions.
  a $f(x) = x^2 + 11x + 30$
  b $f(x) = x^2 + 4x - 12$
  c $f(x) = -x^2 + 8x - 15$
  d $f(x) = x^2 - 36$

5 Solve the following quadratic equations by factorizing.
  a $x^2 - 3x - 4 = 0$
  b $x^2 - 2x - 24 = 0$
  c $-x^2 + 10x - 16 = 0$
  d $x^2 = 11x - 28$

6 The following quadratic equations are of the form $ax^2 + bx + c = 0$.

  Solve them by using the quadratic formula $x = \dfrac{-b \pm \sqrt{b^2 - 4ac}}{2a}$.

  a $x^2 + 5x - 25 = 0$
  b $x^2 + 9x - 24 = 0$
  c $4x^2 + 8x + 3 = 0$
  d $-x^2 + 9x - 15 = 0$

7 For each of the following:
  i form an equation in x
  ii solve the equation to find the possible value(s) of x.

  a

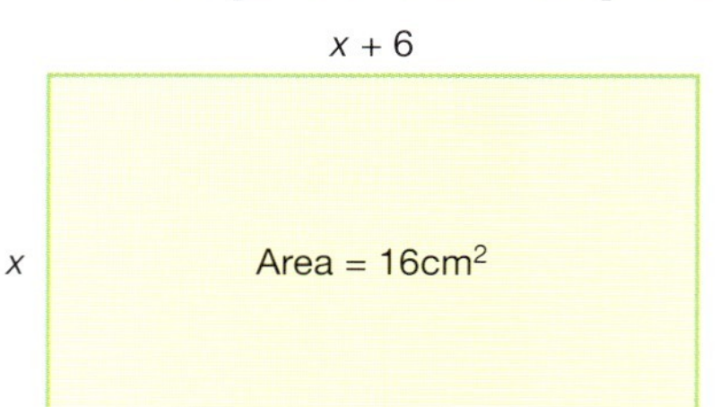

  b

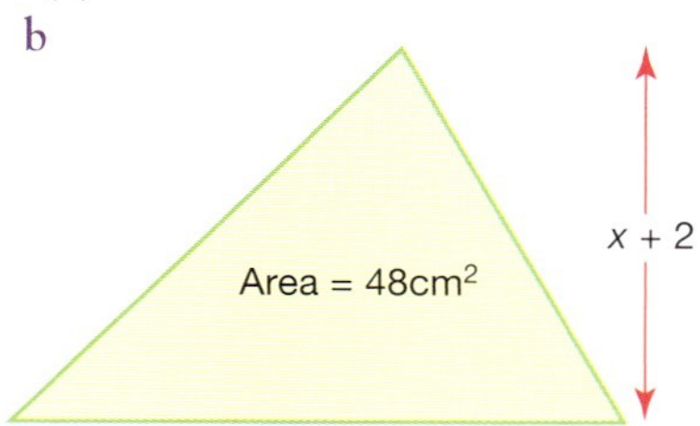

4.4 Exponential functions and their graphs

The exponential expression: a^b; $b \in \mathbb{Q}$
Graphs and properties of exponential functions.
$f(x) = a^x$; $f(x) = a^{\lambda x}$; $f(x) = ka^{\lambda x} + c$; $k, a, c, \lambda \in \mathbb{Q}$.
Growth and decay; basic concepts of asymptotic behaviour.

1 i Plot the following exponential functions.
ii State the equation of any asymptotes.
a $f(x) = 2^x + 1$ b $f(x) = -2^x + 2$ c $f(x) = 3^x - 3$

2 A tap is dripping at a constant rate into a container. The level (l cm) of water in the container is given by the equation $l = 3t + 5$, where t is the time in hours.
a Calculate the level of water in the container at the start.
b Calculate the level of water in the container after 4 hours.
c Calculate the time taken for the level of the water to reach 248 cm.
d Plot a graph to show the level of water over the first 6 hours.
e Use your graph to estimate the time taken for the water to reach a level of 1 m.

3 a Plot a graph of $y = 5^x$ for values of x between -1 and 3.
b Use your graph to find approximate solutions to the following equations.
i $5^x = 100$ ii $5^x = 50$

4 The graph below shows a graph of the function $f(x) = 2^x$.

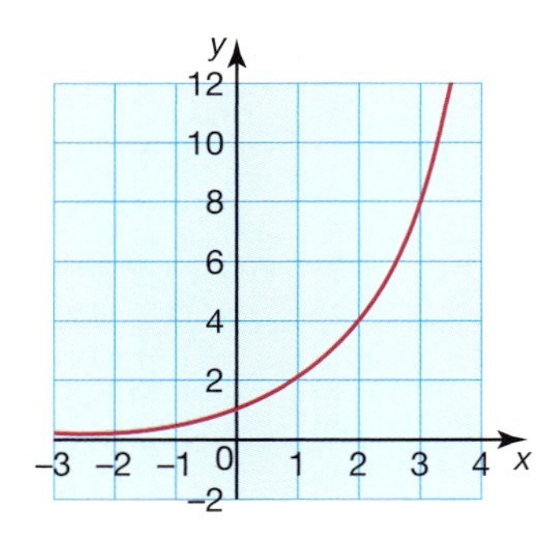

a Copy the graph and on the same axes sketch the graph of $f(x) = 2^x + 3$. Label it clearly.
b On the same axes sketch the graph of $f(x) = -2^x + 6$. Label it clearly.

5 The half-life of plutonium 239 is 24 000 years. How long will 2 g of plutonium 239 take to decay to 62.5 mg?

4.5 Trigonometric functions and their graphs

Graphs and properties of the sine and cosine functions:
$f(x) = a\sin bx + c$; $f(x) = a\cos bx + c$; $a, b, c \in \mathbb{Q}$.
Amplitude and period.

1 Sketch the graph of the trigonometric function $f(x) = \sin x$ for the values of x between 0° and 360°.

2 Copy and complete the following sentences.
 a The sine curve has a period of .... and has an amplitude of ....
 b The cosine curve has a period of .... and oscillates between ... and ... .

3 The graph of the function $f(x) = \cos x$ is shown below.

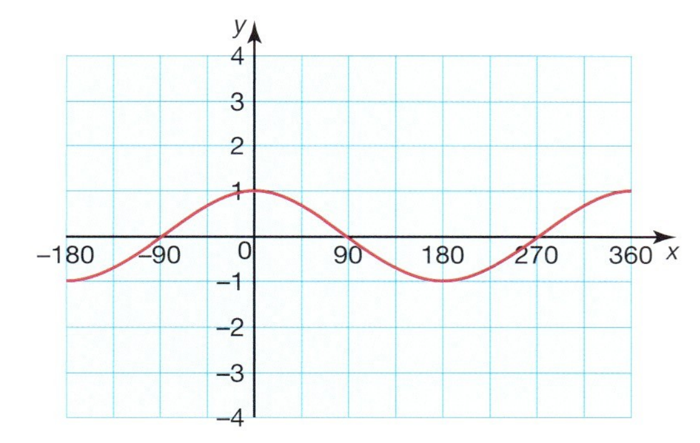

 a Copy the graph and on the same axes sketch the graph of $f(x) = \cos x + 2$. Label it clearly.
 b Sketch the graph of $f(x) = 3\cos x$ on the same axes and label it clearly.

4 The graph of $f(x) = \cos x$ is shown below. Point A(180, −1) is marked on it.

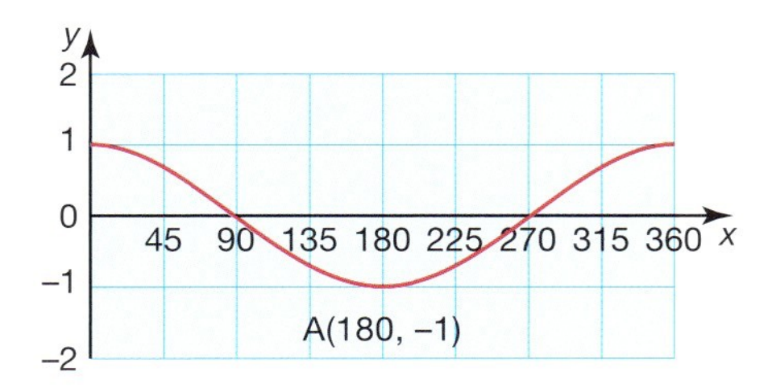

Give the coordinates of point A after each of the following transformations.
 a $f(x) = 2\cos x$
 b $f(x) = \cos x + 3$
 c $f(x) = \cos\frac{1}{2}x$
 d $f(x) = -\cos x - 1$

5 Using a GDC, solve the following equations graphically, finding all the solutions in the range $-180° \leq x \leq 180°$.
 a $\sin x = 0.4$
 b $-\sin x = 0.5$
 c $3\cos x - 1 = -2.5$
 d $\cos x = \frac{1}{2} + \sin x$

5.2 Straight lines

Equation of a line in two dimensions: the forms $y = mx + c$ and $ax + by + d = 0$.
Gradient; intercepts.
Points of intersection of lines; parallel lines; perpendicular lines.

1 Calculate the gradient of the line passing through the points with the following coordinates.
a (3, 5) and (5, 13)
b (6, 1) and (10, −9)

2 The gradients (m) of straight lines are given. Calculate the gradient of a line perpendicular to each of the ones stated.
a $m = 4$
b $m = -6$
c $m = \frac{2}{3}$
d $m = -1\frac{3}{4}$

3 Find the length of the following line segments.

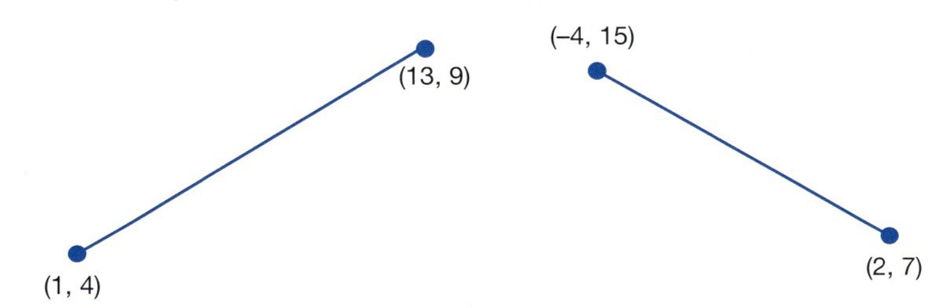

4 Calculate the coordinates of the midpoint of the line segments in question 3 above.

5 The following tables give the x- and y-coordinates of several points on a line. Deduce from the coordinates the equation of the line in the form $y = mx + c$.

a

x	y
−2	−7
−1	−4
0	−1
1	2
2	5

b

x	y
−2	0
−1	0.5
0	1
1	1.5
2	2

6 Deduce the gradient and y-intercept of the following straight lines from their equations.
a $y = x - 2$
b $y = -3x + 1$
c $2y - 4x + 6 = 0$
d $3y + 5x - 12 = 0$

7 Calculate the equation of the straight line passing through the following pairs of points. Give your answers in the form

i $y = mx + c$

ii $ax + by + d = 0$

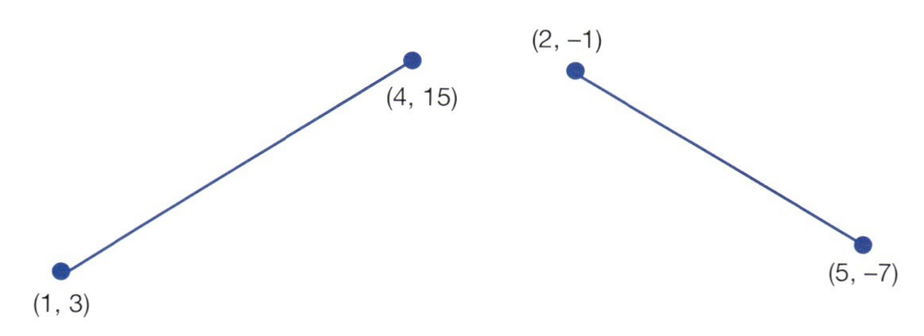

8 Plot the following straight lines. Indicate clearly where they intercept both axes.

a $y = 2x - 4$

b $2x + 2y - 1 = 0$

9 Solve the following simultaneous equations either algebraically or graphically.

a $y - 2x = 1$
$5y + 2x = 17$

b $y = 5 - 2x$
$x + 3y = 5$

10 The cost of buying three cups of tea (t) and one cake (c) is £4.40. The cost of buying two cups of tea and three cakes is £4.80.

a Construct two equations from the information given.

b Solve the two equations simultaneously and work out the cost of:

i one cup of tea

ii one cake.

5.3 Right-angled trigonometry

Right-angled trigonometry.
Use of the ratios of sine, cosine and tangent.

1 Copy the right-angled triangle below. The angle x is shown.
In relation to the angle x label the sides 'opposite', 'adjacent' and 'hypotenuse'.

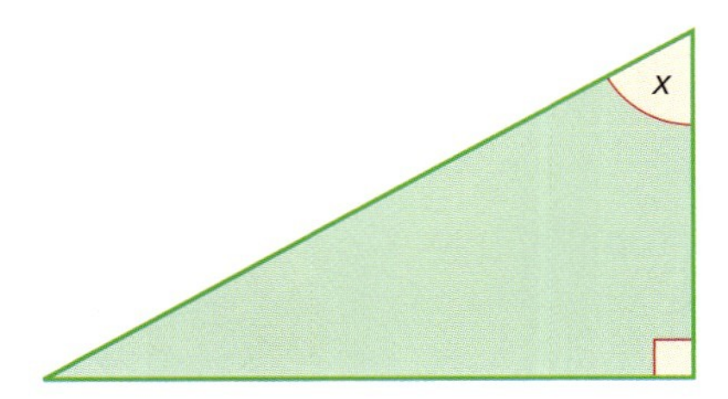

2 Calculate the size of the angle marked x in each of the following right-angled triangles.

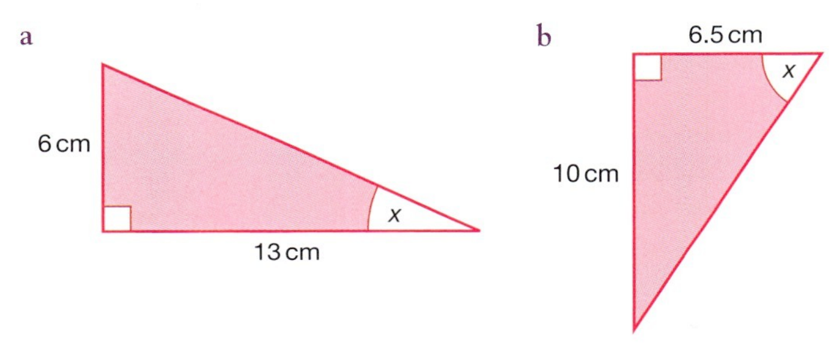

3 Calculate the length of the side marked a in each of the following right-angled triangles.

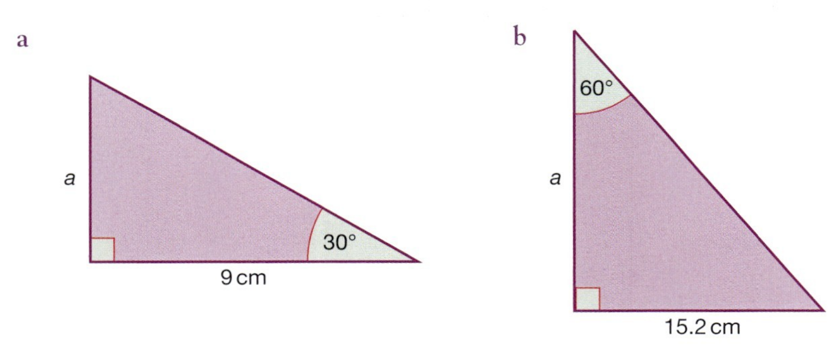

4 Calculate the length of the sides marked a and the size of the angles marked x in each of the following diagrams.

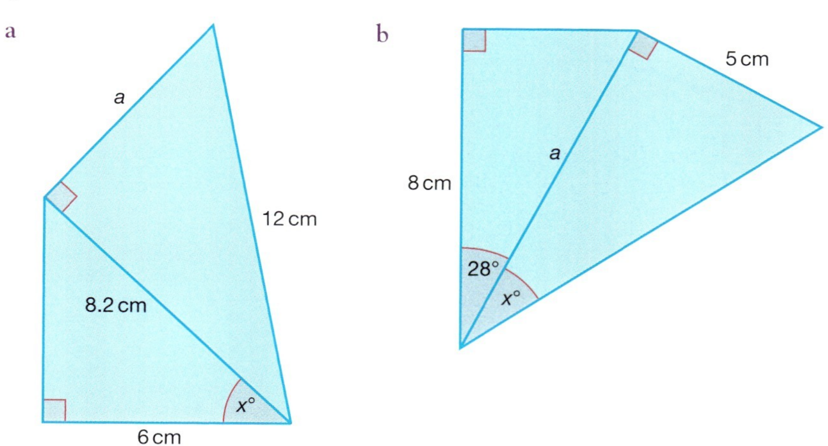

5 Towns A, B and C are situated as shown. B is due North of A, whilst C is due East of A.

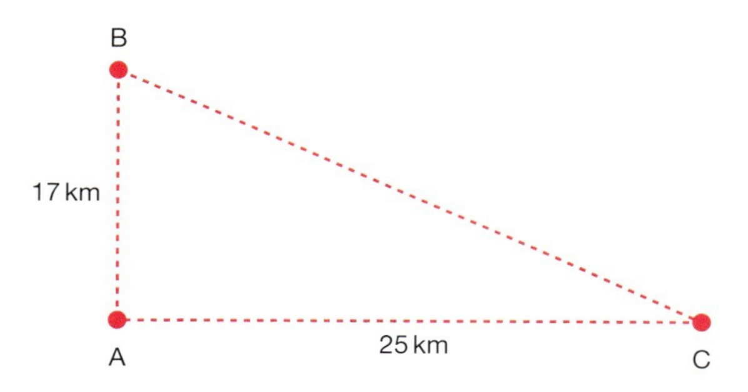

a Calculate the distance BC, giving your answer to three significant figures.
b Calculate the bearing of C from B.

continued on the next page…

6 Two towers X and Y are 120 m apart. The height of tower X is 58 m and the height of tower Y is 72 m. A tightrope walker attaches a rope from X to Y as shown.

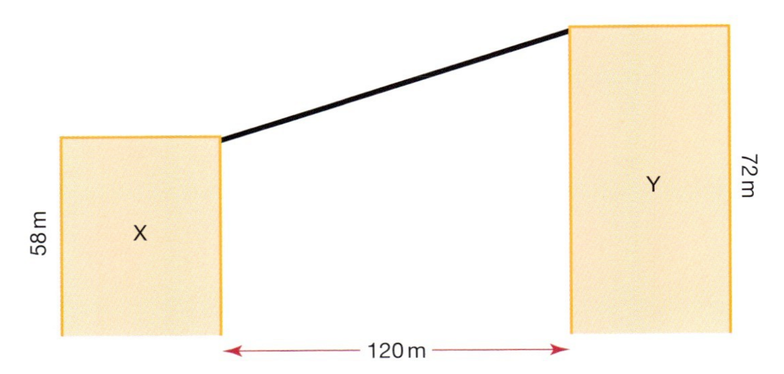

a Calculate the angle of elevation of the tightrope.
b Calculate the distance walked by the tightrope walker getting from X to Y.

7 Three villages X, Y and Z are shown in the diagram below. X is Northwest of Y and 20 km away. Z is Northeast of Y. The distance XZ is 56 km.
Calculate the distance between villages Y and Z.

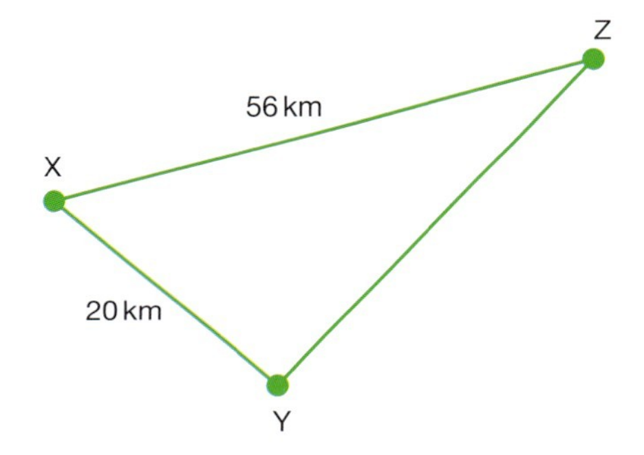

5.4 Trigonometry and non-right-angled triangles

The sine rule: $\frac{a}{\sin A} = \frac{b}{\sin B} = \frac{c}{\sin C}$.

The cosine rule: $a^2 = b^2 + c^2 - 2bc\cos A$; $\cos A = \frac{b^2 + c^2 - a^2}{2bc}$.

Area of a triangle $= \frac{1}{2}ab\sin C$.

Construction of labelled diagrams from verbal statements.

1 Give all the possible solutions for the angle x, where $0 \leq x \leq 180°$ for each of the following.
 a $\sin x = 0.65$ b $\sin x = 0.25$ c $\sin x = 1$ d $\sin x = 0$

2 Give all the possible solutions for the angle θ, where $0 \leq \theta \leq 360°$ for each of the following.
 a $\cos\theta = 0.5$ b $\cos\theta = -0.4$ c $\cos\theta = -1$

3 Use the sine rule to find the length of the side marked x below.

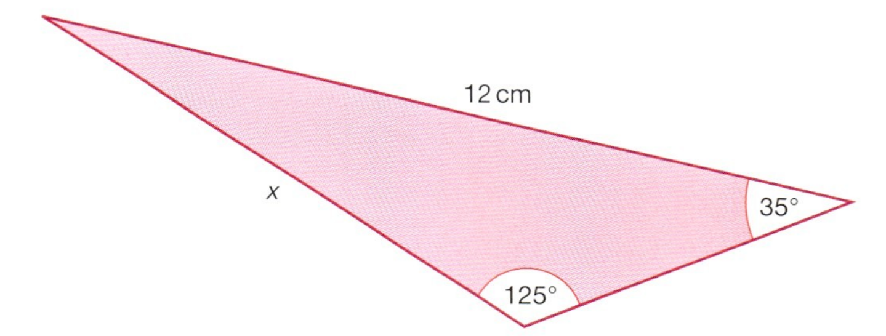

4 Find the size of angle A in the triangle below.

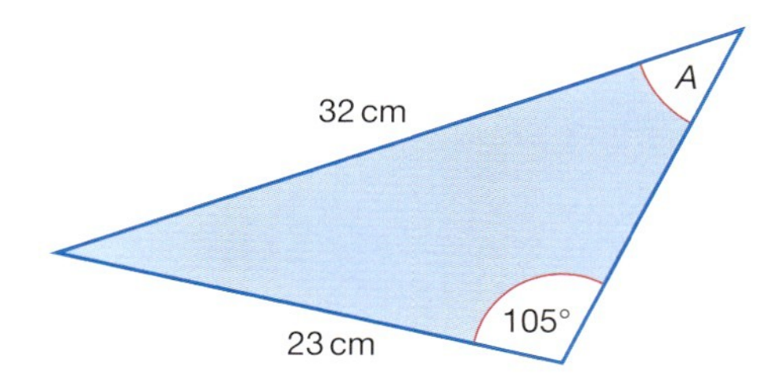

continued on the next page…

5 Use the cosine rule to answer the following questions.

a Find the length of the side labelled a.

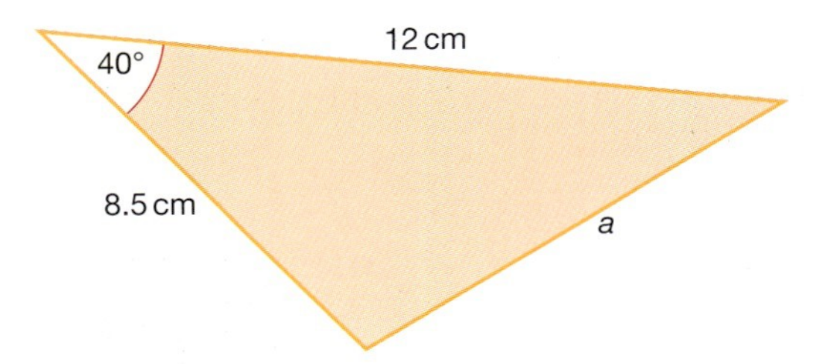

b Calculate the size of angle a.

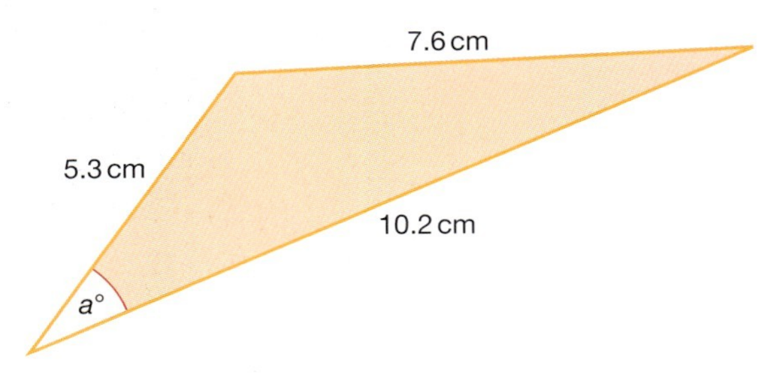

6 Calculate the area of the following triangle.

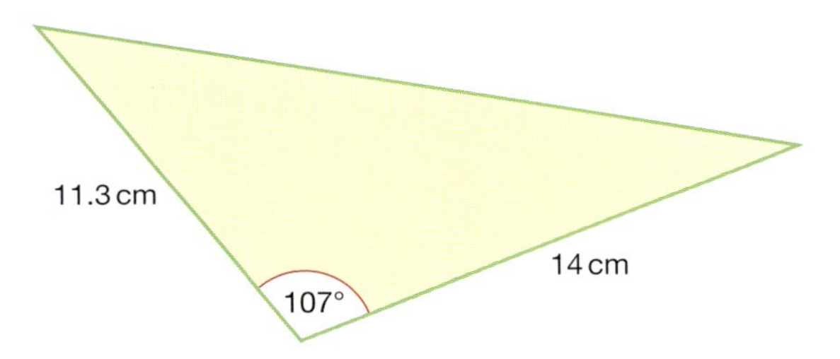

7 Calculate the size of angle C in the triangle below.

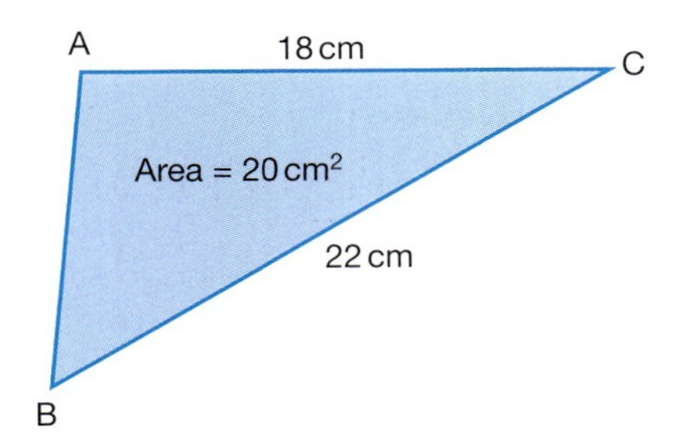

8 a Construct the following triangle with a ruler and a pair of compasses.
Triangle PQR with PQ = 6 cm, PR = 7 cm and QR = 9 cm

b Calculate the size of angle PRQ.

5.5 Geometry of three-dimensional shapes

Geometry of three-dimensional shapes: cuboid; prism; pyramid; cylinder; sphere; hemisphere; cone. Lengths of lines joining vertices with vertices, vertices with midpoints and midpoints with midpoints; sizes of angles between two lines and between lines and planes.

1 The cube in the diagram has edge lengths of 10 cm as shown. Calculate:

a the length BD

b the length of the body diagonal BH

c the size of angle DBH.

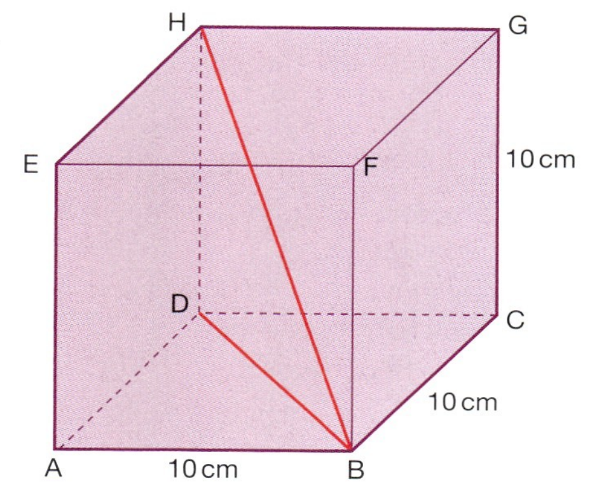

2 The diagram shows a square-based pyramid. The centre of the base, X, is directly beneath the pyramid's apex E. M is the midpoint of CB. Calculate:

a the length CX

b the height of the pyramid EX

c the distance XM

d the angle the face BCE makes with the base of the pyramid.

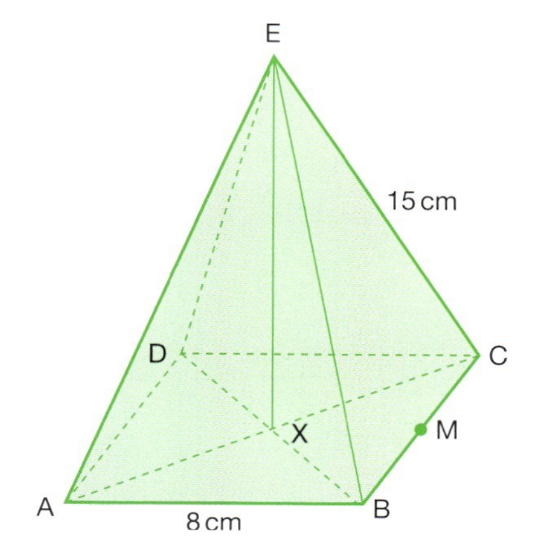

3 Calculate the total surface area of the shapes below.

a

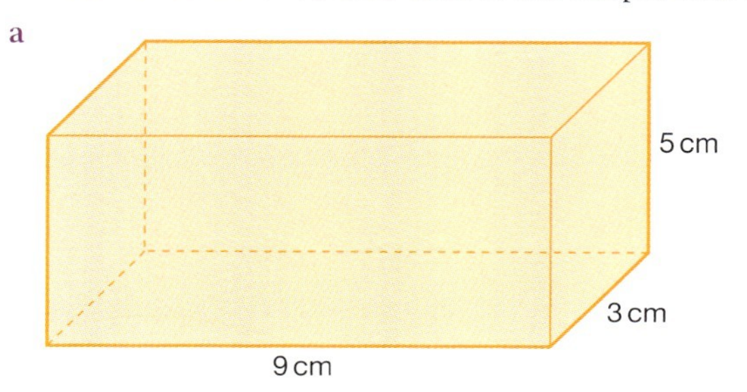

b

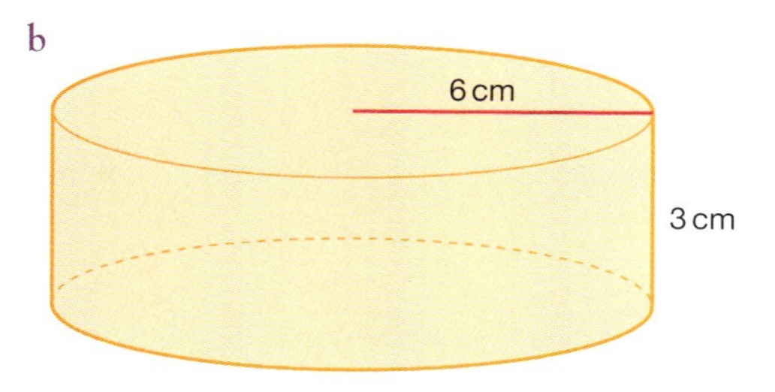

continued on the next page…

4 The diagram shows a triangular prism. Its cross-section is a right-angled isosceles triangle.
Calculate:
a the length of the edge OJ
b the total surface area of the prism
c the volume of the prism
d the angle OL makes with the base JKLM.

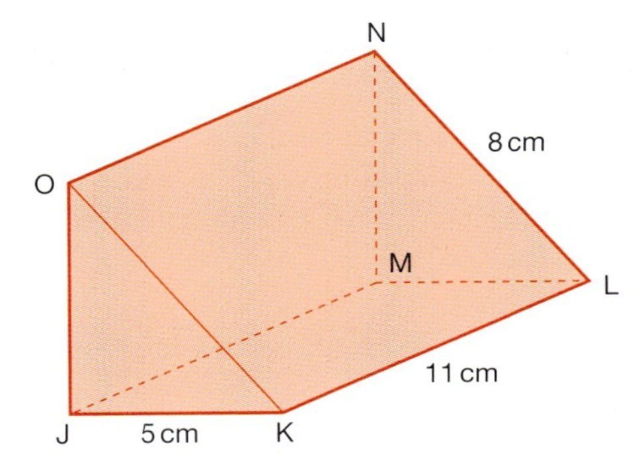

5 The sector shown has a radius of 14 cm and an angle at its centre of 260°.
Calculate:
a the length of the arc
b the area of the sector.

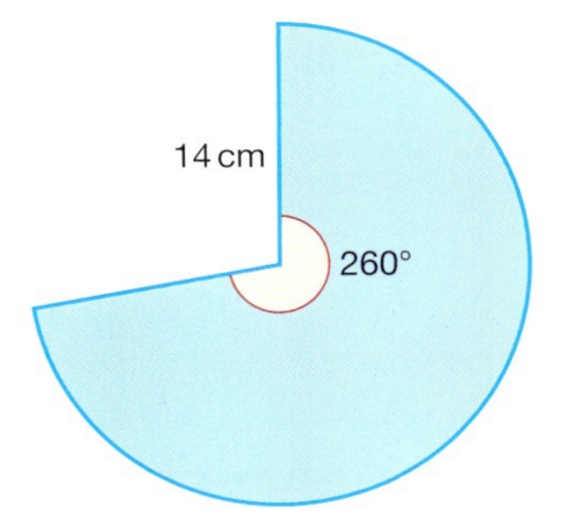

6 The sphere and cube below have the same volume.

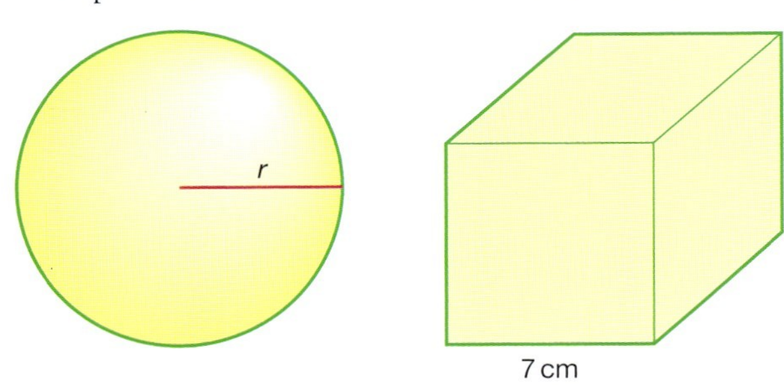

Calculate the radius, r, of the sphere.

7 Two points A and B are directly opposite each other on the surface of a cylinder as shown.

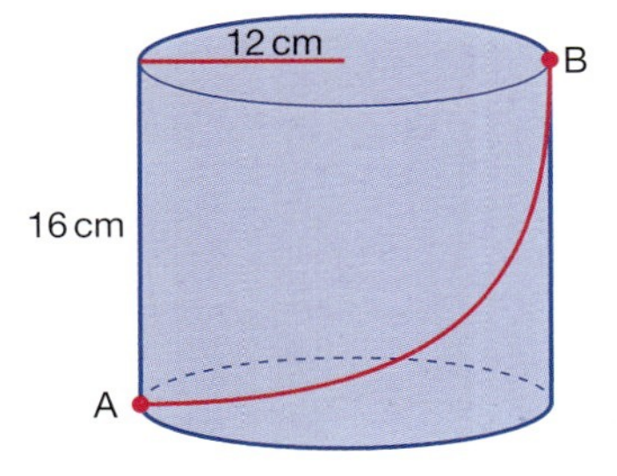

a Calculate the circumference of the top of the cylinder.
b A line is drawn on the surface of the cylinder. It represents the shortest distance from A to B on the cylinder's surface. Calculate the length of this line from A to B.

6.3 and 6.5 Grouped discrete or continuous data and Measures of central tendency

Grouped discrete or continuous data: frequency tables; mid-interval values; upper and lower boundaries.
Frequency histograms.
Stem and leaf diagrams (stem plots).
Measures of central tendency.
For simple discrete data: mean; median; mode.
For grouped discrete and continuous data: approximate mean; modal group; 50th percentile.

1 The amount of milk (in litres) drunk by a group of students in a week is given in the table below.

Number of litres	0	1	2	3	4	5	6	7	8
Frequency	6	1	4	9	22	16	2	4	1

a Draw a frequency histogram for this data.
b State the modal value.
c Calculate the mean number of litres drunk per student.
d Calculate the median number of litres drunk per student.

2 The masses M kg of suitcases being checked-in for a flight at an airport are recorded. The results are shown below.

Mass (kg)	$0 \le M < 5$	$5 \le M < 10$	$10 \le M < 15$	$15 \le M < 20$	$20 \le M < 25$	$25 \le M < 30$
Frequency	6	18	64	105	94	18

a State the modal group.
b Estimate the mean mass of the suitcases.
c State, giving reasons, which group the median mass belongs to.

3 An ornithologist records the wingspan (in centimetres) of a particular type of bird. He measures 20 adult birds. The results are listed below.

15.1 15.8 14.6 14.7 15.7 15.6 15.6 15.7 15.6 15.2
14.1 14.9 15.8 15.7 16.1 16.2 14.4 14.8 15.5 15.4

a Draw a stem and leaf diagram of the results.
b Using the stem and leaf diagram, state the median wingspan.

4 The mass M kg of football players in a team is recorded.
For the team of 11 players, $\Sigma M = 836$ kg.
a Calculate the mean mass, $\overline{M}$, of the 11 players.
b The mean of 11 players and 1 substitute is 76.75 kg. Calculate the mass of the substitute.

continued on the next page…

5 The cost (C euros) of a litre of unleaded petrol at different petrol stations is shown in the frequency histogram below.

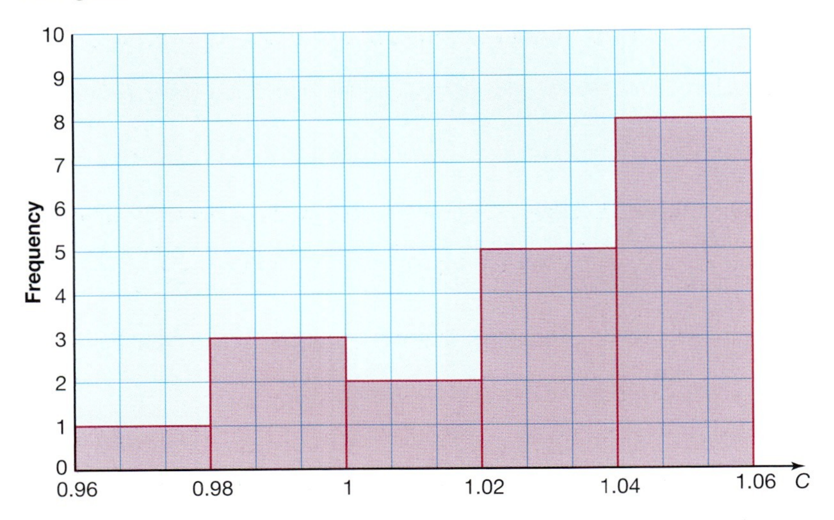

a How many petrol stations were surveyed?
b Estimate the mean price of unleaded petrol.

6.4 Cumulative frequency

Cumulative frequency tables for grouped discrete data and for grouped continuous data; cumulative frequency curves.
Box and whisker plots (box plots).
Percentiles; quartiles.

1 The percentage test result (x) for 20 students in a class are given in the grouped frequency table below.

Percentage	$0 \le x < 20$	$20 \le x < 40$	$40 \le x < 60$	$60 \le x < 80$	$80 \le x < 100$
Frequency	2	4	6	7	1

a Calculate the cumulative frequency for the data.
b Draw a cumulative frequency curve for the results.
c From your graph estimate the median test result.

2 A business surveys its employees to find how long it takes them to travel to work each morning (t minutes). The results are displayed below.

Time (min)	$0 \le t < 15$	$15 \le t < 30$	$30 \le t < 45$	$45 \le t < 60$	$60 \le t < 75$
Frequency	1	3	12	22	12

a Calculate the cumulative frequency for the data.
b Draw a cumulative frequency curve for the results.
c From your graph estimate the median time taken to travel to work.
d From your graph estimate the time taken to travel to work by the middle 50% of the employees.
e Draw a box plot to summarize the data.

3 150 students enter an international maths competition. The scores are out of a maximum 300 points. The scores (x) for the 150 students are summarized in the table below.

Score	$0 \le x < 50$	$50 \le x < 100$	$100 \le x < 150$	$150 \le x < 200$	$200 \le x < 250$	$250 \le x < 300$
Frequency	7	25	56	32	21	9

a Draw a cumulative frequency curve of the scores.
b From your graph estimate the median score.
c Students in the top 20% are invited to take part in the next round of the competition. From your graph estimate the score needed for a student to be in the top 20%.

continued on the next page…

4 The mass (m grams) of forty apples of the same variety was recorded at a market and at a supermarket. The results are given in the table below.

Mass (g)	$60 \le m < 100$	$100 \le m < 140$	$140 \le m < 180$	$180 \le m < 220$	$220 \le m < 260$
Frequency (market)	5	8	12	9	6
Frequency (supermarket)	0	3	28	9	0

a On the same axes draw two cumulative frequency graphs, one for the mass of the apples at the market and the other for the mass of apples at the supermarket.

b Using your graph complete the table below.

	Minimum value	**Lower quartile**	**Median**	**Upper quartile**	**Maximum value**
Market					
Supermarket					

c Use your results table above to draw a box and whisker plot for each set of apple data.

6.6 Measures of dispersion

Measures of dispersion: range; interquartile range; standard deviation.

1 For the following lists of numbers calculate:
  i the range   ii the interquartile range.
  a 2 6 12 14 15 15 17 21 22 22
  b 26 27 1 14 18 7 19 3 12

2 The number of goals conceded by a football team is recorded. The results are given in the frequency table below.

Number of goals conceded	0	1	2	3	4	5
Frequency	7	6	2	5	2	1

Calculate:
  a the range in the number of goals scored
  b the interquartile range of the number of goals scored.

3 Calculate the standard deviation using the formula $s_n = \sqrt{\frac{\sum x^2}{n} - \bar{x}^2}$ of the following lists of numbers.
  a 3 4 7 7 7 8 8 8 8 9
  b 1 4 6 8 10 12 15 18 24 30

4 The temperatures (in °C) at two holiday resorts are recorded every other day during the month of June. The results are given in the table below.

Day	1	3	5	7	9	11	13	15	17	19	21	23	25	27	29
Temperature Resort A	23	24	22	24	25	26	25	23	23	24	22	24	24	26	27
Temperature Resort B	15	17	24	28	33	33	26	22	22	19	16	16	15	26	31

  a Calculate the mean temperature for resort A and B.
  b Calculate the range of temperatures at both resorts.
  c Calculate the interquartile range of temperatures at both resorts.
  d Calculate the standard deviation of the temperature at both resorts.
  e Explain the meaning of your answer to part **d**.

5 The times (in seconds) taken by two sprinters to run 100 m during their training sessions are recorded and given below.

Sprinter A 11.2 10.9 11.0 10.8 10.9 11.0 11.1 11.1 10.9
Sprinter B 10.2 9.9 10.1 11.8 11.2 10.1 10.1 10.3 10.4

  a Calculate the mean sprint time for each runner.
  b Calculate the standard deviation of the sprint times for each runner.
  c Which runner is faster? Justify your answer.
  d Which runner is more consistent? Justify your answer.

continued on the next page…

6 Two maths classes sit the same maths exam. One of the classes has students of similar mathematical ability, the other has students of different abilities. A summary of their percentage scores is presented below.

	Mean	Standard deviation
Class A	65	8
Class B	50	2

From the results table deduce which class is likely to have students of similar mathematical ability. Give reasons for your answer, which refer to the table above.

6.7 and 6.8 Scatter diagrams, bivariate data and linear correlation and The regression line for *y* on *x*

Scatter diagrams; line of best fit, by eye, passing through the mean point.
Bivariate data: the concept of correlation.

Pearson's product–moment correlation coefficient: use of the formula $r = \frac{s_{xy}}{s_x s_y}$.

Interpretation of positive, zero and negative correlations.

The regression line for y on x: use of the formula $y - \bar{y} = \frac{s_{xy}}{(s_x)^2}(x - \bar{x})$
Use of the regression line for prediction purposes.

1 Decide on the possible correlation (if any) between the following variables. Justify your answers.
  a The height of a child and the child's mass
  b A student's result in a maths exam and the same student's result in a science exam
  c A student's result in a maths exam and the same student's result in an art exam
  d The outside temperature and the number of umbrellas sold
  e The number of cigarettes a woman smokes during pregnancy and the mass of her baby
  f The number of people living in a household and the amount of water which the household uses
  g A person's height and their intelligence
  h The number of DVDs sold and the attendance at cinemas

2 Describe the correlation, if any, depicted in the following scatter diagrams.

a

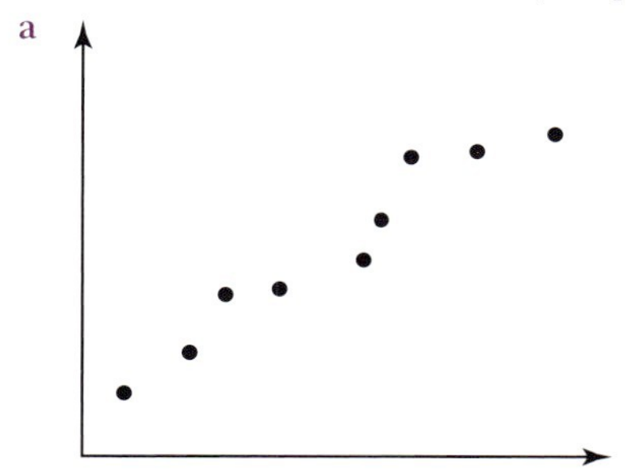

b

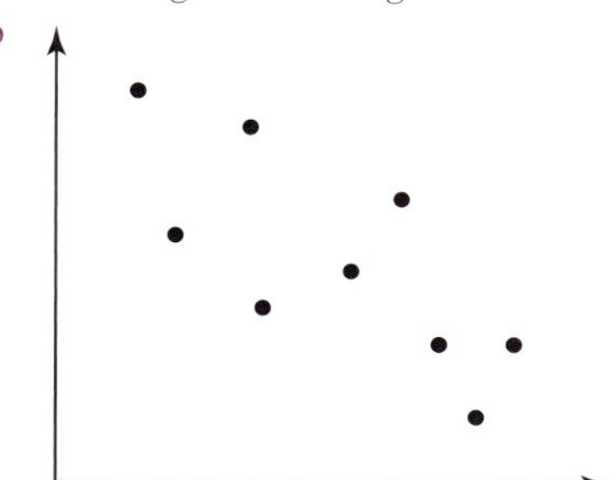

3 A farmer wishes to find a way of increasing the milk yield of his herd of cows. He decides to mix a special feed with the ordinary feed to see whether it has any effect on the yield. The results are shown in the table below.

Special feed (%)	0	2	4	6	8	10	12	14
Milk yield (litres)	2050	2100	2180	2230	2300	2360	2390	2470
Special feed (%)	16	18	20	22	24	26	28	30
Milk yield (litres)	2540	2600	2650	2720	2800	2830	2850	2860

  a Plot a scatter graph of the results, with the yield on the y-axis.
  b Calculate the mean special feed % ($\bar{x}$) and the mean milk yield ($\bar{y}$).
  c Plot the point ($\bar{x}$, $\bar{y}$) on the graph and label it clearly.

continued on the next page…

d Assuming the relationship between the two variables is linear, draw a line of best fit passing through the point $(\bar{x}, \bar{y})$.
e Use your line of best fit to predict the yield the farmer would get if the percentage of special feed was 15%.
f Calculate the equation of the line of best fit in the form $y = mx + c$.
g Use your equation of the line of best fit to extrapolate the results and predict the milk yield if the percentage of special feed was 100%.
h Do you think your answer to part **h** is valid? Justify your answer.

4 A supermarket counts the average number of people entering the store during a period of time and the amount of money collected at the tills. The results are displayed below.

Average number of people in store	Amount collected at tills (£)
72	3006
51	2021
12	812
108	3102
156	4671
92	4092
26	1125
48	1995
52	1991
61	2082
17	742
5	306
88	4128
16	738

a Using your GDC, calculate Pearson's product–moment correlation coefficient.
b Explain the meaning of your answer to part **a**.

5 A lorry driver starts a journey with 1000 litres of fuel in his fuel tank. During his journey he records the amount of fuel left in the tank and the number of kilometres he has travelled. The results are displayed below.

Distance travelled (km)	0	50	150	300	700	1000	2000	4000
Fuel in tank (litres)	1000	985	970	930	850	785	575	215

a Using your GDC, calculate the equation of the y on x regression line, giving it in the form $y = mx + c$.
b Explain the meaning of the value of m in the context of this problem
c Use your equation to predict how much fuel he has left in the tank if he has travelled 3000 km.
d Use your equation to estimate how far he can travel on 1000 litres of fuel.

6.9 The χ^2 test for independence

The χ^2 test for independence: formulation of null and alternative hypotheses; significance levels; contingency tables; expected frequencies; use of the formula $\chi^2 = \sum \frac{(f_o - f_e)^2}{f_e}$; degrees of freedom; use of tables for critical values; p-values.

1 The numbers of male and female students getting either a distinction, pass or fail for a course are given in the contingency table below.

	Distinction	Pass	Fail	Total
Male	18	84	28	130
Female	20	132	18	170
Total	38	216	46	300

a State the number of degrees of freedom of the table.
b Assuming the results are independent, construct a contingency table to show the expected frequencies.
c It is thought that the likelihood of failing is dependent on the gender of the student.
  i State the null hypothesis (H_0).
  ii State the alternative hypothesis (H_1).
d Using the formula $\chi^2 = \sum \frac{(f_o - f_e)^2}{f_e}$ for testing independence, calculate the value of χ^2.
e Decide whether failing is dependent on gender at the 5% level of significance. Justify your answer clearly.

2 Researchers wish to know if students prefer a different type of film genre according to their gender. The results of their survey are given in the contingency table below.

	Romance	Horror	Action	Comedy	Total
Male	3	14	15	18	50
Female	25	15	10	30	80
Total	28	29	25	48	130

a State the number of degrees of freedom of the table.
b Assuming the results are independent, construct a contingency table to show the expected frequencies.
c It is thought that there is a difference in the viewing preferences of males and females.
  i State the null hypothesis (H_0) .
  ii State the alternative hypothesis (H_1).
d Calculate the value of χ^2.
e State whether viewing preferences are dependent on gender at a significance level of:
  i 10% ii 5% iii 1%.

continued on the next page…

3 A holiday resort wishes to survey its customers as to their level of satisfaction with the resort. Customers were asked to rate the resort either as 'excellent', 'good', 'satisfactory' or 'poor'. In particular they wish to see whether the level of satisfaction is dependent on age. The results are displayed below.

	Excellent	Good	Satisfactory	Poor
Under 16	10	21	23	6
16–25	6	11	12	4
26–55	7	25	12	20
Over 55	8	38	40	19

a State the null hypothesis.
b Calculate the value of χ^2.
c State whether the null hypothesis is rejected at a significance level of:
i 1% ii 5% iii 10%.

7.2 and 7.3 Differentiation and The gradient of a curve at a given point

The principle that $f(x) = ax^n$, $f'(x) = anx^{n-1}$ and $f''(x) = an(n-1)x^{n-2}$.
The derivative of functions of the form $f(x) = ax^n + bx^{n-1} + \ldots, n \in \mathbb{Z}$.
Gradients of curves for given values of x.
Values of x where $f'(x)$ is given.
Equation of the tangent at a given point.

1 Differentiate the following functions with respect to x.
   a $f(x) = x^2 + 3x - 4$
   b $f(x) = \frac{1}{2}x^2 - 5x + 4$
   c $f(x) = 2x^3 - 4x^2$
   d $f(x) = \frac{1}{3}x^6 - \frac{1}{2}x^4 - 1$

2 Find the derivative of the following expressions.
   a x^{-1}
   b $2x^{-3}$
   c $x^{-2} + 2x^{-1} - 3$
   d $\dfrac{3}{x^2}$

3 For each of the following functions
   i find the derivative $f'(x)$ with respect to x
   ii find the second derivative $f''(x)$ with respect to x.
   a $f(x) = x(x - 3)$
   b $f(x) = 2x^2(x + 2)$
   c $f(x) = (x - 2)(x + 3)$
   d $f(x) = (x^2 - 3x)(x + 4)$

4 For each of the following expressions:
   i find the derivative
   ii find the second derivative.
   a $\dfrac{2x^3 - x^2}{x}$
   b $\dfrac{2x^5 - x^3}{3x^2}$
   c $\dfrac{x^3 - 2x^2}{x^4}$
   d $\dfrac{(x - 6)(2x^2 - 1)}{x}$

5 The graph of the function $f(x) = x^2 - 4x + 1$ is shown below.

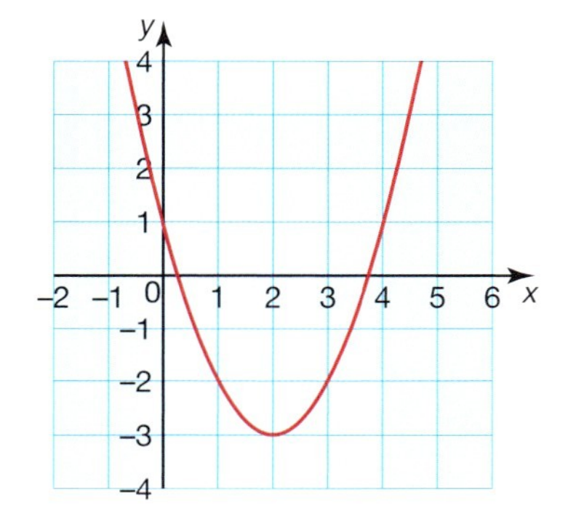

   a Calculate the gradient function $f'(x)$.
   b Calculate the gradient of the graph when:
      i $x = 3$
      ii $x = 2$
      iii $x = 0$.

continued on the next page...

6 The graph of the function $f(x) = \frac{1}{2}x^2 - 4x + 2$ is shown below.

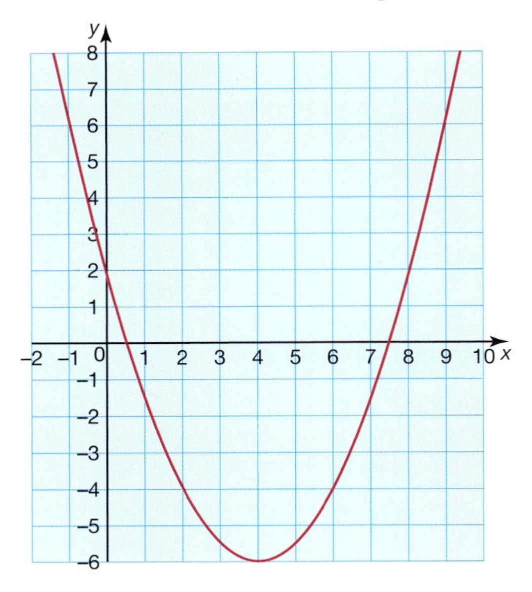

a Calculate the gradient function $f'(x)$.

b Calculate the values of x at the points on the graph where the gradient is:

i 0 ii 2 iii -5.

7 A curve has the equation $y = x^3 + 4x + 2$.

a Find $\frac{dy}{dx}$.

b Deduce from your answer to part **a** the least possible value of $\frac{dy}{dx}$. Justify your answer.

c Calculate the value(s) of x where $\frac{dy}{dx}$ is:

i 7 ii 4 iii 31.

8 The function $f(x) = x^3 - 13x + 12$ is shown below.

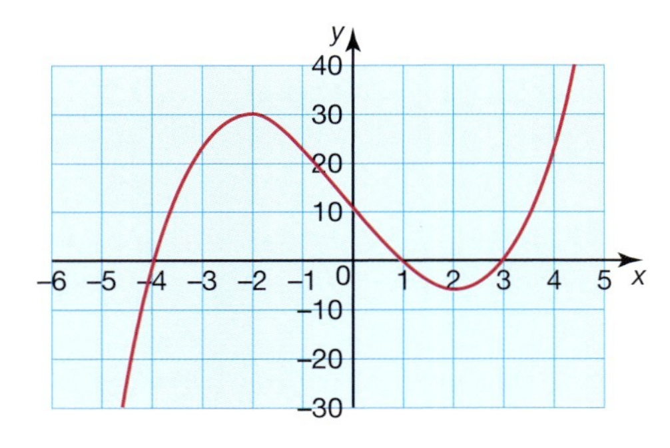

a Calculate the gradient function $f'(x)$.

b Calculate the gradient of the curve when $x = 3$.

c Give the gradient of the tangent to the curve at the point (3, 0).

d Calculate the equation of the tangent to the curve at the point (3, 0).

9 The function $f(x) = -x^2 - 2x + 8$ is shown below.

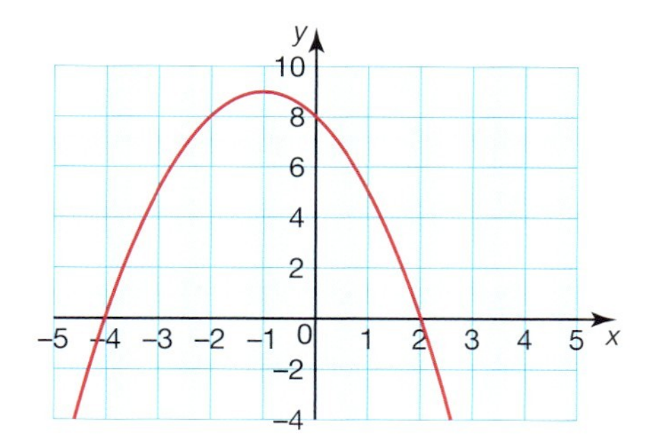

a Calculate the gradient function.
b Show that the points A(−2, 8) and B(1, 5) lie on the curve.
c Calculate the gradient of the curve at points A and B.
d Calculate the equation of the tangent to the curve at A.
e Calculate the equation of the tangent to the curve at B.
f Calculate the coordinates of the point of intersection of the two tangents.

7.4 and 7.5 Increasing and decreasing functions and Stationary points

Increasing and decreasing functions.
Graphical interpretation of $f'(x) > 0$, $f'(x) = 0$, $f'(x) < 0$.
Values of x where the gradient of a curve is 0 (zero): solution of $f'(x) = 0$.
Local maximum and minimum points.

1 The function $f'(x) = x^2 + 6x + 7$, is shown.
From the graph deduce the range of values of x for which $f(x)$ is a decreasing function.

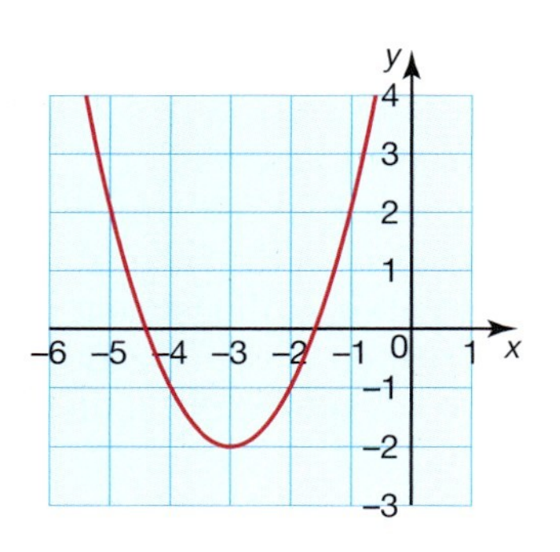

2 For each of the following calculate:
i $f'(x)$
ii the range of values of x for which $f'(x)$ is increasing.
a $f(x) = x^2 - 18$
b $f(x) = x^2 - 10x + 27$
c $f(x) = -x^2 + 8x - 10$
d $f(x) = x^3 - 2x^2 - 8x$

3 Prove that $f(x) = \frac{1}{3}x^3 + x$ is an increasing function for all values of x.

4 Calculate the range of values of k in the function $f(x) = \frac{1}{3}x^3 + x^2 + kx$, given that $f(x)$ is an increasing function for all values of x.

5 A function is of the form $f(x) = x^2 + bx + c$. State:
a the number of stationary points
b the nature of the stationary point(s).

6 For the function $f(x) = \frac{1}{3}x^3 + \frac{1}{2}x^2 - 20x + 2$:
a calculate $f'(x)$
b solve the equation $f'(x) = 0$
c explain the significance of your answer(s) to part **b** in relation to stationary points
d from the equation of the function, deduce the nature of the stationary point(s). Give reasons for your answers.

7 For the function $f(x) = x^3 - 12x - 5$:
a calculate $f'(x)$
b solve the equation $f'(x) = 0$
c determine the coordinates of the stationary points
d determine the nature of the stationary points
e calculate the y-intercept of the graph of the function
f sketch the graph.

8.2 and 8.3 Simple interest and Compound interest

Simple interest: use of the formula $I = \frac{Crn}{100}$ where C = capital, r = % rate, n = number of time periods, I = interest.

Compound interest: use of the formula $I = C \times \left(1 + \frac{r}{100}\right)^n - C$.
Depreciation.
The value of r can be positive or negative.

1 Find the simple interest paid on the following capital sums C, deposited in a savings account for n years at a fixed rate of interest of r%.
   a C = £550 n = 5 years r = 3%
   b C = $8000 n = 10 years r = 6%
   c C = €12 500 n = 7 years r = 2.5%

2 A capital sum of £25 000 is deposited in a bank. After 8 years, the simple interest gained was £7000. Calculate the annual rate of interest on the account assuming it remained constant over the 8 years.

3 A bank lends a business $250 000. The annual rate of interest is 8.4%. When paying back the loan, the business pays an amount of $105 000 in simple interest. Calculate the number of years the business took out the loan for.

4 $15 000 is deposited in a savings account. The following arithmetic sequence represents the total amount of money in the savings account each year. Assume that no further money is either deposited or taken out of the account.

Number of years	0	1	2	3	4	5		n
Total savings in account ($)	15 000	15 375	15 750	16 125	16 500	16 875		

   a Explain, giving reasons, whether the sequence above simulates simple interest or compound interest.
   b Calculate the interest rate.
   c State the formula for calculating the total amount of money (T) in the account after n years.
   d State the formula for calculating the total amount of interest (I) gained after n years.

continued on the next page...

5 $15 000 is deposited in a savings account. The following geometric sequence represents the total amount of money in the savings account each year. Assume that no further money is either deposited or taken out of the account.

Number of years	Total savings in account ($)
0	15 000
1	16 500
2	18 150
3	19 965
4	21 961.50
5	24 157.65
n	

a Explain, giving reasons, whether the sequence above simulates simple interest or compound interest.
b Calculate the interest rate.
c State the formula for calculating the total amount of money (T) in the account after n years.
d State the formula for calculating the total amount of interest (I) gained after n years.

6 Find the compound interest paid on the following capital sums C, deposited in a savings account for n years at a fixed rate of interest of r% per year.
a C = £400 $n = 2$ years $r = 3\%$
b C = $5000 $n = 8$ years $r = 6\%$
c C = €18 000 $n = 10$ years $r = 4.5\%$

7 A car is bought for €12 500. Its value depreciates by 15% per year.
a Calculate its value after:
i 1 year ii 2 years.
b After how many years will the car be worth less than €1000?

8 €4000 is invested for three years at 6% per year. What is the interest paid if the interest rate is compounded:
a yearly?
b half-yearly?
c quarterly?
d monthly?
e daily?

Answers to exercises and student assessments

Presumed knowledge assessments

Student assessment 1

1 a = b < c >
 d >

2 a

 b

 c

 d

3 a $x \geqslant -1$ b $x < 2$
 c $-2 \leqslant x < 2$ d $-1 \leqslant x \leqslant 1$

4 a

 b

 c

 d

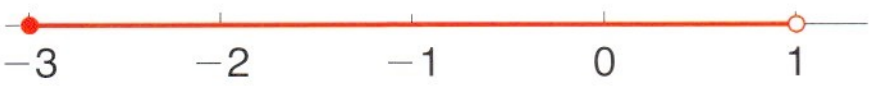

5 $\frac{3}{14}, \frac{2}{5}, \frac{1}{2}, \frac{4}{7}, \frac{9}{10}$

Student assessment 2

1 a 23 b 18

2 6, 15

3 9000

4 22 977

5 360.2

6 $\frac{8}{18} = \frac{4}{9} = \frac{16}{36} = \frac{56}{126} = \frac{40}{90}$

7 a $2\frac{1}{16}$ b $3\frac{3}{8}$

8 a 0.4 b 1.75
 c $0.\dot{8}\dot{1}$ d $1.\dot{6}$

9 a $4\frac{1}{5}$ b $\frac{3}{50}$
 c $1\frac{17}{20}$ d $2\frac{1}{200}$

Student assessment 3

1 a $\frac{7}{10}$ b 45 cm

2 a 375 g b 625 g

3 a 450 m b 80 cm

4 a 1 : 25 b 1.75 m

5 300 : 750 : 1950

6 60°, 90°, 90°, 120°

7 150°

8 a 13.5 h b 12 pumps

9

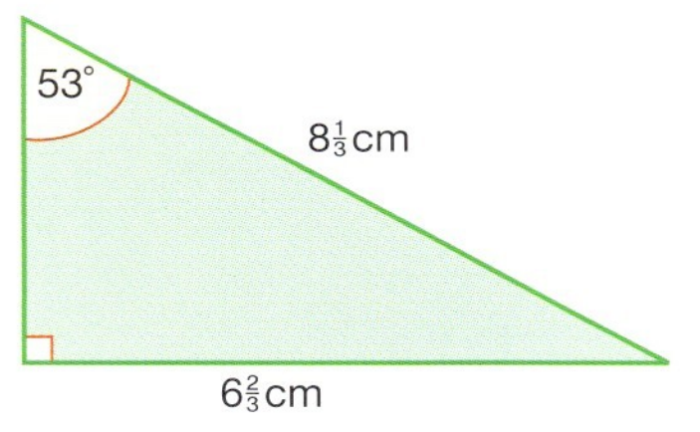

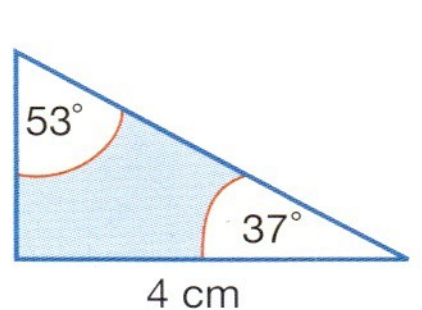

10 a 4 min 48 s b 1.6 litres/min

Student assessment 4

1

Fraction	Decimal	Percentage
$\frac{1}{4}$	0.25	25%
$\frac{3}{5}$	0.6	60%
$\frac{5}{8}$	0.625	62.5%
$\frac{2}{3}$	$0.\dot{6}$	$66\frac{2}{3}$%
$2\frac{1}{4}$	2.25	225%

2 750 m

3 525€

4 £97 200

5 a 29.2% b 21.7% c 125%
 d $8.\dot{3}$% e 20% f 10%

6 8.3%

7 a ¥6500 b 61.8%

8 \$200 \$25 \$524 \$10

9 \$462 \$4000 \$4500 \$5500

10 15 marks

11 35 000

12 25 000 units

13 470 tonnes

Student assessment 5

1 a $6x - 9y + 15z$ b $8pm - 28p$
c $-8m^2n + 4mn^2$ d $20p^3q - 8p^2q^2 - 8p^3$
e $-2x - 2$ f $22x^2 - 14x$
g 2 h $\frac{5}{2}x^2 - x$

2 a $8(2p - q)$ b $p(p - 6q)$
c $5pq(p - 2q)$ d $3pq(3 - 2p + 4q)$

3 a 0 b −7 c 29
d 7 e 7 f 35

4 a $n = p - 4m$ b $y = \frac{4x - 5z}{3}$
c $y = \frac{10px}{3}$ d $y = \frac{3w}{m} - x$
e $r = \frac{pqt}{4mn}$ f $q = r(m - n) - p$

5 a $(q + r)(p - 3r)$ b $(1 + t)(1 - t)(1 + t^2)$
c 750 000 d 50

6 a $x^2 - 2x - 8$ b $x^2 - 16x + 64$
c $x^2 + 2xy + y^2$ d $x^2 - 121$
e $6x^2 - 13x + 6$ f $9x^2 - 30x + 25$

7 a $(x - 11)(x + 7)$ b $(x - 3)(x - 3)$
c $(x - 12)(x + 12)$ d $3(x - 2)(x + 3)$
e $(2x - 3)(x + 4)$ f $(2x - 5)^2$

8 a $f = \sqrt{\frac{p}{m}}$ b $t = \sqrt{\frac{m}{5}}$
c $p = \left(\frac{A}{\pi r}\right)^2 - q$ d $x = \frac{ty}{y - t}$

9 a x^4 b nq
c y^3 d $\frac{4}{q}$

Student assessment 6

1 a 9 b 11 c −4
d 6

2 a 1.5 b 7 c 4
d 3

3 a −10 b 12 c 10
d $11\frac{1}{4}$

4 a 16 b $-8\frac{2}{3}$ c 2
d 3.5

5 a $x = 5$ $y = 2$ b $x = 3\frac{1}{3}$ $y = 4\frac{1}{3}$
c $x = 5$ $y = 4$

6 a $4x + 40 = 180$ b $x = 35°$
c 35°, 70°, 75°

7 9

8 30°, 30°, 30°, 30°, 30°, 30°, 30°, 45°, 45°, 45°, 45°

9 a

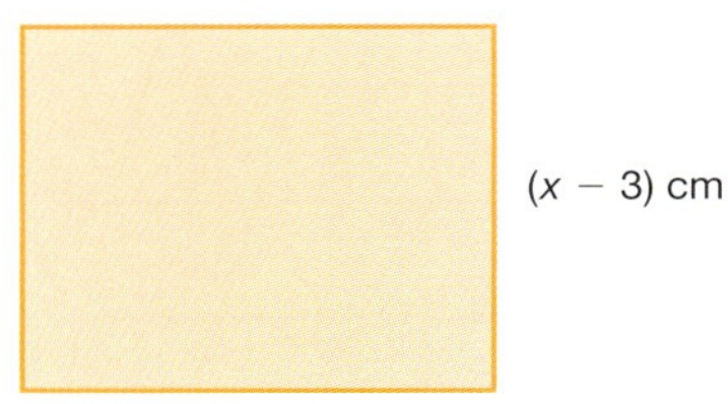

Perimeter = 54 cm

b $4x - 6 = 54$
c Length = 15 cm Width = 12 cm

10 a $x, x - 8, x - 23$ b 55, 47, 32

Student assessment 7

1 a $2^3 \times 5^2$ b $2^2 \times 3^5$

2 a $4 \times 4 \times 4$ b $6 \times 6 \times 6 \times 6$

3 a 800 b 27

4 a 7 b −2 c −1
d $\frac{1}{3}$

5 a $2^2 \times 3^5$ b 2^{14}

6 a $6 \times 6 \times 6 \times 6 \times 6$
b $\frac{1}{2 \times 2 \times 2 \times 2 \times 2}$

Student assessment 8

1 a Circumference = 34.6 cm
Area = 95.0 cm^2
b Circumference = 50.3 mm
Area = 201.1 mm^2

2 9.9 cm^2

3 a 39.3 cm^2 b 34 cm^2 c 101.3 cm^2

4 a Circumference = 27.0 cm
Area = 58.1 cm^2
b Circumference = 47.1 mm
Area = 176.7 mm^2

5 325.8 cm^2

6 a 56.5 cm^2 b 108 cm^2 c 254.5 cm^2

Topic 1

Exercise 1.1.1

1 a 25 b 18 c 40

2 a 12 b 6 c 5

3 a 169 b 4 c 13

4 a 216 b 9 c 4

5 a 3510 b 149 530 c 0

Exercise 1.1.2

1 a 11 b −1 c 79
d 144 e 6 f $18\frac{2}{11}$

2 a 100π cm^2
b 36π cm^2
c C = 16π cm^2, D = π cm^2
d 534 cm^2
e the area of the circle pattern

3 a 18 cm^2 b 24 cm^2 c 106 cm^2
d 254 cm^2

4 a 523.6 cm^3 b 2356.2 cm^3 c 785.4 cm^3

Exercise 1.1.3

1 a Graph vi) b Graph vii)
c Graph ii) d Graph iii)
e Graph x) f Graph i)
g Graph v) h Graph ix)
i Graph iv) j Graph viii)

2 a $y = -x - 5$ b $y = 2x - 4$
c $y = -(x + 5)^2$ d $y = -(x - 5)^2 - 3$

3 a $y = -x + 5$ b $y = 2x + 4$
c $y = (x - 5)^2$ d $y = (x + 5)^2 + 3$

4 Pupil's screens

Exercise 1.1.4

1 a y-intercept = −3 x-intercept ≈ ± 1.7
b y-intercept = 11 x-intercept None
c y-intercept = 1 x-intercept ≈ −0.5, 1.3, 3.2
d y-intercept = 3.5 x-intercept ≈ −1.2

2 a $\left(-1\frac{2}{3}, 1\frac{1}{3}\right)$
b (−3.45, 4.45) and (1.45, −0.45)
c (−1.29, −0.67) and (1.29, −0.67)
d (−0.77, −1.71), (0.88, −1.61) and (5.89, 15.3)

Exercise 1.1.5

1 a

x	−3	−2	−1	0	1	2	3
y	2	−2	−4	−4	−2	2	8

b

x	−3	−2	−1	0	1	2	3
y	−28	−14	−10	−10	−8	2	26

c

x	0	0.5	1	1.5	2	2.5	3
y	–	8	4	2.67	2	1.6	1.33

d

x	−1	−0.5	0	0.5	1	1.5	2	2.5	3
y	0	0.71	1	1.22	1.41	1.58	1.73	1.87	2

2 a 0, 3.6, 7.2, 10.8, 14.4, 18, 21.6, 25.2, 28.8, 32.4, 36 $m\,s^{-1}$
b 18 $m\,s^{-1}$

3 a 0, 6.25, 10, 11.25, 10, 6.25, 0, −8.75, −20 m
b 11.25 m
c 1.5 s
d 3 s
e because they give negative heights

Exercise 1.1.6

1 a mean = 7.58, median = 8.5, mode = 10
b mean = 14.7, median = 14, mode = 12.2

2 mean = 3.63, median = 3, mode = 2

3 a Test A: mean = 5.33, median = 6, mode = 6
Test B: mean = 5.33, median = 6, mode = 6
b Student's answers
c Student's answers

Topic 2

Exercise 2.1.1

1 a $\mathbb{N}, \mathbb{Z}, \mathbb{Q}, \mathbb{R}$ b $\mathbb{Z}, \mathbb{Q}, \mathbb{R}$
c $\mathbb{R}$ d $\mathbb{Q}, \mathbb{R}$

2 a Rational b Rational c Irrational

3 a Rational b Rational c Rational

4 a Irrational b Rational c Rational

5 a Irrational b Irrational c Rational

6 a Rational b Rational c Rational

7 Rational

8 Irrational

9 Rational

10 Rational

Exercise 2.2.1

1 a 69 000 b 74 000
c 89 000 d 4000
e 100 000 f 1 000 000

2 a 78 500 b 6900
c 14 100 d 8100
e 1000 f 3000

3 a 490 b 690
c 8850 d 80
e 0 f 1000

Exercise 2.2.2

1 a 5.6 b 0.7
c 11.9 d 157.4
e 4.0 f 15.0
g 3.0 h 1.0
i 12.0

2 a 6.47 b 9.59
c 16.48 d 0.09
e 0.01 f 9.30
g 100.00 h 0.00
i 3.00

Exercise 2.2.3

1 a 50 000 b 48 600 c 7000
d 7500 e 500 f 2.57
g 1000 h 2000 i 15.0

2 a 0.09 b 0.6 c 0.94
d 1 e 0.95 f 0.003
g 0.0031 h 0.0097 i 0.01

3 a 420 b 5.05 c 166
d 23.8 e 57.8 f 4430
g 1.94 h 4.11 i 0.575

Exercise 2.2.4

Answers may vary slightly from those given.

1 a 1200 b 3000 c 3000
d 150 000 e 0.8 f 100

2 a 200 b 200 c 30
d 550 e 500 f 3000

3 a 130 b 80 c 1
d 4 e 200 f 250

4 c because $689 \times 400 \approx 700 \times 400 = 280\,000$
e because $77.9 \times 22.6 \approx 80 \times 20 = 1600$
f because $\frac{8.4 \times 46}{0.2} \approx \frac{80}{20} \times 50 = 2000$

5 a $120\,\text{m}^2$ b $40\,\text{m}^2$ c $400\,\text{cm}^2$

6 a $200\,\text{cm}^3$ b $4000\,\text{cm}^3$ c $2000\,\text{cm}^3$

Exercise 2.2.5

1 a 0.4% b 2.04% c 0.8%

2 The second player as his percentage error is 3.2% and the first player's is 4%.

3 a 9737.5 m b 9262.5 m

4 a $118.2\,\text{km}\,\text{h}^{-1}$ b 2.9%

Exercise 2.3.1

1 d and e

2 a 6×10^5 b 4.8×10^7 c 7.84×10^{11}
d 5.34×10^5 e 7×10^6 f 8.5×10^6

3 a 6.8×10^6 b 7.2×10^8 c 8×10^5
d 7.5×10^7 e 4×10^9 f 5×10^7

4 a 6×10^5 b 2.4×10^7 c 1.4×10^8
d 3×10^9 e 1.2×10^{13} f 1.8×10^7

5 $1.44 \times 10^{11}\,\text{m}$

6 a 8.8×10^8 b 2.04×10^{11} c 3.32×10^{11}
d 4.2×10^{22} e 5.1×10^{22} f 2.5×10^{25}

7 a 2×10^2 b 3×10^5 c 4×10^6
d 2×10^4 e 2.5×10^6 f 4×10^4

8 a 4.26×10^5 b 8.48×10^9
c 6.388×10^7 d 3.157×10^9
e 4.5×10^8 f 6.01×10^7
g 8.15×10^{10} h 3.56×10^7

9 Mercury 5.8×10^7 km
Venus 1.08×10^8 km
Earth 1.5×10^8 km
Mars 2.28×10^8 km
Jupiter 7.78×10^8 km
Saturn 1.43×10^9 km
Uranus 2.87×10^9 km
Neptune 4.5×10^9 km

Exercise 2.3.2

1 a 6×10^{-4} b 5.3×10^{-5}
c 8.64×10^{-4} d 8.8×10^{-8}
e 7×10^{-7} f 4.145×10^{-4}

2 a 6.8×10^{-4} b 7.5×10^{-7}
c 4.2×10^{-10} d 8×10^{-9}
e 5.7×10^{-11} f 4×10^{-11}

3 a -4 b -3 c -8
d -5 e -7

4 6.8×10^5 6.2×10^3 8.414×10^2
6.741×10^{-4} 3.2×10^{-4} 5.8×10^{-7}
5.57×10^{-9}

Exercise 2.4.1

1 a one hundred b a hundredth
c one thousand d a thousandth
e one thousand f a thousandth
g a thousandth h one thousand
i a millilitre j one million

2 a kilogram b centimetre
c metre or centimetre d millilitre
e tonne f metre
g litre h km
i litre j centimetre

3 Student's lines

4 Student's estimates – answers may vary considerably

Exercise 2.4.2

1 a 40 mm b 62 mm c 280 mm
d 1200 mm e 880 mm f 3650 mm
g 8 mm h 2.3 mm

2 a 2.6 m b 89 m c 2300 m
d 750 m e 2.5 m f 400 m
g 3800 m h 25000 m

3 a 2 km b 26.5 km c 0.2 km
d 0.75 km e 0.1 km f 5 km
g 15 km h 75.6 km

Exercise 2.4.3

1 a 2000 kg b 7200 kg c 2.8 kg
d 0.75 kg e 450 kg f 3 kg
g 6.5 kg h 7000 kg

2 a 2600 ml b 700 ml c 40 ml
d 8 ml

3 a 1.5 litres b 5.28 litres c 0.75 litres
d 0.025 litres

4 a 138.3 tonnes b 1.383×10^5 kg

5 a 720 ml b 0.53 litres

Exercise 2.4.4

1 a 68 °F b 176 °F c 392 °F

2 a 10 °C b 135 °C c 475 °C

3 a 303 K b 153 K c 308 K

4 a 70 °F b 190 °F c 430 °F

5 a 2.9% b 8.0% c 9.7%

6 10 °C, 50 °F

Exercise 2.5.1

1 a 8, 13, 18; arithmetic
b −2, −8, −20; not arithmetic
c 1, −3, 13; not arithmetic
d −2, 5, −2; not arithmetic
e 4, 0, −4; arithmetic
f 9, 3, 5; not arithmetic

2 a i) $3n + 2$ ii) 32
b i) $4n - 4$ ii) 36
c i) $n - 0.5$ ii) 9.5
d i) $-3n + 9$ ii) −21
e i) $3n - 10$ ii) 20
f i) $-4n - 5$ ii) −45

3 a

Position	1	2	5	12	50	n
Term	1	5	17	45	197	$4n - 3$

b

Position	1	2	5	10	75	n
Term	5	11	29	59	449	$6n - 1$

c

Position	1	3	8	50	100	n
Term	2	0	−5	−47	−97	$-n + 3$

d

Position	1	2	3	10	100	n
Term	3	0	−3	−24	−294	$-3n + 6$

e

Position	2	5	7	10	50	n
Term	1	10	16	25	145	$3n - 5$

f

Position	1	2	5	20	50	n
Term	−5.5	−7	−11.5	−34	−79	$-1.5n - 4$

4 a i) +4 ii) $4n + 1$ iii) 201
b i) +1 ii) $n - 1$ iii) 49
c i) +3 ii) $3n - 13$ iii) 137
d i) +0.5 ii) $0.5n + 5.5$ iii) 30.5
e i) +4 ii) $4n - 62$ iii) 138
f i) −3 ii) $-3n + 75$ iii) −75

5 8 years

Exercise 2.5.2

1 a 308 b 488 c −187
d 0

2 a −15 b −95 c −55
d $\frac{n(n + 17)}{4}$

3 a 3 b −5 c 52
d 470

4 a 0.5 b 2 c 2475

5 a +8 b −44 c −80

6 a $\frac{x}{3}$ b 6 c 150

7 11

8 a Student's proof b 12
c 19

Exercise 2.6.1

1 a Geometric b Geometric
c Not geometric d Geometric
e Not geometric f Not geometric

2 a i) 3 ii) 162, 486
iii) $u_n = 2(3)^{n-1}$
b i) $\frac{1}{5}$ ii) $\frac{1}{25}$,
iii) $u_n = 25\left(\frac{1}{5}\right)^{n-1}$
d i) −3 ii) −243, 729
iii) $u_n = -3^n$

3 a −6, −12, −24 b 8

4 a −4 b $\frac{1}{4}$ c −65 536

5 1338.23€

Exercise 2.6.2

1 a i) 2 ii) $\frac{1023}{8}$
b i) −3 ii) $\frac{14762}{9}$
c i) 1.5 ii) 566.65 (2 d.p)
d i) $\frac{1}{10}$ ii) 11.111 111 11

2 a i) 8 ii) 3280
b i) 7 ii) $\frac{127}{5}$
c i) 9 ii) $\frac{171}{32}$
d i) n ii) $\frac{a(r^n - 1)}{r - 1}$

3 a 1364 b $728\frac{2}{3}$ c 62

4 a 6 b $\frac{1}{18}$ c $\frac{9331}{18}$

5 a 1 or 3 b $-\frac{1}{2}$ c $-\frac{85}{4}$

6 a 823 543 b 960 800

Exercise 2.6.3

1 a 27 b $-\frac{16}{3}$ c $\frac{10}{9}$
d $\frac{49}{5}$

2 a $\frac{1}{3}$ b 4 c $\frac{32}{81}$

 d 0.000 381

3 a 3 b $\frac{1}{2}$

4 $\frac{27}{2}$

Exercise 2.7.1

1 a (3, 2) b (5, 2) c (2, 1)
 d (−4, 1) e (4, −2) f (3, −2)
 g (−1, −1) h (−3, −3)
 i Infinite solutions
 j No solution

2 The lines in part **i** are the same line.
 The lines in part **j** are parallel.

Exercise 2.7.2

1 a i

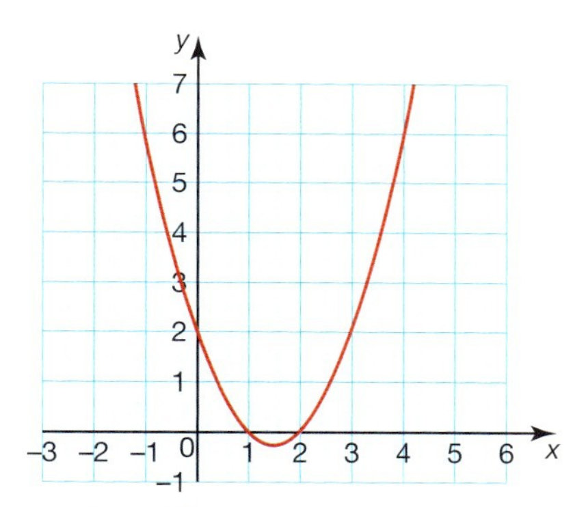

 ii $x = 1$ and 2

 b i

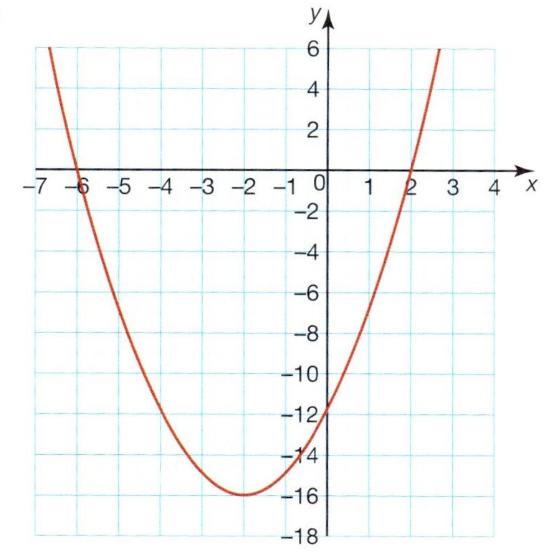

 ii $x = -6$ and 2

 c i

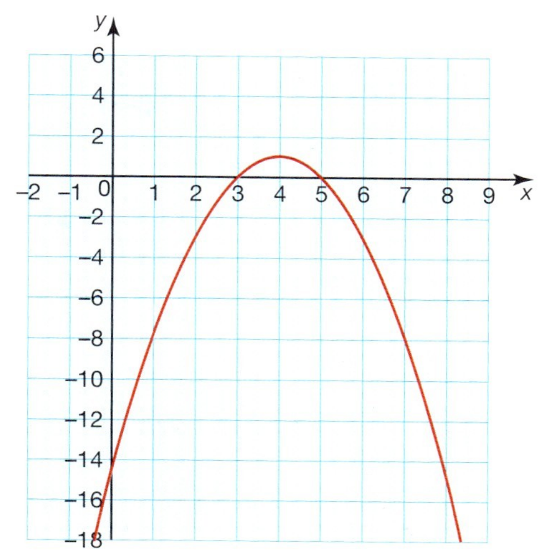

 ii $x = 3$ and 5

 d i

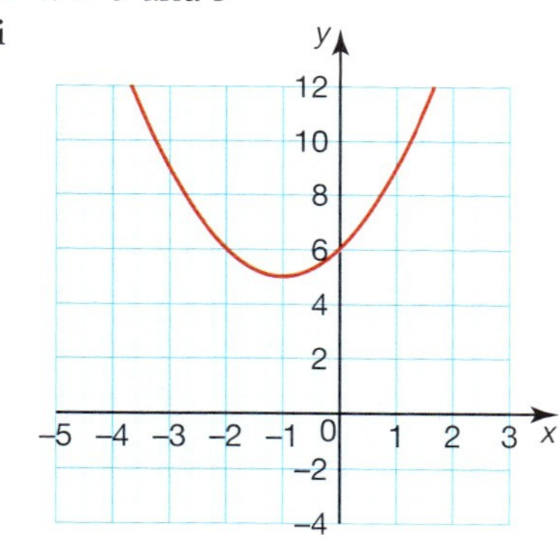

 ii No solution

 e i

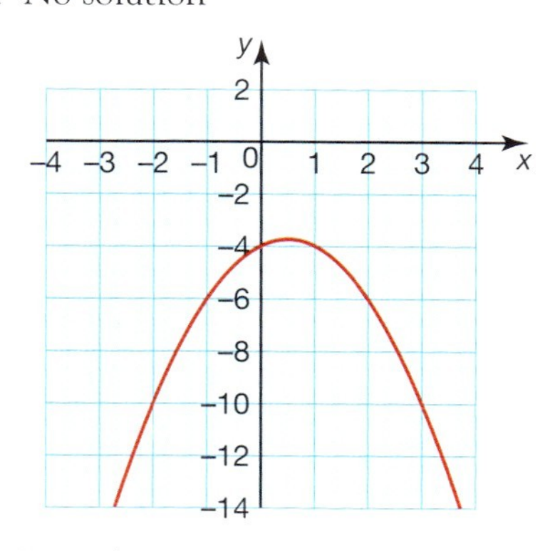

 ii No solution

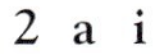
2 a i

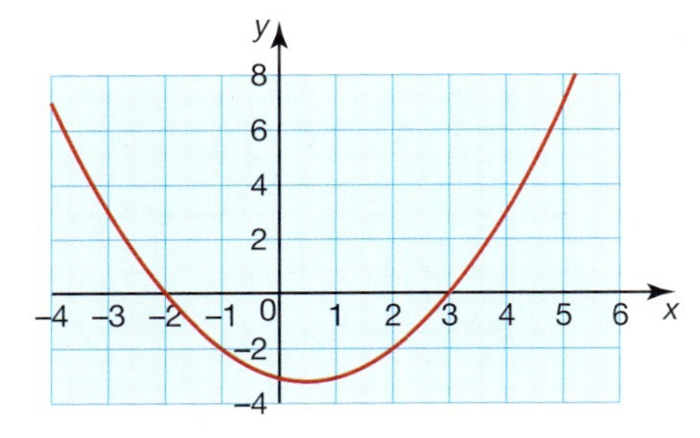

ii $x = -2$ and 3

b i

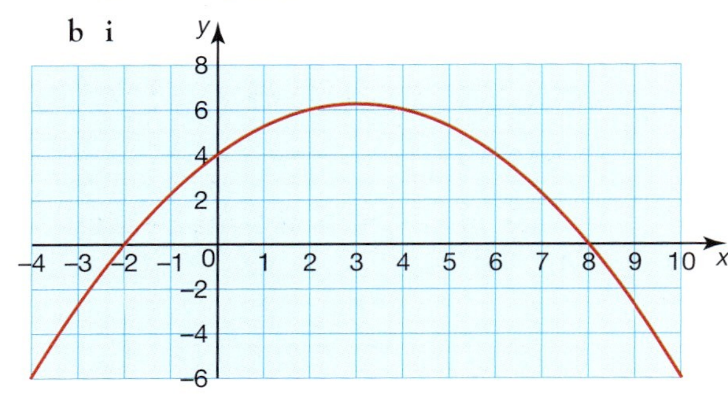

ii $x = -2$ and 8

c i

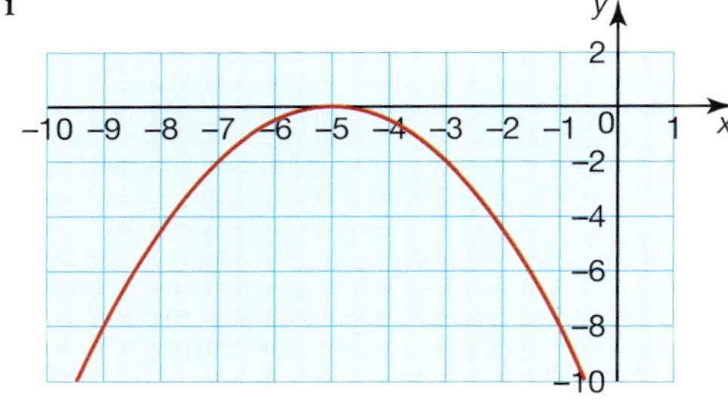

ii $x = -5$

Student assessment 1

1 a Rational b Irrational c Rational
d Rational e Rational f Irrational

2 a 6470 b 88 500 c 65 000
d 10

3 a 3.8 b 6.8 c 0.85
d 1.58 e 10.0 f 0.008

4 a 40 b 5.4 c 0.06
d 49 000 e 700 000 f 687 000

5 92.3 cm^3

Answers to questions 6–8 may vary from those given below.

6 18 000 yards

7 a 25 b 4 c 4

8 170 cm^2

9 a 168.02 cm^2 b Students's calculation

10 1.63%

Student assessment 2

1 a 6×10^6 b 4.5×10^{-3}
c 3.8×10^9 d 3.61×10^{-7}
e 4.6×10^8 f 3×10^0

2 7.41×10^{-9} 3.6×10^{-5} 5.5×10^{-3}
4.21×10^7 6.2×10^7 4.9×10^8

3 a 6×10^6 8.2×10^5 4.4×10^{-3}
8×10^{-1} 5.2×10^4
b 6×10^6 8.2×10^5 5.2×10^4
8×10^{-1} 4.4×10^{-3}

4 a 3 b 9 c −3
d 6 e −1 f 8

5 a 1.2×10^8 b 5.6×10^8
c 2×10^5 d 2.5×10^5

6 43.2 minutes

7 4.73×10^{15} km

8 1.62×10^{11} mm

9 7 kg

10 a 104 °F b 932 °F

Student assessment 3

1 a i) $4n - 3$ ii) 37
b i) $-3n + 4$ ii) −26

2 a $u_5 = 27$, $u_{100} = 597$ b $u_5 = 1.5$, $u_{100} = -46$

3 a

Position	1	2	3	10	25	n
Term	17	14	11	−10	−55	$-3n + 20$

b

Position	2	6	10	80	n
Term	−4	−2	0	35	$\frac{1}{2}n - 5$

4 \$405

5 a $-\frac{1}{3}$ b 243 c 10

6 a 70 b 595

7 a +3 b −12 c 330

8 a +4 b 4 c 220

9 a i) 11 ii) 20470

b i) 13 ii) $\frac{2731}{32}$

10 a 363 b $\frac{3279}{5}$

11 a (2, 1) b (−2, 3)

12 a i)

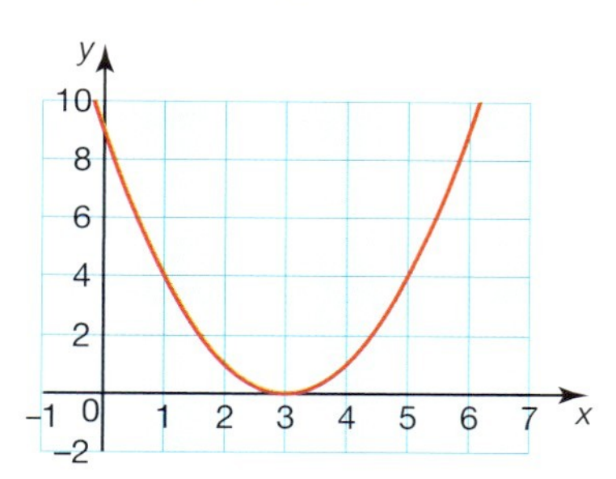

ii) $x = 3$

b i)

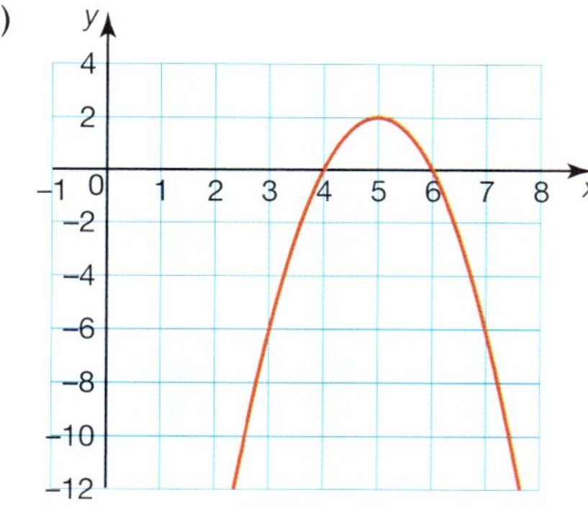

ii) $x = 4$ and 6

Topic 3

Exercise 3.1.1

1 a i) Continents of the world
ii) Student's own answers
b i) Even numbers
ii) Student's own answers
c i) Days of the week
ii) Student's own answers
d i) Months with 31 days
ii) Student's own answers
e i) Triangle numbers
ii) Student's own answers
f i) Boy's names beginning with the letter m
ii) Student's own answers
g i) Odd numbers
ii) Student's own answers
h i) Vowels
ii) o, u
i i) Planets of the solar system
ii) Student's own answers
j i) Numbers between 3 and 12 inclusive
ii) Student's own answers
k i) Numbers between −5 and 5 inclusive
ii) Student's own answers

2 a 7 c 7 d 7
f Unquantifiably finite, though theoretically infinite
h 5 i 9

Exercise 3.1.2

1 a $Q = \{2, 4, 6, 8, 10, 12, 14, 16, 18, 20, 22, 24, 26, 28\}$
b $R = \{1, 3, 5, 7, 9, 11, 13, 15, 17, 19, 21, 23, 25, 27, 29\}$
c $S = \{2, 3, 5, 7, 11, 13, 17, 19, 23, 29\}$
d $T = \{1, 4, 9, 16, 25\}$
e $U = \{1, 3, 6, 10, 15, 21, 28\}$

2 a $B = \{55, 60, 65\}$
b $C = \{51, 54, 57, 60, 63, 66, 69\}$
c $D = \{64\}$

3 a {p, q, r}, {p, q}, {p, r}, {q, r}, {p}, {q}, {r}, { }
b {p, q}, {p, r}, {q, r}, {p}, {q}, {r}

4 a True b True c True
d False e False f True
g True h False

Exercise 3.2.1

1 a True b True c False
d False e False f True

2 a $A \cap B = \{4, 6\}$ b $A \cap B = \{4, 9\}$
c $A \cap B$ = {yellow, green}

3 a $A \cup B = \{2, 3, 4, 6, 8, 9, 10, 13, 18\}$
b $A \cup B = \{1, 4, 5, 6, 7, 8, 9, 16\}$
c $A \cup B$ = {red, orange, blue, indigo, violet, yellow, green, purple, pink}

4 a $U = \{a, b, p, q, r, s, t\}$
b $A' = \{a, b\}$

5 a $U = \{1, 2, 3, 4, 5, 6, 7, 8\}$
b $A' = \{1, 4, 6, 8\}$
c $A \cap B = \{2, 3\}$
d $A \cup B = \{1, 2, 3, 4, 5, 7, 8\}$
e $(A \cap B)' = \{1, 4, 5, 6, 7, 8\}$
f $A \cap B' = \{5, 7\}$

6 a i) A = {even numbers from 2 to 14}
ii) B = {multiples of 3 from 3 to 15}
iii) C = {multiples of 4 from 4 to 20}
b i) $A \cap B = \{6, 12\}$
ii) $A \cap C = \{4, 8, 12\}$
iii) $B \cap C = \{12\}$
iv) $A \cap B \cap C = \{12\}$
v) $A \cup B = \{2, 3, 4, 6, 8, 9, 10, 12, 14, 15\}$
vi) $C \cup B = \{3, 4, 6, 8, 9, 12, 15, 16, 20\}$

7 a i) $A = \{1, 2, 4, 5, 6, 7\}$
ii) $B = \{3, 4, 5, 8, 9\}$
iii) $C' = \{1, 2, 3, 4, 5, 8, 9\}$
iv) $A \cap B = \{4, 5\}$
v) $A \cup B = \{1, 2, 3, 4, 5, 6, 7, 8, 9\}$
vi) $(A \cap B)' = \{1, 2, 3, 6, 7, 8, 9\}$
b $C \subset A$

8 a i) $W = \{1, 2, 4, 5, 6, 7, 9, 10\}$
ii) $X = \{2, 3, 6, 7, 8, 9\}$
iii) $Z' = \{1, 4, 5, 6, 7, 8, 10\}$
iv) $W \cap Z = \{2, 9\}$
v) $W \cap X = \{2, 6, 7, 9\}$
vi) $Y \cap Z = \{\ \}$ or $\varnothing$
b Z

9 a

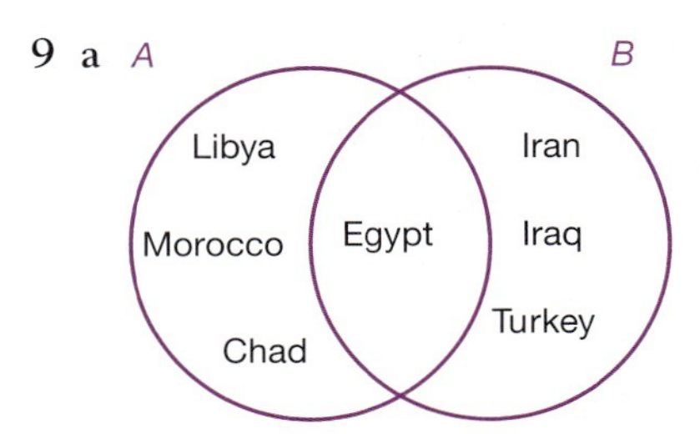

b i) $A \cap B$ = {Egypt}
ii) $A \cup B$ = {Libya, Morocco, Chad, Egypt, Iran, Iraq, Turkey}

10 a

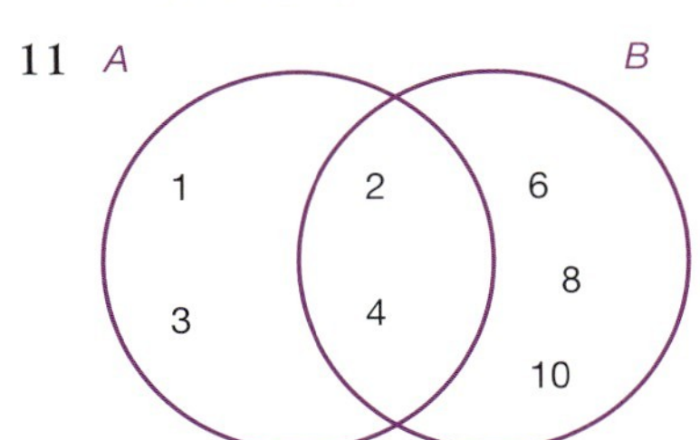

b i) $P \cap Q = \{11, 13, 17\}$
ii) $P \cup Q = \{2, 3, 5, 7, 11, 13, 15, 17, 19\}$

11

12

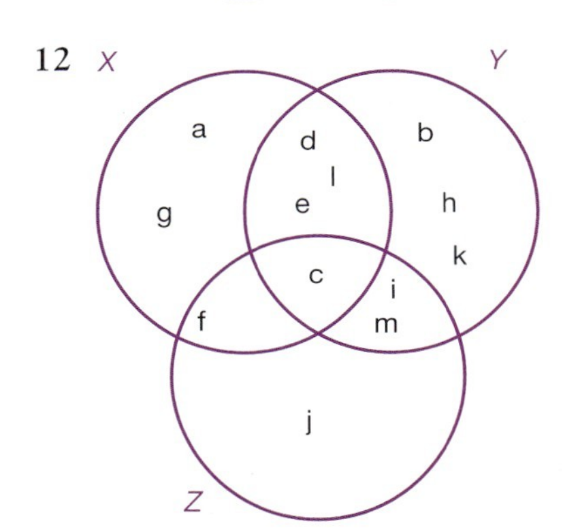

13

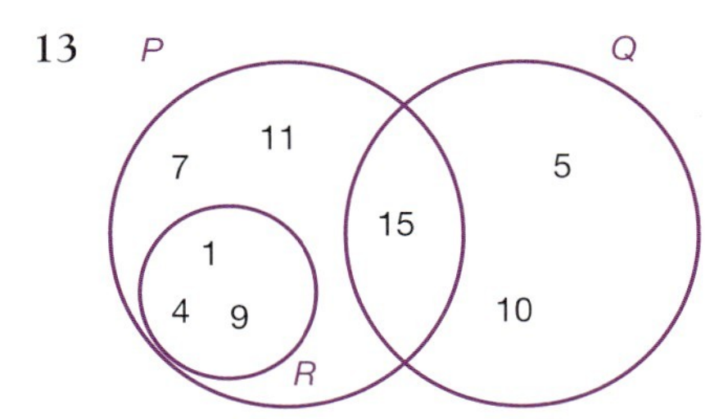

Exercise 3.2.2

1 a 5 **b** 14 **c** 13

2 45

3 a 10 **b** 50

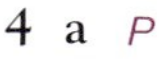

4 a

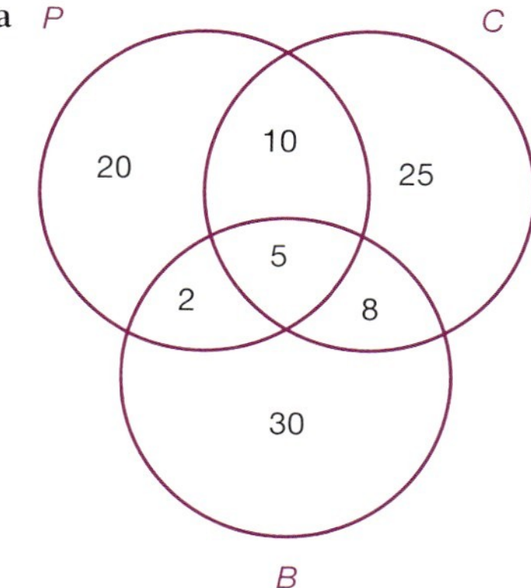

b 100 c 15 d 13

e 75

5 a 16 b 10 c 56

6 a 45 b 65 c 90

Exercise 3.3.1

1 {HHH, THH, THT, TTH, HHT, HTH, HTT, TTT} 8 events

2 {(1, 1)(1, 2)(1, 3)(1, 4)(1, 5)(1, 6)
(2, 1)(2, 2)(2, 3)(2, 4)(2, 5)(2, 6)
(3, 1)(3, 2)(3, 3)(3, 4)(3, 5)(3, 6)
(4, 1)(4, 2)(4, 3)(4, 4)(4, 5)(4, 6)
(5, 1)(5, 2)(5, 3)(5, 4)(5, 5)(5, 6)
(6, 1)(6, 2)(6, 3)(6, 4)(6, 5)(6, 6)} 36 events

3 {(H, 1)(H, 2)(H, 3)(H, 4)(H, 5)(H, 6)
(T, 1)(T, 2)(T, 3)(T, 4)(T, 5)(T, 6)} 12 events

4 {(M M)(M F)(F M)(F F)} 4 events

5 {(M M)(F F)} 2 events

6 a Pass or fail
b {(P P)(P F)(F P)(F F)}

7 a Win or lose
b {(2, 0)(2, 1)(1, 2)(0, 2)}

8 a Win or lose
b {(3, 0)(3, 1)(3, 2)(2, 3)(1, 3)(0, 3)}

Exercise 3.4.1

1 Letter B

2 Card E

Exercise 3.5.1

1 a No b Yes c Yes
d No e Yes f Yes
g Yes h No i Yes
j Yes

2 a Teresa is a girl and Abena is a girl. (True)
b $-1 < x < 8$ (True)
c A pentagon has five sides and a triangle has 4 sides. (False)
d London is in England and England is in Europe. (True)
e $x < y < z$ (True)
f 5 is a prime number and 4 is an even number. (True)
g A square is a rectangle and a triangle is a rectangle. (False)
h Paris is the capital of France and Ghana is in Asia. (False)
i 37 is a prime number and 51 is a prime number. (False)
j Parallelograms are rectangles and trapeziums are rectangles. (True)

Exercise 3.5.2

1

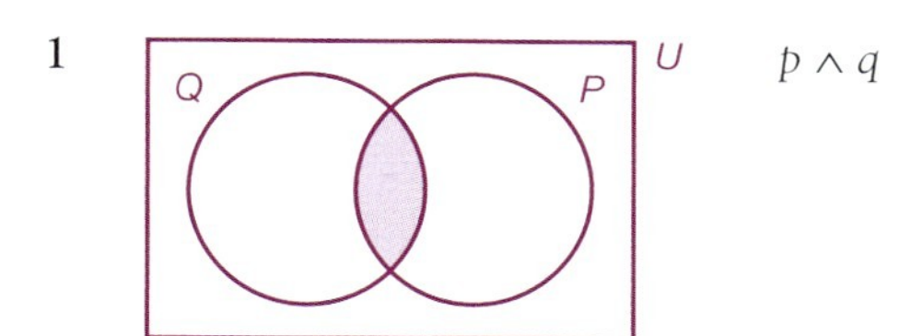

$p \wedge q$

2

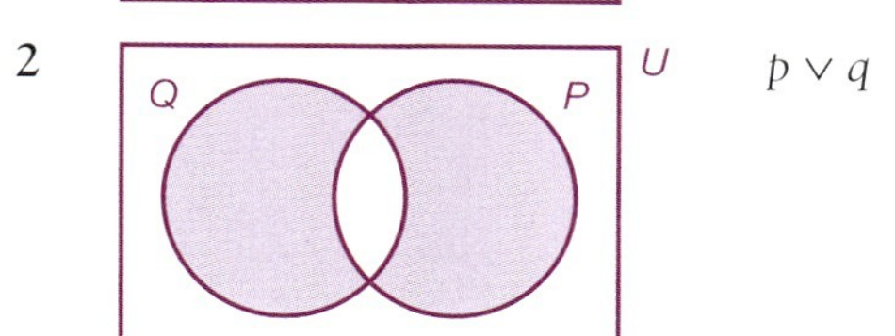

$p \vee q$

3

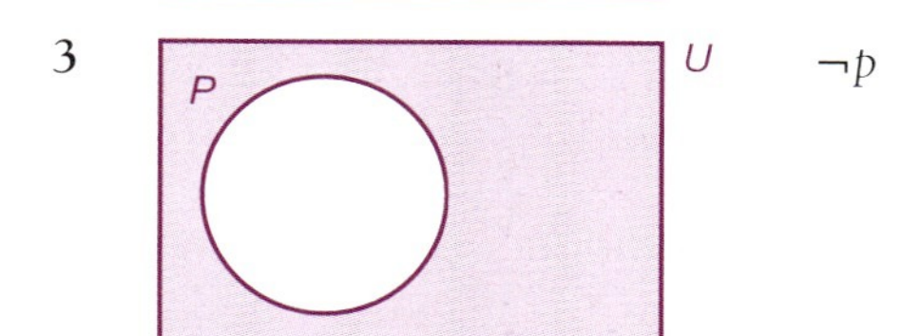

$\neg p$

4

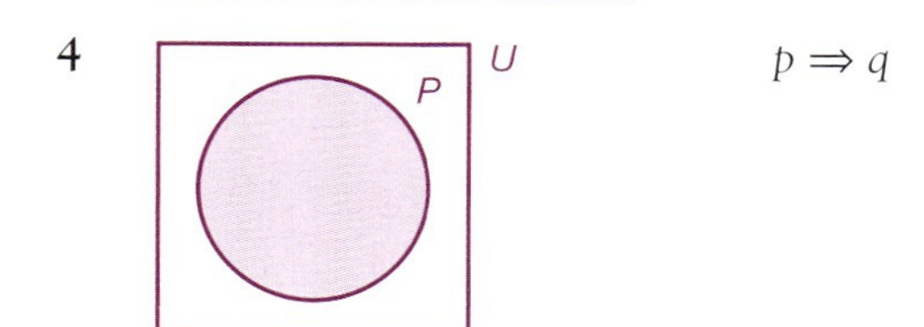

$p \Rightarrow q$

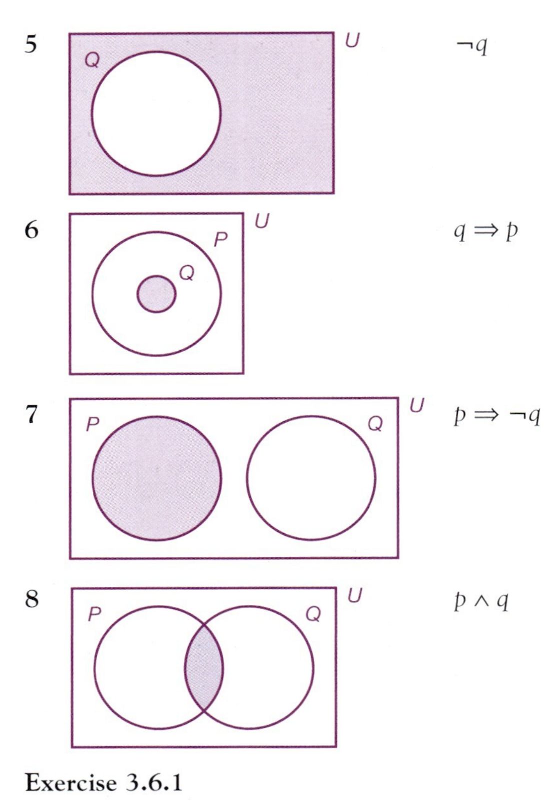

Exercise 3.6.1

1

p	q	r
T	T	T
T	T	F
T	F	T
T	F	F
F	T	T
F	T	F
F	F	T
F	F	F

2

p	q	r	$\neg p$	$p \vee q$	$\neg p \vee r$	$(p \vee q) \wedge (\neg p \vee r)$
T	T	T	F	T	T	T
T	T	F	F	T	F	F
T	F	T	F	T	T	T
T	F	F	F	T	F	F
F	T	T	T	T	T	T
F	T	F	T	T	T	T
F	F	T	F	T	T	F
F	F	F	F	T	T	F

Exercise 3.6.2

1 a Neither b Contradiction c Neither
d Tautology e Tautology

2 a Neither

p	q	$\neg p$	$\neg q$	$\neg p \wedge \neg q$
T	T	F	F	F
T	F	F	T	F
F	T	T	F	F
F	F	T	T	T

b Neither

p	$\neg p$	$\neg(\neg p)$	$\neg(\neg p) \vee p$
T	F	T	T
F	T	F	F

c Neither

q	r	$\neg r$	$q \wedge \neg r$
T	T	F	F
T	F	T	T
F	T	F	F
F	F	T	F

d Neither

p	q	r	$p \wedge q$	$(p \wedge q) \wedge r$
T	T	T	T	T
T	T	F	T	F
T	F	T	F	F
F	T	T	F	F
T	F	F	F	F
F	T	F	F	F
T	F	F	F	F
F	F	F	F	F

e Neither

p	q	r	$p \wedge q$	$(p \wedge q) \vee r$
T	T	T	T	T
T	T	F	T	T
T	F	T	F	T
F	T	T	F	T
T	F	F	F	F
F	T	F	F	F
T	F	F	F	F
F	F	F	F	F

Exercise 3.7.1

1 a True　b False　c True
d True　e False

2 a I think so it follows that I am.
b …does not exist.

Exercise 3.7.2

Note: The students may use different words but the sense will be the same.

1 a If you do not have your mobile phone then you cannot send a text.
If you have your phone you can send a text.

b If you cannot travel a long way on 20 euros then you do not have a small car.
If you do not have a small car then you cannot not travel a long way on 20 euros.

c If you can speak French then you will enjoy France more
If you do enjoy France more then you do speak French.

d If it rains then I do not play tennis.
If it does not rain then I do play tennis.

e We stop playing golf if there is a threat of lightening.
We do not stop playing golf if there is not a threat of lightening.

f A tennis serve is easy if you practice it.
If you do not practice it then a tennis serve is hard.

g If a polygon has six sides then it is a hexagon.
If a polygon is not a hexagon then it does not have six sides.

h If you are less than 160 cm tall then I am taller than you.
If you are not less than 160 cm tall then you are taller than me.

i If the bus was full then I was late.
If I was not late then the bus was not full.

j If the road was greasy then the car skidded.
If the car skidded then the road was greasy.

2 a Statement: If a number is an odd number then it is a prime number. (False)
Converse: If a number is a prime number then it is an odd number. (False)
Inverse : If a number is not an odd number then it is not a prime number. (False)
Contrapositive: If a number is not a prime number then it is not an odd number. (False)

b Statement. If a polygon has six sides then it is called an octagon. (False)
Converse. If a polygon is called an octagon then it has six sides. (False)
Inverse. If a polygon does not have six sides it is not called an octagon. (False)
Contrapositive. If it is not called an octagon then it does not have six sides. (False)

c Statement. If it is an acute angled triangle then it has three acute angles. (True)
Converse. If it has three acute angles then it is called an acute angled triangle. (True)
Inverse. If it is not called an acute angled triangle then it does not have three acute angles. (True)
Contrapositive. If it does not have three acute angles then it is not called an acute angled triangle. (True)

d Statement. If two triangles are similar then they are congruent. (False)
Converse. If two triangles are congruent then they are similar. (True)
Inverse. If two triangles are not similar then they are not congruent. (True)
Contrapositive. If two triangles are not congruent then they are not similar. (False)

e Statement. If two triangles are congruent then they are similar. (True)
Converse. If two triangles are similar then they are congruent. (False)
Inverse. If two triangles are not congruent then they are not similar. (False)
Contrapositive. If two triangles are not similar then they are not congruent. (True)

f Statement. If a solid is a cuboid then it has six faces. (True)
Converse. If a solid has six faces then it is a cuboid. (True)
Inverse. If a solid is not a cuboid then it does not have six faces. (True)
Contrapositive. If a solid does not have six faces then it is not a cuboid. (True)

g Statement. If a solid has eight faces then it is called a regular octahedron. (False)
Converse. If a solid is called a regular octahedron then it has eight faces. (True)
Inverse. If a solid does not have eight faces then it is not called a regular octahedron. (True)
Contrapositive. If a solid is not called a regular octahedron then it does not have eight faces. (False)

h Statement. If a number is a prime number then it is an even number. (False)
Converse. If a number is even then it is a prime number. (False)
Inverse. If a number is not a prime number then it is not even. (False)
Contrapositive. If a number is not even then it is not prime. (False)

Exercise 3.8.1

1 **a** $\frac{1}{6}$ **b** $\frac{1}{6}$ **c** $\frac{1}{2}$
d $\frac{5}{6}$ **e** 0 **f** 1

2 **a** i) $\frac{1}{7}$ ii) $\frac{6}{7}$ **b** Total = 1

3 **a** $\frac{1}{250}$ **b** $\frac{1}{50}$ **c** 1
d 0

4 **a** $\frac{5}{8}$ **b** $\frac{3}{8}$

5 **a** $\frac{1}{13}$ **b** $\frac{5}{26}$ **c** $\frac{21}{26}$
d $\frac{3}{26}$ **e** Student's own answers

6 $\frac{1}{6}$

7 **a** i) $\frac{1}{10}$ ii) $\frac{1}{4}$
b i) $\frac{1}{19}$ ii) $\frac{3}{19}$

8 **a** $\frac{1}{37}$ **b** $\frac{18}{37}$ **c** $\frac{18}{37}$
d $\frac{1}{37}$ **e** $\frac{21}{37}$ **f** $\frac{12}{37}$
g $\frac{17}{37}$ **h** $\frac{11}{37}$

9 **a** RCA RAC CRA CAR ARC ACR
b i) $\frac{1}{6}$ ii) $\frac{1}{3}$ iii) $\frac{1}{2}$ iv) $\frac{1}{24}$

10 **a** $\frac{1}{4}$ **b** $\frac{1}{2}$ **c** $\frac{1}{13}$ **d** $\frac{1}{26}$
e $\frac{3}{13}$ **f** $\frac{1}{52}$ **g** $\frac{5}{13}$ **h** $\frac{4}{13}$

Exercise 3.9.1

1 a

	Dice 1			
Dice 2	1	2	3	4
1	1, 1	2, 1	3, 1	4, 1
2	1, 2	2, 2	3, 2	4, 2
3	1, 3	2, 3	3, 3	4, 3
4	1, 4	2, 4	3, 4	4, 4

b $\frac{1}{4}$ **c** $\frac{1}{4}$ **d** $\frac{9}{16}$

2

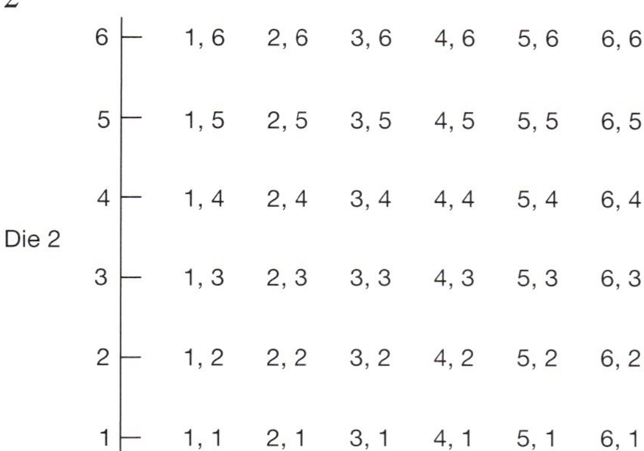

a $\frac{1}{36}$ **b** $\frac{1}{6}$ **c** $\frac{1}{18}$

d $\frac{1}{6}$ **e** $\frac{1}{4}$ **f** $\frac{3}{4}$

g $\frac{11}{36}$ **h** $\frac{1}{6}$ **i** $\frac{11}{18}$

Exercise 3.9.2

1 a

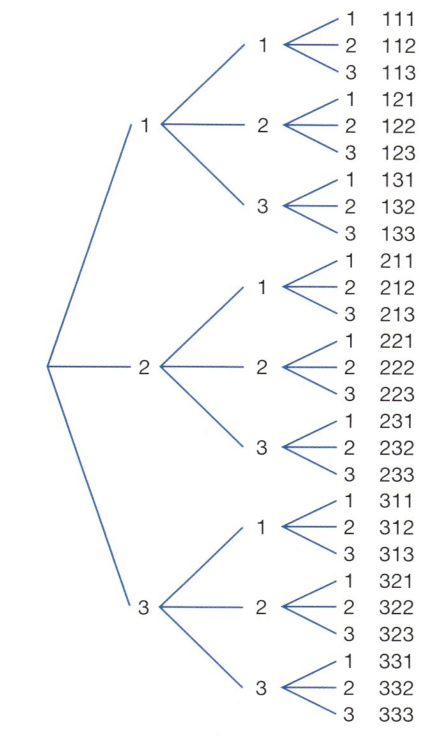

b **i)** $\frac{1}{27}$ **ii)** $\frac{1}{3}$ **iii)** $\frac{1}{9}$

iv) $\frac{1}{3}$ **v)** $\frac{5}{9}$ **vi)** $\frac{1}{3}$

2 a

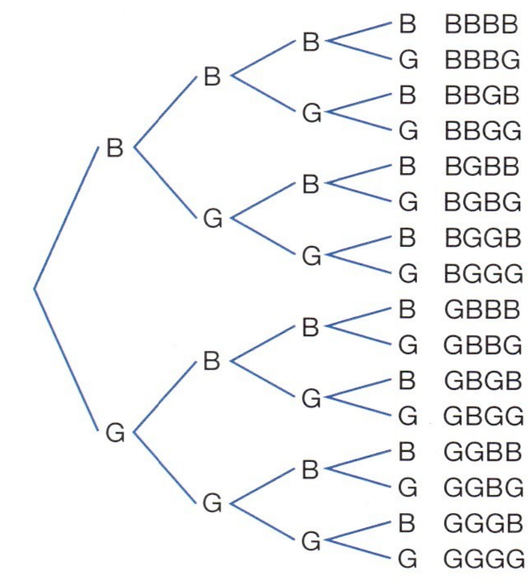

b **i)** $\frac{1}{16}$ **ii)** $\frac{3}{8}$ **iii)** $\frac{15}{16}$

iv) $\frac{5}{16}$

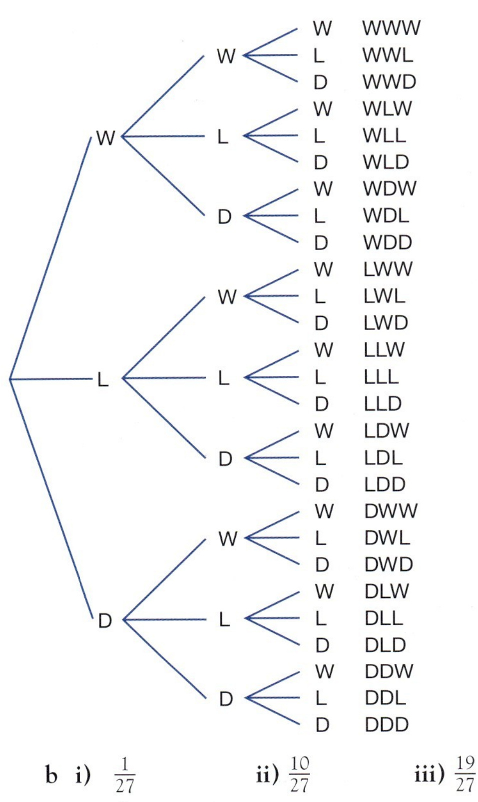

b i) $\frac{1}{27}$ ii) $\frac{10}{27}$ iii) $\frac{19}{27}$
iv) $\frac{8}{27}$

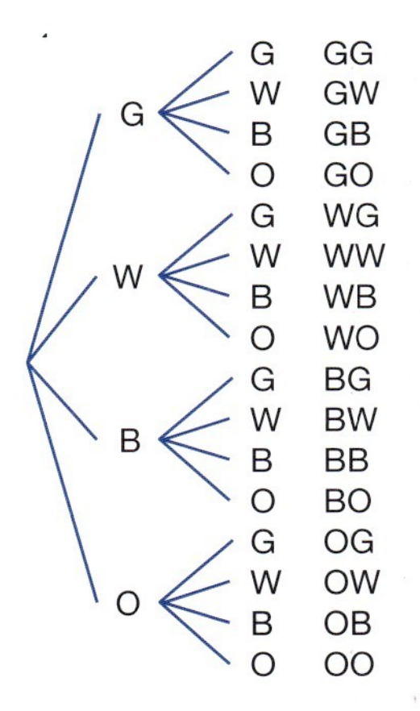

b i) $\frac{1}{16}$ ii) $\frac{1}{4}$ iii) $\frac{1}{8}$

Exercise 3.9.3

1 a

Roll 1	Roll 2	Outcomes	Probability
Six ($\frac{1}{6}$)	Six ($\frac{1}{6}$)	Six, Six	$\frac{1}{6}\times\frac{1}{6}=\frac{1}{36}$
	Not six ($\frac{5}{6}$)	Six, Not six	$\frac{1}{6}\times\frac{5}{6}=\frac{5}{36}$
Not six ($\frac{5}{6}$)	Six ($\frac{1}{6}$)	Not six, Six	$\frac{5}{6}\times\frac{1}{6}=\frac{5}{36}$
	Not six ($\frac{5}{6}$)	Not six, Not six	$\frac{5}{6}\times\frac{5}{6}=\frac{25}{36}$

b i) $\frac{1}{6}$ ii) $\frac{11}{16}$ iii) $\frac{5}{36}$
iv) $\frac{125}{216}$ v) $\frac{91}{216}$

c They add up to 1, because either iv or v.

2 a $\frac{4}{25}$ b $\frac{54}{125}$ c $\frac{98}{125}$

3 a

1st	2nd	3rd	Outcome
L (0.35)	L (0.35)	L (0.35)	L, L, L
		OT (0.65)	L, L, OT
	OT (0.65)	L (0.35)	L, OT, L
		OT (0.65)	L, OT, OT
OT (0.65)	L (0.35)	L (0.35)	OT, L, L
		OT (0.65)	OT, L, OT
	OT (0.65)	L (0.35)	OT, OT, L
		OT (0.65)	OT, OT, OT

b i) 0.275 ii) 0.123 iii) 0.444
iv) 0.718

4 0.027

5 a 0.75^2 (0.56) b 0.75^3 (0.42)
c 0.75^{10} (0.06)

Exercise 3.9.4

1 a $\frac{25}{81}$ b $\frac{16}{81}$ c $\frac{20}{81}$
d $\frac{40}{81}$

2 a $\frac{5}{18}$ b $\frac{1}{6}$ c $\frac{5}{18}$
d $\frac{5}{9}$

3 a $\frac{2}{50}$ b $\frac{3}{10}$

4 a $\frac{1}{45}$ b $\frac{1}{3}$

5 a $\frac{920}{9312}$ b $\frac{8372}{9312}$

6 a

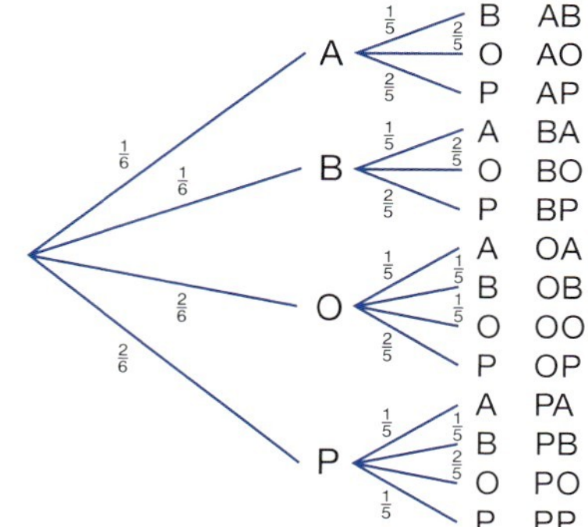

b i) $\frac{1}{15}$ ii) $\frac{1}{15}$ iii) $\frac{3}{5}$

Exercise 3.9.5

1 a

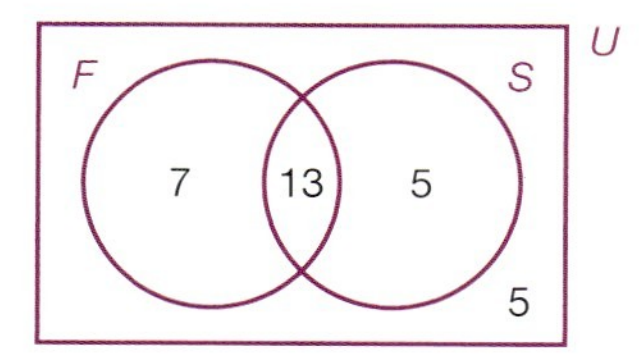

b $\frac{13}{30}$

2 a $\frac{5}{35} = \frac{1}{7}$ b $\frac{14}{35} = \frac{2}{5}$ c $\frac{13}{35}$

3 $\frac{45}{108}$

4 a

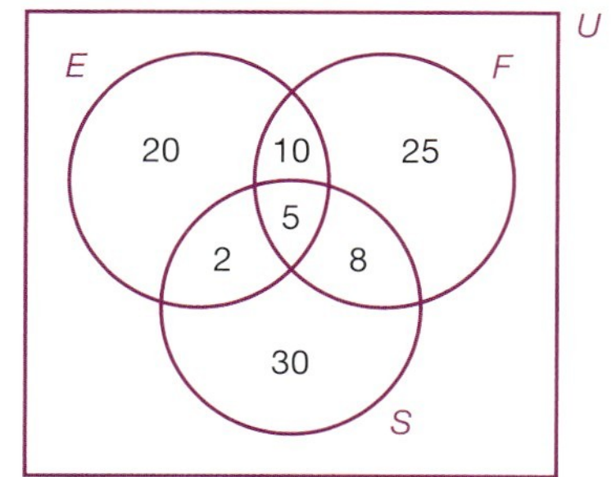

b i) $\frac{5}{100}$ ii) $\frac{20}{100}$ iii) $\frac{25}{100}$

Exercise 3.10.1

1 a No (mutually exclusive)
b 0.8

2 a Yes
b i) $\frac{1}{6}$ ii) $\frac{2}{3}$ iii) $\frac{1}{2}$

3 $\frac{1}{144}$ if each month is equally likely, or $\frac{36}{5329}$ if taken as 30 days out of 365. (Leap years excluded – though this could be an extension.)

4 a Yes
b i) 0.1 ii) 0.9 iii) 0.2
iv) 0.4

5 a 0.675 b 0.2 c 0.875
d 0.77

6 0.93

7 0.66

8 e Only 23 people are needed before the probability of two sharing the same birthday is greater than 50%.

Student assessment 1

1 a {odd numbers from 1 to 7}
b {odd numbers}
c {square numbers}
d {oceans}

2 a 7 b 2 c 7
d Student's own answer

3 a

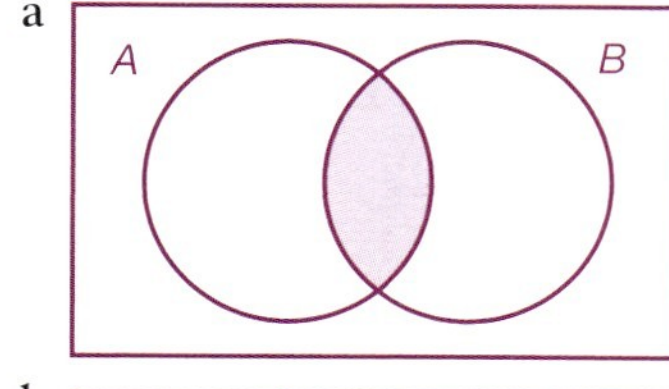

b

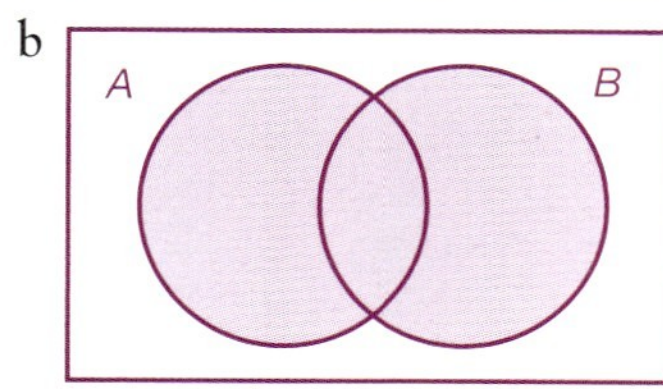

c

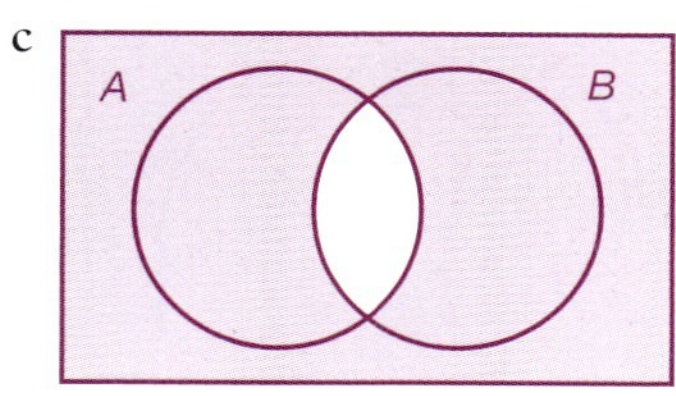

4 {o, r, k}, {w, r, k}, {w, o, k}, {w, o, r}, {w, o, r, k}

5 $P' = \{1, 3, 5, 7\}$

6 {(r r r)(b r r)(r b r)(r r b)(b b r)(b r b)(r b b)(b b b)}

7 The Amazon river is not in Africa.

8 $p \vee q$ means either p or q or both must be true for the statement to be true.
$p \vee q$ means either p or q but not both must be true for the statement to be true.

9 a $\frac{1}{7}$ b $\frac{12}{365}$ c $\frac{1}{365}$
d $\frac{1}{1461}$

10 a

	1	2	3	4	5	6
H	1H	2H	3H	4H	5H	6H
T	1T	2T	3T	4T	5T	6T

b i) $\frac{1}{12}$ ii) $\frac{1}{4}$ iii) $\frac{1}{4}$

Student assessment 2

1 {2, 4}, {2, 6}, {2, 8}, {4, 6}, {4, 8}, {6, 8}, {2, 4, 6}, {2, 4, 8}, {2, 6, 8}, {4, 6, 8}

2
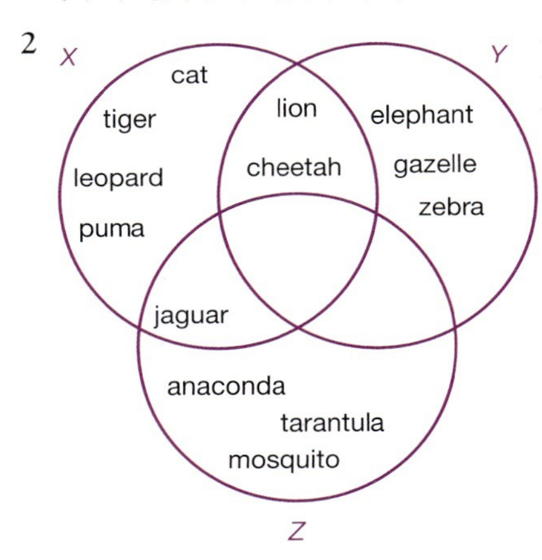

b {lion, cheetah} c ∅ d ∅

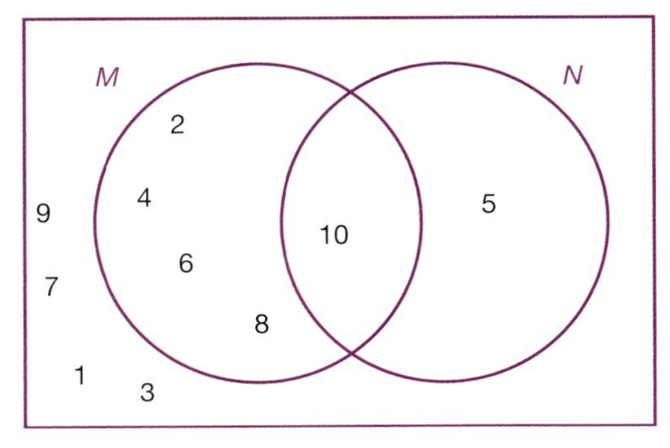

b X = {multiples of 10}

4 a Let the number liking only tennis be x.

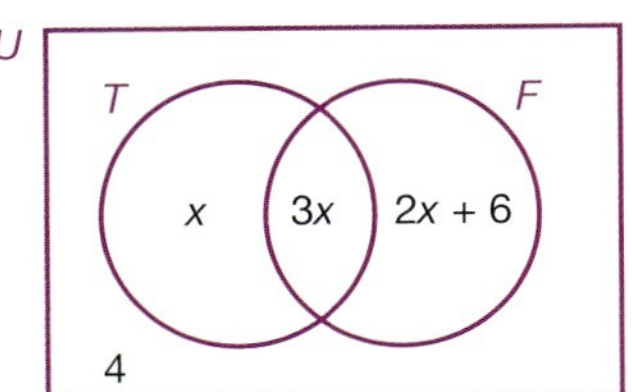

b 15 c 5 d 16

5 a 5 b 35 c 40
d 50 e 15 f 12
g 10 h 78 i 78

6 p and q.

7

p	q	$p \wedge q$	$p \vee q$
T	T	T	T
T	F	F	T
F	T	F	T
F	F	F	F

8 A statement that is always true. Student's example.

9 a $\frac{12}{27}$ b $\frac{8}{27}$ c $\frac{6}{27}$

Student assessment 3

1 0.31

2
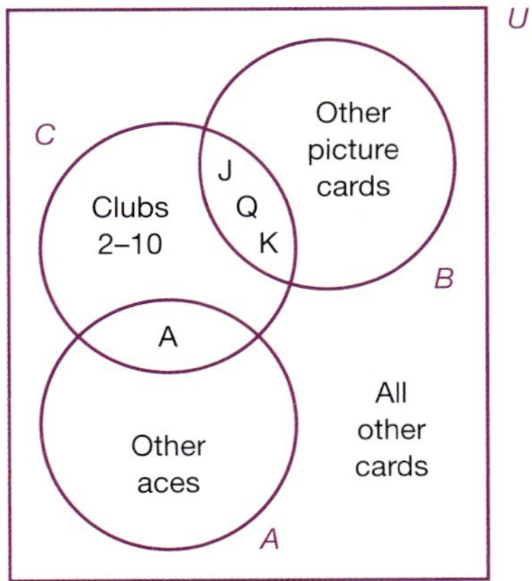

b i) $\frac{16}{52}$ ii) $\frac{36}{52}$ iii) $\frac{16}{52}$
iv) $\frac{1}{52}$ v) 0

3 a

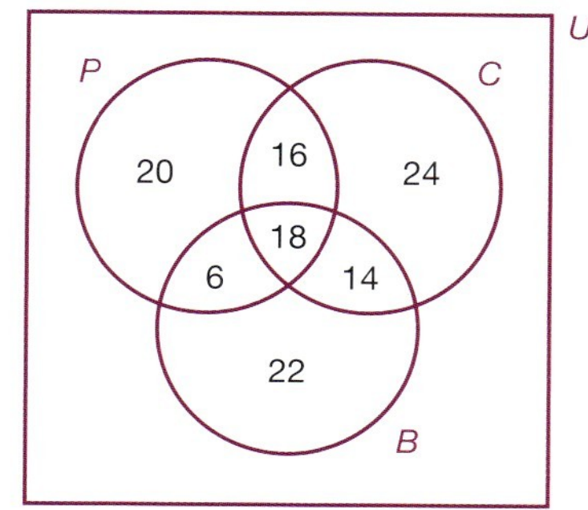

b i) $\frac{66}{120}$ ii) $\frac{60}{120}$

4 0.53

5 0.88

6 0.4

7 a i) 0.8 ii) 0.7 iii) 0.3
b i) 0.06 ii) 0.56 iii) 0.38

Topic 4

Exercise 4.1.1

1 a, b and c

2 a domain: $-1 \leq x \leq 3$ range: $-3 \leq f(x) \leq 5$
b domain: $-4 \leq x \leq 0$ range: $-10 \leq f(x) \leq 2$
c domain: $-3 \leq x \leq 3$ range: $2 \leq f(x) \leq 11$
d domain: $y > 3$ range: $0 < g(y) < \frac{1}{3}$
e domain: $t \in \mathbb{R}$ range: $h(t) \in \mathbb{R}$
f domain: $y \in \mathbb{R}$ range: $f(y) = 4$
g domain: $n \in \mathbb{R}$ range: $f(n) \leq 2$

Exercise 4.2.1

1 a 2.375 b 0.5 c 0.125
d −4

2 a −1 b −6 c −3.5
d −16

3 a −4 b −0.25 c 5
d 2.75

4 a 2.75 b 0.25 c −3.5
d 0.5

5 a $f(x) = 50 + 100x$
b 300€

6 a

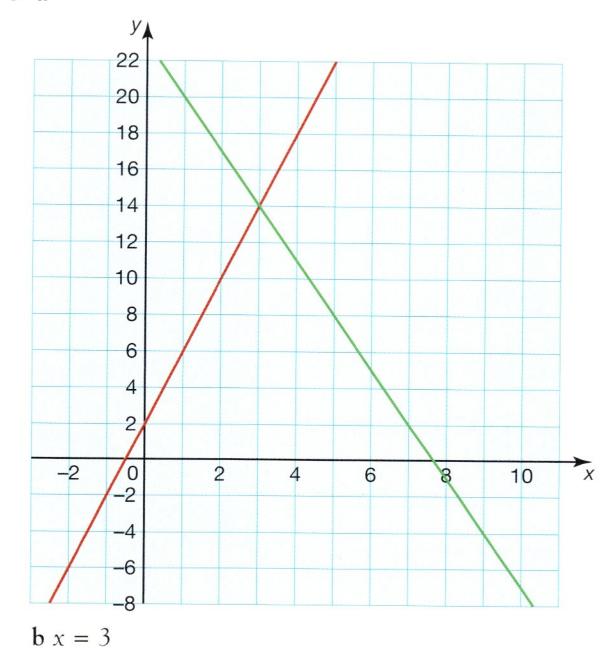

b $x = 3$

Exercise 4.2.2

1 a $x - 3$ b $x + 5$ c $3x$
d $\frac{x}{4}$ e $\frac{x-5}{2}$ f $\frac{x+6}{3}$
g $2x - 4$ h $2(x - 3)$ i $\frac{x+24}{12}$
j $\frac{x+4}{6}$

2 a 6 b 4 c −1

3 a 2 b −0.5 c −6

4 a 3 b 1.5 c 2

5 a 4 b −2 c −11

Exercise 4.3.1

1 a

x	−4	−3	−2	−1	0	1	2	3
y	10	4	0	−2	−2	0	4	10

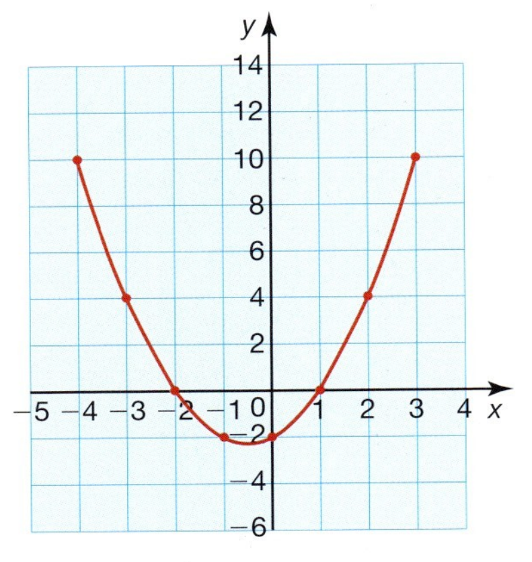

b $x = -\frac{1}{2}$, $(-0.5, -2.25)$, minimum

2 a

x	−3	−2	−1	0	1	2	3	4	5
y	−12	−5	0	3	4	3	0	−5	−12

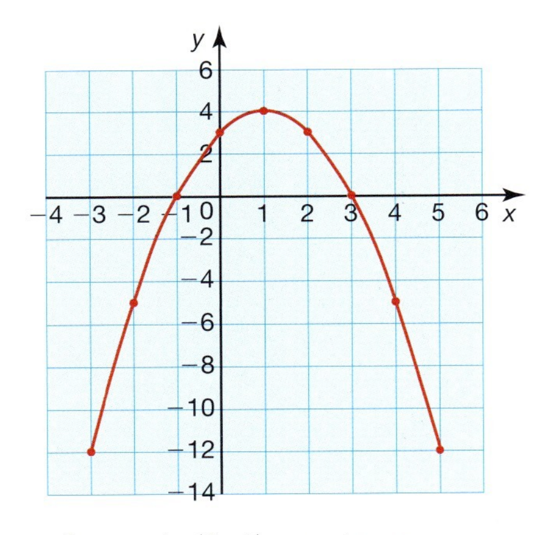

b $x = 1$, $(1, 4)$, maximum

3 a

x	−1	0	1	2	3	4	5
y	9	4	1	0	1	4	9

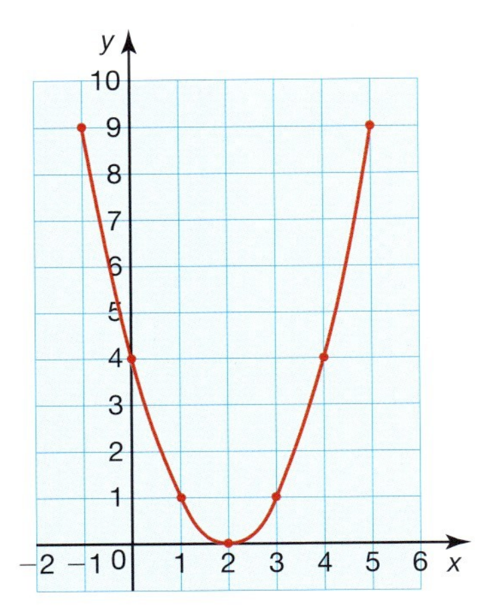

b $x = 2$, $(2, 0)$, minimum

4 a

x	−4	−3	−2	−1	0	1	2
y	−9	−4	−1	0	−1	−4	−9

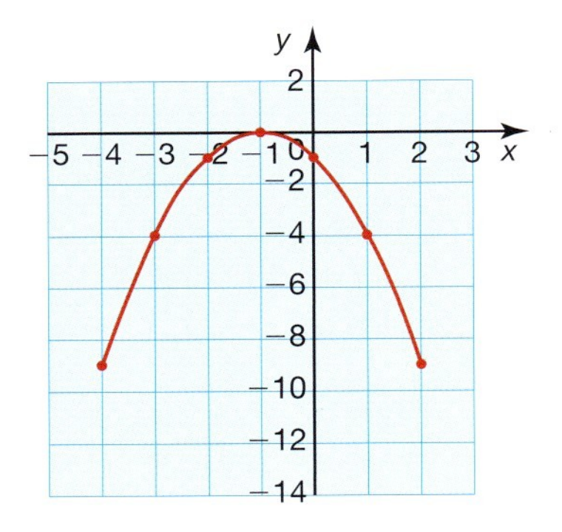

b $x = -1$, $(-1, 0)$, maximum

5 a

x	−4	−3	−2	−1	0	1	2	3	4	5	6
y	9	0	−7	−12	−15	−16	−15	−12	−7	0	9

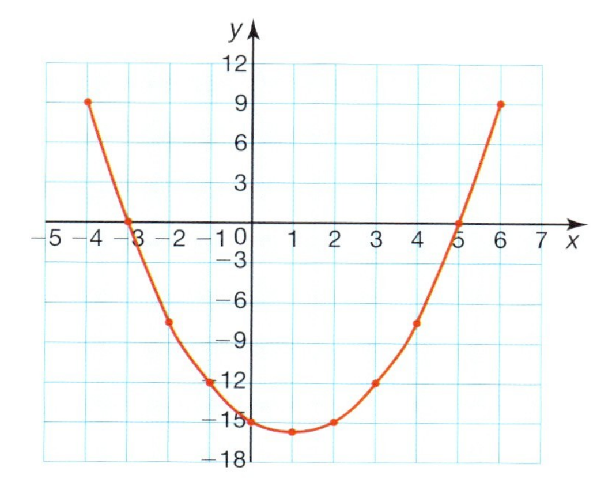

b $x = 1$, $(1, -16)$, minimum

6 a

x	−2	−1	0	1	2	3
y	9	1	−3	−3	1	9

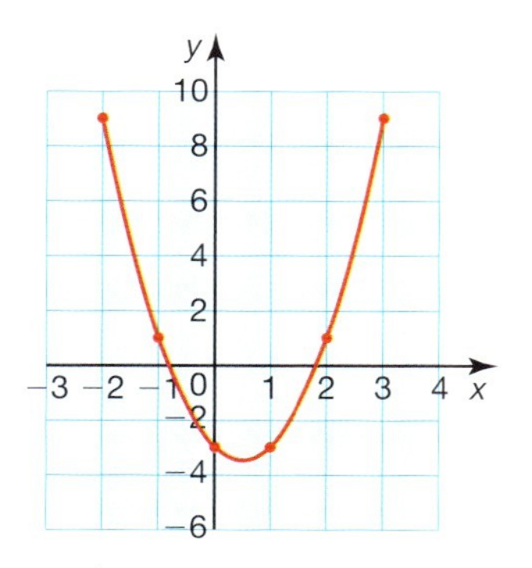

b $x = \frac{1}{2}$, $(0.5, -3.5)$, minimum

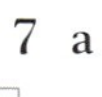

7 a

x	−3	−3	−1	0	1	2	3
y	−15	−4	3	6	5	0	−9

b $x = \frac{1}{4}$, $(0.25, 6.125)$, maximum

8 a

x	−2	−1	0	1	2	3
y	12	0	−6	−6	0	12

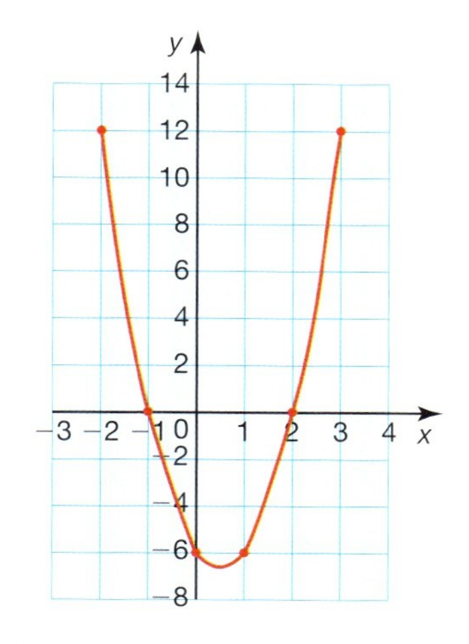

b $x = \frac{1}{2}$, $(0.5, -6.75)$, minimum

9 a

x	−1	0	1	2	3
y	7	−4	−7	−2	11

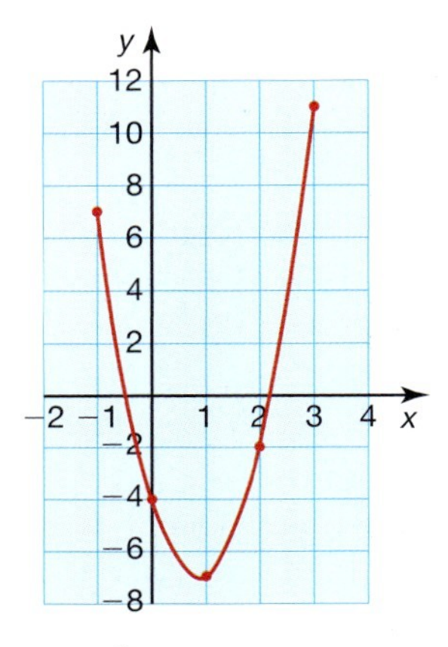

b $x = \frac{7}{8}$, (0.875, −7.0625), minimum

10 a

x	−2	−1	0	1	2	3
y	−25	−9	−1	−1	−9	−25

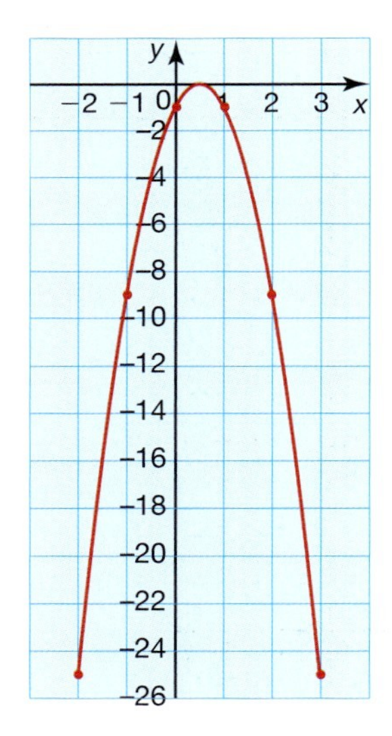

b $x = \frac{1}{2}$, (0.5, 0), maximum

Exercise 4.3.2

1 −2 and 3

2 −1 and 1

3 3

4 −4 and 3

5 2

6 0.5 and 3

7 1

8 $-\frac{1}{3}$ and 2

Exercise 4.3.3

1 −1.6 and 2.6

2 No solution

3 2 and 4

4 −3.5 and 2.5

5 0.3 and 3.7

6 0 and 3.5

7 −0.2 and 2.2

8 $-\frac{1}{3}$ and 2

Exercise 4.3.4

1 a $(x + 4)(x + 3)$ b $(x + 6)(x + 2)$
c $(x + 12)(x + 1)$ d $(x - 3)(x - 4)$
e $(x - 6)(x - 2)$ f $(x - 12)(x - 1)$

2 a $(x + 5)(x + 1)$ b $(x + 4)(x + 2)$
c $(x + 3)^2$ d $(x + 5)^2$
e $(x + 11)^2$ f $(x - 6)(x - 7)$

3 a $(x + 12)(x + 2)$ b $(x + 8)(x + 3)$
c $(x - 6)(x - 4)$ d $(x + 12)(x + 3)$
e $(x + 18)(x + 2)$ f $(x - 6)^2$

4 a $(x + 5)(x - 3)$ b $(x - 5)(x + 3)$
c $(x + 4)(x - 3)$ d $(x - 4)(x + 3)$
e $(x + 6)(x - 2)$ f $(x - 12)(x - 3)$

5 a $(x - 4)(x + 2)$ b $(x - 5)(x + 4)$
c $(x + 6)(x - 5)$ d $(x + 6)(x - 7)$
e $(x + 7)(x - 9)$ f $(x + 9)(x - 6)$

6 a $2(x + 1)^2$ b $(2x + 3)(x + 2)$
c $(2x - 3)(x + 2)$ d $(2x - 3)(x - 2)$
e $(3x + 2)(x + 2)$ f $(3x - 1)(x + 4)$
g $(2x + 3)^2$ h $(3x - 1)^2$
i $(3x + 1)(2x - 1)$

Exercise 4.3.5

1 a −4 and −3
b −2 and −6
c −5 and 2
d −2 and 5
e −3 and −2
f −3
g −2 and 4
h −4 and 5
i −6 and 5
j −6 and 7

2 a −3 and 3
b −5 and 5
c −12 and 12
d −2.5 and 2.5
e −2 and 2
f $-\frac{1}{3}$ and $\frac{1}{3}$
g −4 and −2
h 2 and 4
i −4 and 6
j −6 and 8

3 a −9 and 4
b −1
c 0 and 8
d 0 and 7
e −1.5 and −1
f −1 and 2.5
g −12 and 0
h −9 and −3
i −6 and 6
j 10 and 10

4 −4 and 3

5 2

6 4

7 $x = 6$, height = 3 cm, base length = 12 cm

8 $x = 10$, height = 20 cm, base length = 2 cm

9 $x = 6$, base = 6 cm, height = 5 cm

10 11 m × 6 m

Exercise 4.3.6

1 a −3.14 and 4.14
b −5.87 and 1.87
c −6.14 and 1.14
d −4.73 and −1.27
e −6.89 and 1.89
f 3.38 and 5.62

2 a −5.30 and −1.70
b −5.92 and 5.92
c −3.79 and 0.79
d −1.14 and 6.14
e −4.77 and 3.77
f −2.83 and 2.83

3 a −0.73 and 2.73
b −1.87 and 5.87
c −1.79 and 2.79
d −3.83 and 1.83
e 0.38 and 2.62
f 0.39 and 7.61

4 a −0.85 and 2.35
b −1.40 and 0.90
c 0.14 and 1.46
d −2 and −0.5
e −0.39 and 1.72
f −1.54 and 1.39

Exercise 4.4.1

1 a i

x	−3	−2	−1	0	1	2	3
$f(x)$	0.04	0.11	0.33	1	3	9	27

ii

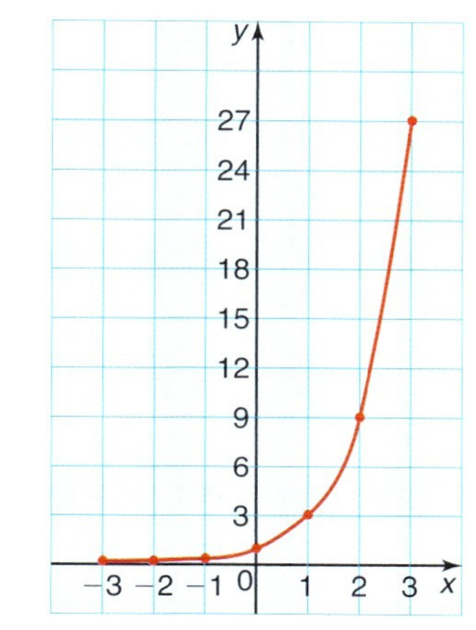

b i

x	−3	−2	−1	0	1	2	3
$f(x)$	1	1	1	1	1	1	1

ii

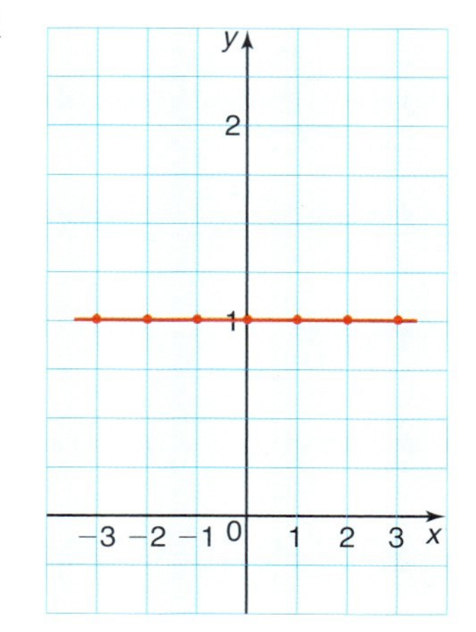

c i

x	−3	−2	−1	0	1	2	3
$f(x)$	3.125	3.25	3.5	4	5	7	11

ii

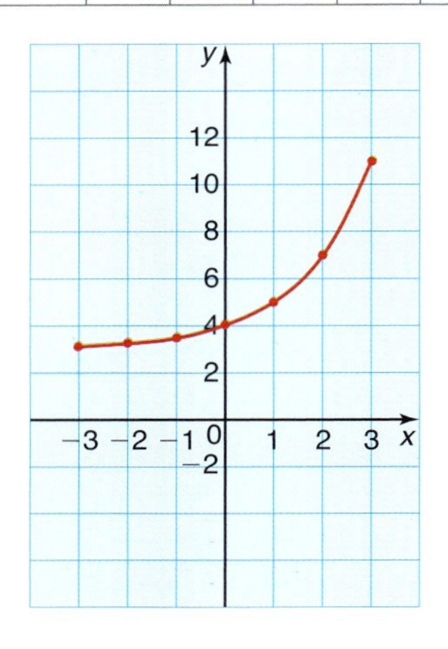

d i

x	−3	−2	−1	0	1	2	3
$f(x)$	−2.875	−1.75	−0.5	1	3	6	11

ii

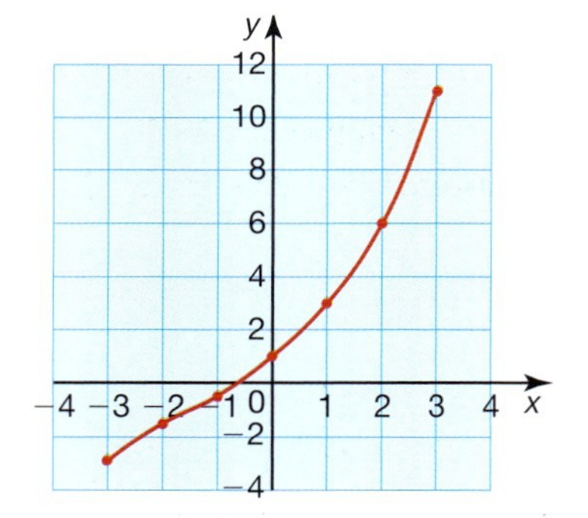

e i

x	−3	−2	−1	0	1	2	3
$f(x)$	3.125	2.25	−1.5	1	1	2	5

ii

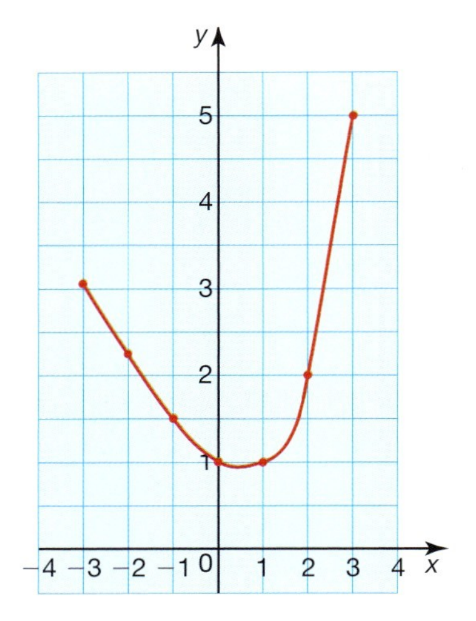

f i

x	−3	−2	−1	0	1	2	3
$f(x)$	−8.96	−3.89	−0.67	1	2	5	18

ii

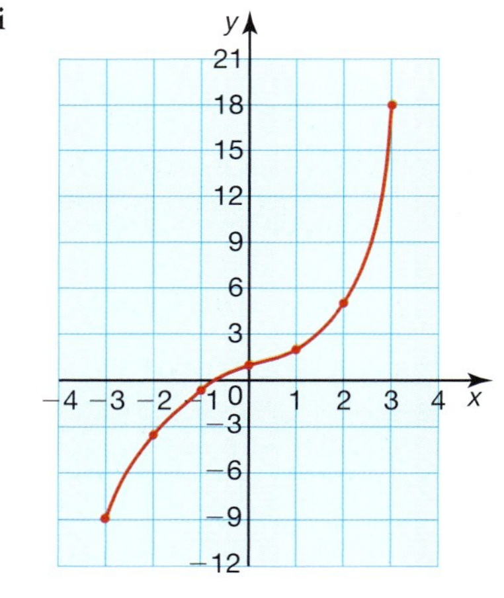

2 **a** 7 cm **b** 0 cm **c** 5 hours
e $\approx 5\frac{1}{2}$ hours

3 **a** ≈ 2.5 **b** $-\frac{1}{2}$

4 ≈ 4.3

Exercise 4.4.2

1 Translation $\begin{pmatrix} 0 \\ c \end{pmatrix}$

2 Stretch parallel to y-axis of scale factor k.

3 Changes the slope of the curve.

4 Reflection in the y-axis.

Exercise 4.4.3

1 16 777 216

2 240 000 years

3 4900 million

4 0.098 g

5 8

6 a 358 000 km^2
  b 7 years

7 2.0

8 9%

Exercise 4.5.1

1 b The maximum and minimum height of the curve (the amplitude) is altered. The y-coordinate of the maximum and minimum points is multiplied by a.

2 b The frequency with which the graph completes one cycle (a period) is $\frac{360}{b}$.

3 b The graph is translated by an amount, c, vertically up or down, i.e. it is translated by $\begin{pmatrix} 0 \\ c \end{pmatrix}$.

4 b The maximum and minimum height of the curve (the amplitude) is altered. The y-coordinate of the maximum and minimum points is multiplied by a.

5 b The frequency with which the graph completes one cycle (a period) is $\frac{360}{b}$.

6 b The graph is translated by an amount, c, vertically up or down, i.e. it is translated by $\begin{pmatrix} 0 \\ c \end{pmatrix}$.

Exercise 4.5.2

1 a 0.5, 360° b 2, 360° c 1, 72°
  d 1, 1440° e 3, 180° f −1, 120°
  g −2, 720° h −3, 180°

2 a 1, 360° b 3, 360° c 2, 360°
  d 1, 720° e 1, 180° f −1, 360°
  g −1, 1080° h −1, 180°

Exercise 4.5.3

1 a $x = 30°, 150°$
  b $x = 29.0°, 331°$
  c $x = 10°, 50°, 130°, 170°, 250°, 290°$
  d $x = 0°, 98.8°$

2 a $x = 60°, 300°$
  b $x = 45°, 225°$
  c $x = 22.5°, 45°, 112.5°, 202.5°, 225°, 292.5°$
  d $x = 30°, 60°, 150°, 180°, 270°, 300°$

Exercise 4.7.1

1 The graph is stretched by a scale factor 'a' parallel to the y-axis.

2 The graph is translated $\begin{pmatrix} -b \\ 0 \end{pmatrix}$.

Exercise 4.7.2

1 a i) Vertical asymptote: $x = -1$
  Horizontal asymptote: $y = 0$
  ii) Graph intercepts y-axis at (0, 1)
  iii)

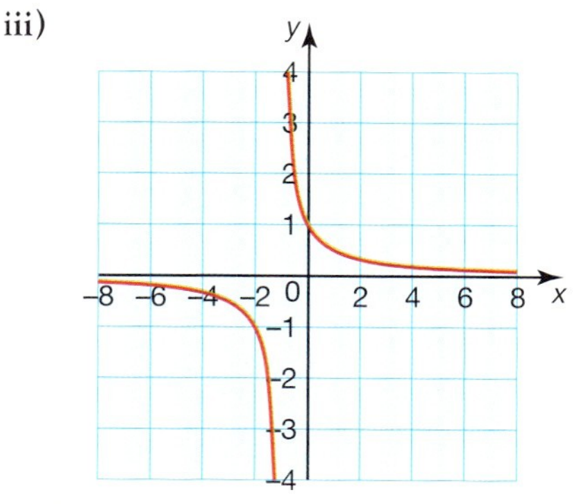

  b i) Vertical asymptote: $x = -3$
  Horizontal asymptote: $y = 0$
  ii) Graph intercepts y-axis at $\left(0, \frac{1}{3}\right)$

iii)

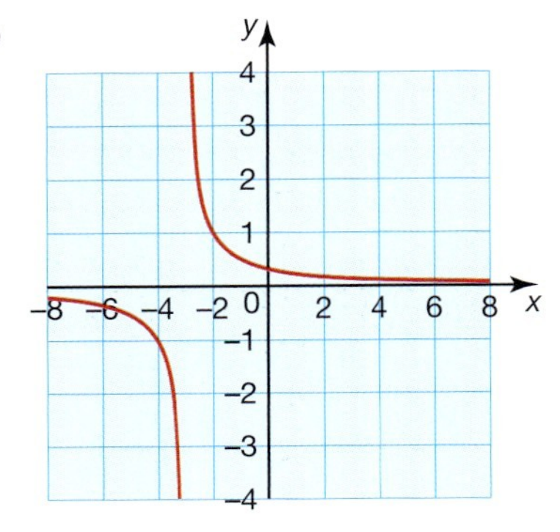

c **i)** Vertical asymptote: $x = 4$

Horizontal asymptote: $y = 0$

ii) Graph intercepts y-axis at $\left(0, -\frac{1}{2}\right)$

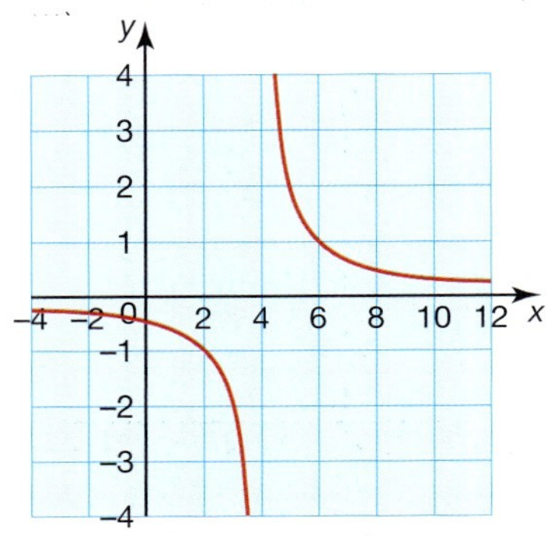

d **i)** Vertical asymptote: $x = 3$

Horizontal asymptote: $y = 0$

ii) Graph intercepts y-axis at $\left(0, \frac{1}{3}\right)$

iii)

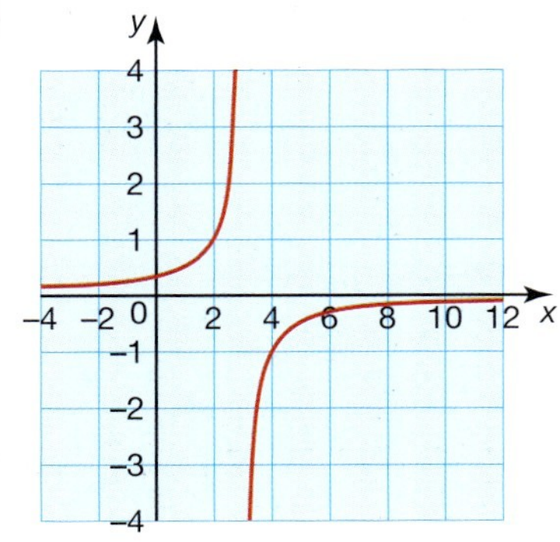

2 **a** **i)** Vertical asymptote: $x = 0$

Horizontal asymptote: $y = 2$

ii) Graph intercepts x-axis at $\left(-\frac{1}{2}, 0\right)$

iii)

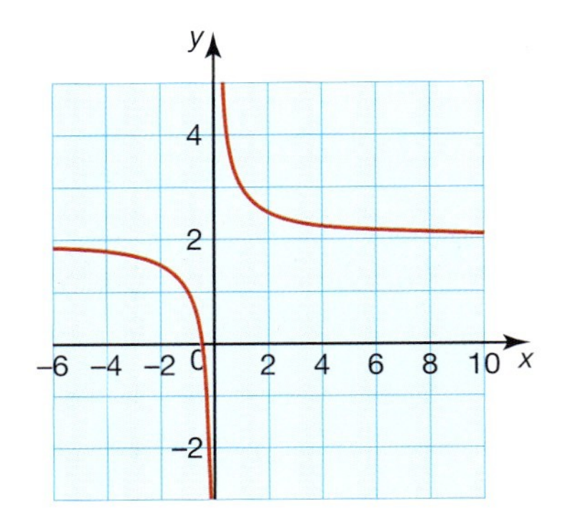

b **i)** Vertical asymptote: $x = 0$

Horizontal asymptote: $y = -3$

ii) Graph intercepts x-axis at $\left(\frac{1}{3}, 0\right)$

iii)

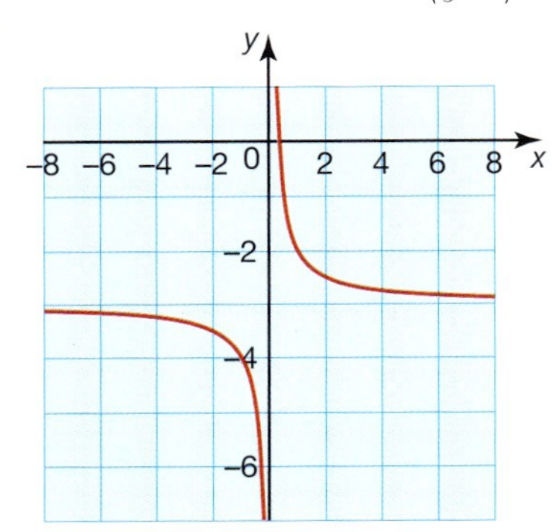

c **i)** Vertical asymptote: $x = 1$

Horizontal asymptote: $y = 4$

ii) Graph intercepts x-axis at $\left(\frac{3}{4}, 0\right)$ and y-axis at $(0, 3)$

iii)

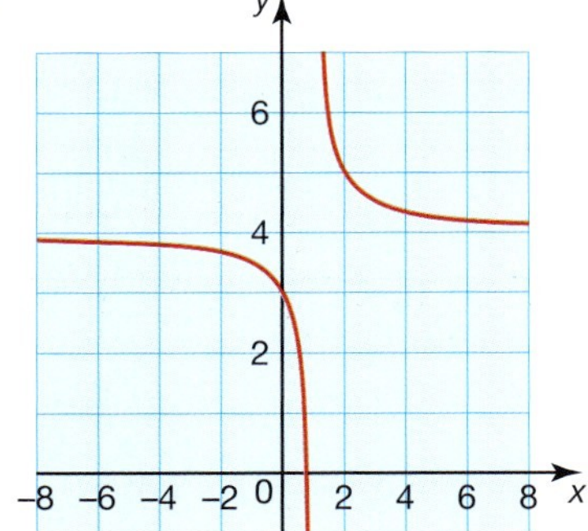

d **i)** Vertical asymptote: $x = -4$

Horizontal asymptote: $y = -1$

ii) Graph intercepts x-axis at $(-3, 0)$ and y-axis at $\left(0, -\frac{3}{4}\right)$

iii)

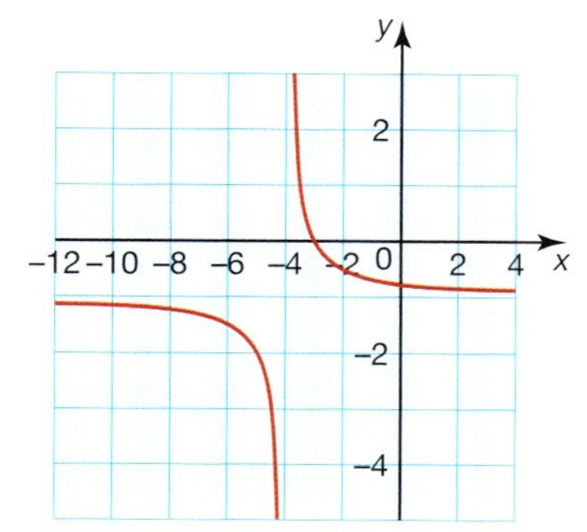

3 a i) Vertical asymptote: $x = -\frac{1}{2}$

Horizontal asymptote: $y = 0$

ii) Graph intercepts y-axis at $(0, 1)$

iii)

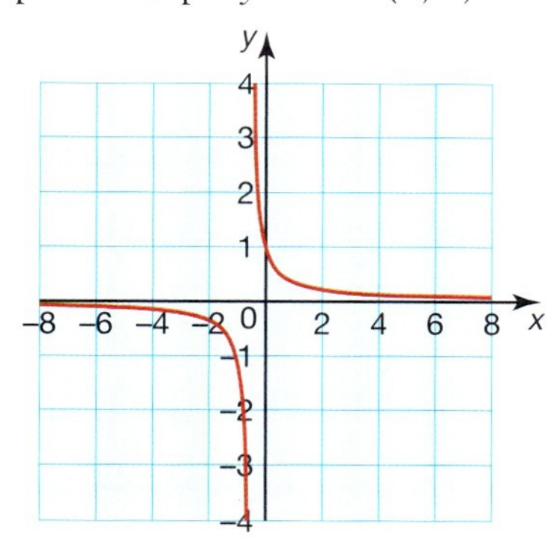

b i) Vertical asymptote: $x = \frac{1}{2}$

Horizontal asymptote: $y = 1$

ii) Graph intercepts axes at $(0,0)$

iii)

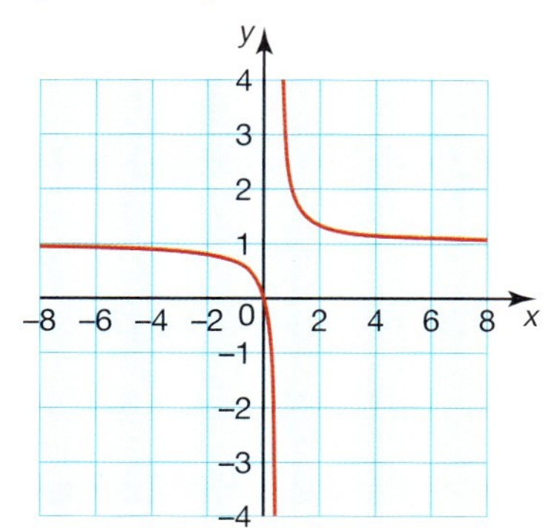

c i) Vertical asymptote: $x = \frac{1}{3}$

Horizontal asymptote: $y = 0$

ii) Graph intercepts y-axis at $(0, -2)$

iii)

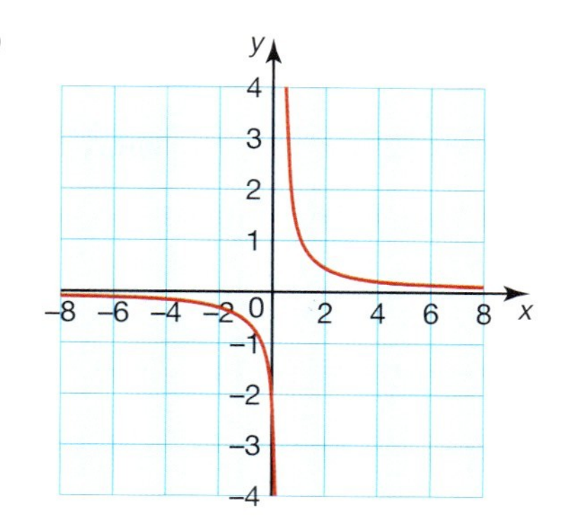

d i) Vertical asymptote: $x = \frac{1}{4}$

Horizontal asymptote: $y = 2$

ii) Graph intercepts x-axis at $\left(-\frac{3}{8}, 0\right)$ and the y-axis at $(0, 3)$

iii)

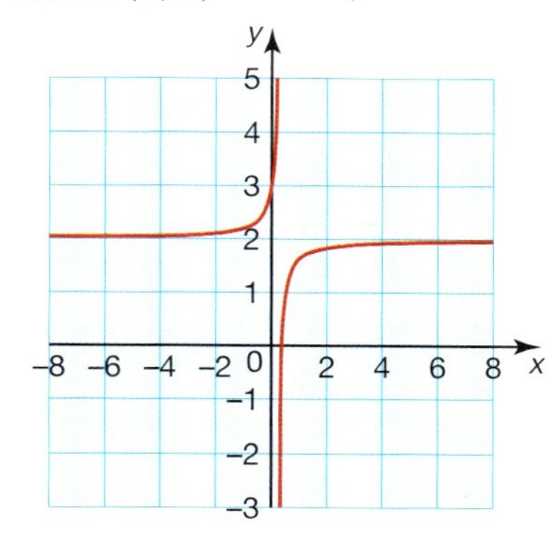

Exercise 4.7.3

1 a i) Vertical asymptotes: $x = 1$ and 2

Horizontal asymptote: $y = 0$

ii) Graph intercepts y-axis at $\left(0, \frac{1}{2}\right)$

iii)

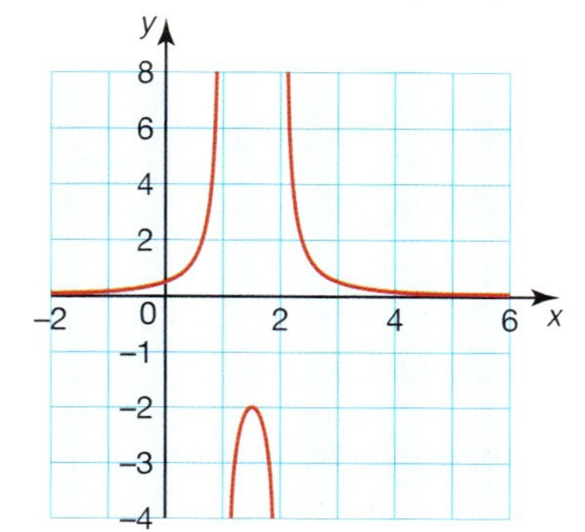

b **i)** Vertical asymptotes: $x = -3$ and 4
Horizontal asymptote: $y = 0$
ii) Graph intercepts y-axis at $\left(0, -\frac{1}{12}\right)$
iii)

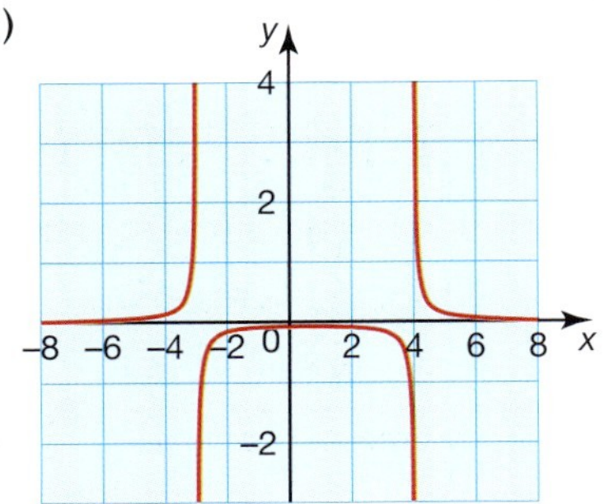

c **i)** Vertical asymptotes: $x = 0$ and 5
Horizontal asymptote: $y = 1$
ii) Graph intercepts x-axis at $\left(\frac{5 \pm \sqrt{21}}{2}, 0\right)$
iii)

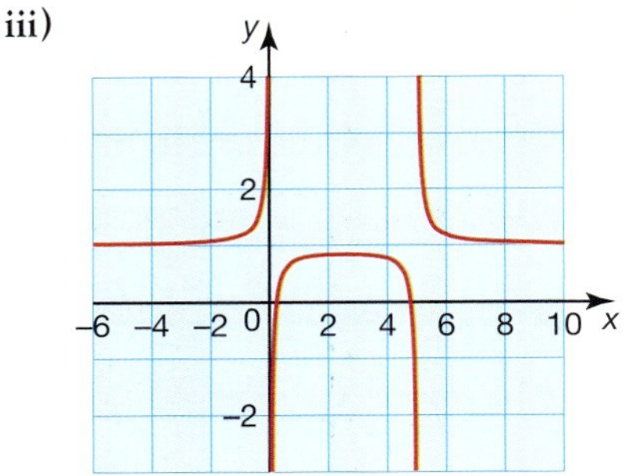

d **i)** Vertical asymptote: $x = -2$
Horizontal asymptote: $y = -3$
ii) Graph intercepts the x-axis at $\left(-2 \pm \sqrt{\frac{1}{3}}, 0\right)$ and the y-axis at $\left(0, -2\frac{3}{4}\right)$
iii)

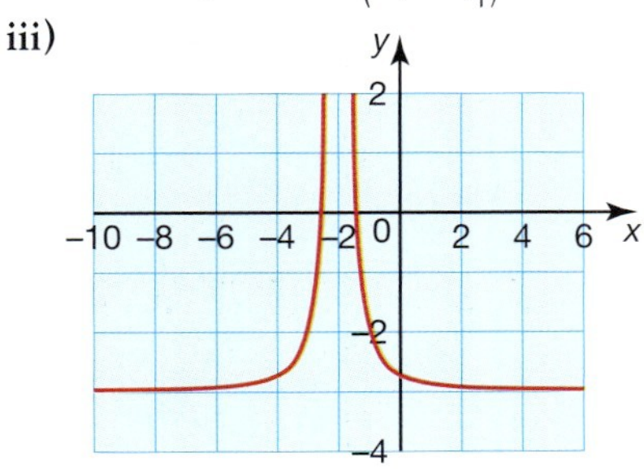

2 **a** **i)** Vertical asymptotes: $x = -4$ and 1
Horizontal asymptote: $y = 0$
ii) Graph intercepts the y-axis at $\left(0, -\frac{1}{4}\right)$
iii)

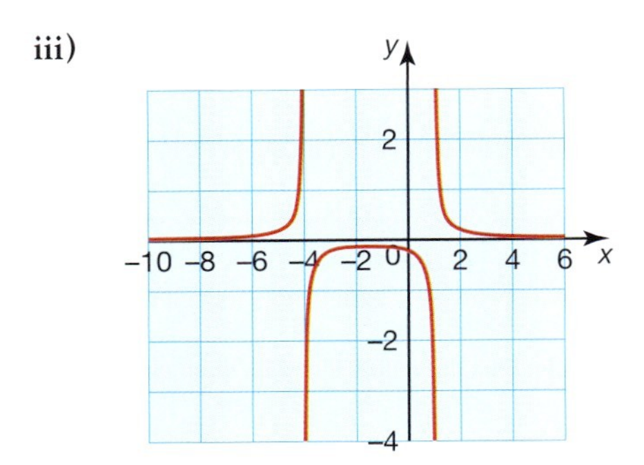

b **i)** Vertical asymptotes: $x = -5$ and 2
Horizontal asymptote: $y = 0$
ii) Graph intercepts the y-axis at $\left(0, -\frac{1}{10}\right)$
iii)

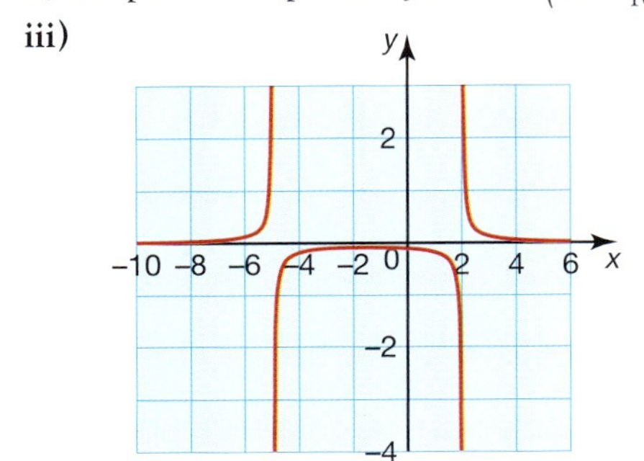

c **i)** Vertical asymptote: $x = -1$
Horizontal asymptote: $y = 3$
ii) Graph intercepts the y-axis at $(0, 4)$
iii)

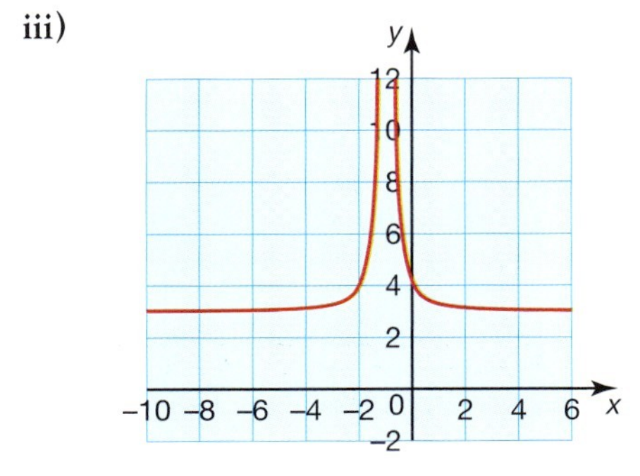

d **i)** Vertical asymptotes: $x = -\frac{1}{2}$ and 4
Horizontal asymptote: $y = -1$
ii) Graph intercepts the x-axis at $\left(\frac{7 \pm \sqrt{89}}{4}, 0\right)$ and the y-axis at $\left(0, -\frac{5}{4}\right)$

iii)

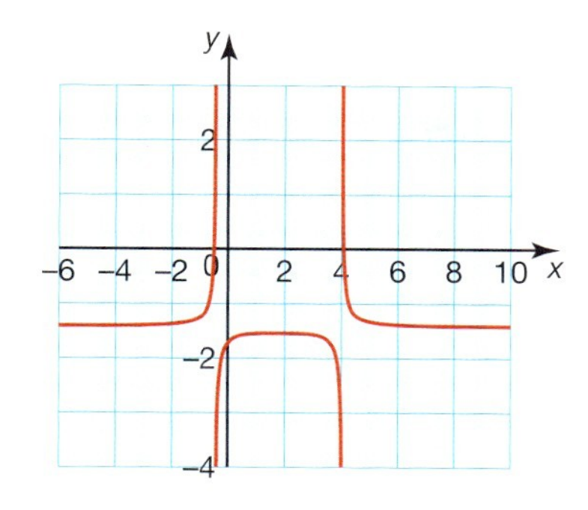

Exercise 4.7.4

1 a i)

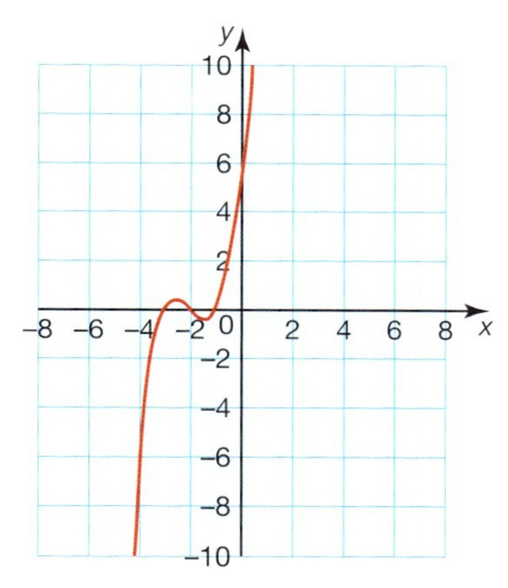

ii) Graph intercepts the y-axis at (0, 6)

iii) $(x + 3)(x + 2)(x + 1)$

b i)

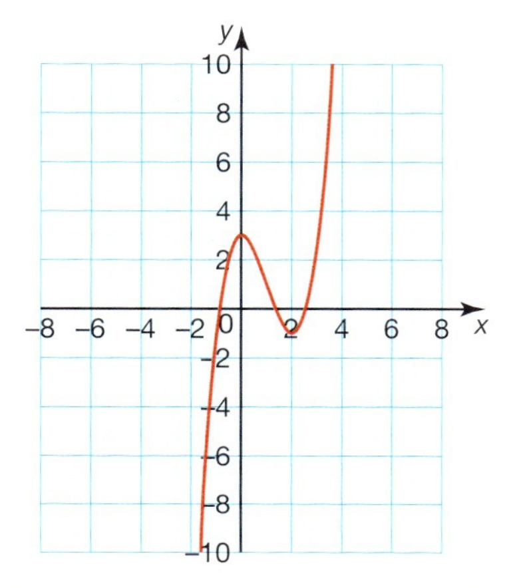

ii) Graph intercepts the y-axis at (0, 3)

iii) $(x + 0.88)(x - 1.35)(x - 2.53)$

c i)

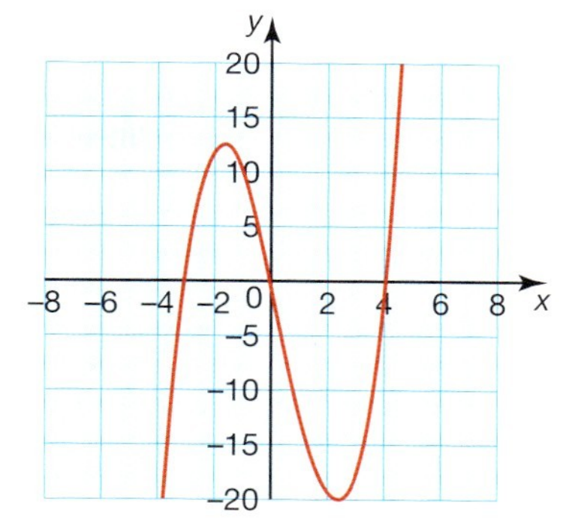

ii) Graph intercepts the y-axis at (0, 0)

iii) $x(x + 3)(x - 4)$

d i)

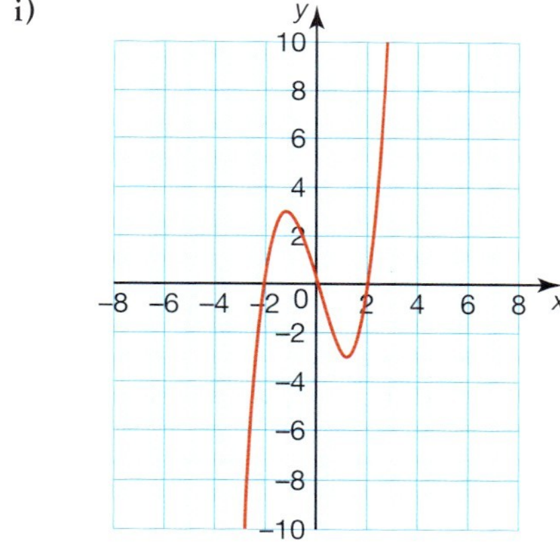

ii) Graph intercepts the y-axis at (0, 0)

iii) $x(x + 2)(x - 2)$

2 a i)

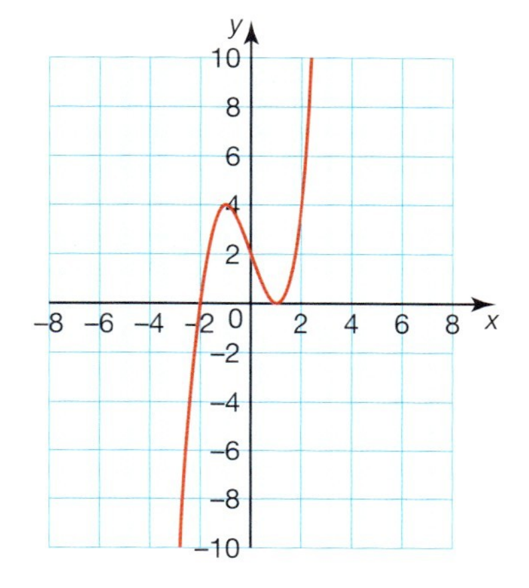

ii) Graph intercepts the y-axis at (0, 2)

iii) $(x + 2)(x - 1)(x - 1)$

b i)

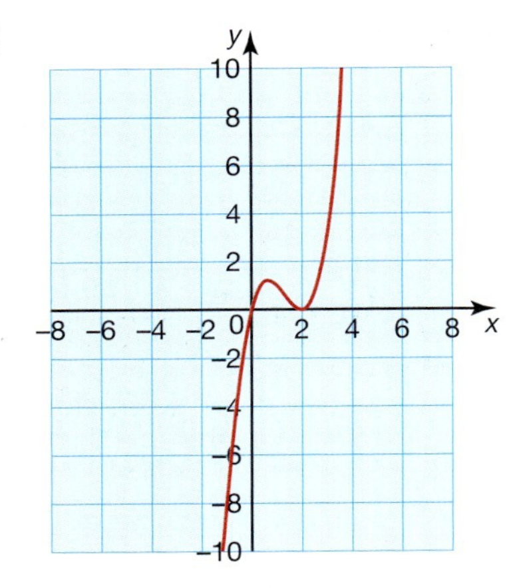

ii) Graph intercepts the y-axis at $(0, 0)$

iii) $x(x - 2)(x - 2)$

c i)

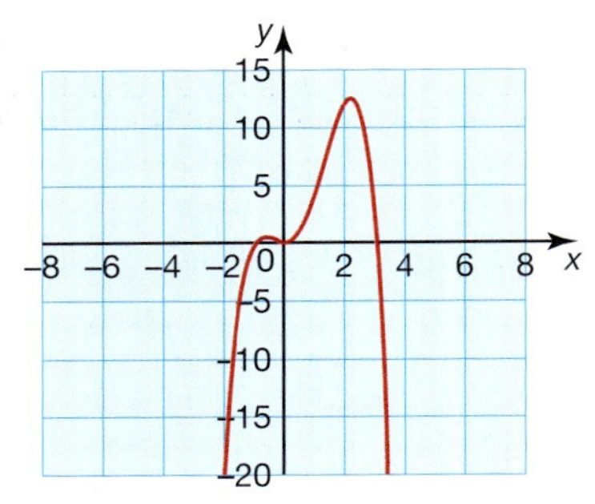

ii) Graph intercepts the y-axis at $(0, 0)$

iii) $-x^2(x + 1)(x - 3)$

d i)

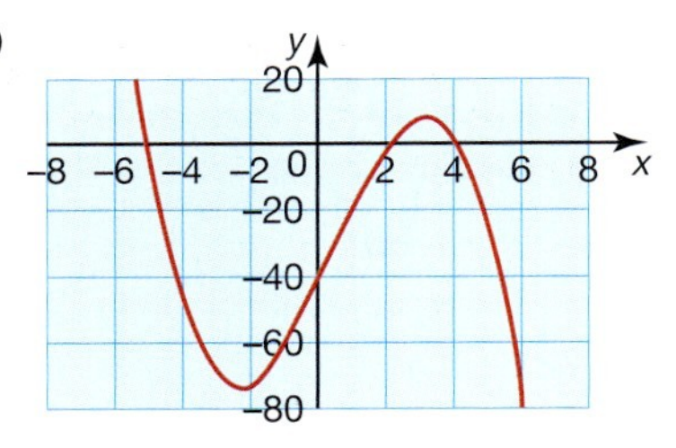

ii) Graph intercepts the y-axis at $(0, -40)$

iii) $-(x + 5)(x - 2)(x - 4)$

Exercise 4.8.1

1 a $x = 4$ and -5
  b $x = -3$ and $\frac{1}{2}$
  c $x = -\frac{1}{3}$ and $\frac{1}{2}$
  d $x = -\frac{5}{2}$ and 7
  e $x = 1$ and $\frac{1}{4}$
  f $x = -2$, 1 and 4

2 a $x = 0.2$ and 4.9
  b $x = -1.6$ and 1.0
  c $x = -0.5$, 0.7 and 5.9
  d $x = -0.8$ and 1.3
  e $x = 0.4$ and 2.2
  f $x = 17.6°$ and $162.4°$; $0 \leq x \leq 360$

Student assessment 1

1 b is not a function as each input value should only produce one output value.

2 a domain: $-3 \leq x \leq 1$ range: $-11 \leq f(x) \leq 1$
  b domain: $x \in R$ range: $f(x) \geq 0$

3 $3 \leq f(q) \leq 27$

4 a 5 b -1 c $-\frac{11}{2}$

5 a $f^{-1}(x) = 3 - \frac{x}{3}$
  b $g^{-1}(x) = \frac{2}{3}x + 6$

6 a 5 b 2

7 a $(-3, 4)$ b $(3, 5)$ c $\left(0, \frac{7}{2}\right)$

8 a

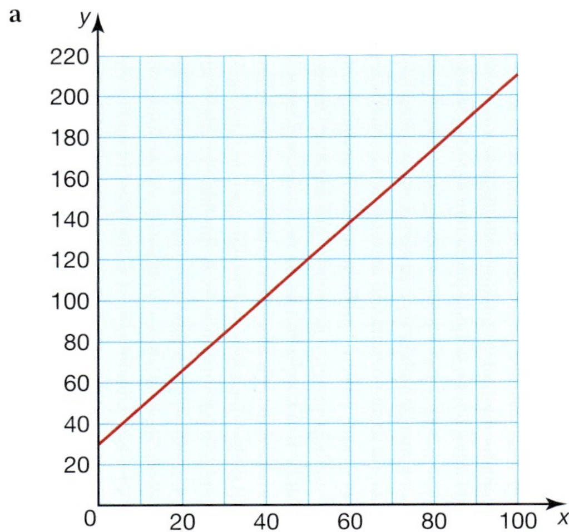

  b i) ≈ 50 °F ii) ≈ 104 °F iii) ≈ 176 °F

9 a $R(n) = 12n$

b

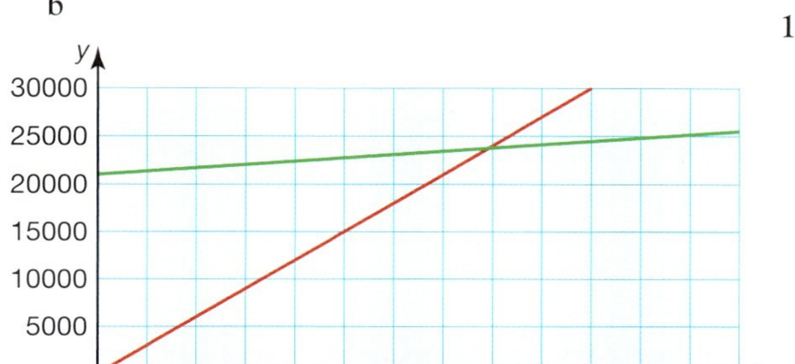

c 20 000 copies

d ≈ 2.4 million euros (Exact 2 415 000€)

Student assessment 2

1

x	−3	−2	−1	0	1	2	3
$f(x) = x^2 + 3x - 9$	−9	−11	−11	−9	−5	1	9

b

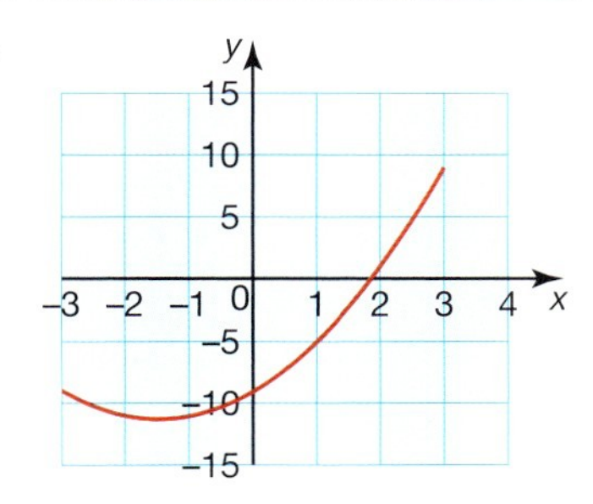

c Range: $-11.25 \leq f(x) \leq 9$

2 a $f(x) = (x - 3)(x - 6)$

b $h(y) = (3y - 2)(y + 1)$

c $f(x) = (x - 5)(x + 2)$

d $h(x) = (2x - 1)(x + 4)$

3 a $x = -4$ and -2 b $x = 1$ and 5

c $x = -5$ d $x = 1$ and $\frac{4}{3}$

4 a $x = 0.191$ and 1.31

b $x = 2.62$ and -0.228

5 8244.13€

6 $211.8x$

7 22 years

8 17 m

Student assessment 3

1 a b

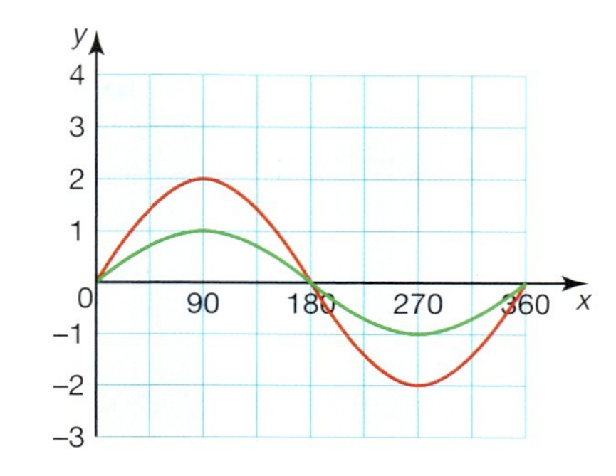

2 a b

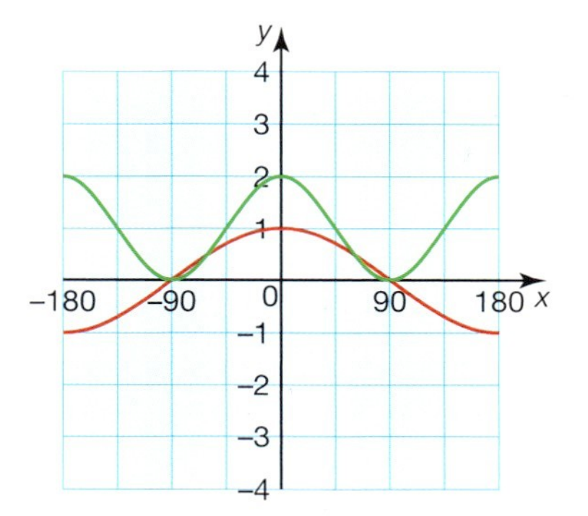

3 a b

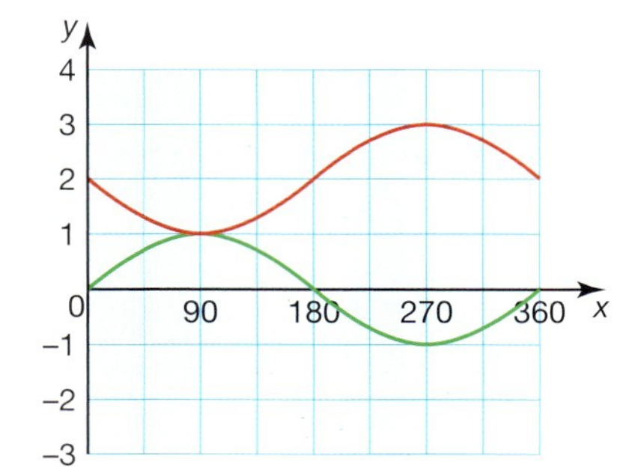

4 a b

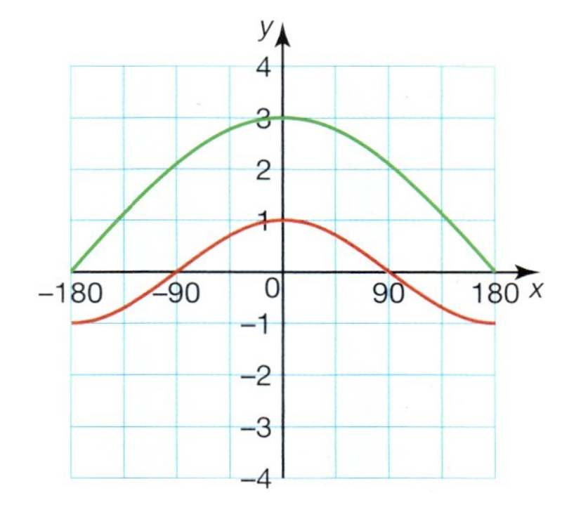

5 $x = 20.9°$ and $69.1°$

6 $x = 109.5°$ and $250.5°$

7 a Vertical asymptote $x = 3$
Horizontal asymptote $y = 1$

b $\left(0, \frac{1}{3}\right)$ and $(1, 0)$

c

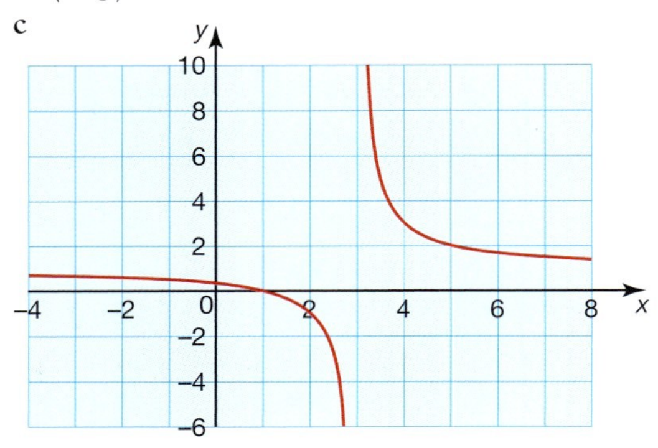

8 a Vertical asymptote $x = 2$
Horizontal asymptote $y = -1$

b $\left(0, -\frac{5}{4}\right)$ and $\left(\frac{5}{2}, 0\right)$

c

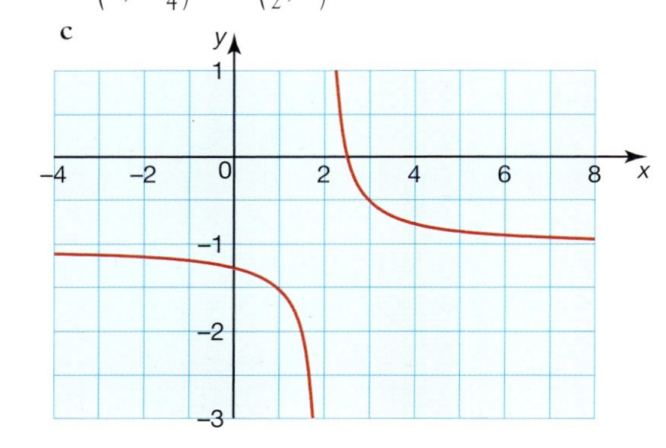

9 a Vertical asymptotes $x = -1$ and 5
Horizontal asymptote $y = 0$

b $\left(0, -\frac{1}{5}\right)$

c

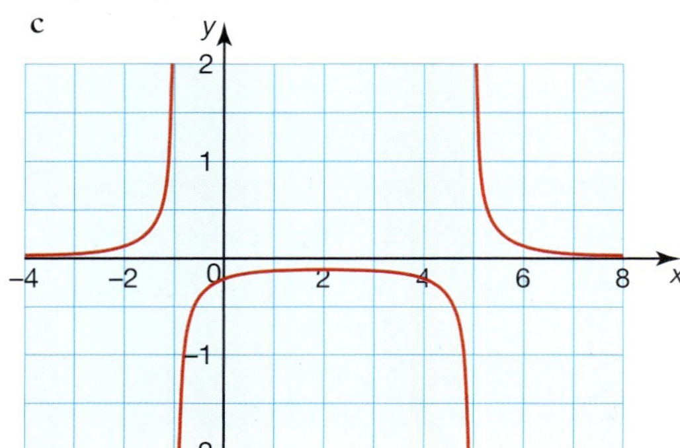

10 a Vertical asymptotes $x = 2$ and 3
Horizontal asymptote $y = 2$

b $\left(0, 2\frac{1}{6}\right)$

c

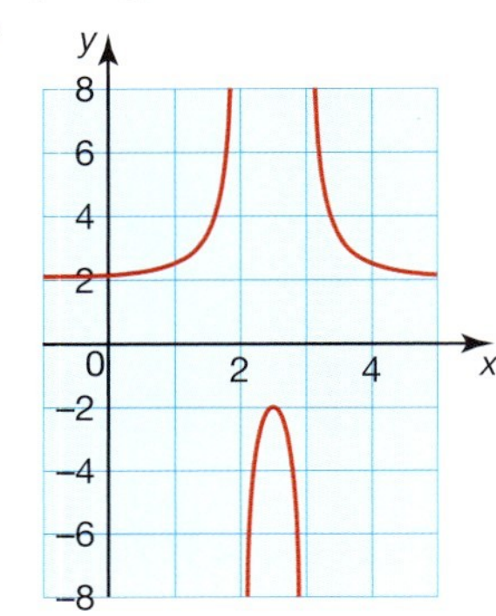

Topic 5

Exercise 5.1.1

1 a ii) 5.66 units **iii)** (3, 4)
b ii) 4.24 units **iii)** (4.5, 2.5)
c ii) 5.66 units **iii)** (3, 6)
d ii) 8.94 units **iii)** (2, 4)
e ii) 6.32 units **iii)** (3, 4)
f ii) 6.71 units **iii)** (−1.5, 4)
g ii) 8.25 units **iii)** (−2, 1)
h ii) 8.94 units **iii)** (0, 0)
i ii) 7 units **iii)** (0.5, 5)
j ii) 6 units **iii)** (2, 3)
k ii) 8.25 units **iii)** (0, 4)
l ii) 10.82 units **ii)** (0, 1.5)

2 a i) 4.25 units **ii)** (2.5, 2.5)
b i) 5.66 units **ii)** (5, 4)
c i) 8.94 units **ii)** (4, 2)
d i) 8.94 units **ii)** (5, 0)
e i) 4.24 units **ii)** (−1.5, 4.5)
f i) 4.47 units **ii)** (−4, −3)
g i) 7.21 units **ii)** (0, 3)
h i) 7.21 units **ii)** (5, −1)
i i) 12.37 units **ii)** (0, 2.5)
j i) 8.49 units **ii)** (1, −1)
k i) 11 units **ii)** (0.5, −3)
l i) 8.25 units **ii)** (4, 2)

Exercise 5.2.1

1 a i) 1 **ii)** −1
b i) 1 **ii)** −1
c i) 1 **ii)** −1

d i) 2 ii) $-\frac{1}{2}$
e i) 3 ii) $-\frac{1}{3}$
f i) 2 ii) $-\frac{1}{2}$
g i) 4 ii) $-\frac{1}{4}$
h i) $\frac{1}{2}$ ii) -2
i i) 0 ii) infinite
j i) infinite ii) 0
k i) $\frac{1}{4}$ ii) -4
l i) $\frac{3}{2}$ ii) $-\frac{2}{3}$

2 a i) -1 ii) 1
b i) -1 ii) 1
c i) -2 ii) $\frac{1}{2}$
d i) $-\frac{1}{2}$ ii) 2
e i) -1 ii) 1
f i) -2 ii) $\frac{1}{2}$
g i) $-\frac{3}{2}$ ii) $\frac{2}{3}$
h i) $\frac{2}{3}$ ii) $-\frac{3}{2}$
i i) $-\frac{1}{4}$ ii) 4
j i) -1 ii) 1
k i) 0 ii) infinite
l i) -4 ii) $\frac{1}{4}$

Exercise 5.2.2

1 a $y = 7$ b $y = 2$
c $x = 7$ d $x = 3$
e $y = x$ f $y = \frac{1}{2}x$
g $y = -x$ h $y = -2x$

2 a $y = x + 1$ b $y = x + 3$
c $y = x - 2$ d $y = 2x + 2$
e $y = \frac{1}{2}x + 5$ f $y = \frac{1}{2}x - 1$

3 a $y = -x + 4$ b $y = -x - 2$
c $y = -2x - 2$ d $y = -\frac{1}{2}x + 3$
e $y = -\frac{3}{2}x + 2$ f $y = -4x + 1$

4 a 2 a 1 b 1 c 1 d 2 e $\frac{1}{2}$ f $\frac{1}{2}$
3 a -1 b -1 c -2 d $-\frac{1}{2}$ e $-\frac{3}{2}$ f -4
b The gradient is equal to the coefficient of x.
c The constant being added/subtracted indicates where the line intersects the y-axis.

5 b Only the intercept c is different.

6 The lines are parallel.

Exercise 5.2.3

1 a $m = 2$ $c = 1$ b $m = 3$ $c = 5$
c $m = 1$ $c = -2$ d $m = \frac{1}{2}$ $c = 4$
e $m = -3$ $c = 6$ f $m = -\frac{2}{3}$ $c = 1$
g $m = -1$ $c = 0$ h $m = -1$ $c = -2$
i $m = -2$ $c = 2$

2 a $m = 3$ $c = 1$ b $m = -\frac{1}{2}$ $c = 2$
c $m = -2$ $c = -3$ d $m = -2$ $c = -4$
e $m = \frac{1}{4}$ $c = 6$ f $m = 3$ $c = 2$
g $m = 1$ $c = -2$ h $m = -8$ $c = 6$
i $m = 3$ $c = 1$

3 a $m = 2$ $c = -3$ b $m = \frac{1}{2}$ $c = 4$
c $m = 2$ $c = -4$ d $m = -8$ $c = 12$
e $m = 2$ $c = 0$ f $m = -3$ $c = 3$
g $m = 2$ $c = 1$ h $m = -\frac{1}{2}$ $c = 2$
i $m = 2$ $c = -\frac{1}{2}$

4 a $m = 2$ $c = -4$ b $m = 1$ $c = 6$
c $m = -3$ $c = -1$ d $m = -1$ $c = 4$
e $m = 10$ $c = -2$ f $m = -3$ $c = \frac{3}{2}$
g $m = -9$ $c = 2$ h $m = 6$ $c = -14$
i $m = 2$ $c = -\frac{3}{2}$

5 a $m = 2$ $c = -2$ b $m = 2$ $c = 3$
c $m = 1$ $c = 0$ d $m = \frac{3}{2}$ $c = 6$
e $m = -1$ $c = \frac{2}{3}$ f $m = -4$ $c = 2$
g $m = 3$ $c = -12$ h $m = 0$ $c = 0$
i $m = -3$ $c = 0$

6 a $m = 1$ $c = 0$ b $m = -\frac{1}{2}$ $c = -2$
c $m = -3$ $c = 0$ d $m = 1$ $c = 0$
e $m = -2$ $c = -\frac{2}{3}$ f $m = \frac{2}{3}$ $c = -4$
g $m = -\frac{2}{5}$ $c = 0$ h $m = -\frac{1}{3}$ $c = -\frac{7}{6}$
i $m = 3$ $c = 0$ j $m = -\frac{2}{3}$ $c = -\frac{8}{3}$

Exercise 5.2.4

1 a i) $y = 2x - 1$ ii) $2x - y - 1 = 0$
b i) $y = 3x + 1$ ii) $3x - y + 1 = 0$
c i) $y = 2x + 3$ ii) $2x - y + 3 = 0$
d i) $y = x - 4$ ii) $x - y - 4 = 0$
e i) $y = 4x + 2$ ii) $4x - y + 2 = 0$
f i) $y = -x + 4$ ii) $x + y - 4 = 0$
g i) $y = -2x + 2$ ii) $2x + y - 2 = 0$
h i) $y = -3x - 1$ ii) $3x + y + 1 = 0$
i i) $y = \frac{1}{2}x$ ii) $x - 2y = 0$

2 a i) $y = \frac{1}{7}x + \frac{26}{7}$ ii) $x - 7y + 26 = 0$
b i) $y = \frac{6}{7}x + \frac{4}{7}$ ii) $6x - 7y + 4 = 0$
c i) $y = \frac{3}{2}x + \frac{15}{2}$ ii) $3x - 2y + 15 = 0$
d i) $y = 9x - 13$ ii) $9x - y - 13 = 0$
e i) $y = -\frac{1}{2}x + \frac{5}{2}$ ii) $x + 2y - 5 = 0$
f i) $y = -\frac{3}{13}x + \frac{70}{13}$ ii) $3x + 13y - 70 = 0$
g i) $y = 2$ ii) $y - 2 = 0$
h i) $y = -3x$ ii) $3x + y = 0$
i i) $x = 6$ ii) $x - 6 = 0$

Exercise 5.2.5

1 a

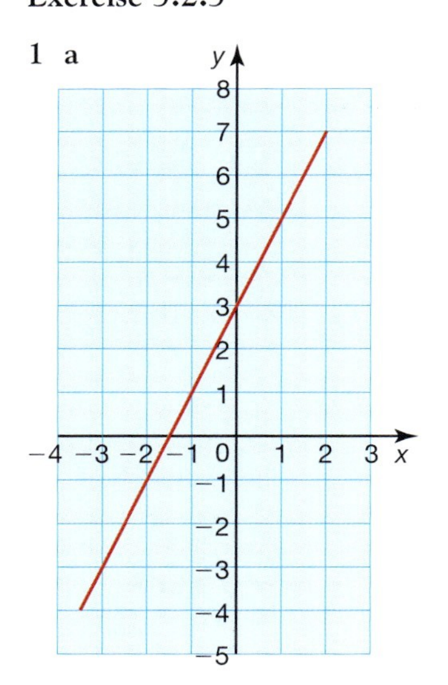

b

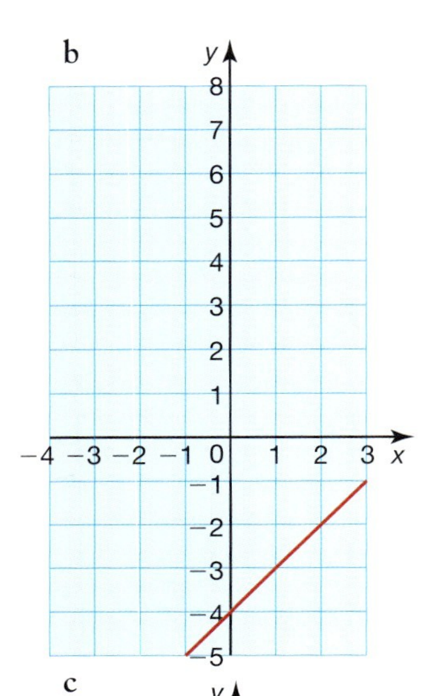

c

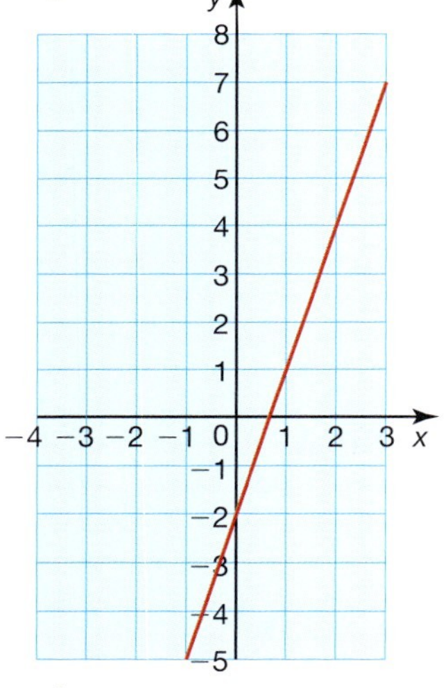

d

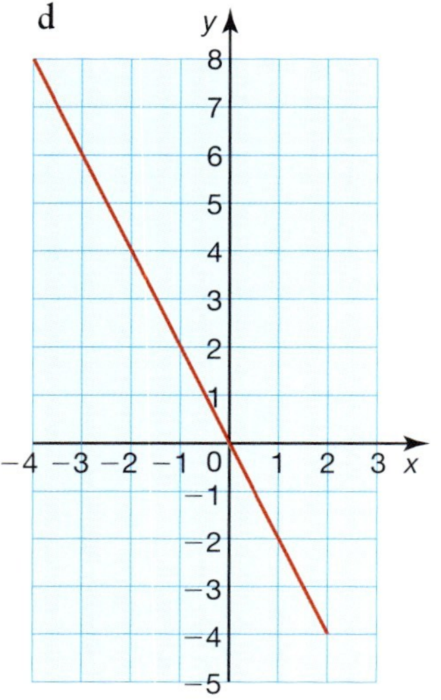

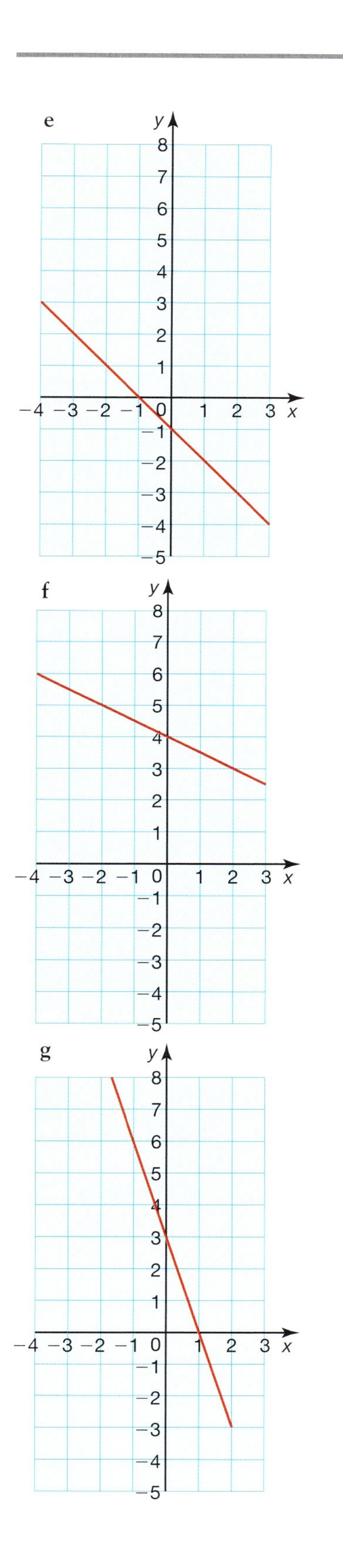
e
y
8
7
6
5
4
3
2
1
−4 −3 −2 −1 0 1 2 3 x
−1
−2
−3
−4
−5
f
y
8
7
6
5
4
3
2
1
−4 −3 −2 −1 0 1 2 3 x
−1
−2
−3
−4
−5
g
y
8
7
6
5
4
3
2
1
−4 −3 −2 −1 0 1 2 3 x
−1
−2
−3
−4
−5

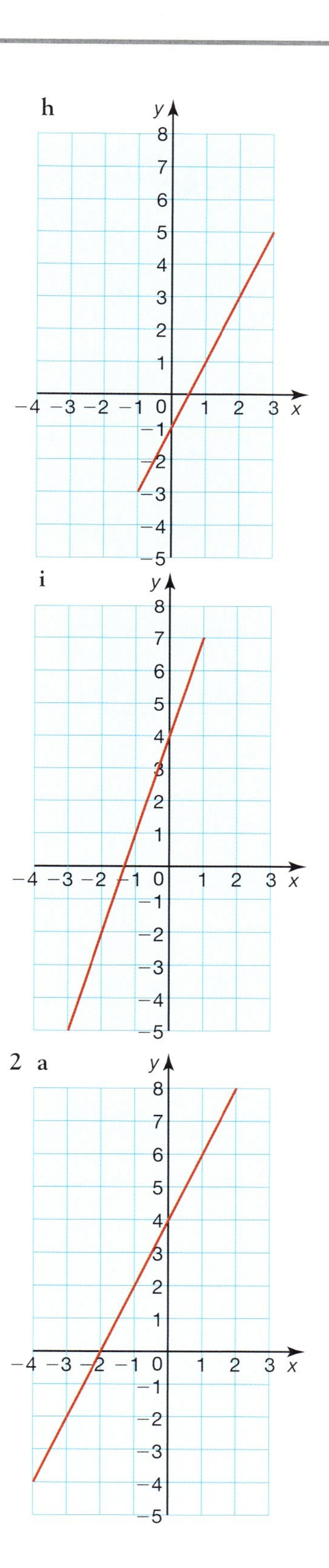
h
y
8
7
6
5
4
3
2
1
−4 −3 −2 −1 0 1 2 3 x
−1
−2
−3
−4
−5
i
y
8
7
6
5
4
3
2
1
−4 −3 −2 −1 0 1 2 3 x
−1
−2
−3
−4
−5
2 a
y
8
7
6
5
4
3
2
1
−4 −3 −2 −1 0 1 2 3 x
−1
−2
−3
−4
−5

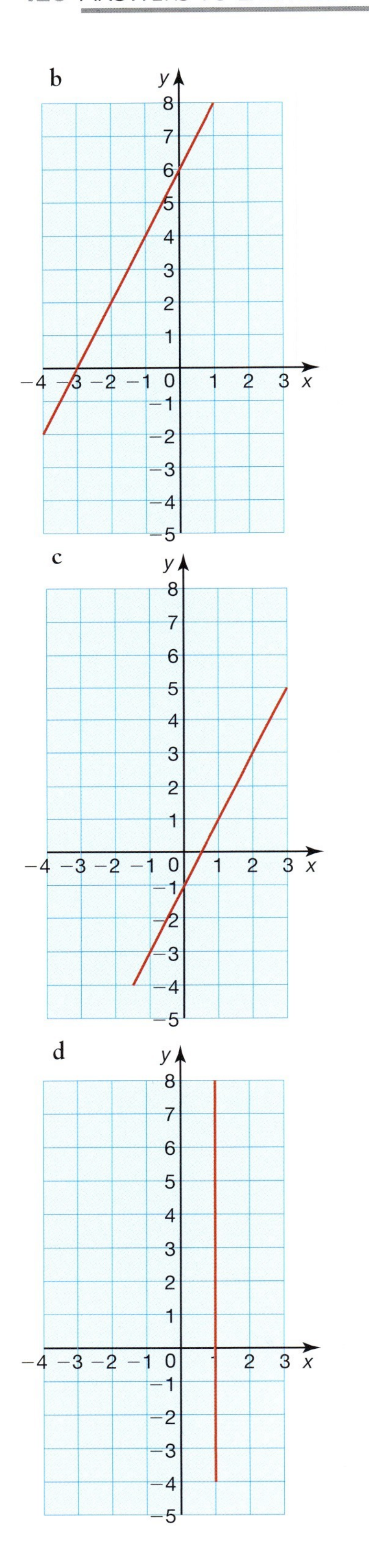
b
y
x
c
y
x
d
y
x

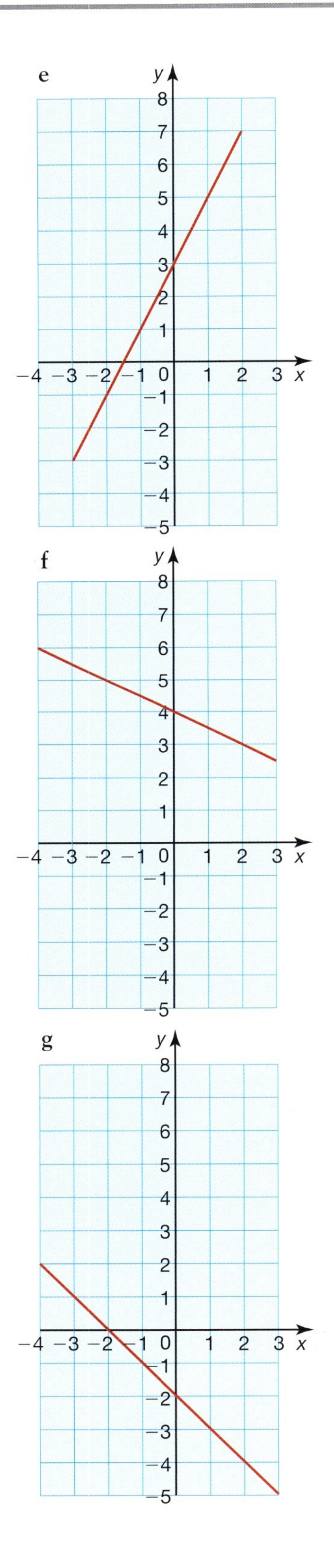
e
y
x
f
y
x
g
y
x

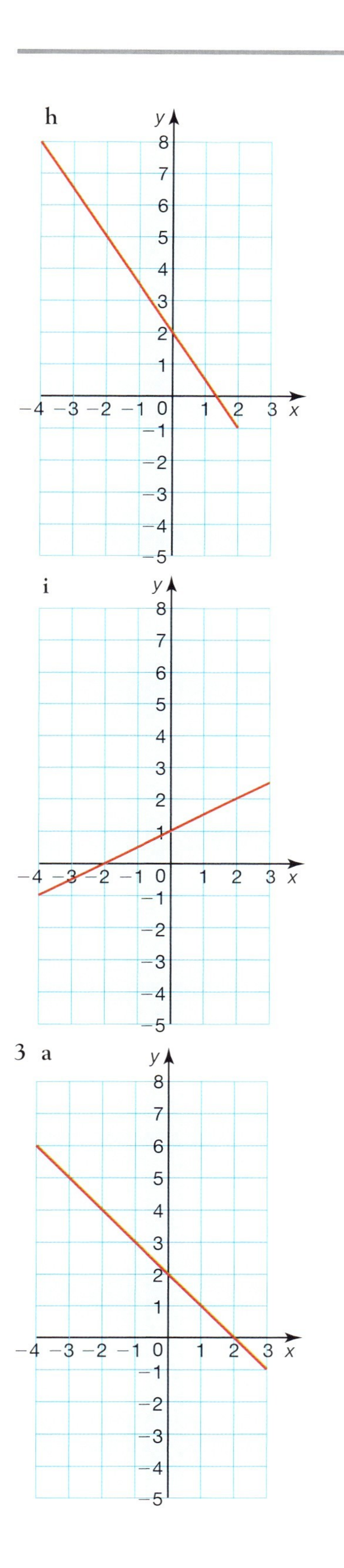
h
y
x
i
y
x
3 a
y
x

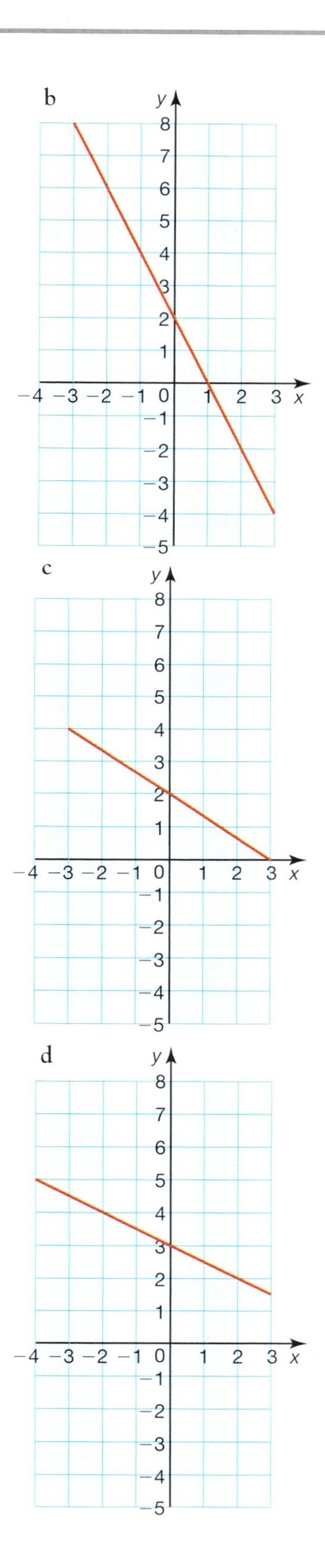
b
y
x
c
y
x
d
y
x

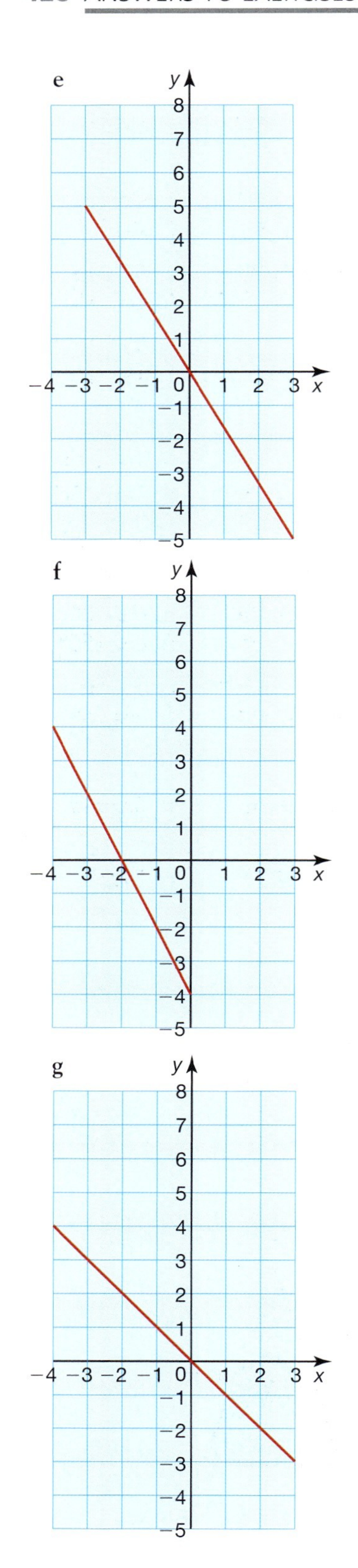

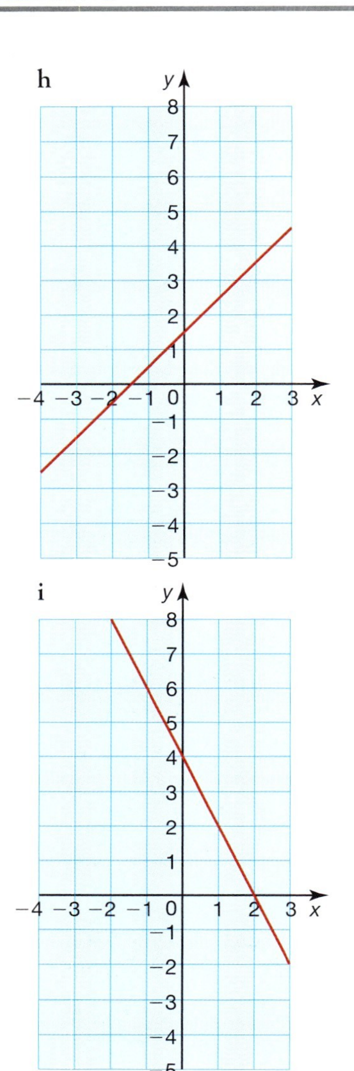

Exercise 5.2.6

1 a $x = 4$ $y = 2$ b $x = 6$ $y = 5$
c $x = 6$ $y = -1$ d $x = 5$ $y = 2$
e $x = 5$ $y = 2$ f $x = 4$ $y = 9$

2 a $x = 3$ $y = 2$ b $x = 7$ $y = 4$
c $x = 1$ $y = 1$ d $x = 1$ $y = 5$
e $x = 1$ $y = 10$ f $x = 8$ $y = 2$

3 a $x = 5$ $y = 4$ b $x = 4$ $y = 3$
c $x = 10$ $y = 5$ d $x = 6$ $y = 4$
e $x = 4$ $y = 4$ f $x = 10$ $y = -2$

4 a $x = 5$ $y = 4$ b $x = 4$ $y = 2$
c $x = 5$ $y = 3$ d $x = 5$ $y = -2$
e $x = 1$ $y = 5$ f $x = -3$ $y = -3$

5 a $x = -5$ $y = -2$ b $x = -3$ $y = -4$
c $x = 4$ $y = 3\frac{2}{3}$ d $x = 2$ $y = 7$
e $x = 1$ $y = 1$ f $x = 2$ $y = 9$

6 a $x = 2$ $y = 3$ b $x = 5$ $y = 10$
c $x = 4$ $y = 6$ d $x = 4$ $y = 4$

7 a $x = 1$ $y = -1$ b $x = 11\frac{2}{3}$ $y = 8$
c $x = 4$ $y = 0$ d $x = 3$ $y = 4$
e $x = 2$ $y = 8$ f $x = 1$ $y = 1$

Exercise 5.2.7

1 a $x = 2$ $y = 3$ b $x = 1$ $y = 4$
c $x = 5$ $y = 2$ d $x = 3$ $y = 3$
e $x = 4$ $y = 2$ f $x = 6$ $y = 1$

2 a $x = 1$ $y = 4$ b $x = 5$ $y = 2$
c $x = 3$ $y = 3$ d $x = 6$ $y = 1$
e $x = 2$ $y = 3$ f $x = 2$ $y = 3$

3 a $x = 0$ $y = 3$ b $x = 5$ $y = 2$
c $x = 1$ $y = 7$ d $x = 6$ $y = 4$
e $x = 2$ $y = 5$ f $x = 3$ $y = 0$

4 a $x = 1$ $y = 0.5$ b $x = 2.5$ $y = 4$
c $x = \frac{1}{5}$ $y = 4$ d $x = \frac{3}{4}$ $y = \frac{1}{2}$
e $x = 5$ $y = \frac{1}{3}$ f $x = \frac{1}{2}$ $y = 1$

Exercise 5.3.1

1 a 19.2 cm b 15.1 cm c 43.8 cm
d 31.8 cm e 6.2 cm f 2.1 cm

2 a 81.1° b 63.4° c 38.7°

3 a 43.6° b 19.5 cm c 42.5°

4 a 20.8 km b 215°

5 a 228 km b 102 km c 103 km
d 147 km e 415 km f 217°

6 a 6.7 m b 19.6 m c 15.3 m

7 a 48.2° b 41.8° c 8 cm
d 8.9 cm e 76.0 cm^2

8 a 34.2 m b 940 m

9 6.9 km

10 a 225.2 m b 48.4°

Exercise 5.4.1

1 a sin 120° b sin 100° c sin 65°
d sin 40° e sin 52° f sin 13°

2 a sin 145° b sin 130° c sin 150°
d sin 132° e sin 76° f sin 53°

3 a 19°, 161° b 82°, 98° c 5°, 175°
d 72°, 108° e 13°, 167° f 28°, 152°

4 a 70°, 110° b 9°, 171° c 53°, 127°
d 34°, 146° e 16°, 164° f 19°, 161°

Exercise 5.4.2

1 a −cos 160° b −cos 95° c −cos 148°
d −cos 85° e −cos 33° f −cos 74°

2 a −cos 82° b −cos 36° c −cos 20°
d −cos 37° e −cos 9° f −cos 57°

3 a cos 80° b −cos 90° c cos 70°
d cos 135° e cos 58° f cos 155°

4 a −cos 55° b −cos 73° c cos 60°
d cos 82° e cos 88° f cos 70°

Exercise 5.4.3

1 a 8.9 cm b 8.9 cm c 6.0 mm
d 8.6 cm

2 a 33.2° b 127.3° c 77.0°
d 44.0°

3 a 25.3°, 154.7°
b

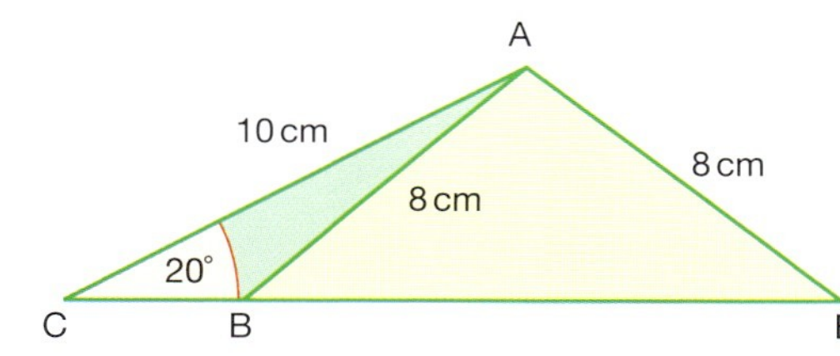

4 a 74.6°, 105.4°
b

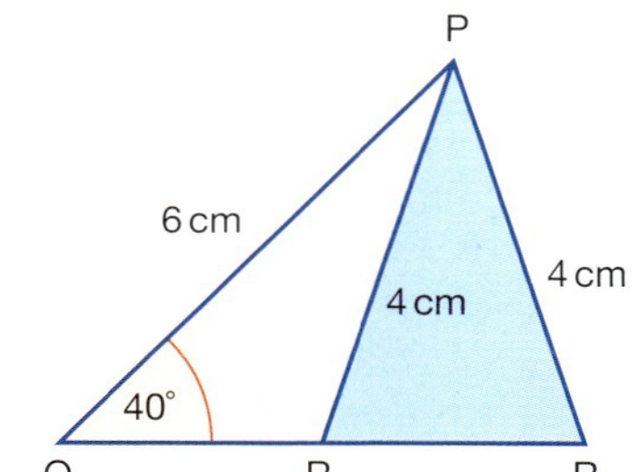

Exercise 5.4.4

1 a 4.7 m b 12.1 cm c 9.1 cm
d 3.1 cm e 10.7 cm

2 a 125.1° b 108.2° c 33.6°
d 37.0° e 122.9°

3 a 42.9 m b 116.9° c 24.6°
d 33.4° e 35.0 m

4 370 m

5 a 669 m b 546 m c 473 m

6 73.9 m

Exercise 5.4.5

1 a $70.0\,cm^2$ b $70.9\,mm^2$ c $121.6\,cm^2$
d $17.0\,cm^2$

2 a 24.6° b 13.0 cm c 23.1 cm
d 63.2°

3 $16\,800\,m^2$

4 a $3.9\,m^2$ b $222\,m^2$

Exercise 5.5.1

1 a 5.7 cm b 6.9 cm c 54.7°
d 2.8 cm

2 a 5.8 cm b 6.2 cm c 18.9°

3 a 6.4 cm b 13.6 cm c 61.9°

4 a 75.3° b 56.3°

5 a i) 7.2 cm ii) 21.1°
b i) 33.7° ii) 68.9°

6 a i) 8.5 cm ii) 28.3°
b i) 20.6° ii) 61.7°

7 a 6.5 cm b 11.3 cm c 70.7 cm

8 a 11.7 cm b 7.6 cm

9 a 25.0 cm b 20.5 cm c 26.0 cm
d 12.5 cm e 22.0 cm

10 a TU = TQ = 10 cm, QU = 8.5 cm
b $Q = U = 64.8°$, $T = 50.4°$
c $38.5\,cm^2$

Exercise 5.5.2

1 a RW b TQ c SQ
d WU e QV f SV

2 a JM b KN c HM
d HO e JO f MO

3 a ∠TPS b ∠UPQ c ∠VSW
d ∠RTV e ∠SUR f ∠VPW

4 a 5.83 cm b 31.0°

5 a 10.2 cm b 29.2° c 51.3°

6 a 6.71 cm b 61.4°

7 a 7.81 cm b 11.3 cm c 12.4°

8 a 14.1 cm b 8.49 cm c 7.5 cm
d 69.3°

9 a 17.0 cm b 5.66 cm c 7.00 cm
d 51.1°

Exercise 5.5.3

1 a Volume = $27.6\,cm^3$, surface area = $60.8\,cm^2$
b Volume = $277.1\,cm^3$, surface area = $235.3\,cm^2$
c Volume = $42\,cm^3$, surface area = $95.5\,cm^2$

2 a 16 cm b $4096\,cm^3$ c $3217\,cm^3$
d 21.5%

3 a $42\,cm^2$ b $840\,cm^3$ c $564\,cm^2$

4 6.3 cm

5 $2.90\,m^3$

6 a $24\,cm^2$ b 2 cm

7 a $216\,cm^2$ b 15.2 cm c $25.0\,cm^3$

8 a $94.2\,cm^2$ b 14 cm c $12.6\,cm^3$

9 4.4 cm

Exercise 5.5.4

1 a Volume = $905\,cm^3$, surface area = $452\,cm^2$
b Volume = $3591\,cm^3$, surface area = $1134\,cm^2$
c Volume = $2309.6\,cm^3$ (1 d.p.), surface area = $845\,cm^2$
d Volume = $1.4\,cm^3$ (1 d.p.), surface area = $6.16\,cm^2$

2 a 5.6 cm b 0.4 m

3 a 1.15 cm b 3.09 mm

4 6.30 cm

5 86.7 cm^3

6 11.9 cm

7 a 4190 cm^3 b 8000 cm^3 c 47.6%

8 10.0 cm

9 A = 4.1 cm, B = 3.6 cm, C = 3.1 cm

10 3 : 2

11 1 : 4

12 a 157 cm^2 b 15.0 cm c 707 cm^2

13 a 804.2 cm^2 b 5.9 cm

Exercise 5.5.5

1 a 40 cm^3 b 133 cm^3
c 64 cm^3 d 70 cm^3

2 Volume = 147.1 cm^3, surface area = 157.4 cm^2

3 Volume = 44.0 cm^3, surface area = 73.3 cm^2

4 a 8 cm b 384 cm^3 c 378 cm^3

5 1121.9 cm^2

6 7 cm

7 5 cm

8 a 3.6 cm b 21.7 cm^3 c 88.7 cm^3

9 6.93 cm^2

10 a 693 cm^2 b 137 cm^2 c 23.6 cm

Exercise 5.5.6

1 a Arc length = 6.3 cm, sector area = 25.1 cm^2
b Arc length = 2.1 cm, sector area = 15.7 cm^2
c Arc length = 11.5 cm, sector area = 34.6 cm^2
d Arc length = 23.6 cm, sector area = 58.9 cm^2

2 a 4.19 cm b 114 cm^2 c 62.8 cm^3

3 a 56.5 cm^3 b 264 cm^3 c 1.34 cm^3
d 166 cm^3

4 6.91 cm

5 a 15.9 cm b 8.41 cm c 2230 cm^3

6 a Volume = 559.2 cm^3, surface area = 414.7 cm^2
b Volume = 3117.0 cm^3, surface area = 1649.3 cm^2

7 3.88 cm

8 a 33.0 cm b 5.25 cm c 7.31 cm
d 211 cm^3 e 148 cm^2

9 1131 cm^2

10 a 314 cm^2 b 12.5

11 a 2304 cm^3 b 603.2 cm^3 c 1700.8 cm^3

12 a 81.6 cm^3 b 275 cm^3 c 8 : 27

13 a 81.8 cm^3 b 101 cm^2

14 a 771 m^3 b 487 m^2

15 3166.7 cm^3

16 a 654.5 cm^3 b 12.5 cm c 2945.2 cm^3

Student assessment 1

1 a i) 13 units ii) (0, 1.5)
b i) 10 units ii) (4, 6)
c i) 5 units ii) (0, 4.5)
d i) 26 units ii) (−5, 2)

2 a b c d

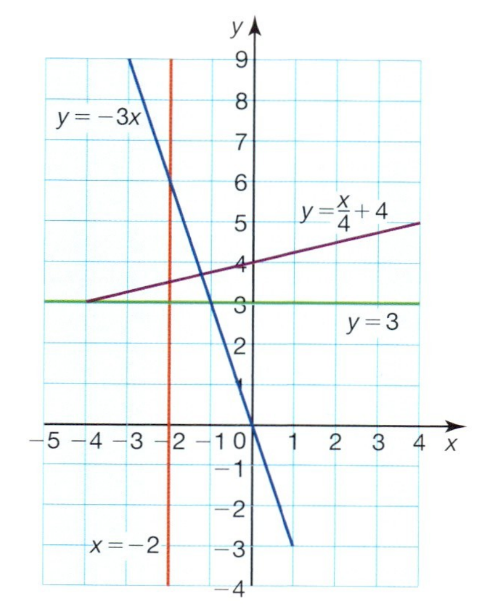

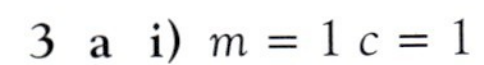

3 a i) $m = 1$ $c = 1$

ii)

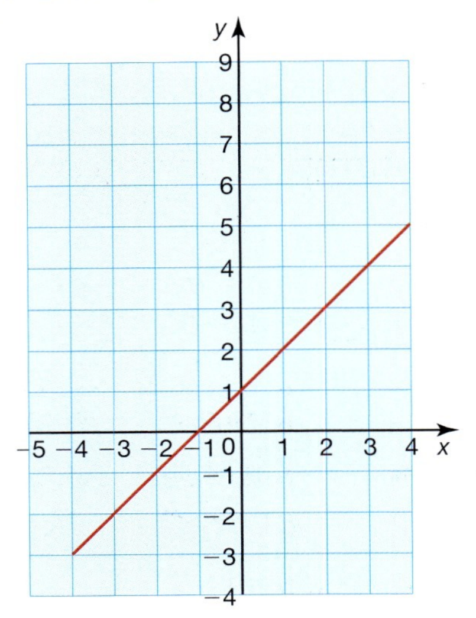

b i) $m = -3$ $c = 3$

ii)

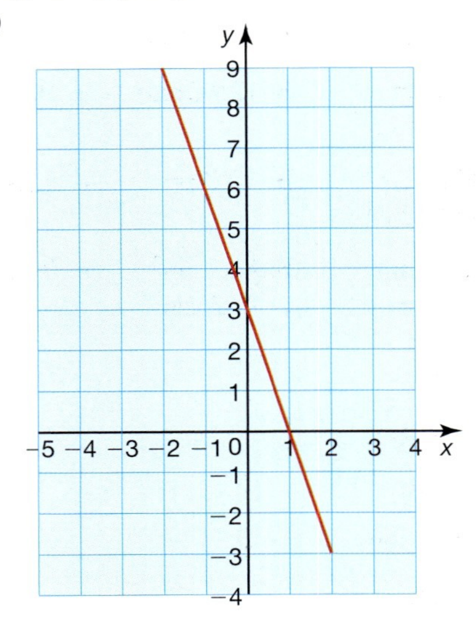

c i) $m = 2$ $c = 4$

ii)

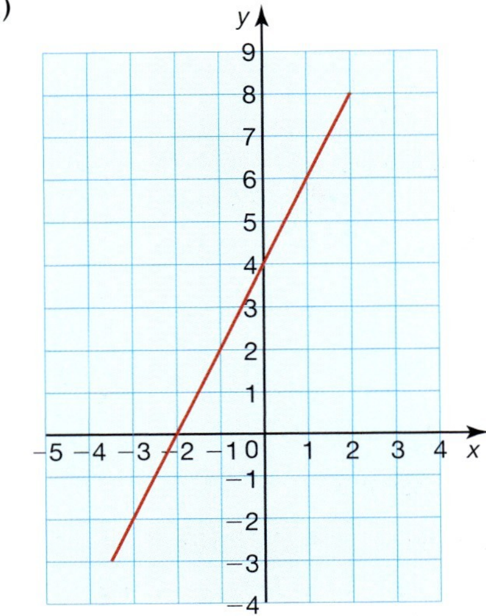

d i) $m = \frac{5}{2}$ $c = 4$

ii)

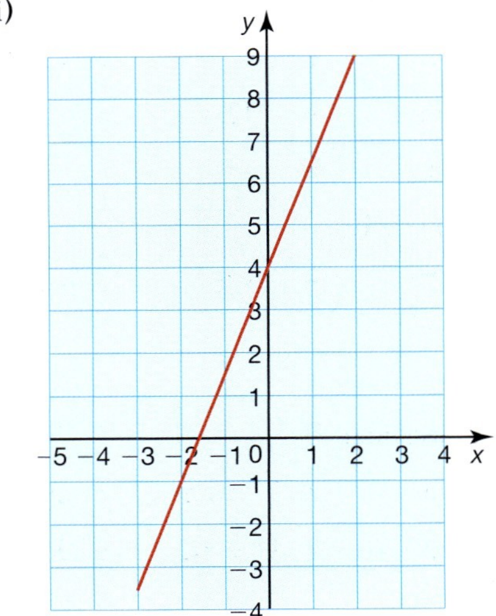

4 a $y = 3x - 4$ **b** $y = -2x + 7$
c $y = 2x - 5$ **d** $y = -4x + 3$

5 a $x = 2$ $y = 2$
b $x = 1$ $y = -1$
c $x = -2$ $y = 4$
d $x = -2$ $y = 0$

Student assessment 2

1 a 4.0 cm b 43.9 cm c 20.8 cm
d 3.9 cm

2 a 37° b 56° c 31°
d 34°

3 a 5.0 cm b 6.6 cm c 9.3 cm
d 28.5 cm

Student assessment 3

1 a 160.8 km b 177.5 km

2 a $\tan\theta = \frac{5}{x}$ b $\tan\theta = \frac{7.5}{(x + 16)}$
c $\frac{5}{x} = \frac{7.5}{(x + 16)}$ d 32 m e 8.88°

3 a 285 m b 117° c 297°

4 a 1.96 km b 3.42 km c 3.57 km

5 a 4003 m b 2.35°

6 Student's graph

7 a sin 130° b sin 30° c −cos 135°
d −cos 60°

8 134°

Student assessment 4

1 a 11.7 cm b 12.3 cm c 29°

2 a 10.8 cm b 11.9 cm c 30°
d 49°

3 Student's graph

4 a −cos 52° b cos 100°

5 a 9.8 cm b 30° c 19.6 cm

6 a 678.4 m b 11.6° c 718.0 m

Student assessment 5

1 a 18.0 m b 27° c 28.8 m
d 278 m²

2 a 12.7 cm b 67° c 93.4 cm²
d 14.7 cm

3 a 38°, 322° b 106°, 254°

4 a

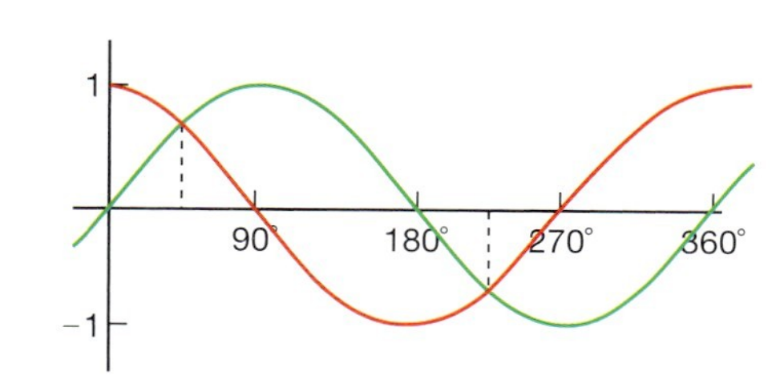

b $\theta = 45°, 225°$

5 a 5.8 cm b 6.7 cm c 7.8 cm
d 47° e 19 cm² f 37°

6 a 31.2 cm b 25.6 cm c 32.6 cm
d 14.6 cm e 27.6 cm

Student assessment 6

1 a 530.9 cm² b 1150.3 cm³

2 a 1210.7 cm² b 2897.9 cm³

3 a 22.9 cm b 229.2 cm² c 985.1 cm²
d 1833.5 cm³

4 a 904.8 cm³ b 12.0 cm c 13.4 cm
d 958.2 cm²

5 a 10.0 cm b 82.1 cm³ c 71.8 cm³
d 30.8 cm³ e 41.1 cm³

Topic 6

Exercise 6.1.1

1 Discrete
2 Continuous
3 Discrete
4 Continuous
5 Discrete
6 Discrete
7 Continuous
8 Discrete
9 Continuous
10 Continuous

Exercise 6.2.1

1 a

Number of chocolates	Tally	Frequency
35	||||	4
36	~~||||~~ ||	7
37	||||	3
38	~~||||~~ |	6

b c

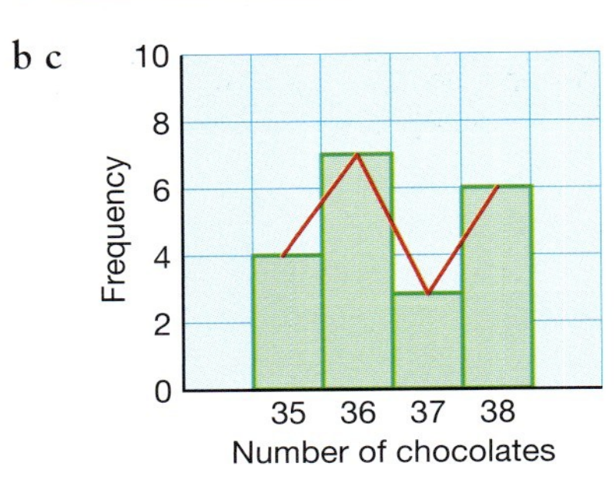

2 Student's own results

Exercise 6.3.1

1 a Class 12X

Score	Frequency
31–40	1
41–50	5
51–60	3
61–70	4
71–80	8
81–90	5
91–100	5

Class 12Y

Score	Frequency
31–40	3
41–50	8
51–60	6
61–70	3
71–80	2
81–90	4
91–100	5

b Student's own response

2

Number of apples	Frequency
1–20	9
21–40	6
41–60	7
61–80	11
81–100	7
101–120	4
121–140	4
141–160	2

Exercise 6.3.2

1

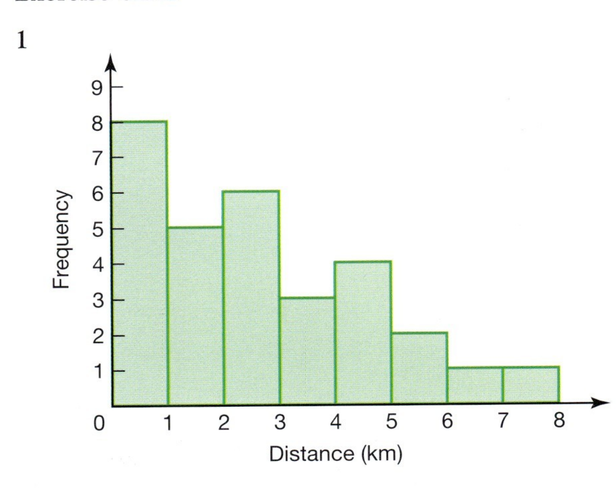

2

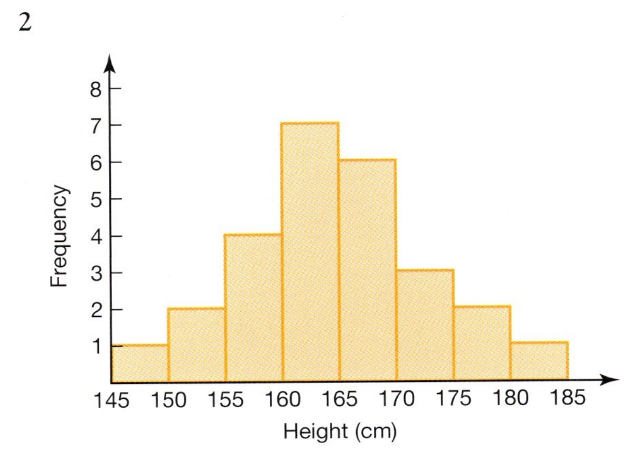

3 a

Time (min)	Frequency	Frequency density
$0 \le t < 10$	6	0.6
$10 \le t < 15$	3	0.6
$15 \le t < 20$	13	2.6
$20 \le t < 25$	7	1.4
$25 \le t < 30$	3	0.6
$30 \le t < 40$	4	0.4
$40 \le t < 60$	4	0.2

b

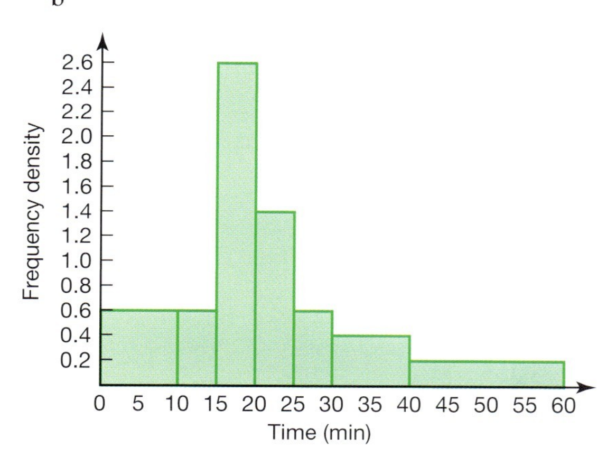

4 a

Age (years)	Frequency	Frequency density
$0 \le a < 1$	35	35
$1 \le a < 5$	48	12
$5 \le a < 10$	140	28
$10 \le a < 20$	180	18
$20 \le a < 40$	260	13
$40 \le a < 60$	280	14
$60 \le a < 90$	150	5

b

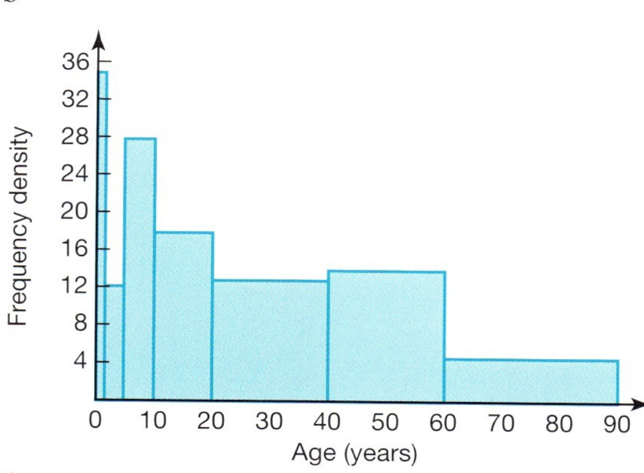

5 a

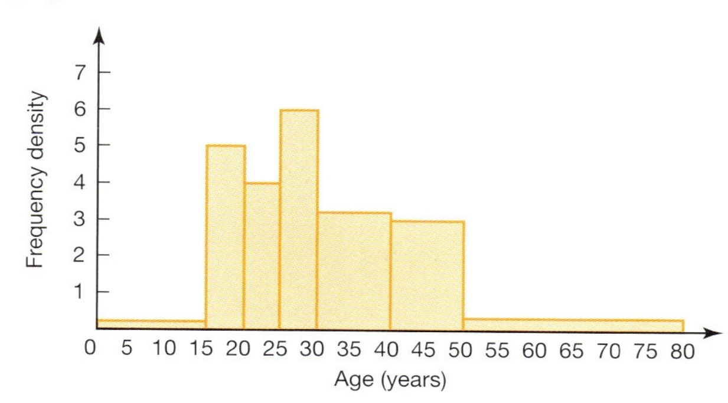

b Student's own answers

Exercise 6.3.3

1

0	9
1	1 4 5 6 6 7 8
2	1 1 2 2 3 4 6 7 7 8 8 8
3	0 3 3 3 3 4 6 6 7 8 8 9

Key **2** | 5 means 25

2

4	6 6 6 8
5	2 2 4 4 5 8 8
6	2 3 3 5 7 8
7	1 2 4 6
8	3 5 8

Key **5** | 2 means 52

3

5	4 8 9 9
6	1 2 5 8
7	2 3 3 7 9 9
8	0 0 3 3 3 8 8 9
9	4 7
0	0 4

Key **5** | 8 means 5.8

Exercise 6.3.4

1 Student's own response

2

2008		2009
8 6 4	**2**	
6 5 5 3 2	**3**	
8 2 1	**4**	6 6 6 8
4	**5**	2 2 4 4 5 8 8
4 3	**6**	2 3 3 5 7 8
8 2	**7**	1 2 4 6
8 8 8 6 2	**8**	3 5 8
	9	
5 0	**10**	

Key **5** | 2 means 52

3

A		B
8 9	**0**	9
1 2 2 2 4 4 6 6 7 8 8	**1**	1 4 5 6 6 7 8
1 1 2 2 2 3 3 5 8 9	**2**	1 1 2 2 3 4 6 7 7 8 8 8
2 3 3 4 5 6 7 8 9 9	**3**	0 3 3 3 3 4 6 6 7 8 8 9

Key **2** | 5 means 25

Exercise 6.4.1

1 a

Finishing time (h)	0–	0.5–	1.0–	1.5–	2.0–	2.5–	3.0–3.5
Frequency	0	0	6	34	16	3	1
Cumulative frequency	0	0	6	40	56	59	60

b

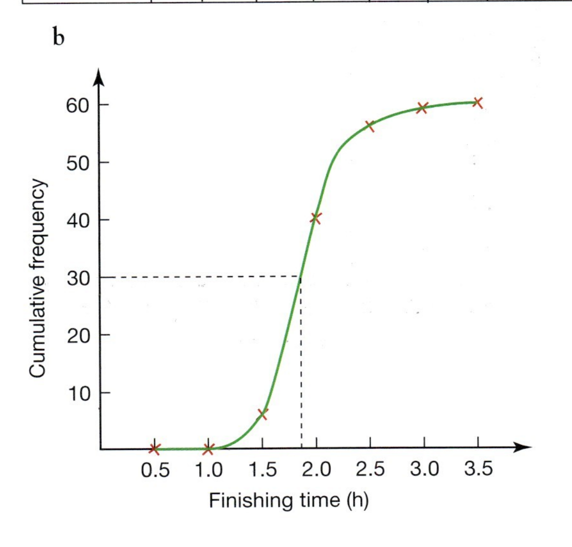

c Median ≈ 1.8 hours

d As many runners finished before as after the median.

2 a

	Class A		Class B		Class C	
Score	**Freq.**	**Cum. freq.**	**Freq.**	**Cum. freq.**	**Freq.**	**Cum. freq.**
$0 \le x < 20$	1	1	0	0	1	1
$20 \le x < 40$	5	6	0	0	2	3
$40 \le x < 60$	6	12	4	4	2	5
$60 \le x < 80$	3	15	4	8	4	9
$80 \le x < 100$	3	18	4	12	8	17

b

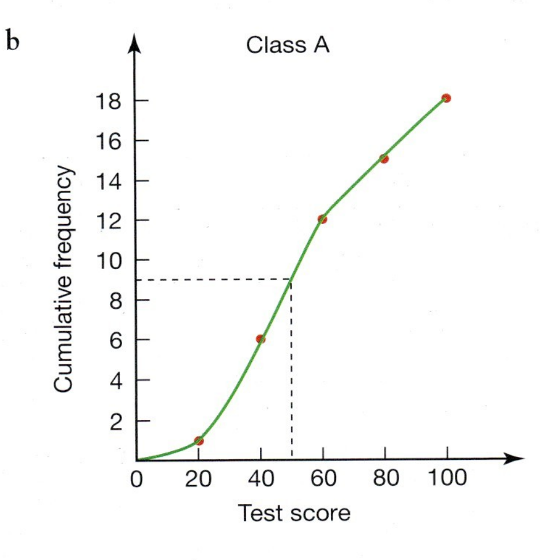

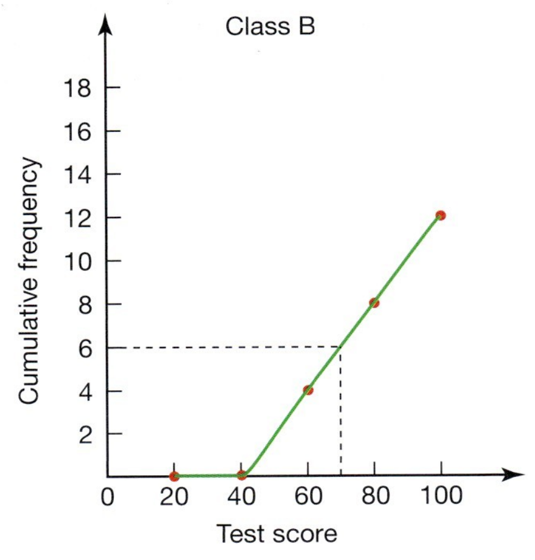

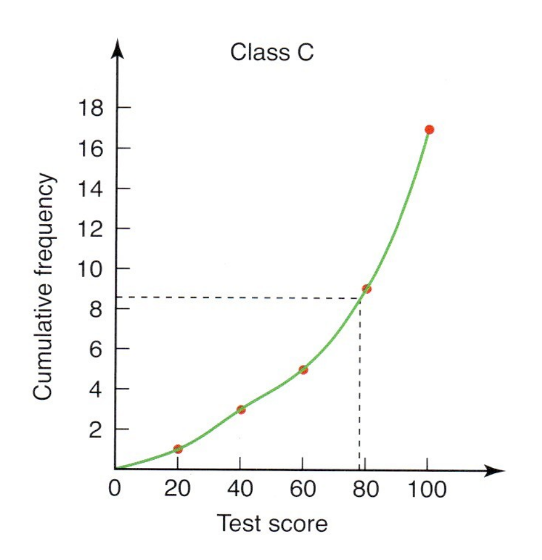

c Class A median ≈ 50
Class B median ≈ 70
Class C median ≈ 78

d As many students were above as below the median.

3 a

	2007		2008		2009	
Height (cm)	**Freq.**	**Cum. freq.**	**Freq.**	**Cum. freq.**	**Freq.**	**Cum. freq.**
$150 \le h < 155$	6	6	2	2	2	2
$155 \le h < 160$	8	14	9	11	6	8
$160 \le h < 165$	11	25	10	21	9	17
$165 \le h < 170$	4	29	4	25	8	25
$170 \le h < 175$	1	30	3	28	2	27
$175 \le h < 180$	0	30	2	30	2	29
$180 \le h < 185$	0	30	0	30	1	30

b

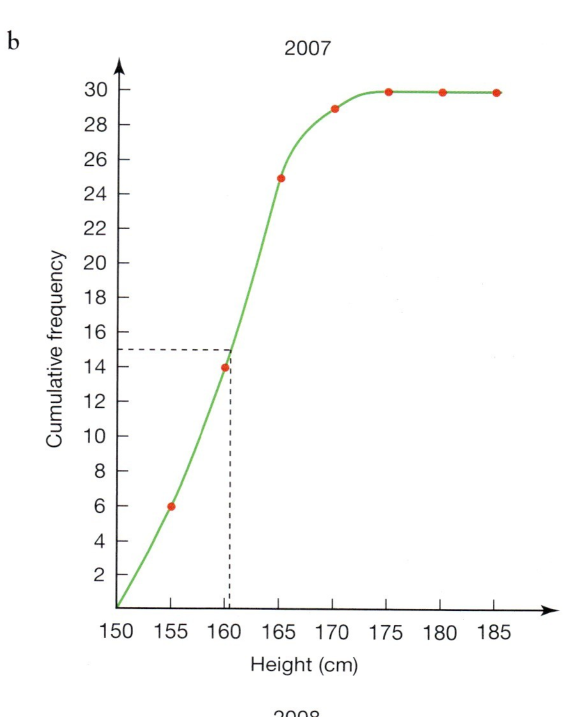

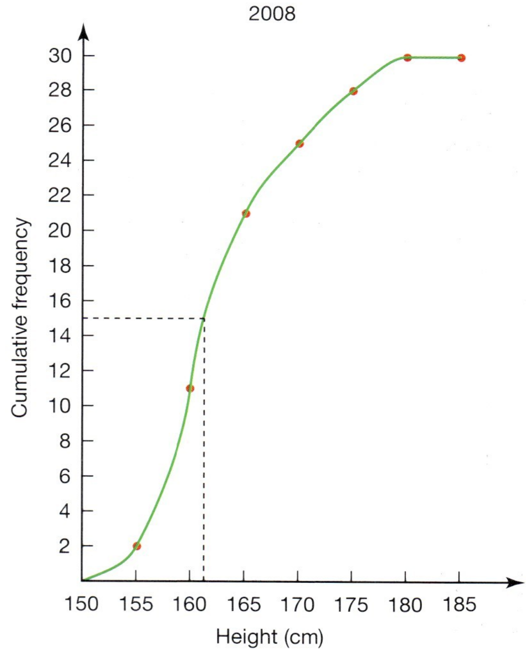

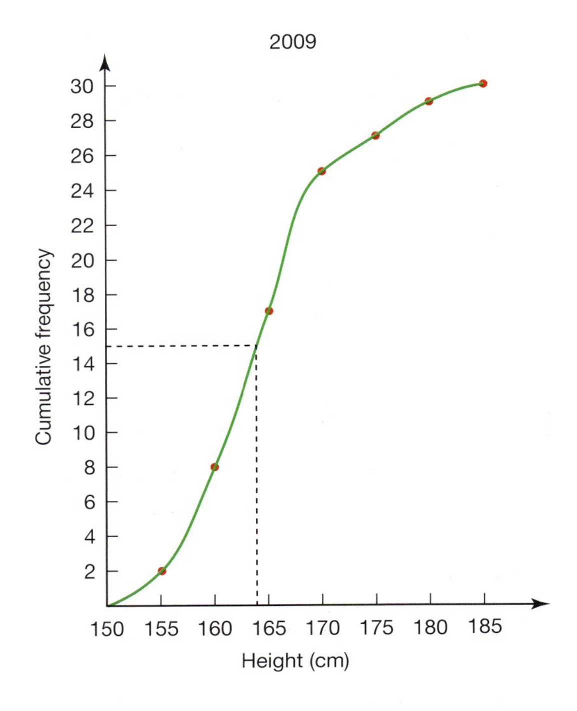

c Median (2007) ≈ 161 cm
Median (2008) ≈ 162 cm
Median (2009) ≈ 164 cm

d As many students are taller than the median as shorter than the median.

Exercise 6.4.2

1 a Class A ≈ 30
Class B ≈ 30
Class C ≈ 40

b Student's own responses

2 a 2007 ≈ 7 cm
2008 ≈ 8 cm
2009 ≈ 8 cm

b Student's own responses

3 a

Distance thrown (m)	0–	20–	40–	60–	80–100
Frequency	4	9	15	10	2
Cumulative frequency	4	13	28	38	40

b

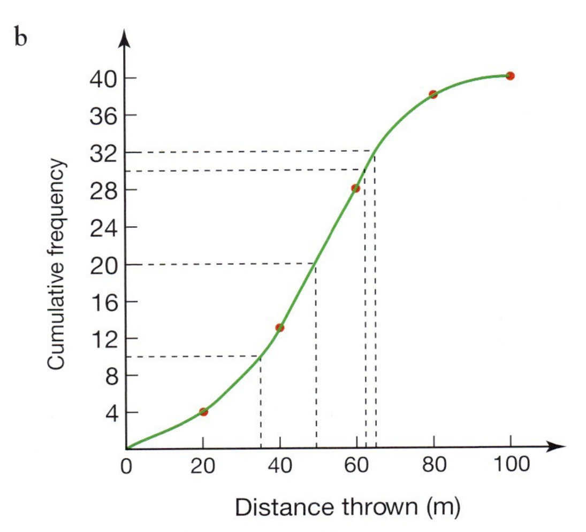

c Qualifying distance ≈ 66 m

d Interquartile range ≈ 28 m

e Median ≈ 50 m

4 a

Type A		
Mass (g)	**Frequency**	**Cum. freq.**
75–	4	4
100–	7	11
125–	15	26
150–	32	58
175–	14	72
200–	6	78
225–250	2	80

Type B		
Mass (g)	**Frequency**	**Cum. freq.**
75–	0	0
100–	16	16
125–	43	59
150–	10	69
175–	7	76
200–	4	80
225–250	0	80

b

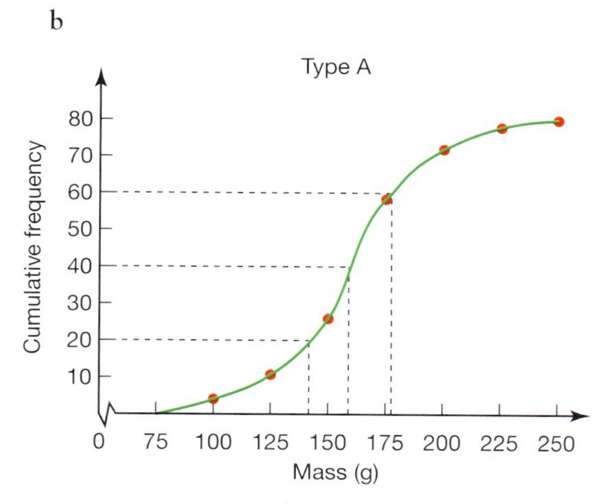

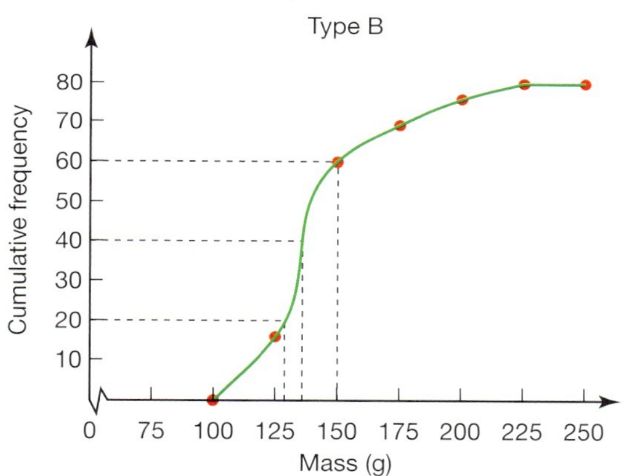

c Median type A ≈ 157 g
Median type B ≈ 137 g

d i) Lower quartile type A ≈ 140 g
Lower quartile type B ≈ 127 g

ii) Upper quartile type A ≈ 178 g
Upper quartile type B ≈ 150 g

iii) Interquartile type range type A ≈ 38 g
Interquartile type range type B ≈ 23 g

e Student's own report

5 a Student's own explanation

b Student's own explanation

Exercise 6.4.3

1 a

	Goals scored	Goals let in
i) Mean	1.15	2.00
ii) Median	1	2
iii) q_1	0	1
iv) q_3	1.5	3
v) IQR	1.5	2

b

Goals let in
Goals scored
1 2 3 4 5

c Student's own report

2 a

	Resort A	Resort B
i) Mean	8.5	8.5
ii) Median	8	8
iii) q_1	7	8
iv) q_3	10	9
v) IQR	3	1

b

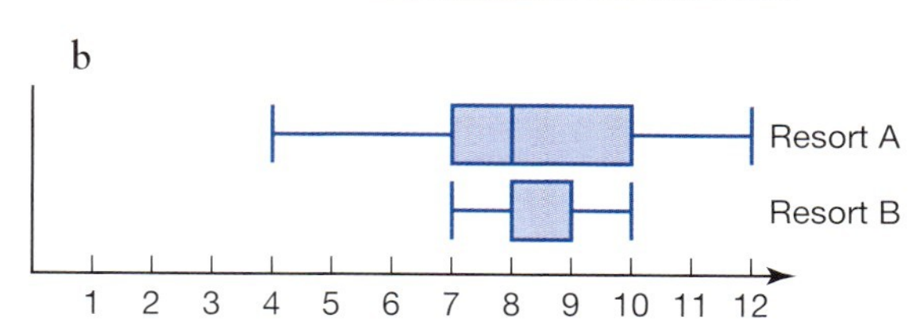

3 Student's own explanation using the box and whisker plots below.

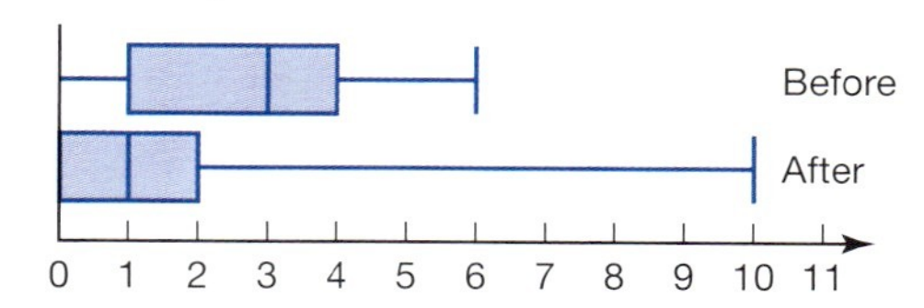

Exercise 6.5.1

1 a Mean = 1.67 Median = 1
Mode = 1
b Mean = 6.2 Median = 6.5
Mode = 7
c Mean = 26.4 Median = 27
Mode = 28
d Mean = 13.95 s Median = 13.9 s
Mode = 13.8 s

2 91.1 kg

3 103 points

Exercise 6.5.2

1 Mean = 3.35 Median = 3
Mode = 1 and 4

2 Mean = 7.03 Median = 7
Mode = 7

3 a Mean = 6.33 Median = 7
Mode = 8
b The mode as it gives the highest number of flowers per bush.

Exercise 6.5.3

1 a 29.1
b 30–39

2 a 60.9
b 60–69

3 a 5 mins 50 secs
b 0–4
c Student's own comments

Exercise 6.6.1

1 a i) 5.5 ii) 7 iii) 5
iv) 2.58
b i) 78.75 ii) 16 iii) 9
iv) 5.40
c i) 3.85 ii) 3.9 iii) 1.6
iv) 1.05

2 i) 2.31 ii) 6 iii) 2
iv) 1.43

3 i) 71.53 ii) 8 iii) 1
iv) 1.44

4 i) 2.72 ii) 6 iii) 3
iv) 1.69

5 a 80 b 11.92

6 a 6.5 b 0.18

Exercise 6.7.1

1 Students' answers may differ from those given below.
a Possible positive correlation (strength depending on topics tested)
b No correlation
c Positive correlation (likely to be quite strong)
d Negative correlation (likely to be strong)
e Depends on age range investigated. 0–16 years likely to be a positive correlation. Ages 16+ little correlation.
f Strong positive correlation

2 a

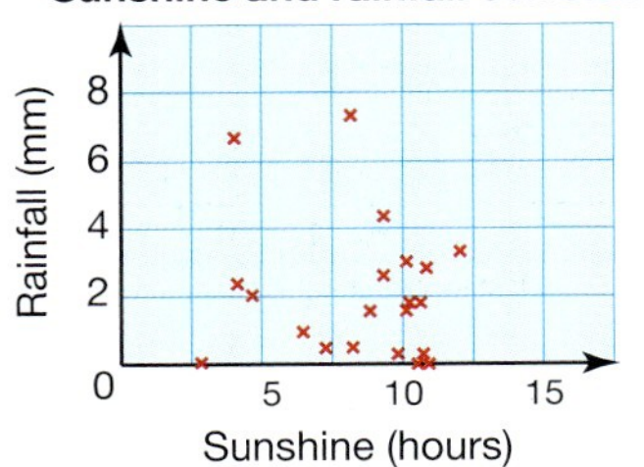

b Graph shows a very weak negative correlation. Student's answers as to whether this is what they expected.

3 a

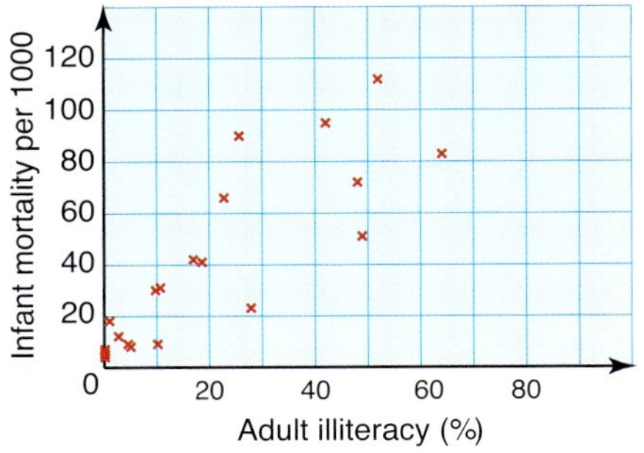

b Student's answer, however it is important to stress that although there is a correlation, it doesn't imply that one variable affects the other.

c Students's own explanations

d

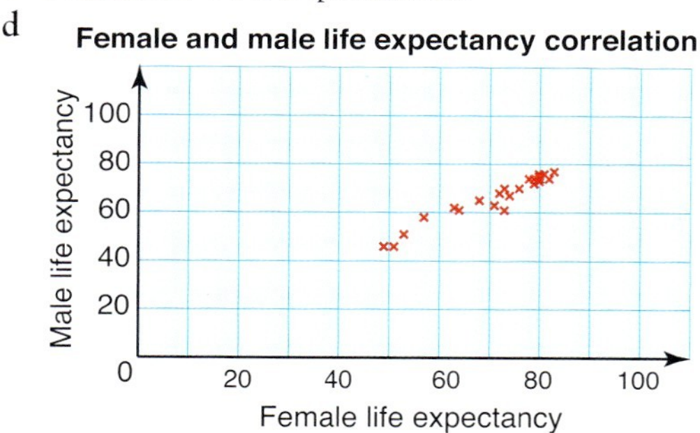

4 a d

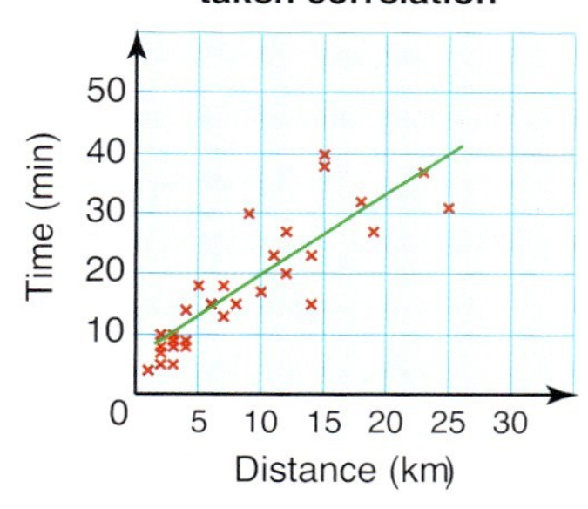

b (Strong) positive correlation

c Students's own explanations

e $\approx$ 30 mins

Exercise 6.7.2

1 a c

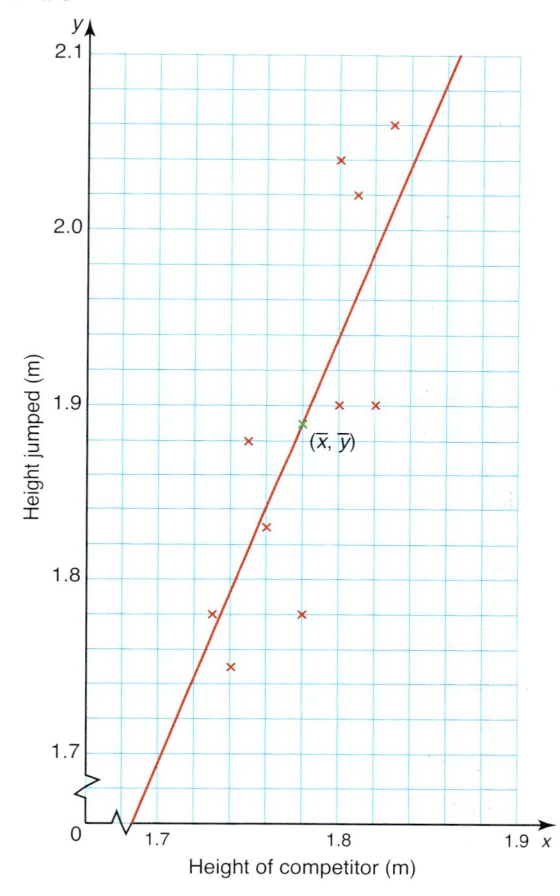

b $\bar{x} = 1.78\,\text{m}$ $\quad \bar{y} = 1.89\,\text{m}$

d $r = 0.79$

This implies a fairly good correlation, i.e. the taller the competitor, generally the greater the height jumped.

2 a c

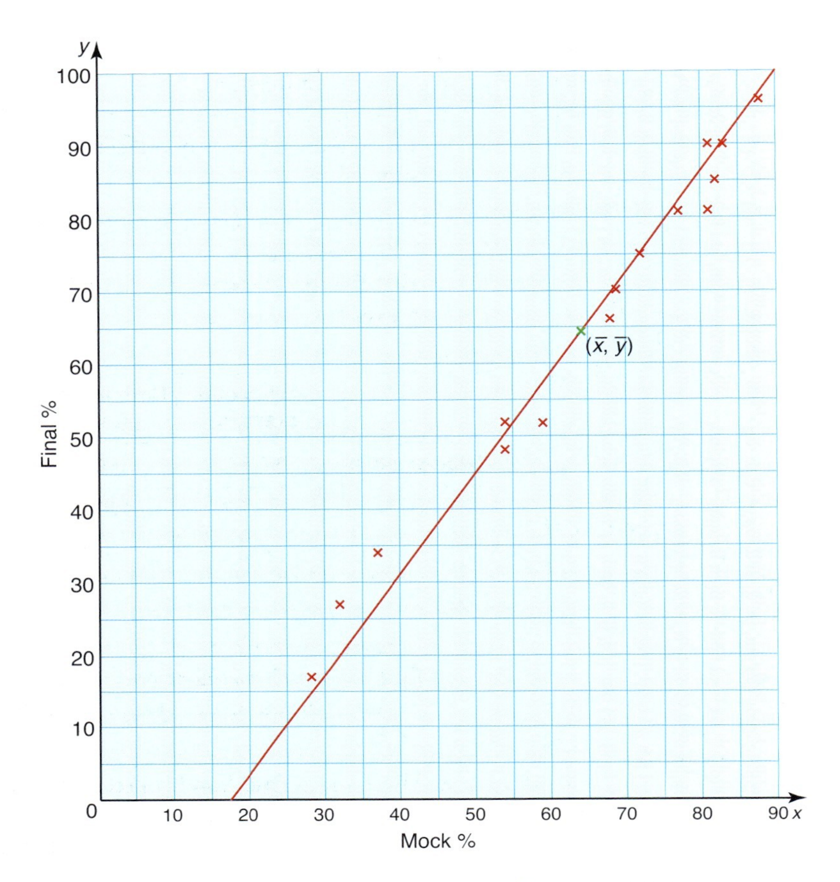

b $\bar{x} = 64.3\%$ $\bar{y} = 64.3\%$

d $r = 0.99$

This implies a very strong correlation between the mock % and the final % for English.

3 a $r = 0.89$

This implies a strong correlation between the mock % and final % for mathematics.

b Although both show a strong correlation between mock % and final %, the results appear to suggest that it is stronger for English.

4 a b Student's solutions and justifications

5 a–d Student's own data, graph and analysis.

e r is likely to be lower for 11-year-old students than for 17-year-old students as children of 11 are at different stages of development.

Exercise 6.8.1

1 a $r = -0.959$

b $y = -1.468x + 92.742$

c $y = 66$ minutes

d A valid estimate as 18 hours is within the data range collected.

2 a $r = 0.970$ b $y = 0.866x - 24.433$

c IQ: 95 → IB: 58% IQ: 155 → IB: 110%

d The estimate for the IQ of 95 is valid as it falls in the range of the data collected. The estimation for the IQ of 155 is an extrapolation producing an IB result greater than 100% and is therefore invalid.

3 a $r = 0.973$ indicates a strong positive correlation between the salary and the number of years of experience.

b $y = 0.00116x - 28.552$

c $x = \$33\,200$ – this is a valid estimate as 10 years falls within the data range collected.

d 87 years experience – this value falls outside the data range collected. The result implies a fire fighter approximately 100 years old, and is therefore invalid.

4 a $r = -0.946$ indicates a strong negative correlation, i.e. as height increases, temperature decreases.
b $y = -0.00189x + 7.4$
c Height = 41 000 m – although −70°C is slightly outside the data range, this is marginal and therefore the answer is likely to be valid.

Exercise 6.9.1

1 a H_0 : A person's opinion regarding the wearing of safety helmets is independent of whether they are a cyclist or not.
H_1 : A person's opinion regarding the wearing of safety helmets is dependent of whether they are a cyclist or not.
b

	Helmet compulsory	Helmet voluntary	Total
Cyclist	125	175	300
Non-cyclist	125	175	300
Total	250	350	600

c $X^2 = 98.743$
d 1 degree of freedom
e The table gives a critical value of 3.841. $98.743 > 3.841$, therefore the null hypothesis is rejected. The opinions are dependent on whether they are cyclists or not.

2 a H_0 : Being given the drug and living for more than 3 months are independent events.
H_1 : Being given the drug does affect the chance of surviving for longer than 3 months.

b

	Alive after 3 months	Not alive after 3 months	Total
Given drug	78.81	61.19	140
Given placebo	73.19	56.81	130
Total	152	118	270

c $X^2 = 4.040$
d 1 degree of freedom
e $4.040 < 6.635$, therefore the null hypothesis is valid. The drug does not affect the chances of survival.

3 a H_0 : Being a smoker does not cause high blood pressure.
H_1 : Being a smoker does cause high blood pressure.
b

	High blood pressure	Normal blood pressure	Total
Non-smoker	96.63	33.37	130
Smoker	349.37	120.63	470
Total	446	154	600

c $X^2 = 8.215$
d 1 degree of freedom
e $8.215 > 6.635$ therefore the null hypothesis is rejected. Smoking does cause high blood pressure.

4 a H_0 : Gender and holiday preference are independent events.
H_1 : Holiday preference is affected by gender.
b

	Beach	Walking	Cruise	Sail	Ski	Total
Male	68.93	40.17	30.59	26.93	43.37	210
Female	82.07	47.83	36.41	32.07	51.63	250
Total	151	88	67	59	95	460

c $X^2 = 12.233$
d 4 degrees of freedom
e $12.233 > 7.779$ therefore the null hypothesis is rejected. Holiday preference is dependent upon gender.

5 a H_0 : Age and musical preference are independent.
H_1 : Musical preference is dependent on age.
b $X^2 = 49.077$
c 8 degrees of freedom
d $49.077 > 15.507$ therefore the null hypothesis is rejected. Musical preference is dependent upon age.

Student assessment 1

1 a Discrete b Discrete c Continuous
d Discrete e Continuous f Continuous
g Continuous

2 a

Time (min)	Frequency	Frequency density
$0 \leq t < 30$	8	0.3
$30 \leq t < 45$	5	0.3
$45 \leq t < 60$	8	0.5
$60 \leq t < 75$	9	0.6
$75 \leq t < 90$	10	0.7
$90 \leq t < 120$	12	0.4

b

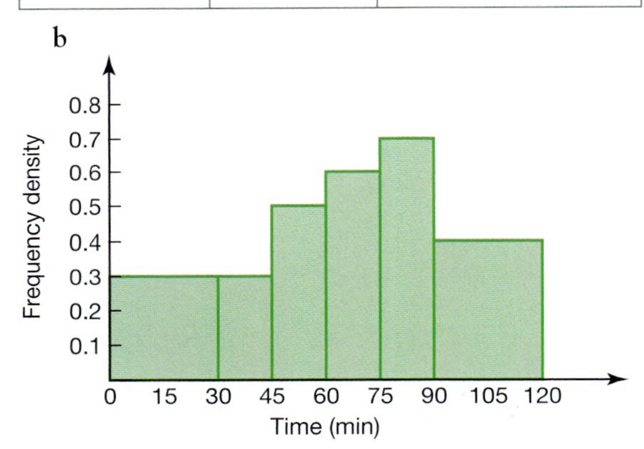

3

Key 4 | 3 means 4.3

```
Before training          After training
                2  6  3  8
8  6  5  4  2  1  5  2  2  3  3  4  4  5  5  6  8  9
7  7  7  3  3  0  4  3  5  6  7  8  9  9
   9  8  8  6  3  3
            9  4  3
```

4 a

Mark (%)	Frequency	Cumulative frequency
31–40	21	21
41–50	55	76
51–60	125	201
61–70	74	275
71–80	52	327
81–90	45	372
91–100	28	400

b

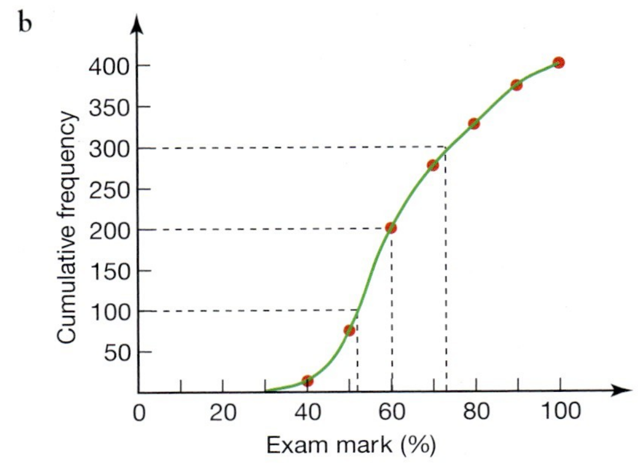

c i) Median ≈ 60%
ii) Lower quartile ≈ 52%
Upper quartile ≈ 74%
iii) IQR ≈ 22%

5

Mark (%)	Frequency	Cumulative frequency
1–10	10	10
11–20	30	40
21–30	40	80
31–40	50	130
41–50	70	200
51–60	100	300
61–70	240	540
71–80	160	700
81–90	70	770
91–100	30	800

b

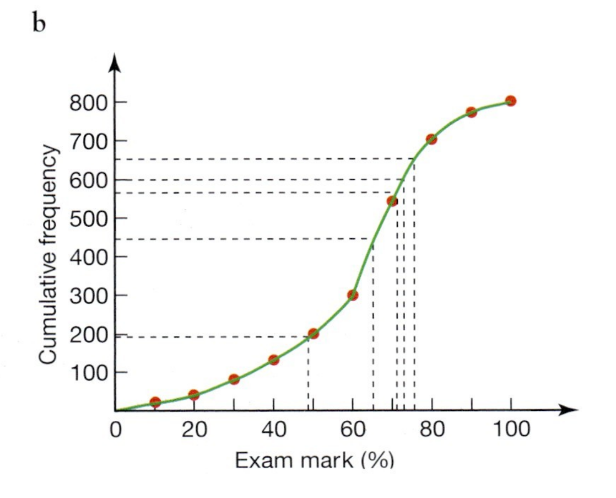

c A grade ≥ 72%
d Fail < 50%
e 200 pupils failed.
f 200 pupils achieved an A grade.

6 a

Time (mins)	$10 \leq t < 15$	$15 \leq t < 20$	$20 \leq t < 25$	$25 \leq t < 30$	$30 \leq t < 35$	$35 \leq t < 40$	$40 \leq t < 45$
Motorway frequency	3	5	7	2	1	1	1
Motorway cumulative frequency	3	8	15	17	18	19	20
Country lanes frequency	0	0	9	10	1	0	0
Country Lanes cumulative frequency	0	0	9	19	20	20	20

b

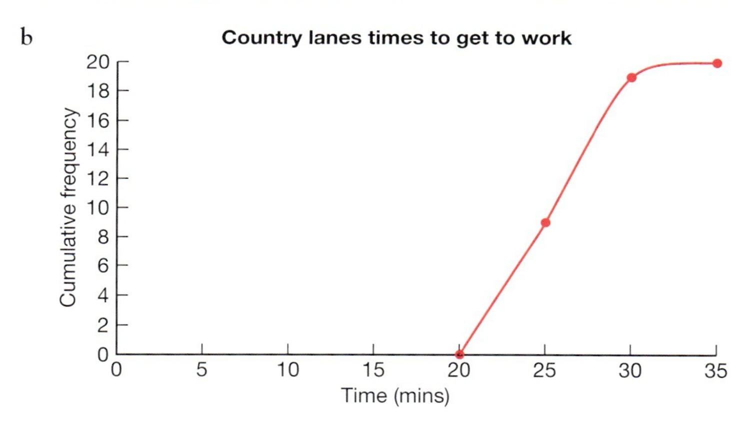

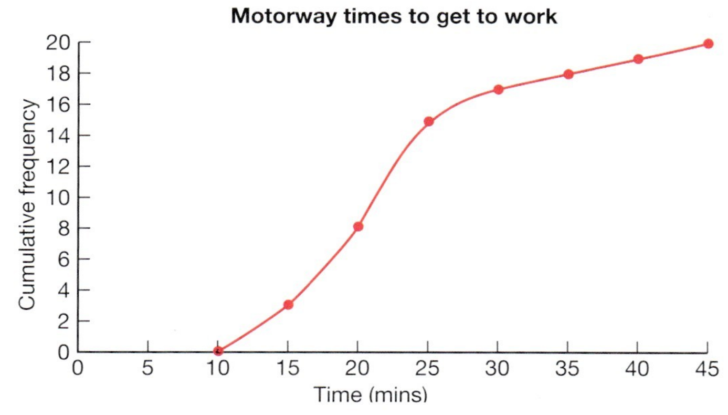

c i) Motorway median ≈ 21mins
Country Lanes median ≈ 26 mins

ii) Motorway lower quartile ≈ 17 mins
Motorway upper quartile ≈ 25 mins
Country lanes lower quartile ≈ 23 mins
Country lanes upper quartile ≈ 28 mins

iii) Motorway IQR ≈ 8 mins
Country lanes IQR ≈ 5 mins

d Student's explanations

e Student's explanations

7 Box and whisker plot A is likely to belong to mixed ability class + student's explanation. Box and whisker plot B is likely to belong to maths set + student's explanation.

8 **a** Mean = 5.4 Median = 5
Mode = 5
b Mean = 75.4 Median = 72
Mode = 72
c Mean = 13.8 Median = 15
Mode =18
d Mean = 6.1 Median = 6
Mode = 3

9 61 kg

10 **a** 2.83 **b** 3 **c** 3

Frequency
Number of pets

11 **a** 2.4 **b** 2 **c** 3

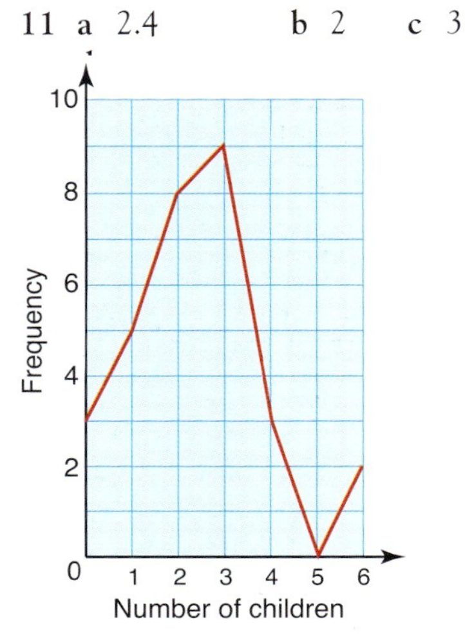

12 **a**

Group	Mid-interval value	Frequency
0–19	9.5	0
20–39	29.5	4
40–59	49.5	9
60–79	69.5	3
80–99	89.5	6
100–119	109.5	4
120–139	129.5	4

b 75.5

Student assessment 2

1 **a**

Group	Mid-interval value	Frequency
10–19	14.5	3
20–29	24.5	7
30–39	34.5	9
40–49	44.5	2
50–59	54.5	4
60–69	64.5	3
70–79	74.5	2

b 39.2

2 3.8

3 **a** 2.05 **b** 5 **c** 1.26

4 Student's answers may differ from those give below.
a Negative correlation (likely to be strong). Assume that motorcycles are not rare or vintage.
b Factors such as social class, religion and income are likely to affect results. Therefore little correlation is likely.

5 a

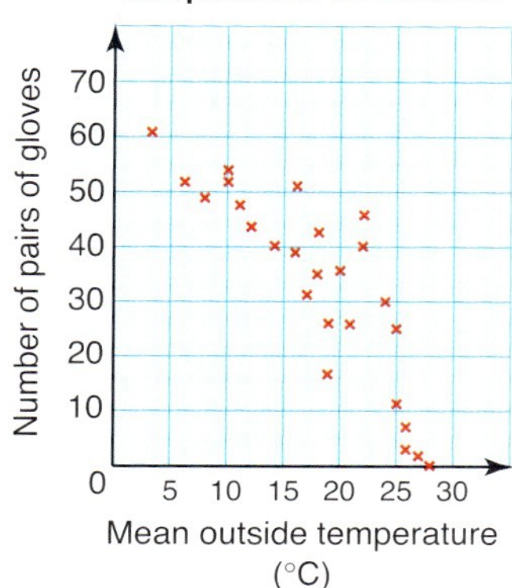

b The graph indicates a negative correlation.
c Student's own explanation

6 a $y = 76241x + 3.76 \times 10^6$
b $r = 0.277$
c There doesn't appear to be a correlation between a footballer's salary and his popularity. This gives some weight to the newspaper report.

7 a H_0 : The drug has no effect on the dog's condition.
H_1 : The drug improves the dog's condition.
b

	Improved	Did not improve	Total
Given drug	168.68	91.32	260
Not given drug	97.32	52.68	150
Total	266	144	410

c $X^2 = 2.061$
d 1 degree of freedom
e $2.061 < 3.841$ therefore the null hypothesis is supported, i.e. the drug does not significantly improve the dog's condition.

8 a H_0 : The opinion on whether to ban fox hunting is independent of whether you live in the city or the country.
H_1 : The opinion on whether to ban fox hunting is dependent of whether you live in the city or the country.
b $X^2 = 1683.01$
c $1683.01 > 6.635$ therefore the null hypothesis is rejected, i.e. the opinion to ban fox hunting is dependent on whether you live in the country or the city.

Topic 7

Exercise 7.1.1

1 a 4 b 6 c −2
d $2x$

2 a 4 b 8 c −8
d $4x$

3 a 1 b 2 c 3
d x

Exercise 7.1.2

1 a $\frac{dy}{dx} = 3x^2$ b $\frac{dy}{dx} = 6x$
c $\frac{dy}{dx} = 2x + 2$ d $\frac{dy}{dx} = 2x$
e $\frac{dy}{dx} = 3$ f $\frac{dy}{dx} = 4x - 1$

2

Function $f(x)$	Gradient function $f'(x)$
x^2	$2x$
$2x^2$	$4x$
$\frac{1}{2}x^2$	x
$x^2 + x$	$2x + 1$
x^3	$3x^2$
$3x^2$	$6x$
$x^2 + 2x$	$2x + 2$
$x^2 - 2$	$2x$
$3x - 3$	3
$2x^2 - x + 1$	$4x - 1$

4 If then $f(x) = ax^n$ then $\frac{dy}{dx} = anx^{n-1}$.

Exercise 7.1.3

1 a $4x^3$ b $5x^4$ c $6x$
d $15x^2$ e $18x^2$ f $56x^6$

2 a x^2 b x^3 c $\frac{1}{2}x$
d $2x^3$ e $\frac{6}{5}x^2$ f $\frac{2}{3}x^2$

Exercise 7.2.1

1 a $15x^2$ b $14x$ c $24x^5$
d $\frac{1}{2}x$ e $4x^5$ f $\frac{15}{4}x^4$
g 0 h 6 i 0

2 a $6x + 4$ b $15x^2 - 4x$
c $30x^2 - x$ d $18x^2 - 6x + 1$
e $48x^3 - 4x$ f $x^2 - x + 1$
g $-12x^3 + 8x$ h $-30x^4 + 12x^3 - 1$
i $-\frac{9}{2}x^5 + 2x^2$

3 a $2x + 1$ b 4 c $6x + 1$
d $\frac{1}{2}x + \frac{1}{2}$ e $6x + 3$ f $6x^2 - 8x$
g $2x + 10$ h $4x + 7$ i $3x^2 - 4x - 3$

Exercise 7.2.2

1 a $-x^{-2}$ b $-3x^{-4}$ c $-4x^{-3}$
d $2x^{-3}$ e x^{-4} f $2x^{-6}$

2 a x^{-1} b $2x^{-1}$ c $3x^{-2}$
d $\frac{2}{3}x^{-3}$ e $\frac{3}{7}x^{-2}$ f $\frac{2}{9}x^{-3}$

3 a $-3x^{-2} + 2$ b $4x - x^{-2}$
c $-3x^{-2} + 2x^{-3} + 2$ d $-3x^{-4} + 3x^2$
e $-8x^{-5} + 3x^{-4}$ f $x^{-3} - x^{-4}$

Exercise 7.2.3

1 a $6t + 1$ b $6t^2 + 2t$ c $15t^2 - 2t$
d $-2t^{-2}$ e $-t^{-3}$ f $4t^3 + 2t^{-3}$

2 a $-3x^{-2}$ b $-2t^{-2} - 1$ c $-2r^{-3} + r^{-2}$
d $-2l^{-5} + 2$ e $\frac{1}{2} + n^{-4}$ f $-\frac{4}{5}t^{-3} - 3t^2$

3 a $2x + 4$ b $1 - 2t$ c $3t^2$
d $2 - 6r$ e $x - x^{-2}$ f $3t^{-4}$

4 a $2t$ b $4t$
c $-2t^{-3}$ d $3t^2 - 2 + 4t^{-3}$
e $2t^2 - \frac{4}{3}t + 1$ f $-9t^{-4} + \frac{45}{2}t^{-6} - 4t$

Exercise 7.2.4

1 a $12x$ b $12x^2 - 1$ c $10x^4$
d 6 e $\frac{1}{2}$ f 0

2 a $2x^{-3}$ b $6x^{-4} + 6x$
c $2x^{-5}$ d $4x^{-4}$
e $6x^{-3} - 30x^{-4}$ f $12x^{-4} - 2x^{-3}$

3 a $6x - 6$ b $6x^2 + 3x$
c $2x^{-3} + 2$ d $12x^{-4} - 2$
e $20x^3$ f $12x^{-4}(1 - x^{-1})$

Exercise 7.3.1

1 a 6 b -3 c 0
d 0 e $-\frac{1}{2}$ and -5 f 6

2 a $-\frac{1}{4}$ b -2
c $-3\frac{3}{16}$ d -3
e $3\frac{31}{32}$ f -36 and -68

3 a i) $4\frac{1}{2}$ ii) $31\frac{1}{2}$ iii) 72
iv) 0
b $\frac{dN}{dt} = 10t - \frac{3}{2}t^2$
c i) $8\frac{1}{2}$ ii) $16\frac{1}{2}$ iii) 6
iv) -50
e The graph increases during the first 6–7 seconds, hence number of new infections increases. When $t = 10$ the graph is at zero, hence number of new infections is zero.
f The rate is initially increasing, i.e. the gradient of the graph is increasing. After approximately 4 seconds, the rate of increase starts to decrease, i.e. the gradient of curve is less steep. After 10 seconds the gradient is negative, hence the rate of increase is negative too.

4 a i) 243 m ii) 2000 m
b $\frac{dh}{dt} = 60t - 3t^2$
c i) $108\,\text{m h}^{-1}$ ii) $225\,\text{m h}^{-1}$ iii) $0\,\text{m h}^{-1}$
e The steepness of the graph indicates the rate at which the balloon is ascending. After 20 hours the graph has peaked therefore the rate of ascent is $0\,\text{m h}^{-1}$.
f The steepest part of the graph occurs when $t = 10$ hours. Therefore this represents when the balloon is climbing at its fastest.

Exercise 7.3.2

1 a (3, 6) b (2.5, 19.75)
c (2, 10) d (0, −1)
e (3, 15) f (−5, −14)

2 a $\left(1, 4\frac{5}{6}\right)$ and $\left(-2, -8\frac{2}{3}\right)$
b $\left(-1, -4\frac{1}{3}\right)$ and (−3, −9)
c $\left(2, -5\frac{1}{3}\right)$
d (1, 4) and $\left(-\frac{1}{3}, -1\frac{13}{27}\right)$

3 a $\frac{ds}{dt} = 4 + 10t$ b $\frac{1}{2}$ seconds
c 3 seconds d 57 m

4 a 20°C b $\frac{dT}{dt} = 24t - 3t^2$
c i) 21 °C/min
ii) 48 °C/min
iii) 0 °C/min
d t = 2 or 6 minutes e 236 °C

Exercise 7.3.3

1 a $f'(x) = 2x - 3$ b 1
c 1 d $y = x - 1$

2 a 4 b $y = 4x - 10$

3 a −4 b $y = -4x - 2$

4 a −4 b $y = -4x + 17$
c $y = 5$

5 a $-2x + 2$ b (−6, 10)
c (−4, 10)

6 a $f'(x) = -x - 1$ b −3
c (2, −8)

Exercise 7.4.1

1 a i) $f'(x) = 2x$ ii) $x > 0$
b i) $f'(x) = 2x - 3$ ii) $x > \frac{3}{2}$
c i) $f'(x) = -2x + 10$ ii) $x < 5$
d i) $f'(x) = 3x^2 - 24x + 48$
ii) $x < 4$ and $x > 4$
e i) $f'(x) = -3x^2 + 25$
ii) $-\frac{5}{\sqrt{3}} < x < \frac{5}{\sqrt{3}}$
f i) $f'(x) = x^3 - x$
ii) $-1 < x < 0$ and $x > 1$

2 a $x < 0$ b $x < \frac{3}{2}$ c $x < 5$
d never e $x < -\frac{5}{\sqrt{3}}$ and $x > \frac{5}{\sqrt{3}}$
f $x < -1$ and $0 < x < 1$

3 a $f'(x) = x^2 + \frac{1}{3}$
$x^2 \geqslant 0$ for all values of x, $\rightarrow x^2 + \frac{1}{3} > 0$ for all values of x
Therefore $f'(x)$ is an increasing function for all values of x.
b $f'(x) = x^2 - 1 - x^4 = x^2(1 - x^2) - 1$
When $x < -1$ or $x > 1 \rightarrow (1 - x^2) < 0$
therefore $f'(x) < 0$
When $-1 < x < 1 \rightarrow 0 < x^2(1 - x^2) < 1$
therefore $f'(x) < 0$
Hence $f'(x)$ is a decreasing function for all values of x.

4 $k < -\frac{1}{3}$

Exercise 7.5.1

1 a i) $f'(x) = 2x - 3$ ii) (3, 4)
b i) $f'(x) = 2x + 12$ ii) (−6, −1)
c i) $f'(x) = -2x + 8$ ii) (4, 3)
d i) $f'(x) = -6$ ii) no stationary points

2 a i) $f'(x) = 3x^2 - 24x + 48$
ii) (4, 6)
b i) $f'(x) = 3x^2 - 12$
ii) (−2, 16) and (2, −16)
c i) $f'(x) = 3x^2 - 6x - 45$
ii) (−3, 89) and (5, −167)
d i) $f'(x) = x^2 + 3x - 4$
ii) $\left(-4, 13\frac{2}{3}\right)$ and $\left(1, -7\frac{1}{6}\right)$

3 a i) $f'(x) = -4 - 2x$
ii) (−2, 5)
iii) (−2, 5) is a maximum
iv) (0, 1)

v)

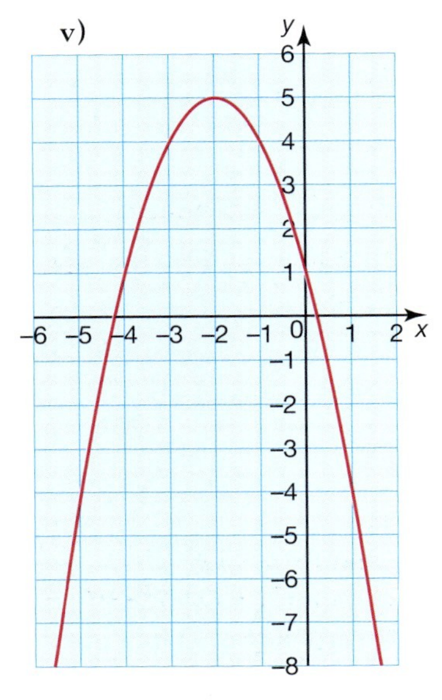

b i) $f'(x) = x^2 - 8x + 12$

ii) $\left(2, 7\frac{2}{3}\right)$ and $(6, -3)$

iii) $\left(2, 7\frac{2}{3}\right)$ is a maximum, $(6, -3)$ is a minimum

iv) $(0, -3)$

v)

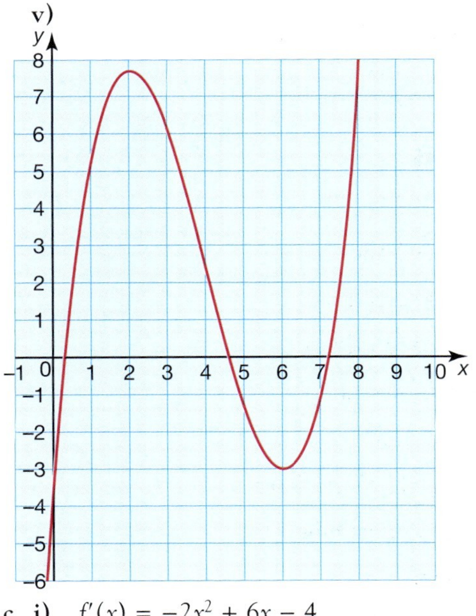

c i) $f'(x) = -2x^2 + 6x - 4$

ii) $\left(1, -1\frac{2}{3}\right)$ and $\left(2, -1\frac{1}{3}\right)$

iii) $\left(1, -1\frac{2}{3}\right)$ is a minimum point, $\left(2, -1\frac{1}{3}\right)$ is a maximum point

iv) $(0, 0)$

v)

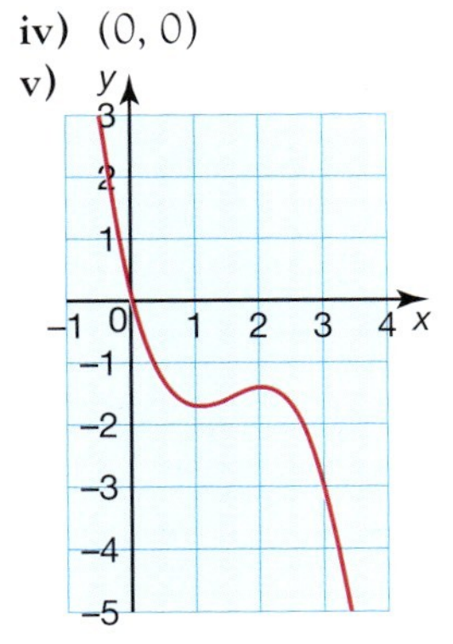

d i) $f'(x) = 3x^2 - 9x - 30$

ii) $(-2, 38)$ and $\left(5, -133\frac{1}{2}\right)$

iii) $(-2, 38)$ is a maximum point, $\left(5, -133\frac{1}{2}\right)$ is a minimum point

iv) $(0, 4)$

v)

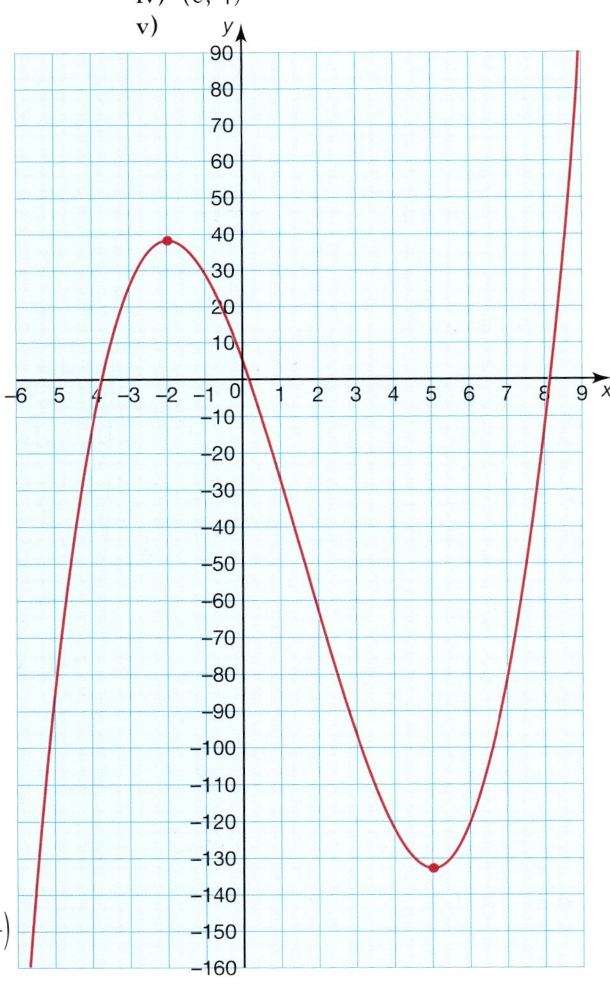

4 a i) $f'(x) = 3x^2 - 18x + 27$ **ii)** $(3, -3)$

iii) $(3, -3)$ is a point of inflexion

iv) $(0, -30)$

v)

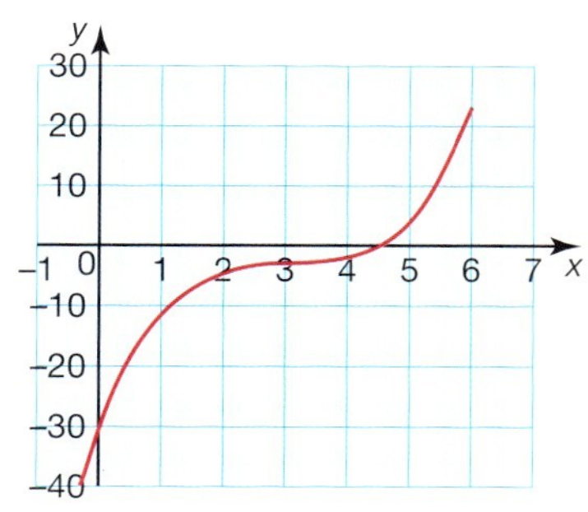

b i) $f'(x) = 4x^3 - 12x^2 + 16$

ii) $(-1, -11)$ and $(2, 16)$

iii) $(-1, -11)$ is a minimum point, $(2, 16)$ is a point of inflexion

iv) $(0, 0)$

v)

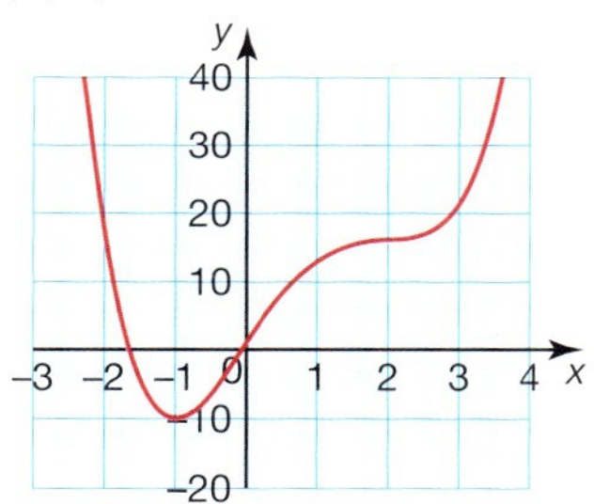

Student assessment 1

1 a $\dfrac{dy}{dx} = 3x^2$ **b** $\dfrac{dy}{dx} = 4x - 1$

c $\dfrac{dy}{dx} = -x + 2$ **d** $\dfrac{dy}{dx} = 2x^2 + 8x - 1$

2 a $f'(x) = 2x + 2$ **b** $f'(x) = 2x - 1$

c $f'(x) = 2x$ **d** $f'(x) = x + 1$

e $f'(x) = -3x^{-2}$ **f** $f'(x) = 1 - 2x^{-2}$

3 a $\dfrac{d^2y}{dx^2} = 12x^2 - 6$

b $\dfrac{d^2s}{dt^2} = 40t^3 - 6t$

4 a $f'(1) = 2$ **b** $f'(0) = 1$

c $f'\left(-\frac{1}{2}\right) = 9$ **d** $f'\left(\frac{1}{4}\right) = 5\frac{1}{2}$

5 a $v = \dfrac{ds}{dx} = 10t$ **b** $v = 30\,\text{m s}^{-1}$

c i) 4.2 seconds **ii)** 88.2 m

Student assessment 2

1 a $f'(x) = 3x^2 + 2x$

b $P\left(-\frac{2}{3}, -\frac{23}{27}\right)$

c $Q(0, -1)$

d P is a maximum and Q a minimum

2 a Substituting (1, 1) into the equation gives $1 = 1^3 - 1^2 + 1 \rightarrow 1 = 1$

b 2

c $y = 2x - 1$

3 a $f'(x) = 2x - 4$

b $x < 2$

4 a $f'(x) = 4x^3 - 4x$

b $(0, 0)$, $(1, -1)$ and $(-1, -1)$

c $(0, 0)$ is a maximum, $(1, -1)$ and $(-1, -1)$ are both minimum points

d i) $(0, 0)$

ii) $(0, 0)$, $(\sqrt{2}, 0)$ and $(-\sqrt{2}, 0)$

e

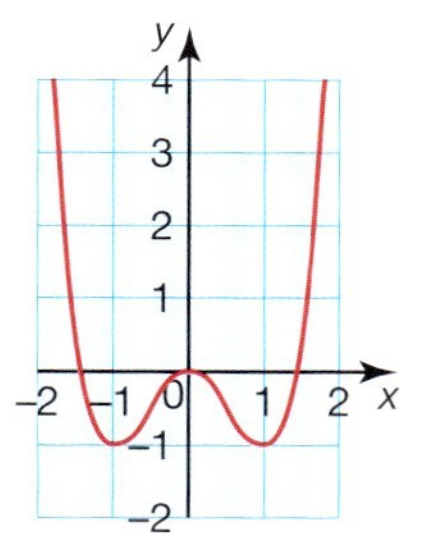

Topic 8

Exercise 8.1.1

1. 334.69 francs
2. 77 500 drachmas
3. 141.86 schillings
4. 9036.14 escudos
5. 164 francs
6. 7 531 645.57 markkas
7. 17 113.40 pesetas
8. 1266.60 punts
9. 4 948 980 lire
10. 15 636.36 schillings

Exercise 8.1.2

1 a £7.81 b 7711 rand

2 a £282 258 b £256 598

3 a 1¥ = $0.01 b $1 = 100¥

4 a 1 rouble = 0.15 shekel
b 1 shekel = 6.86 roubles

5 a $33 331 200 b 1 366 579 200 rupees
c 20 665 344€ d 1734 ounces

Exercise 8.2.1

1 a NZ$34.85 b £420 c 102¥
d 252 baht e HK$369.60

2 a 5 years b 41 years c 5 years
d 8 years e 6 years f 7 years

3 a 7% b 4% c 3.5%
d 7.5% e 8% f 11%

4 a 400 Ft b NZ$200 c 850€
d 1200 baht e 4000€ f US$1200

5 4%

6 2 years

7 4.5%

8 9.5%

9 AU$315

10 6%

Exercise 8.3.1

1 £11 033 750

2 520 875€

3 $10 368

4 1331 students

5 3 276 800 tonnes

6 2 years

7 5 years

8 3 years

9 a $525 b $537.75 c $549.02

10 a €149.40 b €152.30 c €154.84

Exercise 8.4.1

1 $447 600

2 7%

3 £250 000

4 $250 000

5 a 21 years
b 11 years

6 ≈ 15%

7 $8027

Student assessment 1

1 ≈16.4%

2 £561 167

3 59%

4 2001

5 a 480 000€ b 499 200€

6 4 years

7 8 days

8 955 080€

9 5 years

10 a $2012.50 b $483 000 c $1 062 000

Student assessment 2

1 13.6%

2 $1 782 000

3 81%

4 10 years

5 a 300 000€ b 900 000€

6 6 years

7 11 years

8 a $1 000 020 b $1 109 000 c $1 148 000

9 11 years

10 a $3146.50 b $566 370 c $1 675 000

Index